Lecture Notes in Electrical Engineering

Volume 657

The book series *Lecture Notes in Electrical Engineering* (LNEE) publishes the latest developments in Electrical Engineering—quickly, informally and in high quality. While original research reported in proceedings and monographs has traditionally formed the core of LNEE, we also encourage authors to submit books devoted to supporting student education and professional training in the various fields and applications areas of electrical engineering. The series cover classical and emerging topics concerning:

- Communication Engineering, Information Theory and Networks
- Electronics Engineering and Microelectronics
- Signal, Image and Speech Processing
- Wireless and Mobile Communication
- Circuits and Systems
- Energy Systems, Power Electronics and Electrical Machines
- Electro-optical Engineering
- Instrumentation Engineering
- Avionics Engineering
- Control Systems
- Internet-of-Things and Cybersecurity
- Biomedical Devices, MEMS and NEMS

For general information about this book series, comments or suggestions, please contact leontina.dicecco@springer.com.

To submit a proposal or request further information, please contact the Publishing Editor in your country:

China

Jasmine Dou, Associate Editor (jasmine.dou@springer.com)

India, Japan, Rest of Asia

Swati Meherishi, Executive Editor (Swati.Meherishi@springer.com)

Southeast Asia, Australia, New Zealand

Ramesh Nath Premnath, Editor (ramesh.premnath@springernature.com)

USA, Canada:

Michael Luby, Senior Editor (michael.luby@springer.com)

All other Countries:

Leontina Di Cecco, Senior Editor (leontina.dicecco@springer.com)

**** Indexing: The books of this series are submitted to ISI Proceedings, EI-Compendex, SCOPUS, MetaPress, Web of Science and Springerlink ****

More information about this series at http://www.springer.com/series/7818

Liheng Wang · Yirong Wu ·
Jianya Gong
Editors

Proceedings of the 6th China High Resolution Earth Observation Conference (CHREOC 2019)

Springer

Editors
Liheng Wang
China Aerospace Science and Technology Corporation
Beijing, China

Yirong Wu
Aerospace Information Research Institute
Chinese Academy of Sciences
Beijing, China

Jianya Gong
Wuhan University
Wuhan, China

ISSN 1876-1100 ISSN 1876-1119 (electronic)
Lecture Notes in Electrical Engineering
ISBN 978-981-15-3946-6 ISBN 978-981-15-3947-3 (eBook)
https://doi.org/10.1007/978-981-15-3947-3

This Springer imprint is published by the registered company Springer Nature Singapore Pte Ltd.
The registered company address is: 152 Beach Road, #21-01/04 Gateway East, Singapore 189721, Singapore

Contents

A New Multidimensional Spectral and Polarization Information Detection Technology 1
Yechao Wang, Xiaoli Chen, Xiaoming Zhong, and Haibo Zhao

Optimizing Pose of UAV Image Based on PPK Technology 11
Guangrui Yu, Yuncai Su, Lili Yu, Lianbing Gong, and Danyang Zhao

Extraction of Helicopter Rotor Physical Parameters Based on Time-Frequency Image Processing 37
Chenxiao Lai and Daiying Zhou

Design of an Airborne DIAL Measurement System for Measuring Concentrations of Atmospheric Pollutants 51
Yong Chen, Ding-fu Zhou, Ze-hou Yang, Chun-li Chen, Yong-ke Zhang, Xiang-hua Niu, Jing Li, Xiao-feng Li, Guo-juan Zhang, and Guo-hua Jin

A Multiple Moving Ships Detection Method for GF-4 Satellite Image in Thin-Cloud Environment 61
Peng Lv, Yuxin Hu, Qianqian Li, Yangshuan Hou, Xiaohui Wang, and Bin Lei

Identification and Extraction of Nutrient Content in Hyperspectral Black Soil in Frequency Domain 73
Dong-hui Zhang, Ying-jun Zhao, Kai Qin, Dong-hua Lu, Cheng-kai Pei, Ning-bo Zhao, Yue-chao Yang, and Ming Li

Evaluation of Geometric Performances of the Gaofen-6 PMS Camera 85
Liping Zhao and Hongzhou Li

Analysis of Geometric Performances of the Gaofen-6 WFV Camera . . . 105
Liping Zhao, Xingke Fu, and Xianhui Dou

Evaluation and Analysis of Geometric Performances of the Gaofen-1 B/C/D Satellite 125
Liping Zhao and Xingke Fu

Preliminary Analysis of Gaofen-1 B/C/D Satellite Stereo Mapping Performance 141
Liping Zhao and Xianhui Dou

Ecological Vulnerability Assessment and Cause Analysis of the Farming–Pastoral Zone in Northern China—Taking Yulin City as an Example 157
Yanmei Zhong, Jianguo Cheng, Lingkui Meng, and Wen Zhang

Airport and Ship Target Detection on Satellite Images Based on YOLO V3 Network 167
Ren Ying

River Segmentation Based on Visual Salience Calculation of Spectral Residual and Region Growing in SAR Images 175
Guiming Zhang, Gong Zhang, Liyan Luo, Jiantao Wang, and Qing Ding

Remote Sensing Road Extraction by Refining Road Topology 187
Huiqin Gao, Yuan Yuan, and Xiangtao Zheng

Cloud Detection Method in GaoFen-2 Multi-spectral Imagery 199
Zhaocong Wu, Lin He, Yi Zhang, and Jun Li

High-Resolution Land-Use Mapping in Beijing-Tianjin-Hebei Region Based on Convolutional Neural Network 213
Pan Chen, Zhengchao Chen, Xuan Yang, Baipeng Li, and Bing Zhang

Dichotomy: Trajectory Planning Algorithm Based on Point Group Distribution 223
Naiting Xu, Fan Yang, HaiMing Lian, and Yi Wang

System Design for an Improved SPIDER Imager 241
Guomian Lv, Yueting Chen, Huajun Feng, Zhihai Xu, and Qi Li

Performance of Support Vector Machines, Artificial Neural Network, and Random Forest for Identifying Banana Fusarium Wilt Using UAV-Based Multi-spectral Imagery 261
Huichun Ye, Bei Cui, Shanyu Huang, Yingying Dong, Wenjiang Huang, Anting Guo, Yu Ren, and Yu Jin

Remote Sensing Image On-Board Restoration Based on Adaptive Wiener Filter 271
Yunsen Wang, Wenxiu Mu, Xiaohui Du, Chengzhi Ma, and Xuejin Shen

Design of Space-Based Information Service Architecture Based on Mobile Cloud 283
Ying-chun Hou, Chao Yang, and Yan Hou

Monitoring of a Subgrade Subsidence by Means of Ground-Based SAR Interferometry 295
Pingping Huang, Fang Liu, Weixian Tan, Wei Xu, Qi Lin, and Huifang Ren

A Polarization Calibration Solution Method of Airborne SAR Based on Point Target 303
Zhida Xu, Shucheng Yang, Chunquan Cheng, and Jianwei Tan

Recognition Hydropower Stations from Remote Sensing Images by Multi-stage CNN Detection and Segmentation 313
Xiaowei Tan, Zhifeng Xiao, and Weiping Shao

The Intelligent Planning for Spacecraft Autonomy in On-Orbit Servicing Mission 331
Jing Yu, Dong Hao, Hongyang Liu, and Xiaoqian Chen

Numerical Aerodynamic and Design Analysis of Combined Saucer-Shaped Buoyancy-Lifting Airships 343
Yanxiang Cui, Yanchu Yang, Jinggang Miao, and Xiangqiang Zhang

Monitoring the Surface Subsidence of Liupanshui City Using ALOS-2 Images and NSBAS-InSAR Technology 363
Huan Chen, Guoman Huang, Guoqi Cheng, and Yuting Sheng

L^1-Norm Method for Dual-Wavelength Phase Unwrapping 373
Guoqi Cheng, Jixian Zhang, Huan Chen, Jiaqi Chen, and Lijun Lu

High-Efficiency Flexible GaAs Solar Modules Used for Unmanned Aerial Vehicles and Stratospheric Airships 385
J. K. Yang, X. Y. Jiao, Y. Yang, X. S. Wang, L. L. Song, J. Xue, and Z. C. Chen

Landslide Risk Zoning Method Based on Improved Water Seepage Capacity Model 393
Zhaohua Wang, Jixian Zhang, Zheng Zhao, and Haiying Gao

Semi-supervised Classification of PolSAR Image Based on Self-training Convolutional Neural Network 405
Xianxiang Qin, Wangsheng Yu, Peng Wang, Tianping Chen, and Huanxin Zou

A SAR Image Data Augmentation Method Based on Generative Adversarial Network 419
Qinglin Lu, Guojing Li, and Wei Ye

Research on High-Resolution Synthetic Aperture Radar Image for Varied Antenna Beam Pointing 431
Yinghui Zhao and Xijuan Yue

Data-Driven Fault Detection Methods for Jilin-1 Satellite Attitude Control System 441
Zhi Qu, Kai Xu, Yanhao Xie, Feng Li, Mengmeng Liu, Shuangxue Han, Jiebing Liu, and Zhigang Chen

A Novel Framework of Artificial Intelligent Geologic Hazards Detection Over Comprehensive Remote Sensing 457
Yueying Zhang, Haonan Ran, Yuexing Peng, and Yu Zheng

Automatic Prediction of Landslides Over InSAR Techniques and Differential Detection Using High-Resolution Remote Sensing Images: Application to Jinsha River 471
Yueying Zhang, Haonan Ran, Yuexing Peng, and Yu Zheng

Internal Calibration and Range Replica Extraction Scheme for Ultrahigh-Resolution Spaceborne SAR 487
Fan Feng, Axin Jin, Jia Sun, Ruohan Hou, and Hongxing Dang

Monitoring and Analysis of Surface Deformation and Glacier Motion Along the Sichuan–Tibet Railway: A Case Study of the Lhasa–Nyingchi Railway Section 499
Jinghui Luo, Qing Ding, Gong Zhang, Wei Zhang, Xiaoxia Wang, and Changli Zheng

Automatic Detection of Ship Based on Rotation Invariant RetinaNet 515
Luyang Zan, Kaixuan Lu, and Zhengchao Chen

A High-Resolution Remote Sensing Images Segmentation Algorithm Based on PCA and Fuzzy C-Means 527
Chenchen Jiang, Hongtao Huo, and Qi Feng

Structural Design and Analysis of a Belly Radome 541
Liuqing Xu

Cross-comparison and Analysis of SJ-9A, SPOT5, and THEOS Based on a Satellite Network 551
Fangyan Yuan, Guoqing Li, Zhengli Zuo, Quan Ran, LiDong Guo, and Guangbin Ma

Water Extraction of Airborne Polarimetric SAR by Introducing Eigenvalue Relative Difference 571
Zheng Changli, Zhang Wei, Ding Qing, Wang Xiaoxia, and Luo Jinghui

Research on Dimension Reduction Detection Technology of Space-Borne GPS Receiver Fault 581
Zhang Xiaopeng, Wang Xiaochen, Zhuang Haixiao, Guo Yongfu, Chen Xi, Zuo Miao, Zhang Xiangyan, Wang Weiwei, and Yang Ping

Establishing the Downscaling Model of NDVI Based on the Iterated Function System 589
Haijun Luan, Meng Zhang, Yunya Wan, Yuanrong He, Qin Nie, and Xinxin Zhang

An Autonomous Navigation Method with InSAR-Aided INS 605
Shuai Jiang, Yalong Pang, Luyuan Wang, Jiyang Yu, Bowen Cheng, Zongling Li, Liang Hao, and Cuilian Wang

A New Multidimensional Spectral and Polarization Information Detection Technology

Yechao Wang, Xiaoli Chen, Xiaoming Zhong, and Haibo Zhao

Abstract Simultaneous acquisition of spectral information and polarized information can obtain more feature information to distinguish targets. Based on the computational imaging technology and a pixelized polarization detector, we propose a novel imaging mode that simultaneously acquires 2D spatial information, 1D spectral information and 1D spectral–polarized information of the target. In the framework of compressed sensing, the imaging results are obtained by the pixelized polarization detector after the spatial modulation of coded aperture and the dispersion of prism. The corresponding spectral and polarization images are reconstructed by an optimization algorithm. The method can obtain 25 bands of spectral and polarization images of four angles (0°, 45°, 90° and 135°), ranging in 450–650 nm (spectral resolution less than 10 nm), and the degree of linear polarization and the angle of polarization of each band. The experimental prototype has been developed, and the data acquisition and processing have been completed. Finally, the technology has successfully achieved simultaneous acquisition of multidimensional spatial–spectral–polarized information.

Keywords Computational imaging · Polarization detector · Multidimensional information · Spectral image reconstruction

1 Introduction

Traditional spectropolarimeter can simultaneously acquire three-dimensional spatial–spectral data of each Stokes parameter by scanning specific domains [1], such as the spatial domain in channeled spectropolarimetry [2], the optical path difference domain in Fourier transform imaging spectropolarimetry [3] and polarization domain in other spectropolarimetry [4]. Some new systems obtain polarization spectra by direct measurements such as integrating integral field spectrometry with division-of-aperture imaging polarimetry [5]. However, the above systems have limitations such

Y. Wang (✉) · X. Chen · X. Zhong · H. Zhao
Beijing Institute of Space Mechanics & Electricity, Beijing, China
e-mail: cast_wangyc_508@163.com

L. Wang et al. (eds.), *Proceedings of the 6th China High Resolution Earth Observation Conference (CHREOC 2019)*, Lecture Notes in Electrical Engineering 657,
https://doi.org/10.1007/978-981-15-3947-3_1

as noise sensitivity, channel cross talk and spectral resolution [6, 7]. Some literatures have introduced the latest compressed spectral imaging method, using channel switching method to measure polarization spectrum, but at the expense of time resolution [8]. Compressed spectrum imaging is a new technology that has evolved in recent years. Based on the compressed sensing framework, high-dimensional snapshot acquisition of spectral information becomes possible [9]. The coded aperture snapshot spectral imager (CASSI) receives information from the array detector by random code modulation of the scene and spectral modulation of the dispersive elements [10–12]. Inspired by this, we propose a new imaging mode that a pixelized polarization detector combined with compression spectroscopy to simultaneously detect two-dimensional spatial information (x, y), one-dimensional spectral information (λ) and four polarized components (0°, 45°, 90° and 135°). The random coded aperture enables the system to code of the spatial information efficiently, and the integration of the polarization detector with the CASSI achieves efficient acquisition. The feasibility of the scheme was verified by experiments, and the model of the system was established. Under the theory of compressed sensing, the optimization problem is solved by the iterative algorithm. This model breaks through the principle limitation of the traditional polarization spectrum detection method and satisfies the synchronous acquisition requirements of multidimensional information such as time-space spectrum polarization.

The scene information $h(x, y, \lambda, p)$ is limited by bandpass filter (BPF), and then it is modulated by coded aperture (CA). After double Amici prism (DAP) dispersion modulation, it is imaged on the polarization detector.

2 A New Multidimensional Spectral and Polarization Information Detection Technology

The principle of the multidimensional polarization spectrum detection system is shown in Fig. 1. The scene information $h(x, y, \lambda, p)$ first enters the objective lens, and the bandpass filter of 450–650 nm is used to intercept the spectrum segments.

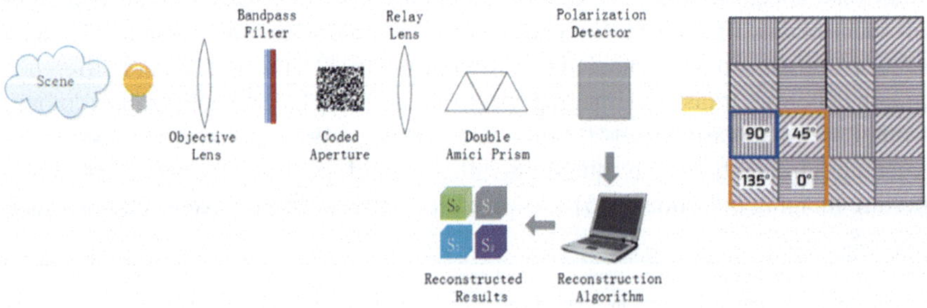

Fig. 1 Schematic of system

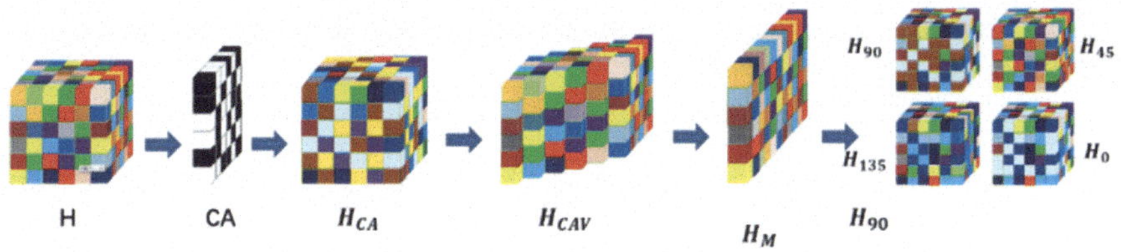

Fig. 2 Data modulation flowchart

Then, the light reaches the coded aperture, and the random coding mask spatially modulates the information $h(x, y, \lambda, p)$. The coded aperture is a random matrix and sets 0 and 1 channels which etched on the glass and made an anti-reflectance process. The light which is modulated by the codes goes through the relay lens and then reaches the dispersive element. Then, a double Amici prism is used to disperse the encoded light from the spectral dimension. It is useful for subsequent reconstruction of the data cube. The coded information passes through the dispersive prism and reaches the polarization detector. A spatially encoded and dispersion-modulated polarization spectrum-compressed image is obtained on the polarization detector (Fig. 2).

The data flowchart indicates that the information H first performs coding of 0 or 1 in space when it passes through the coded aperture (CA). After the dispersion of the double Amici prism, the data cube is expressed as H_{CAV} and it is distributed on both sides of the central wavelength of 550 nm. H_{M} is obtained on the polarization detector, including four polarization components (0°, 45°, 90°, and 135°).

3 System Model

As shown in Fig. 1, the scene information can be expressed as $h(x, y, \lambda, p)$, where x and y represent two-dimensional spatial information, λ represents spectral information, and $p = 0, 1, 2, 3$, representing different polarization information,

$$\boldsymbol{S}_{\mathbf{out}} = \boldsymbol{M}_{\mathbf{pd}} \boldsymbol{S}_{\mathbf{in}} \tag{1}$$

$$\boldsymbol{M}_{\mathbf{pd}}(\theta) = \frac{1}{2} \begin{bmatrix} 1 & \cos 2\theta & \sin 2\theta & 0 \\ \cos 2\theta & \cos^2 2\theta & \cos 2\theta \sin 2\theta & 0 \\ \sin 2\theta & \cos 2\theta \sin 2\theta & \sin^2 2\theta & 0 \\ 0 & 0 & 0 & 0 \end{bmatrix} \tag{2}$$

$$\mathbf{S} = \left[\boldsymbol{S}_0^{\mathrm{T}}, \boldsymbol{S}_1^{\mathrm{T}}, \boldsymbol{S}_2^{\mathrm{T}}, \boldsymbol{S}_3^{\mathrm{T}} \right]^{\mathrm{T}} \tag{3}$$

M_{pd} represents the Mueller matrix of the polarization detector [13]. $\boldsymbol{S}_{\mathbf{in}}$ and $\boldsymbol{S}_{\mathbf{out}}$ represent polarization information of incident and outgoing light. Due to the small field of the lens, the Mueller matrix of other components can be considered as a unit matrix. S is Stokes parameter [14]. Similar to the CASSI, the image acquired by the detector can be expressed as [15]

$$f(x, y) = \sum_{p=0}^{3} \int \psi_P(x, y, \lambda) T(x, y, \lambda) h(x, y, \lambda, p) d\lambda + \omega(x, y) \quad (4)$$

$\psi_P(x, y, \lambda)$ represents the modulation of the polarizer. $T(x, y, \lambda)$ is the modulation introduced by the CASSI, and $\omega(x, y)$ is system noise. Discrete form can be expressed as

$$f(m, n) = \sum_{p=0}^{3} \sum_{k=1}^{25} \psi_P(m, n, \lambda) T(m, n, k) h(m, n, k, p) + \omega(x, y) \quad (5)$$

(m, n) is the discrete coordinate, and k is the number of spectral bands. Expressed as matrix forms,

$$\boldsymbol{F} = \boldsymbol{\Psi T H} + \boldsymbol{\Omega} = \boldsymbol{K H} + \boldsymbol{\Omega} \quad (6)$$

$\boldsymbol{F}$, $\boldsymbol{\Psi}$, $\mathbf{T}$, $\boldsymbol{H}$, $\boldsymbol{\Omega}$ are the matrix forms of $\psi_P(m, n, \lambda)$, $T(m, n, k)$, $h(m, n, k, p)$ and $\omega(x, y)$. $\mathbf{K} = \boldsymbol{\Psi}\mathbf{T}$ is the measurement matrix of the system. Assume that the data cube has L spectral bands, $M \times N$ spatial pixels and 4 Stokes parameters. It can be expressed as

$$\boldsymbol{F} = [\boldsymbol{K}_0, \boldsymbol{K}_1, \boldsymbol{K}_2, \boldsymbol{K}_3]\left[\boldsymbol{H}_0^{\mathrm{T}}, \boldsymbol{H}_1^{\mathrm{T}}, \boldsymbol{H}_2^{\mathrm{T}}, \boldsymbol{H}_3^{\mathrm{T}}\right] + \boldsymbol{\Omega} \quad (7)$$

K_{p} is the measurement matrix of $\boldsymbol{S}_{\mathbf{p}} \in \Re^{MNL \times 1}$,and $\boldsymbol{F} \in \Re^{MN+L-1 \times 1}$ is the vectorized form of f.

4 Reconstruction Algorithm

The original polarization spectrum information is obtained by solving an optimization problem. We need to reconstruct all the original information from an incomplete observation. Due to the sparsity of the spectral data itself, the underdetermined system can find a unique solution by solving the following optimization problem [16],

$$\widehat{\boldsymbol{F}} = \arg\min_{\boldsymbol{F}} \left\| \boldsymbol{G} - \mathbf{HF}_{\mathbf{2}}^{\mathbf{2}} \right\| + \tau \Gamma(\boldsymbol{F}) \quad (8)$$

where $\boldsymbol{G}$ is the observed value, $\boldsymbol{H}$ is the equivalent observation matrix, $\boldsymbol{F}$ is the scene spectral data, and τ is a parameter used to adjust the balance between the two parts.

Formula (8) is the objective function of the optimization and consists of two parts: (a) the system fidelity term, which is used to measure the error between the optimization result and the system observation, and (b) the regularization term, which is generally used to constrain the objective function according to the intrinsic property of the target [17]. This paper selects the total variational (TV) regularizer [18]. Two-step iterative shrinkage/thresholding (TwIST) is an effective algorithm to solve constrained optimization problems, which can realize reconstruction quickly and efficiently [19]. In this paper, TwIST is used to reconstruct the polarization spectrum data cube.

5 Experiment

5.1 Hardware Implementation

The experimental device includes (1) objective lens L1, (2) bandpass filter (BPF), (3) coded aperture (CA), (4) an F/8 relay lens L_2, (5) a double Amici prism (DAP), (6) a monochromatic pixelized polarized CMOS detector (P-detector). The cutoff range of BPF is 450 nm-650 nm. The CA includes 520 × 520 elements of random binary pattern with 13.8 μm× 13.8 μm in size. The central wavelength of the double Amici prism is 550 nm. The resolution of the pixelized polarized CMOS detector is 2448 × 2048, and the pixel size is 3.45 μm× 3.45 μm. Adjacent polarizers of 0°, 45°, 90° and 135° are placed on the array [20, 21] (Fig. 3).

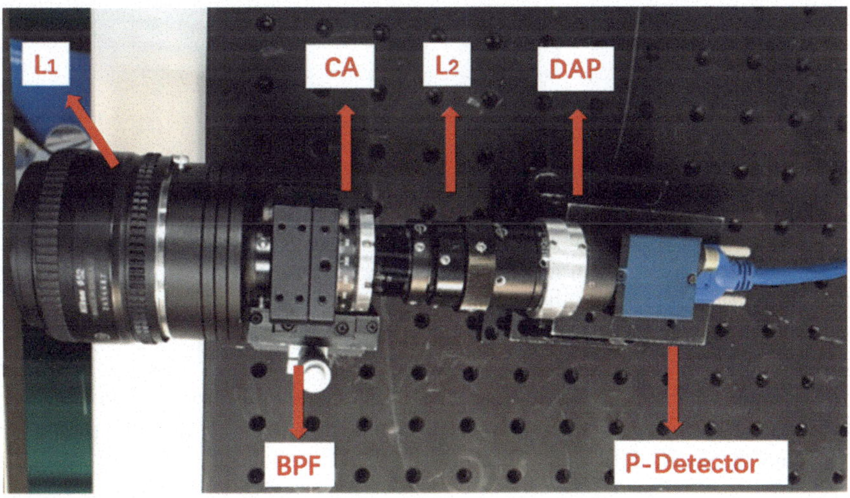

Fig. 3 Experimental prototype developed at laboratory

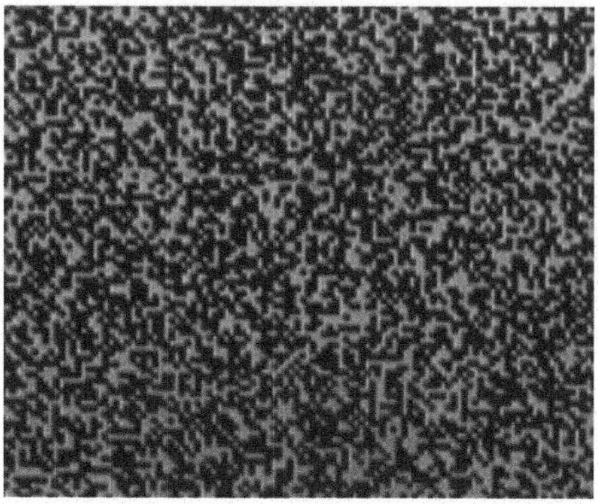

Fig. 4 Coded aperture image obtained at a wavelength of 614 nm

5.2 Calibration

A monochromator is used as the light source. The wavelength ranges in 450–650 nm, and the wavelength interval is 1 nm. The center wavelength is determined, and its position corresponds to the pixels on the detector. Twenty-five wavelengths are selected as the basis of reconstruction, and the coded aperture spectrum image of the corresponding wavelength is obtained. In order to eliminate dark current noise, 10 dark current noise images of each wavelength were acquired. The calibration images at different wavelengths are averaged in energy, and then the exposure time during the calibration process is equivalent to a constant [22–24] (Fig. 4).

5.3 Experiment Results

In Fig. 5, the scene consists of a blue book, a red racket, a color card and a checkerboard. The polarization angles are 0°, 45°, 90° and 135°. An image with no polarization angle was acquired.

Spectral and polarization information can be reconstructed by TwIST. Figure 6 shows the reconstructed results of four polarization angles (0°, 45°, 90° and 135°) ranging in 450–650 nm, including 25 bands, using a color display at the corresponding wavelength.

The Stokes parameters S_0, S_1, S_2, S_3 are calculated and displayed in 3D after correction [13], the x-axis and the y-axis represent the spatial domain, and the z-axis represents the spectral domain.

$$S = \begin{bmatrix} S_0 \\ S_1 \\ S_2 \\ S_3 \end{bmatrix} = \begin{cases} S_0 = I_0 + I_{90} \\ S_1 = I_0 - I_{90} \\ S_2 = I_{45} - I_{135} \\ S_3 = I_R - I_L \end{cases} \tag{9}$$

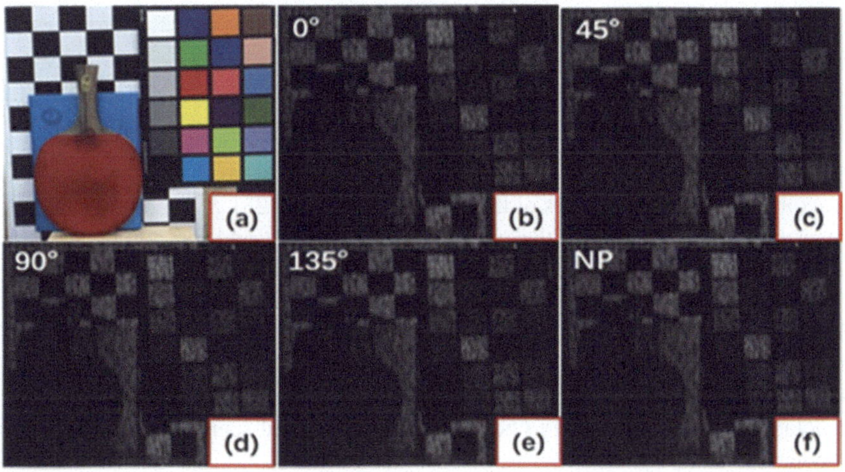

Fig. 5 Experimental scene and measured images. **a** The experimental scene, **b–e** different polarization components (0°, 45°, 90° and 135°), **f** no polarization image

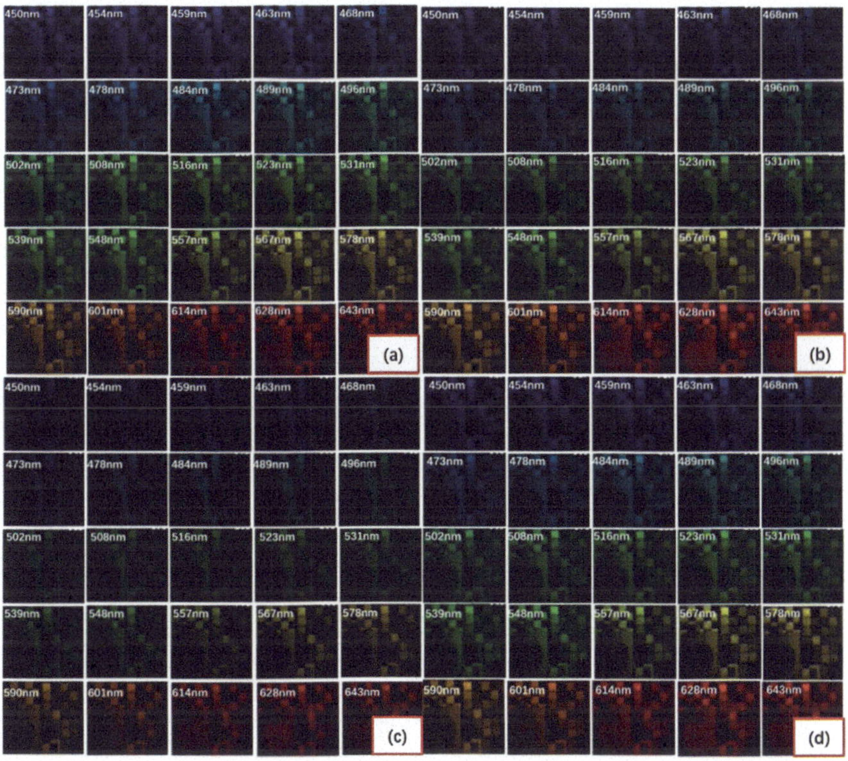

Fig. 6 Reconstructed results. **a–d** represent 0°, 45°, 90° and 135° polarization components

The circularly polarized Stokes parameter is assumes as [8]

$$S_3 = 0.1\left(S_0 - \sqrt{S_1^2 + S_2^2}\right) \tag{10}$$

Figure 7 indicates that S_1, S_2 and S_3 are significantly weaker than S_0, which can be explained by the properties of the Stokes parameters [25]. S_0 represents the total intensity of light, including the polarization component and the non-polarization component. S_1 is the difference between the horizontally polarized light and the vertically polarized light. S_2 is the difference between 45° and 135° polarized lights. Both S_1 and S_2 represent linearly polarized components. S_3 is the difference between the left and right circularly polarized lights and represents a circularly polarized light component. The main component of the light reflected by the natural scene is unpolarized light. The target to be measured is a natural scene, so it is mainly an unpolarized component [1].

Degree of linear polarization (DoLP) and angle of polarization (AoP) are defined as [1]

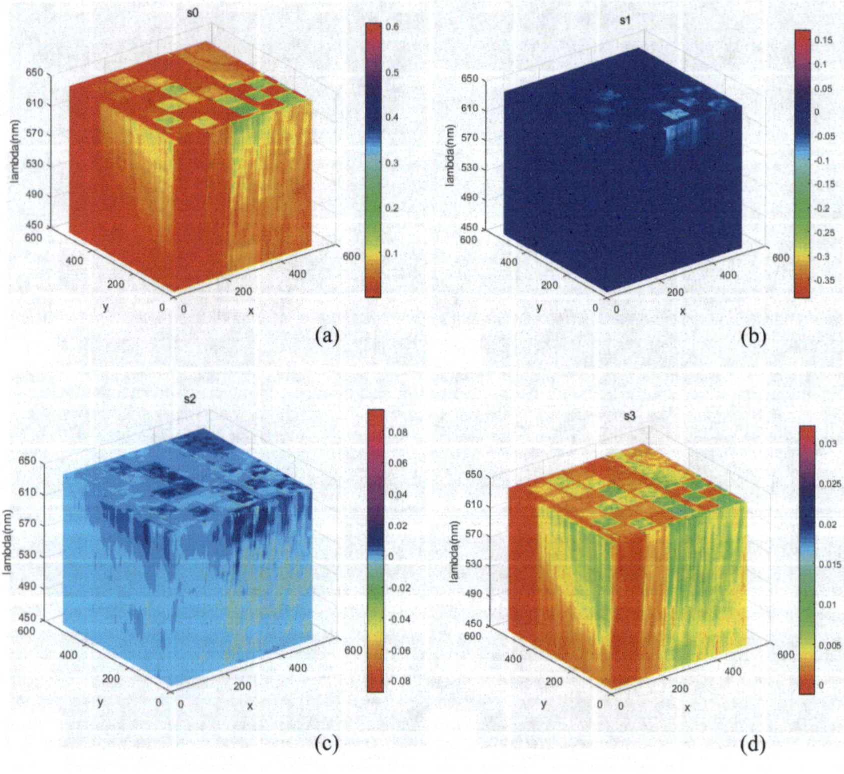

Fig. 7 Reconstructed results. **a–d** The 3D data cubes of S_0, S_1, S_2 and S_3

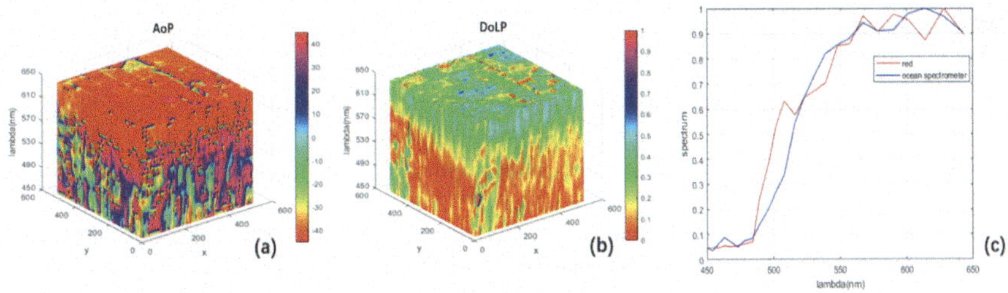

Fig. 8 **a–b** 3D image of AoP and DoLP, **c** reconstructed spectral curve of the red card and the blue curve is obtained by a spectrometer

$$\mathrm{DoLP} = \sqrt{S_1^2 + S_2^2}/S_0 \tag{11}$$

$$\mathrm{AoP} = \arctan(S_1/S_2)/2 \tag{12}$$

Figure 8a, b indicates that the edges of the checkerboard and the racket can be distinguished easily. The weaker area corresponds to a smaller AoP and a smaller DoLP. The area with much polarized characteristics is a larger AoP and medium DoLP. The results indicate that the spectral image with polarization information is easier to distinguish the edge of the object compared with the simple spectral imagery. (c) shows that the reconstructed spectra have good accuracy.

6 Conclusion

This paper proposes a new multidimensional polarization spectrum detection technology based on CASSI and a pixelized polarization detector. The system can simultaneously acquire spatial, spectral and polarization information. The CASSI performs the spatial coding modulation and the spectral dispersion aliasing compression sampling of the scene. The polarization information is acquired by the pixelized polarization detector. The polarization spectrum is reconstructed by TwIST. Compared with the existing polarization spectrum detection technology, the linear compression sampling model does not require Fourier transform and spatial filter. The technology improves the luminous flux and the noise sensitivity, and spectral resolution reduction and channel cross talk caused by traditional polarization spectrum acquisition technology got improved and eliminated. Acquisition of polarization spectrum with 25 bands' range in 450–650 nm using the technology is demonstrated. The feasibility has been successfully verified by the experiment.

References

1. Tyo JS, Goldstein DL, Chenault DB, Shaw JA (2006) Review of passive imaging polarimetry for remote sensing applications. Appl Opt 45:5453–5469
2. Oka K, Kato T (1999) Spectroscopic polarimetry with a channeled spectrum. Opt Lett 24:1475–1477
3. Kudenov MW, Hagen NA, Dereniak EL, Gerhart GR (2007) Fourier transform channeled spectropolarimetry in the mwir. Opt Express 15:12792–12805
4. Chan VC, Kudenov M, Liang C, Zhou P, Dereniak E (2014) Design and application of the snapshot hyperspectral imaging fourier transform (shift) spectropolarimeter for fluorescence imaging. Proc SPIE 8949:894–903
5. Mu T, Zhang C, Li Q, Wei Y, Chen Q, Jia C (2014) Snapshot full-stokes imaging spectropolarimetry based on division-of-aperture polarimetry and integral-field spectroscopy. Proc SPIE 9298:92980D
6. Locke AM, Salyer D, Sabatke DS, Dereniak EL (2003) Design of a SWIR computed tomographic imaging channeled spectropolarimeter. Proc. SPIE 5158:507136
7. Dereniak EL, Craven JM, Kudenov MW (2009) False signature reduction in infrared channeled spectropolarimetry. Proc SPIE 7419:7419
8. Yiren W, Fu C, Wu D, Xie Y (2019) Channeled compressive imaging spectropolarimeter. Opt Exp 27(3|4)
9. Candès EJ, Wakin MB (2008) An introduction to compressive sampling. IEEE Signal Process Mag 25:21–30
10. Wagadarikar A, John R, Willett R, Brady D (2008) Single disperser design for coded aperture snapshot spectral imaging. Appl Opt 47:B44–B51
11. Arce GR, Brady DJ, Carin L, Arguello H, Kittle DS (2014) Compressive coded aperture spectral imaging: an introduction. IEEE Signal Process Mag 31(1):105–115
12. Tan J, Ma Y, Rueda H, Baron D, Arce GR (2016) Compressive hyperspectral imaging via approximate message passing. IEEE J Sel Top Signal Process 10:389–401
13. Chipman RA (2009) Handbook of optics. McGraw-Hill
14. Kalibjian R (2004) Stokes polarization vector and Mueller matrix for a corner-cube reflector. Opt Commun 240:39–68
15. Arce GR, Brady DJ, Carin L, Arguello H, Kittle DS (2014) Compressive coded aperture spectral imaging: an introduction. IEEE Signal Process Mag 31:105–115
16. Zibulevsky M, Elad M (2010) L1-l2 optimization in signal and image processing. IEEE Signal Process Mag 27(3):76–88
17. Daubechies I, DeVore R, Fornasier M, Güntürk CS (2010) Iteratively reweighted least squares minimization for sparse recovery. Commun Pure Appl Math 63(1):1–38
18. Tsai T-H, Brady DJ (2013) Coded aperture snapshot spectral polarization imaging. Appl Opt 52:2153–2161
19. Bioucas-Dias JM, Figueiredo MA (2007) A new twist: two-step iterative shrinkage/thresholding algorithms for image restoration. IEEE Trans Image Process 16:2992–3004
20. Zhao X, Boussaid F, Bermak A, Chigrinov VG (2009) Thin photopatterned micropolarizer array for CMOS image sensors. IEEE Photon Technol Lett 21:805–807
21. Tokuda T, Sato S, Yamada H, Sasagawa K, Ohta J (2009) Polarisation-analysing CMOS photosensor with monolithically embedded wire grid polarizer. Electron Lett 45:228–230
22. Wagadarikar Ashwin A, Pitsianis Nikos P, Sun Xiaobai, Brady David J (2008) Spectral image estimation for coded aperture snapshot spectral imagers. SPIE 7076:707602
23. Ma X, Wang Z, Li Y, Arce GR, Dong L, Garcia-Frias J (2018) Fast optical proximity correction method based on nonlinear compressive sensing. Opt Express 26:14479–14498
24. Powell SB, Gruev V (2013) Calibration methods for division-of-focalplane polarimeters. Opt Express 21:21039–21055
25. Stokes GG (1852) On the composition and resolution of streams of polarized light from different sources. Trans Cambridge Philos Soc 9:399–416

Optimizing Pose of UAV Image Based on PPK Technology

Guangrui Yu, Yuncai Su, Lili Yu, Lianbing Gong, and Danyang Zhao

Abstract Geographic information products of UAV have been developed from single to multiple directions. How to obtain the pose parameters of images with high quality is the key to ensure the accuracy of products. In terms of both production quality and efficiency, the development of RTK-UAVs has gradually become the focus of public attention. In view of the fact that these UAVs are mostly used for navigation, the real-time correction of camera coordinates is not considered, and the accuracy of attitude parameters obtained by the inertial navigation unit of the micro/mini UAV is low. So a small number of GCPs still need to be laid to meet the accuracy requirements of surveying and mapping results. This paper studied a new PPK method, taking the BD930 GNSS module as an example, the UAV system was modified by using the configuration scheme of three non-collinear antennas. A carrier phase dual-difference model combining GPS, BDS and GLONASS was constructed. Then baseline vector's determination was carried out by LAMBDA method. The system errors caused by fixed frequency, camera exposure delay and position difference were eliminated by three spline function GNSS interpolation algorithm and eccentricity measurement technology for space rear intersection. Taking the corrected position of the camera and the calculated attitude parameters as the initial values, GNSS-assisted self-bundle block adjustment was introduced to obtain high-precision pose and distortion parameters of the UAV image. The flight test in a hilly area proved that the accuracy of differential GNSS system could reach 0.01 m. Under dynamic flight, the fixed solution ratio was 58.7% higher than that of the traditional method. Without using any GCPs, the error of CPs was less than 0.3 m, and with four corner GCPs, the median error of CPs was less than 0.2 m. This confirmed that the position and distortion parameters obtained in this paper were correct and could meet the accuracy requirements of geographic information results.

G. Yu · D. Zhao (✉)
Key Laboratory for Precision and Non-traditional Machining Technology of Ministry of Education, Dalian University of Technology, Dalian 116024, People's Republic of China
e-mail: danyangzhao@yeah.net

G. Yu · Y. Su
32023 Troops, Dalian 116023, People's Republic of China

L. Yu · L. Gong
China Nuclear Industry Fifth Construction Co., Ltd, Shanghai 201512, People's Republic of China

L. Wang et al. (eds.), *Proceedings of the 6th China High Resolution Earth Observation Conference (CHREOC 2019)*, Lecture Notes in Electrical Engineering 657,
https://doi.org/10.1007/978-981-15-3947-3_2

Keywords PPK · GNSS interpolation · Eccentricity calibration · Self-BBA · Pose parameters

1 Introduction

Unmanned aerial vehicle (UAV) has been widely used in petroleum geophysical exploration, environmental monitoring, urban planning and border patrol due to its advantages of flexibility, strong current situation, less influence by airspace and weather and high cost performance [1–4]. In particular, UAV plays an irreplaceable role in the emergency support of natural disasters such as earthquake, forest fire, tsunami and debris flow monitoring [5–7]. However, there are several disadvantages in traditional UAV aerial photography, such as poor attitude stability, large image distortion and small image size, and it is difficult to carry the position and orientation system (POS) with high accuracy due to the limitation of load. At present, only the navigation equipment is used to obtain the outline position and attitude of the image. Therefore, in order to ensure the accuracy of bundle block adjustment (BBA) method, the camera needs to be calibrated strictly in advance. At the same time, a large number of ground control points (GCPs) are needed to restore the image pose parameters by aerial triangulation [8, 9]. In this way, it not only increases the workload of field work, but also people entering difficult areas such as mountains, waters and forests, and it is difficult to set up control points, so the accuracy cannot be guaranteed [10, 11]. For this reason, how to obtain high-precision pose data of images at shooting time with little or no help of field GCPs has become the research focus of scholars.

Compared with manned aircraft, UAVs are more cost-effective and responsive. However, the observations acquired by the POS are usually less accurate than those acquired in manned aerial photogrammetry [7]. Global positioning system (GPS) for UAV applications includes single point positioning (SPP) and differential positioning. SPP is mainly used in relatively low precision situations, such as navigation. In order to provide higher accuracy, single-frequency or dual-frequency geodetic-grade GPS receivers for carrier phase measurement must be used, but the price is between $5000 and $15000, and the weight and volume are not suitable for UAV carrying requirements [4]. Differential positioning includes real-time kinematic (RTK) and post-processed kinematic (PPK) technology. RTK-GPS is a typical differential positioning method, and through the measurement of GPS signal carrier phase, the user can obtain a high-precision real-time position, which can meet a variety of measurement and the application of geographic information system requirements [5, 12, 13]. Compared with RTK technology, PPK technology is a post-processing differential of carrier phase observations, which can effectively avoid the interruption of signal transmission while guaranteeing the same accuracy as RTK technology. Therefore, differential positioning can be regarded as a powerful economic alternative to geodetic-grade GPS receivers. However, geodetic-grade receiver and antenna have been still expensive compared to consumer-grade ones. In the low-cost differential system test, it is found that for receivers, the performance difference between consumer-grade and geodetic-grade is small, By contrast, as to antennas, low-cost antennas, the performance degradation is large [4, 14, 15].

With the development of miniaturization of global navigation satellite system (GNSS), RTK and PPK have been continuously applied to micro/mini unmanned aerial vehicles (MUAVs) [16, 17]. The more representative ones are eBee Plus UAV from senseFly, Switzerland, UX5 HP UAV from Trimble, American, CW10 UAV from Chengdu Vertical Automation, China, etc. [18]. Differential GNSS UAV only needs to set up base station on the ground and acquire satellite carrier phase information together with rover station on the UAV. Through PPK algorithm, the accurate picture positions at exposure can be known [12, 19]. On this basis, GNSS-assisted BBA can be used for integrated positioning, which can avoid the complex process of laying a large number of GCPs in traditional UAV photography and improve the accuracy and efficiency of data processing [5, 10, 13]. The commercial UAVs using differential system are mostly applied to navigation, and system integration is high, without considering the dynamic case GNSS satellite signal reception, camera exposure and eccentricity effect between delay and receiver antenna and the camera center, etc. [19, 20]. In order to satisfy the accuracy requirements of topographic mapping and three-dimensional modeling, it is necessary to measure the camera distortion parameters in advance and lay a certain number of GCPs in the later data processing [21–23].

This paper takes BD930 positioning board produced by Trimble company as the research object, completes the UAV system modification, studies PPK algorithm under multi-system and improves the static positioning accuracy and dynamic stability of consumer-grade receivers and antennas. Based on spline curve, a flight model is built, and the influence of system error caused by camera exposure delay and eccentricity is discussed. A flight test was carried out in a hilly area of Henan Province, China. By reasonably setting the lens distortion parameter model, taking corrected GNSS camera station coordinates and calculated image attitude angle as initial values. Though the self-BBA method, the UAV image position and attitude parameters were obtained precisely. The advantages of this method were compared and analyzed based on the statistical results of CPs residuals. Finally, discussion of the results is given, as well as recommendations for further investigations.

2 Measurement Method of UAV Pose Parameters

2.1 Double Difference Model Fused with BDS/GPS/GLONASS

BD930 positioning board supports GPS, GLONASS, Galileo and BDS satellite signal reception. It has fast RTK initialization speed and built-in Kalman filter PVT engine.

It can obtain low multipath, high dynamic and low noise observation data. The low-cost differential GNSS system designed in this paper consists of BD930 board, power supply system, storage system, measurement antenna and post-processing software (see Fig. 1).

Figure 2 depicts the working principle of PPK UAV. The base station on known position receives the carrier and pseudo-range information of satellite together with the antenna on UAV. When more than four satellite signals are received synchronously, the errors such as satellite orbit, cloud occlusion, satellite clock and multi-path effect can be eliminated in GNSS positioning process according to the principle of relative positioning.

Compared with RTK technology, PPK technology carries out post-differential processing of carrier phase observation value. Through the dynamic relative positioning algorithm, the real-time position data of the receiver can be calculated by linear combination of satellite carrier observation information recorded by base station and rover station in the same time period [24]. BD930 receives BDS, GPS and GLONASS satellites at the same time. Figure 3 shows the visibility and reception of the satellite signal drawn by RTKPLOT. It can be found that three kinds of effective satellites can be received at each time, which makes up for the shortcomings of single system in space distribution and integrity.

The pseudo-range and carrier phase observation equations of GPS, BDS and GLONASS can be expressed as follows:

$$\begin{cases} \rho^{G'} = \rho^{G} + c(t_{\mathrm{r}} - t_{\mathrm{s}}^{G}) + L^{G} + D^{G} + M^{G} + T^{G} + v^{G}. \\ \rho^{C'} = \rho^{C} + c(t_{\mathrm{r}} + t^{\mathrm{CG}} - t_{\mathrm{s}}^{C}) + L^{C} + D^{C} + M^{C} + T^{C} + v^{C}. \\ \rho^{R'} = \rho^{R} + c(t_{\mathrm{r}} + t^{\mathrm{RG}} - t_{\mathrm{s}}^{R}) + L^{R} + D^{R} + M^{R} + T^{R} + v^{R}. \end{cases} \tag{1}$$

$$\begin{cases} \lambda^{G}\varphi^{G} = \rho^{G} + \lambda^{G}N^{G} + c(t_{\mathrm{r}} - t_{\mathrm{s}}^{G}) + L^{G} - D^{G} + M^{G} + T^{G} + v^{G}. \\ \lambda^{C}\varphi^{C} = \rho^{C} + \lambda^{C}N^{C} + c(t_{\mathrm{r}} + t^{\mathrm{CG}} - t_{\mathrm{s}}^{C}) + L^{C} - D^{C} + M^{C} + T^{C} + v^{C}. \\ \lambda^{R}\varphi^{R} = \rho^{R} + \lambda^{R}N^{R} + c(t_{\mathrm{r}} + t^{\mathrm{RG}} - t_{\mathrm{s}}^{R}) + L^{R} - D^{R} + M^{R} + T^{R} + v^{R}. \end{cases} \tag{2}$$

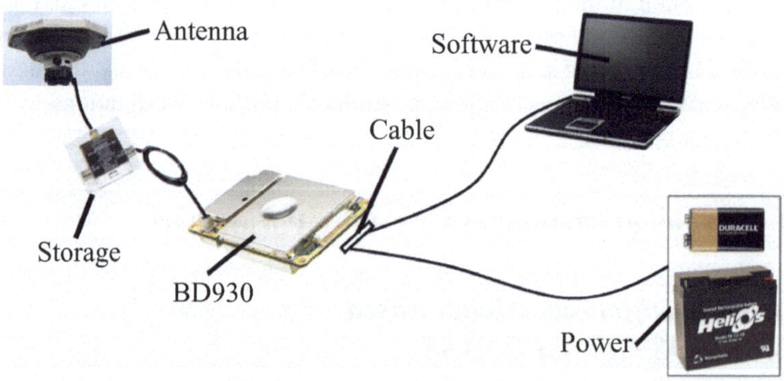

Fig. 1 Low-cost differential GNSS system

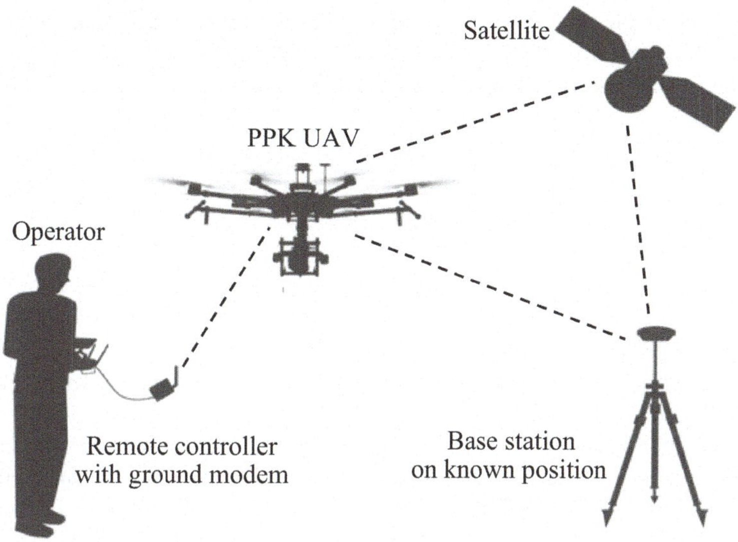

Fig. 2 Working principle of PPK UAV

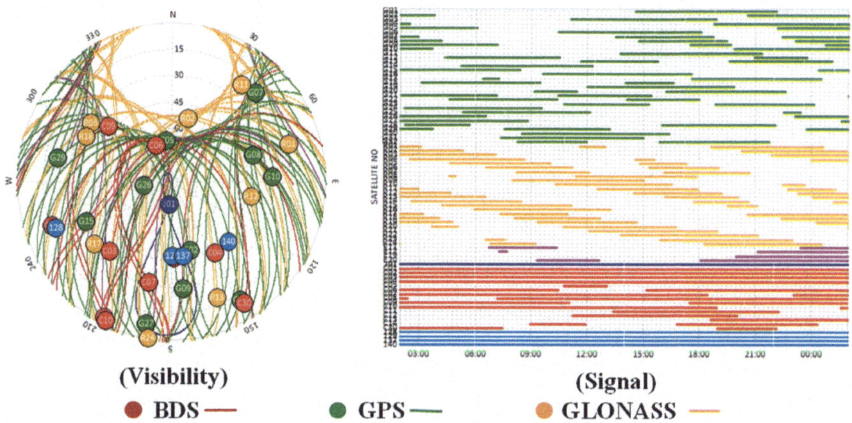

Fig. 3 BD930 receiving satellite drawn by RTKPLOT

where C, G, R are, respectively, BDS, GPS and GLONASS, ρ is the geometric distance between the receiver and the satellite, ρ' is the pseudo-range, λ is the carrier wavelength, c is the electromagnetic wave transmission speed, N is the integer ambiguity, t_r is the receiver clock error, t_s is the satellite clock error, L, D, M, T and V are, respectively, the errors caused by troposphere, ionosphere, multipath, antenna phase center and satellite ephemeris, t^{CG} is the time reference deviation for GPS and BDS, t^{RG} is the time reference deviation for GPS and GLONASS.

The three satellite systems using different time reference and space reference, GPST, BDT and GLONASST correspond to different time reference, and the conversion relationship between them is:

$$\begin{aligned} &\mathrm{GPST} = \mathrm{BDT} + 14s + [\mathrm{UTC(USNO)} - \mathrm{UTC(UTSC)}].\\ &\mathrm{GPST} = \mathrm{GLONASST} - 3h + \tau_{\mathrm{r}} + 1s\times n^{-19}s + [\mathrm{UTC(USNO)} - \mathrm{UTC(SU)}]. \end{aligned} \tag{3}$$

UTC(USNO), UTC(UTSC) and UTC(SU) are, respectively, three coordinated universal time (UTC) reference systems, τ_{r} is the systematic deviation within 1 ms between GLONASST and UTC(SU).

The coordinate datum of the three systems are, respectively, $WGS{-}84$, $CGCS2000$ and $PZ{-}90$.

According to the definition of coordinate system, only two parameters of reference ellipsoid in $WGS{-}84$ and $CGCS2000$ coordinate systems are slightly different, and for the short flight range of UAV, it can be ignored. The transformation relationship between $WGS{-}84$ and $PZ{-}90$ is obtained by Bursa model as follows:

$$\begin{aligned} \begin{bmatrix} X \\ Y \\ Z \end{bmatrix}_{WGS-84} &= \begin{bmatrix} X_0 \\ Y_0 \\ Z_0 \end{bmatrix} + (1+m)\begin{bmatrix} 1 & \beta_Z & -\beta_Y \\ -\beta_Z & 1 & \beta_X \\ \beta_Y & -\beta_X & 1 \end{bmatrix}\begin{bmatrix} X' \\ Y' \\ Z' \end{bmatrix}_{PZ-90} \\ &= \begin{bmatrix} -0.47 \\ -0.51 \\ -1.56 \end{bmatrix} + (1 + 22\times 10^{-9}) \\ &\quad \begin{bmatrix} 1 & -1.728\times 10^{-6} & -0.017\times 10^{-6} \\ 1.728\times 10^{-6} & 1 & 0.076\times 10^{-6} \\ 0.017\times 10^{-6} & -0.076\times 10^{-6} & 1 \end{bmatrix}\begin{bmatrix} X' \\ Y' \\ Z' \end{bmatrix}_{PZ-90}. \end{aligned} \tag{4}$$

The applicability and accuracy of carrier phase difference is better than position and pseudo-range difference [25]. As UAV fly fast, the stability of miniature GNSS receiver is poor. In order to eliminate the common errors of orbit, satellite and atmosphere between base station and rover station, also considering the frequency division multiple access (FDMA) technology adopted by GLONASS, a dual-difference model integrating GPS, BDS and GLONASS is proposed:

$$\begin{cases} \lambda^{iG}\nabla\Delta\varphi_{br}^{ijG} = \nabla\Delta\rho_{br}^{ij} + \lambda^{iG}\nabla\Delta N_{br}^{ijG} + \nabla\Delta L_{br}^{ijG} - \nabla\Delta D_{br}^{ijG} + \nabla\Delta M_{br}^{ijG} \\ \qquad + \nabla\Delta T_{br}^{ijG} + \nabla\Delta v_{br}^{ijG}. \\ \lambda^{iC}\nabla\Delta\varphi_{br}^{ijC} = \nabla\Delta\rho_{br}^{ij} + \lambda^{iC}\nabla\Delta N_{br}^{ijC} + \nabla\Delta L_{br}^{ijC} - \nabla\Delta D_{br}^{ijC} + \nabla\Delta M_{br}^{ijC} \\ \qquad + \nabla\Delta T_{br}^{ijC} + \nabla\Delta v_{br}^{ijC}. \\ \lambda^{iR}\nabla\Delta\varphi_{br}^{ijR} = \nabla\Delta\rho_{br}^{ij} + \lambda^{iR}\nabla\Delta N_{br}^{ijR} + (\lambda^{i} - \lambda^{j})\nabla N_{br}^{jR} + \nabla\Delta L_{br}^{ijR} \\ \qquad - \nabla\Delta D_{br}^{ijR} + \nabla\Delta M_{br}^{ijR} + \nabla\Delta T_{br}^{ijR} + \nabla\Delta v_{br}^{ijR}. \end{cases} \tag{5}$$

Limited by flight control system and endurance time, MUAVs usually fly within 20 km, so the double difference of system errors can be neglected. In a certain epoch, assuming that the differential system receives m GPS, n BDS and q GLONASS satellites at the same time, all of which are referenced by the satellites with the highest altitude angle, and the dual-difference observation model can be combined:

$$\begin{bmatrix} \lambda^{1G}\nabla\Delta\varphi_{br}^{12G} \\ \vdots \\ \lambda^{1G}\nabla\Delta\varphi_{br}^{1mG} \\ \lambda^{1C}\nabla\Delta\varphi_{br}^{12C} \\ \vdots \\ \lambda^{1C}\nabla\Delta\varphi_{br}^{1nC} \\ \lambda^{1R}\nabla\Delta\varphi_{br}^{12R} \\ \vdots \\ \lambda^{1R}\nabla\Delta\varphi_{br}^{1qR} \end{bmatrix} = \begin{bmatrix} a^{G12} & b^{G12} & c^{G12} \\ \vdots & \vdots & \vdots \\ a^{G1m} & b^{G1m} & c^{G1m} \\ a^{C12} & b^{C12} & c^{C12} \\ \vdots & \vdots & \vdots \\ a^{C1n} & b^{C1n} & c^{C1n} \\ a^{R12} & b^{R12} & c^{R12} \\ \vdots & \vdots & \vdots \\ a^{R1q} & b^{R1q} & c^{R1q} \end{bmatrix} \begin{bmatrix} B_x \\ B_y \\ B_z \end{bmatrix}$$
$$+ \begin{bmatrix} \lambda^{1G} & & & & & & & \\ & \ddots & & & & & & \\ & & \lambda^{1G} & & & & & \\ & & & \lambda^{1C} & & & & \\ & & & & \ddots & & & \\ & & & & & \lambda^{1C} & & \\ & & & & & & \lambda^{1R} & & \\ & & & & & & & \ddots & \\ & & & & & & & & \lambda^{1R} \end{bmatrix} \begin{bmatrix} \nabla\Delta N_{br}^{12G} \\ \vdots \\ \nabla\Delta N_{br}^{1mG} \\ \nabla\Delta N_{br}^{12C} \\ \vdots \\ \nabla\Delta N_{br}^{1nC} \\ \nabla\Delta N_{br}^{12R} \\ \vdots \\ \nabla\Delta N_{br}^{1qR} \end{bmatrix}$$
$$+ \begin{bmatrix} 0 & & & & \\ & \ddots & & & \\ & & 0_{m+n-2} & & \\ & & & \lambda^{1R}-\lambda^{2R} & & \\ & & & & \ddots & \\ & & & & & \lambda^{1R}-\lambda^{qR} \end{bmatrix} \begin{bmatrix} 0 \\ \vdots \\ 0_{m+n-2} \\ \nabla N_{br}^{2R} \\ \vdots \\ \nabla N_{br}^{qR} \end{bmatrix}. \quad (6)$$

where $\nabla\Delta$ is the double difference and B is the baseline vector. The integration of BDS, GPS and GLONASS can ensure the effective acquisition of satellite signals. On the premise of neglecting the correlation of the three systems, variance–covariance matrix can be established combining the error propagation law. Using MLAMBDA method can effectively improve the fixed ratio of ambiguity [26]. Therefore, we use the least square method to deal with the redundant carrier observations and solve the three-dimensional coordinates of the UAV measurement antenna.

2.2 Attitude Parameters Calculation of Three-Antenna Configuration

The inertial measurement unit (IMU) equipped with MUAV has low measurement accuracy and is currently only used for navigation. In the data processing process, the measurement results generally participate in iterative calculations in the form of initial values. In this paper, the carrier attitude measurement method based on differential technique is studied.

In order to measure the three-dimensional attitude of UAV, three non-collinear antennas are deployed on the fuselage, in which the antenna plane is parallel to the main plane of UAV and the main baseline is parallel to the axis of the fuselage. Through the translation transformation, the origin of the global coordinate system and the body coordinate system can be unified to the phase center of the main antenna, and the coordinates of the main antenna and the auxiliary antenna in the two coordinate systems can be obtained by static calibration and PPK technology, respectively. The distribution of antennas in the body coordinate system is shown in Fig. 4.

The figure shows that in the body coordinate system, the *Y*-axis is the baseline of the main antenna 1 and the auxiliary antenna 2, the *Z*-axis is perpendicular to the plane where the antenna is located, and the *X*-axis constitutes right-hand coordinate system together.

Sampling b_i and l_i are, respectively, coordinates of the *i*-th antenna in the body coordinate system and the global coordinate system. α, β and γ are the yaw angle, the roll angle and the pitch angle, which represent the rotation parameters of the body coordinate system relative to the global coordinate system. Figure 4 shows the coordinates of the main and auxiliary antennas in the body coordinate system:

Fig. 4 Distribution of antennas in the body coordinate system

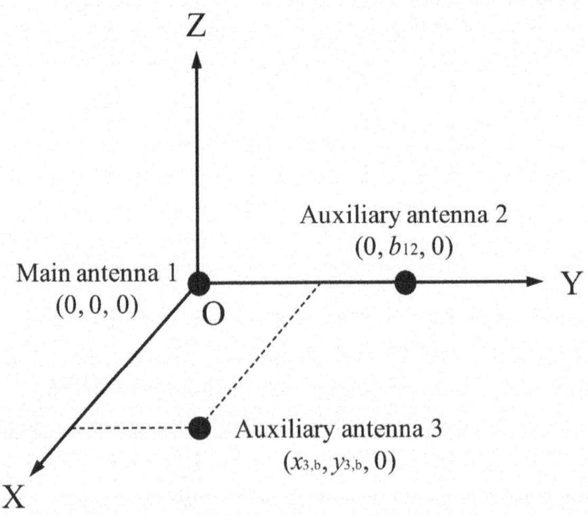

$b_1 = \left[0\,0\,0\right]^{\mathrm{T}}, b_2 = \left[0\ b_{12}\ 0\right]^{\mathrm{T}}, b_3 = \left[x_{3,b}\ y_{3,b}\ 0\right]^{\mathrm{T}}$. The conversion relationship between two coordinate systems can be expressed as:

$$b_i = R_l^b(l_i - l_1) = R_Y(\beta)R_X(\gamma)R_Z(\alpha)(l_i - l_1)$$
$$= \begin{bmatrix} \cos\beta\cos\alpha - \sin\beta\sin\gamma\sin\alpha & \cos\beta sin\alpha + \sin\beta\sin\gamma\cos\alpha & -\sin\beta\cos\gamma \\ -\cos\gamma\sin\alpha & \cos\gamma\cos\alpha & \sin\gamma \\ \sin\beta\cos\alpha + \cos\beta\sin\gamma\sin\alpha & \sin\beta sin\alpha - \cos\beta\sin\gamma\cos\alpha & \cos\beta\cos\gamma \end{bmatrix}(l_i - l_1). \tag{7}$$

According to the orthogonality of the rotation matrix, the conversion relationship of the main baseline between the two coordinate systems is obtained as:

$$l_2 - l_1 = \begin{bmatrix} x_{2,l} - x_{1,l} \\ y_{2,l} - y_{1,l} \\ z_{2,l} - z_{1,l} \end{bmatrix} = b_{12}\begin{bmatrix} -\cos\gamma\sin\alpha \\ \cos\gamma\cos\alpha \\ \sin\gamma \end{bmatrix}. \tag{8}$$

From Eq. (8), the yaw angle α and the pitch angle γ are calculated through the main baseline b_{12}:

$$\alpha = -\arctan\left(\frac{x_{2,l} - x_{1,l}}{y_{2,l} - y_{1,l}}\right). \tag{9}$$

$$\gamma = \arcsin\left(\frac{z_{2,l} - z_{1,l}}{b_{12}}\right) = \arctan\left(\frac{z_{2,l} - z_{1,l}}{\sqrt{\left(x_{2,l} - x_{1,l}\right)^2 + \left(y_{2,l} - y_{1,l}\right)^2}}\right). \tag{10}$$

Next, the roll angle β is calculated according to the relationship of the baseline formed by the auxiliary antenna 3 in the two coordinate systems:

$$b_3 = \begin{bmatrix} x_{3,b} \\ y_{3,b} \\ 0 \end{bmatrix} = R_Y(\beta)R_X(\gamma)R_Z(\alpha)\begin{bmatrix} x_{3,l} - x_{1,l} \\ y_{3,l} - y_{1,l} \\ z_{3,l} - z_{1,l} \end{bmatrix}. \tag{11}$$

The yaw angle α and the pitch angle γ are calculated according to Eqs. (9) and (10). Firstly, the baseline is rotated from the global coordinate system by the α angle around the Z-axis, and then, the γ angle is rotated around the X-axis to obtain a new baseline vector:

$$\begin{bmatrix} x'_{3,l} - x_{1,l} \\ y'_{3,l} - y_{1,l} \\ z'_{3,l} - z_{1,l} \end{bmatrix} = \begin{bmatrix} 1 & 0 & 0 \\ 0 & \cos\gamma & \sin\gamma \\ 0 & -\sin\gamma & \cos\gamma \end{bmatrix}\begin{bmatrix} \cos\alpha & \sin\alpha & 0 \\ -\sin\alpha & \cos\alpha & 0 \\ 0 & 0 & 1 \end{bmatrix}\begin{bmatrix} x_{3,l} - x_{1,l} \\ y_{3,l} - y_{1,l} \\ z_{3,l} - z_{1,l} \end{bmatrix}. \tag{12}$$

Similarly, the new relationship is obtained by the orthogonality of the rotation matrix $R_Y(\beta)$:

$$\begin{bmatrix} x_{3,b}\cos\beta \\ y_{3,b} \\ -x_{3,b}\sin\beta \end{bmatrix} = \begin{bmatrix} x'_{3,l} - x_{1,l} \\ y'_{3,l} - y_{1,l} \\ z'_{3,l} - z_{1,l} \end{bmatrix}. \tag{13}$$

Thus, the roll angle β is:

$$\beta = -\arctan\left(\frac{z'_{3,l} - z_{1,l}}{x'_{3,l} - x_{1,l}}\right). \tag{14}$$

3 Correction of Image Position Parameters

3.1 Cubic Spline Interpolation for Solving Exposure Delay

The sampling frequency of BD930 GNSS differential system carried by UAV is 20 Hz. During the flight process, it is sampled every 0.05 s. Therefore, in order to get the position data of the shooting time, it is necessary to record the flash pulse of the camera and calculate the exposure time coordinates by interpolation algorithm. At present, the curve fitting methods of space trajectory include Lagrange interpolation, Chebyshev polynomial interpolation, spline interpolation, etc. [27]. Some studies have found that the order of interpolation polynomials should not be too high or too low. Lagrange polynomials are prone to "Runge phenomenon" in interpolation calculation. Chebyshev polynomial interpolation requires an order of 11 to achieve millimeter accuracy. As a result, the amount of calculation increases, and the efficiency of operation are affected. UAV flight is easily affected by wind speed, wind direction, navigation error, air temperature, etc. There are some differences in speed at each instant of taking photos. Considering spline function is an important approximation tool in curve fitting, interpolation, numerical differentiation [28]. We propose a GNSS three-coordinate interpolation algorithm based on cubic spline function (see Fig. 5).

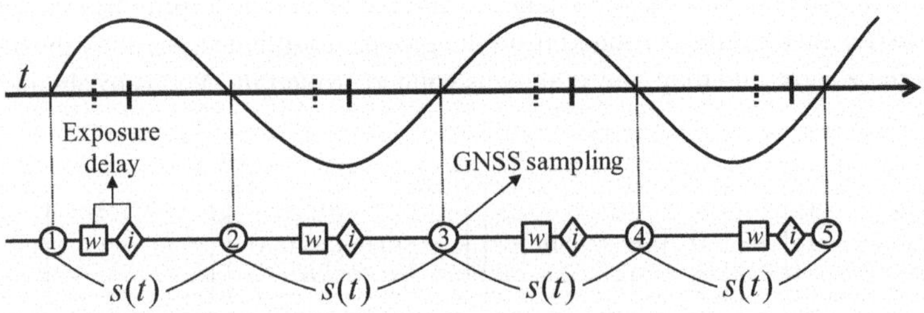

Fig. 5 Exposure delay correction model

Sampling points of UAV on a single route are set as $P = \{a = t_0 < t_1 < \cdots < t_n = b\}$ be the uniform partition over the interval $[a, b]$ with equally spaced grid points $t_i = a + ih, i = 0, 1 \ldots n$. Where $h = (b - a)/n$.

The coordinate obtained by the differential GNSS system at the sampling point is:

$$f(t_i) = y_i, \quad i = 0, 1, \ldots, n. \tag{15}$$

We construct a cubic spline function $s(t) \in S_3(t_1, t_2, \ldots, t_{n-1})$, where $s(t)$ satisfies the following interpolating conditions:

$$s(t_i) = y_i, \quad i = 0, 1, \ldots, n. \tag{16}$$

We denote $s'(t_k) = m_k (k = 0, 1, \ldots, n), h_k = t_{k+1} - t_k = 0.05 (k = 0, 1, \ldots, n-1)$, $s(t)$ in the subinterval $[t_k, t_{k+1}]$ can be expressed as:

$$\begin{aligned} s(t) = {} & \frac{h_k + 2(t - t_k)}{h_k^3}(t - t_{k+1})^2 y_k + \frac{h_k - 2(t - t_{k+1})}{h_k^3}(t - t_k)^2 y_{k+1} \\ & + \frac{(t - t_k)(t - t_{k+1})^2}{h_k^2} m_k + \frac{(t - t_{k+1})(t - t_k)^2}{h_k^2} m_{k+1}. \end{aligned} \tag{17}$$

In order to get constant coefficients $m_0, m_1, \ldots, m_n$, we need to calculate the second derivative of $s(t)$:

$$\begin{aligned} s''(t) = {} & \frac{6t - 2t_k - 4t_{k+1}}{h_k^2} m_k + \frac{6t - 4t_k - 2t_{k+1}}{h_k^2} m_{k+1} \\ & + \frac{6(t_k + t_{k+1} - 2t)}{h_k^2}(y_{k+1} - y_k), \quad x \in [t_k, t_{k+1}]. \end{aligned} \tag{18}$$

and

$$\lim_{t \to t_k^+} s''(t) = -\frac{4}{h_k} m_k - \frac{2}{h_k} m_{k+1} + \frac{6}{h_k^2}(y_{k+1} - y_k). \tag{19}$$

Similarly, we can obtain the expression of $s(t)$ in the interval $[t_{k-1}, t_k]$ and get the equation according to $\lim_{t \to t_k^+} s''(t) = \lim_{t \to t_k^-} s''(t) (k = 1, 2, \ldots, n - 1)$.

$$\lambda_k m_{k-1} + 2m_k + \mu_k m_{k+1} = g_k (k = 1, 2, \ldots, n - 1). \tag{20}$$

where

$$\lambda_k = \frac{h_k}{h_k + h_{k-1}}, \quad \mu_k = \frac{h_{k-1}}{h_k + h_{k-1}}, \quad g_k = 3\left(\mu_k \frac{y_{k+1} - y_k}{h_k} + \lambda_k \frac{y_k - y_{k-1}}{h_{k-1}}\right). \tag{21}$$

By making use of Eq. (21), we obtain a system of n-1 equations in n+1 unknowns, so we next consider a boundary value problems:

$$\begin{aligned} s'(t_0) = f_0', s'(t_n) = f_n'. \\ s''(t_0) = f_0'', s''(t_n) = f_n''. \end{aligned} \tag{22}$$

Therefore, the constant coefficients $m_0, m_1, \ldots, m_n$ can be determined by using (22) to calculate the GNSS antenna position of each exposure time.

3.2 *Real-Time Measurement of Eccentricity Based on Space Resection*

In order to ensure the flight and signal receiving performance, GNSS antenna should be installed on the top of UAV, and affected by the requirement of system gravity center, the phase center of the antenna cannot coincide with the lens center of aerial camera, so there are three eccentric components shown in Fig. 6. Due to the influence of flight control system and airflow changes in the process of photogrammetry, UAV has different angles of tilt and deflection. These changes will make eccentricity affect the accuracy of geographic positioning and later mapping [20]. Thus, we need to initially calibrate the eccentric components and correct them in real time with flight attitude to acquire accurate images position.

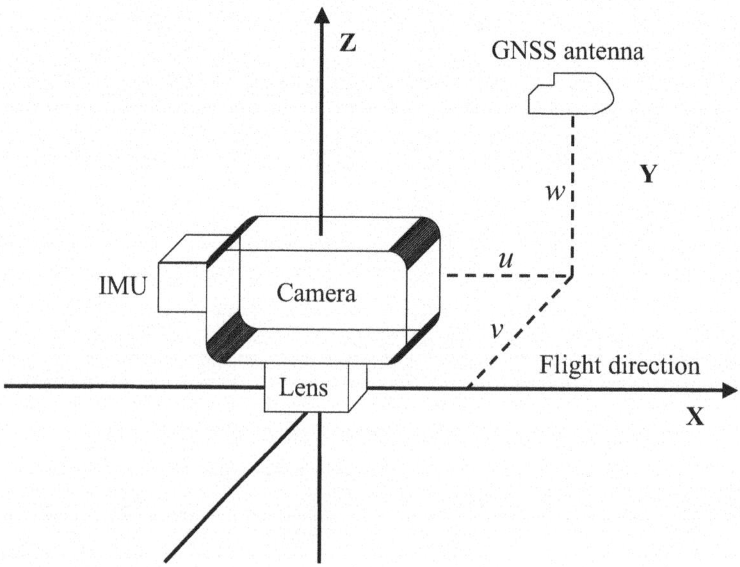

Fig. 6 Eccentric components between antenna and aerial camera

Traditional eccentricity measurement methods include close-range photogrammetry, manual measurement, theodolite measurement and direct projection measurement, which do not take into account the system integration error and real-time correction. In the calibration process, these methods mostly rely on other measurement systems, which need higher site requirements.

In this paper, a real-time eccentricity measurement method based on space rear rendezvous is proposed. Firstly, a global coordinate system $O\text{-}XYZ$ is established by setting a certain number of GCPs on the undulating ground. Then, the UAV is leveled over these points, and the height ensures that all points are in the imaging range. By adjusting the vertical axis level and yaw angle to zero, we get a differential GNSS system with the base station on the ground. Finally, we establish the conversion relationship between image space coordinate system and global coordinate system by space resection algorithm based on single image. The initial value of eccentric components can be obtained by comparing the origin of image space coordinate system with the differential GNSS position. The measurement scheme is shown in Fig. 7.

We set the coordinates of the GCPs in the global coordinate system as $P_i(X_i, Y_i, Z_i)$ $(i = 1, 2, \ldots, i \geq 3)$, and in the image coordinate system as $p_i(x_i, y_i)$ $(i = 1, 2, \ldots, i \geq 3)$. Then, the collinear conditional equation can be established as follows:

$$\begin{cases} x_i - x_0 - \triangle x = -f\dfrac{a_1(X_i - X_S) + b_1(Y_i - Y_S) + c_1(Z_i - Z_S)}{a_3(X_i - X_S) + b_3(Y_i - Y_S) + c_3(Z_i - Z_S)}. \\ y_i - y_0 - \triangle y = -f\dfrac{a_2(X_i - X_S) + b_2(Y_i - Y_S) + c_2(Z_i - Z_S)}{a_3(X_i - X_S) + b_3(Y_i - Y_S) + c_3(Z_i - Z_S)}. \end{cases} \tag{23}$$

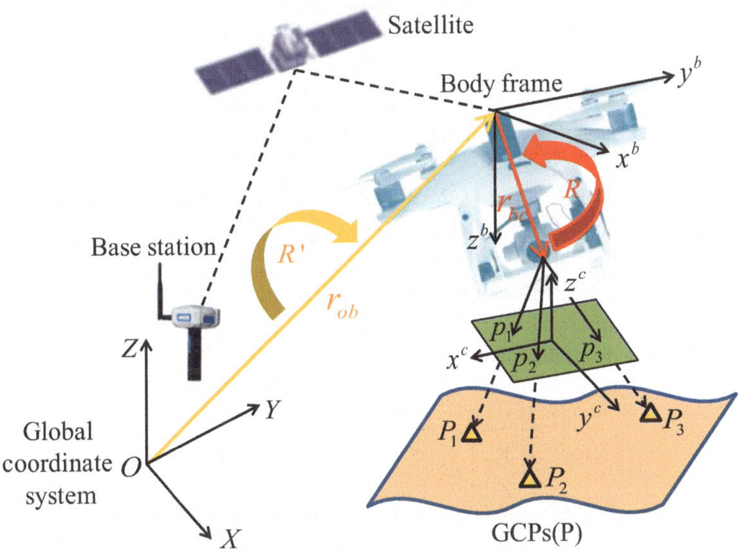

Fig. 7 Eccentric components measurement scheme

and

$$\begin{cases} \triangle x = (x - x_0)(k_1 r^2 + k_2 r^4 + k_3 r^6) - p_1[r^2 + 2(x - x_0)] - 2p_2(x - x_0)(y - y_0). \\ \triangle y = (y - y_0)(k_1 r^2 + k_2 r^4 + k_3 r^6) - p_2[r^2 + 2(y - y_0)] - 2p_1(x - x_0)(y - y_0). \end{cases} \tag{24}$$

In the above Formulas (23) and (24), (x_0, y_0, f) are inner orientation elements, and $(\triangle x, \triangle y)$ are deviations caused by camera distortion parameters. k_1, k_2 and k_3 are radial distortion parameters, and p_1 and p_2 are eccentric distortion parameters. (X_S, Y_S, Z_S) is the coordinate of the camera station S in the global coordinate system. The camera coordinate system is set parallel to the UAV body coordinate system, so the UAV attitude (α, β, γ) is used to represent the image angle elements. The rotation matrix also can be expressed as:

$$R = \begin{bmatrix} \cos\beta\cos\alpha - \sin\beta\sin\gamma\sin\alpha & \cos\beta\sin\alpha + \sin\beta\sin\gamma\cos\alpha & -\sin\beta\cos\gamma \\ -\cos\gamma\sin\alpha & \cos\gamma\cos\alpha & \sin\gamma \\ \sin\beta\cos\alpha + \cos\beta\sin\gamma\sin\alpha & \sin\beta\sin\alpha - \cos\beta\sin\gamma\cos\alpha & \cos\beta\cos\gamma \end{bmatrix}. \tag{25}$$

In theory, six exterior orientation elements of an image can be got by knowing three sets of irrelevant GCPs and image points coordinates. We linearize the redundant observation equation and apply the least square adjustment method to obtain the pose parameters of a single image accurately. Through the coordinate of the phase, center of UAV GNSS antenna is (X_A, Y_A, Z_A), we calculate the coordinate components of the eccentricity in the image space coordinate system as follows:

$$\begin{bmatrix} u \\ v \\ w \end{bmatrix} = R\begin{bmatrix} X_A - X \\ Y_A - Y \\ Z_A - Z \end{bmatrix}. \tag{26}$$

During the flight process of UAV, the attitude changes in real time, which makes the eccentricity, have different effects on image positioning. Combined with the three-antenna configuration method proposed in Sect. 2.2, the differential technique is used to calculate the attitude data $(\alpha_t, \beta_t, \gamma_t)$ at each exposure time. According to Formula (25), the rotation matrix R_t is formed, and the real-time position information of the image is corrected as follows:

$$\begin{bmatrix} X_t \\ Y_t \\ Z_t \end{bmatrix} = \begin{bmatrix} X_{At} \\ Y_{At} \\ Z_{At} \end{bmatrix} - R_t^{\mathrm{T}}\begin{bmatrix} u \\ v \\ w \end{bmatrix}. \tag{27}$$

4 GNSS Assisted Self-BBA

Bundle block adjustment (BBA) is the most rigorous theoretical adjustment method based on the collinear condition equation. It mainly calculates the image exterior orientation elements and the image points global coordinates by the least square principle. In recent years, with the application of POS-assisted aerial triangulation and the structure from motion (SFM) in computer vision technology in aerial photogrammetry, image processing efficiency has been greatly improved. Especially in improving the automation and the robustness of SFM, researchers have invested a lot of energy [29]. But the production cost and the accuracy of the results are considered less.

Most of MUAVs carry non-metric cameras. Before image processing, in order to obtain the corresponding relationship between three-dimensional space and two-dimensional image accurately, we need to calibrate the camera parameters strictly. With the development of self-calibration technology, we consider pose parameters according to PPK and construct a new BBA model with lens distortion parameters, GCPs and image exterior orientation elements. When the inner orientation elements of the camera are known, the observation equation can be expanded to the first term according to Taylor series in the field of unknown initial value. The error equation is as follows:

$$\begin{aligned} V_X &= Bx + A_X t + \quad + Ss - L_X, \ WM : E. \\ V_G &= \quad A_G t \quad\quad - L_G, \ WM : P_G. \\ V_I &= \quad A_I t \quad\quad - L_I, \ WM : P_I. \end{aligned} \tag{28}$$

where

V_X, V_G, V_I are, respectively, the observation correction vector of image points coordinates, differential GNSS camera station coordinates and attitude angles;
$x = [\Delta X \Delta Y \Delta Z]^{\mathrm{T}}$ is the ground point coordinate increment vector;
$t = [\Delta X_s \Delta Y_s \Delta Z_s \Delta \alpha_x \Delta \omega \Delta \kappa]^{\mathrm{T}}$ is the image exterior orientation elements increment vector;
$s = [k_1 k_2 k_3 p_1 p_2]^{\mathrm{T}}$ is the camera distortion parameter vector;
B, A_X, S, A_G, A_I are, respectively, coefficient matrix of observation equation composed of first-order partial derivatives of unknown parameters;
$L_X = \begin{bmatrix} x - x_0 \\ y - y_0 \end{bmatrix}$ is the image point coordinate observation residual vector;
$L_G = \begin{bmatrix} X_A - X_A^0 \\ Y_A - Y_A^0 \\ Z_A - Z_A^0 \end{bmatrix}$ is the differential GNSS camera coordinate residual vector;
$L_I = \begin{bmatrix} \alpha_x - \alpha_{x0} \\ \omega - \omega_0 \\ \kappa - \kappa_0 \end{bmatrix}$ is the attitude angle calculation residual vector;

$P_G = \frac{\sigma_0^2}{\sigma_G^2}E$ is the GNSS camera coordinate weight matrix, $P_I = \frac{\sigma_0^2}{\sigma_I^2}E$ is the attitude angle calculation weight matrix, where σ_0 represents the image coordinate measurement accuracy, σ_G represents the GNSS positioning accuracy, σ_I represents the attitude angle calculation accuracy.

Order:

$$V = \begin{bmatrix} V_X \\ V_G \\ V_I \end{bmatrix}, \quad X = \begin{bmatrix} x \\ t \\ s \end{bmatrix}, \quad L = \begin{bmatrix} L_X \\ L_G \\ L_I \end{bmatrix},$$

$$A = \begin{bmatrix} B & A_X & S \\ 0 & A_G & 0 \\ 0 & A_I & 0 \end{bmatrix}, \quad P = \begin{bmatrix} E & & \\ & P_G & \\ & & P_I \end{bmatrix}. \tag{29}$$

(28) can be written as follows:

$$V = AX - L, \quad WM : P. \tag{30}$$

The norm equation is:

$$(A^{\mathrm{T}}PA)X = A^{\mathrm{T}}PL. \tag{31}$$

The expansion is as follows:

$$\begin{bmatrix} B^{\mathrm{T}}B & B^{\mathrm{T}}A_X & B^{\mathrm{T}}S \\ A_X^{\mathrm{T}}B & A_X^{\mathrm{T}}A_X + A_G^{\mathrm{T}}P_GA_G + A_I^{\mathrm{T}}P_IA_I & A_X^{\mathrm{T}}S \\ S^{\mathrm{T}}B & S^{\mathrm{T}}A_X & S^{\mathrm{T}}S \end{bmatrix} \begin{bmatrix} x \\ t \\ s \end{bmatrix}$$
$$e - \begin{bmatrix} B^{\mathrm{T}}L_X \\ A_X^{\mathrm{T}}L_X + A_G^{\mathrm{T}}P_GL_G + A_I^{\mathrm{T}}P_IL_I \\ S^{\mathrm{T}}L_X \end{bmatrix} = 0. \tag{32}$$

Figure 8 depicts the GNSS-assisted self-BBA processing workflow. In order to improve the accuracy of model calculation, besides setting weights in practice, gross error detection, setting the standardized residual threshold, gross error elimination and iteration calculation are also needed.

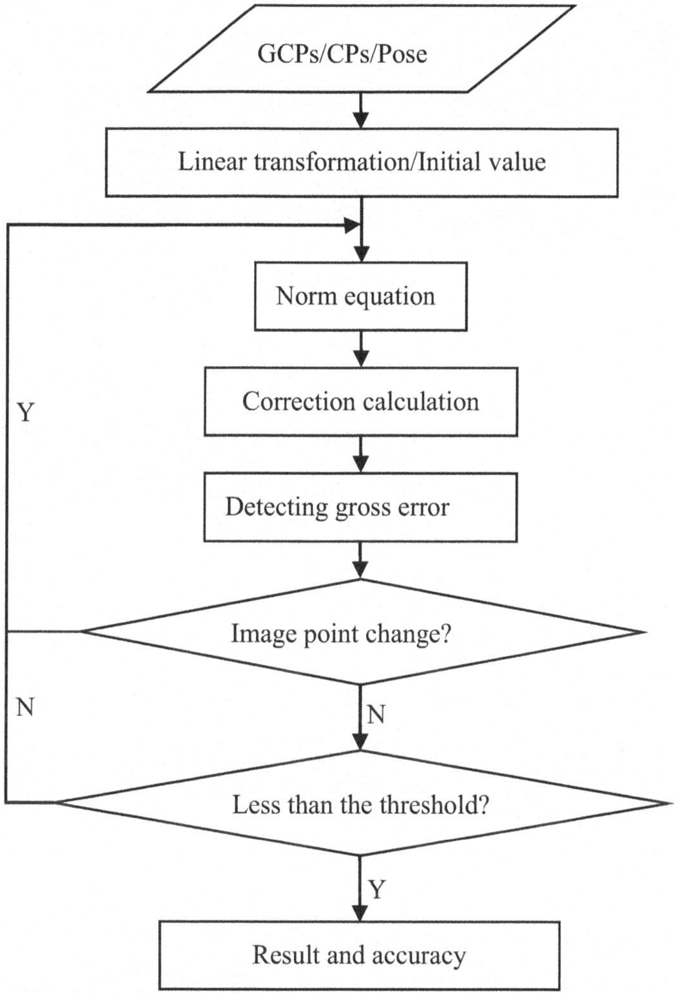

Fig. 8 GNSS-assisted self-BBA processing workflow

5 Experiments

5.1 Precision Testing of Differential GNSS System

In order to explore the accuracy of PPK algorithms fusing BDS/GPS/GLONASS and test the timeliness and feasibility of image position correction method, we have carried out static and dynamic experiments on the differential BD930 GNSS system.

In Fig. 9, the base station is located at a GCP with known position. We get the plane coordinates and altitude values of five feature points in the range of 1 km.

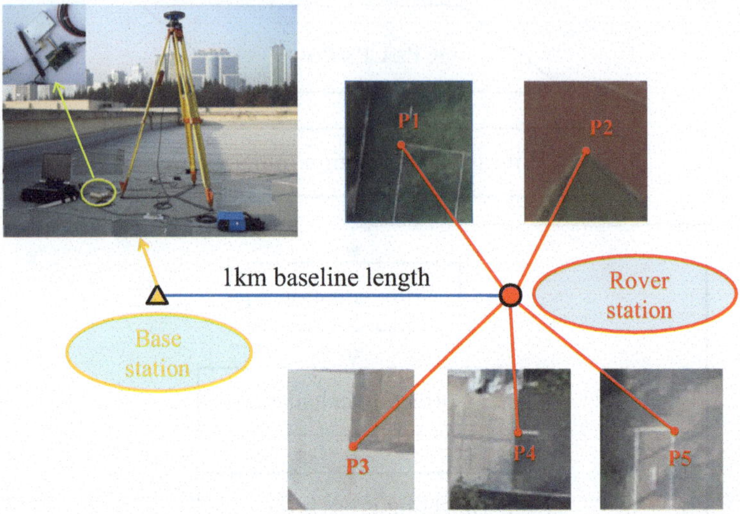

Fig. 9 Static accuracy test

Then, the rover station is, respectively, placed on P_1, P_2, P_3, P_4, P_5, and observes the satellite data for 5 min at the same time with the base station.

In Fig. 10, we get the precision of one point by PPK fusing BDS/GPS/GLONASS under good receiving satellites condition.

From Fig. 9, the calculated GNSS data has a small fluctuation range. We take the average value of each point as the final static positioning result and compare it with the coordinates obtained by geodetic survey. Figure 11 shows the three-direction coordinate residuals of observations at each point.

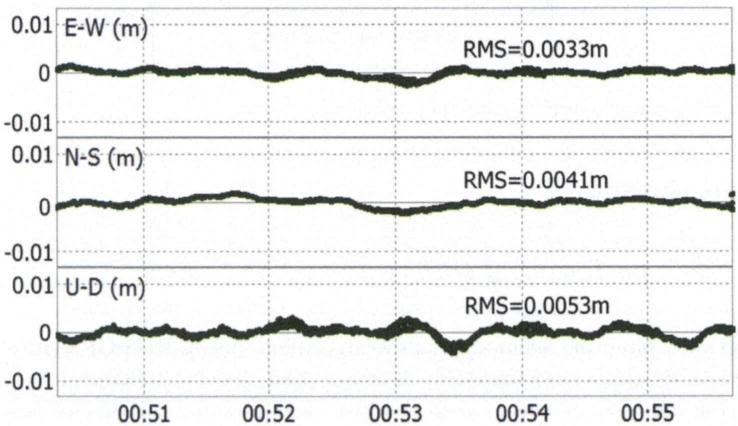

Fig. 10 Data accuracy of 1 km baseline length

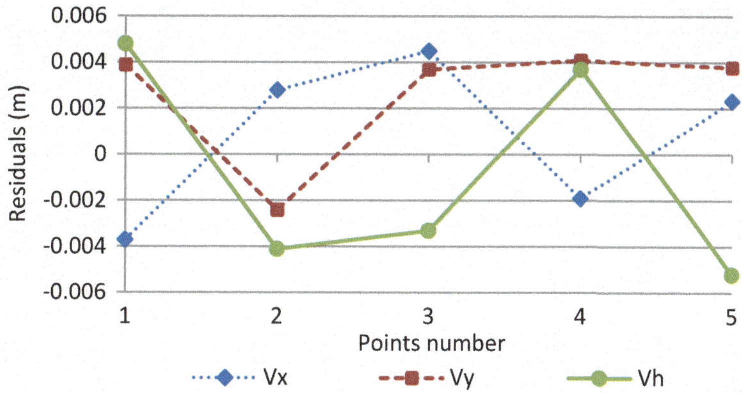

Fig. 11 Residual statistics under static condition

According to the formula $m = \pm\left(\sqrt{\frac{V^T V}{n}}\right)$, the coordinate components median errors of close range observations are calculated as: $m_x = \pm 0.0036\,\text{m}$, $m_y = \pm 0.0041\,\text{m}$; $m_h = \pm 0.0048\,\text{m}$.

The flying range of MUAVs is generally within 20 km from the ground control station. In Fig. 12, we test the static accuracy for baseline length of 5 km, 10 km and 20 km, respectively. The results of different baseline lengths are shown in Table 1.

We can see that the positioning accuracy of the differential GNSS system is higher. Among them, the accuracy and stability of the plane direction are better than that

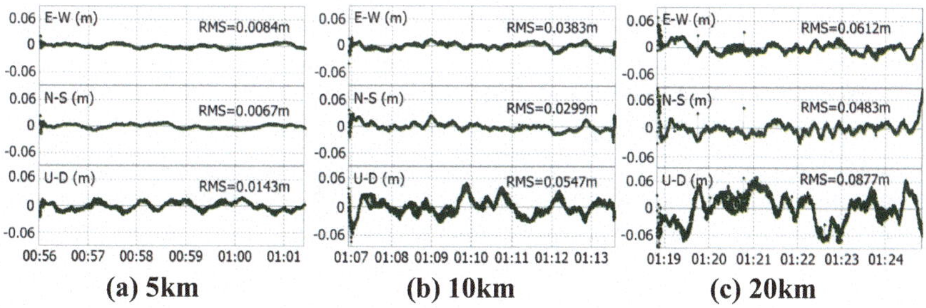

Fig. 12 Data accuracy of different baseline length

Table 1 Data results of different baseline length

Baseline (km)	*RMS-X* (m)	*RMS-Y* (m)	*RMS-Z* (m)
1	0.0033	0.0041	0.0053
5	0.0084	0.0067	0.0143
10	0.0383	0.0299	0.0547
20	0.0612	0.0483	0.0877

Fig. 13 PPK UAV flight test

of the elevation direction. With the increase of the baseline length, the positioning accuracy and stability of the system decrease gradually.

In order to explore the stability of differential GNSS system for dynamic positioning on UAV, and the advantages of the PPK algorithm proposed in this paper, we make fixed-wing UAV FW100 carried BD930 module fly nine north–south direction routes within the A-B-C-D range in Henan Province, China. The layout of the routes is shown in Fig. 13.

After the flight, we use the traditional and new PPK algorithm to calculate the download the GNSS data, respectively, as shown in Fig. 14.

5.2 *Precision Testing of Exterior Orientation Elements*

To verify the accuracy of corrected exterior orientation elements and its advantages in map production, we used FW100 UAV with Sony RX1R II camera to take aerial

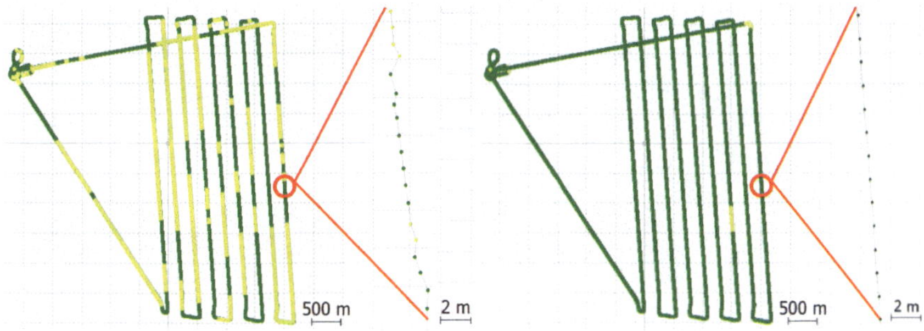

(a) Traditional PPK method: Q=30.4% (b) New PPK method: Q=89.1%

Fig. 14 Comparison of different methods

photography in hilly areas with an area of 2.3 km^2 in Henan Province, China, the flight height is 700 m, the heading overlap is 80%, the side overlap is 60%, the ground resolution is 5 cm, the number of north–south routes is 5, the number of east–west routes is 2, and the number of images is 213. In Fig. 15, 34 GCPs (mark size: 60 cm × 60 cm) are laid on the ground.

After differential post-processing, GNSS data are adjusted by interpolation algorithm and eccentricity correction. In Tables 2 and 3, we get image pose data and camera distortion parameters by GNSS self-BBA without any GCPs. Among them, on the premise of not affecting the data analysis, we consider keeping the geographical environment confidential and hiding the same number of bits in the coordinate values in Table 2.

In Figs. 16 and 17, we adopt three methods, BBA based on lots GCPs (GCP BBA), self-BBA based on POS date (POS self-BBA) and self-BBA based on four corner GCPs and POS date (CGP&POS self-BBA), respectively. By setting 12 CPs, we compare the plane and elevation median errors of the three methods, the calculation results are shown in Table 4.

6 Conclusions

This paper presents a study to use BD930 acquired carrier phase double difference model of BDS, GPS and GLONASS satellites to built one PPK UAV system and improve the accuracy of system pose parameters. In order to eliminate system errors, a cubic spline GNSS interpolation algorithm and a real-time eccentricity correction scheme for space rear intersection are proposed. Through the configuration scheme of the three antennas, the attitude parameters at each exposure time are calculated. Combining with the GNSS self-BBA model, we get the accurate image pose parameters. Through experiments and analysis, we conclude that the static positioning accuracy of the differential GNSS system in a small range ($r < 1$ km) can reach millimeter level. When the UAV performs a wide range of tasks ($r < 20$ km), the plane and elevation positioning accuracy can be achieved within 10 cm, the fixed

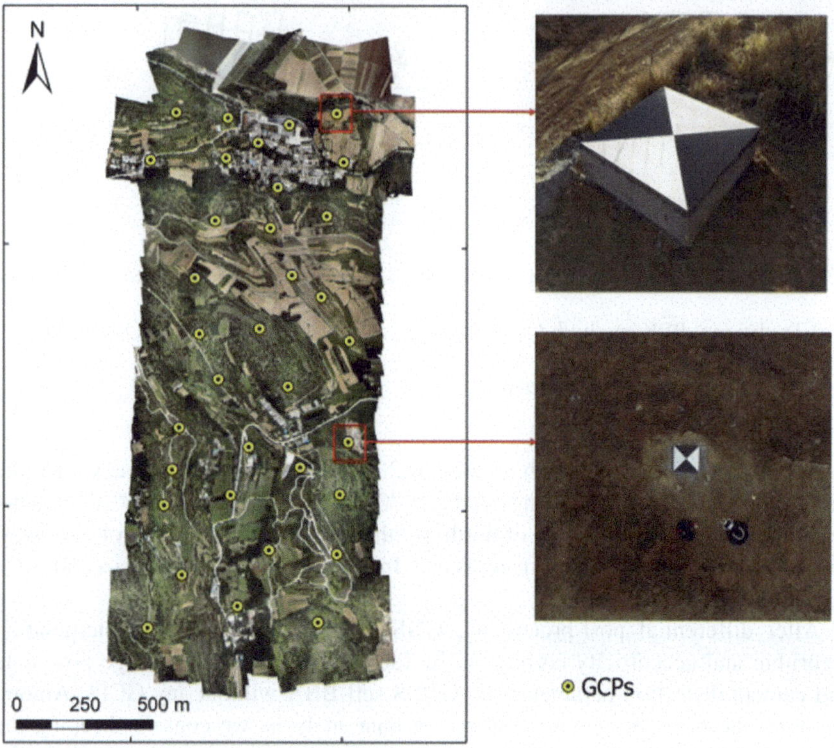

Fig. 15 Aerial range and GCPs layout

Table 2 Pose parameters (parts)

Images	North/m	East/m	Altitude/m	Yaw/degree	Pitch/degree	Roll/degree
DSC08624.JPG	**7459.6267	****888.7313	674.1215	355.8631	3.9429	1.2493
DSC08625.JPG	**7461.8258	****949.1350	671.9626	353.5549	1.2148	−1.4266
DSC08626.JPG	**7464.4908	****010.8944	671.2363	356.1942	2.5114	−0.3901
DSC08627.JPG	**7468.1217	****072.4172	670.7672	355.3299	0.0087	−2.5289
DSC08628.JPG	**7470.7465	****135.7927	670.5383	356.3170	2.0198	−2.3889
DSC08629.JPG	**7472.1258	****193.8808	672.6825	358.9086	−0.0027	−3.6756
DSC08630.JPG	**7473.3177	****255.2846	671.7145	0.6636	−1.3646	−1.7563
DSC08631.JPG	**7474.9277	****311.4490	673.9567	357.7257	−3.1310	−1.2754

solution ratio under dynamic conditions is increased by 58.7% compared with the traditional processing method. In the GNSS assisted self-BBA, we reasonably set the corrected GNSS data weights and compare the GCP BBA, POS self-BBA and CGP&POS self-BBA image processing methods to confirm the advantages of this paper method. In the case of not using any GCPs, the median error in the CPs is less than 0.3 m. In the case of using 4 corner GCPs, the median error in the CPs is less

Table 3 Camera distortion parameters

Distortion parameters	Result
Focus: f(mm)	21.0191
Pixel size: u(mm)	0.004
Principal point: x_p(mm)	−0.1694
Principal point: y_p(mm)	−0.1410
Radial distortion: k_1	3.57893e−004
Radial distortion: k_2	−3.27600e−007
Radial distortion: k_3	−1.67821e−009
Tangential distortion: p_1	6.50577e−005
Tangential distortion: p_2	4.47213e−005

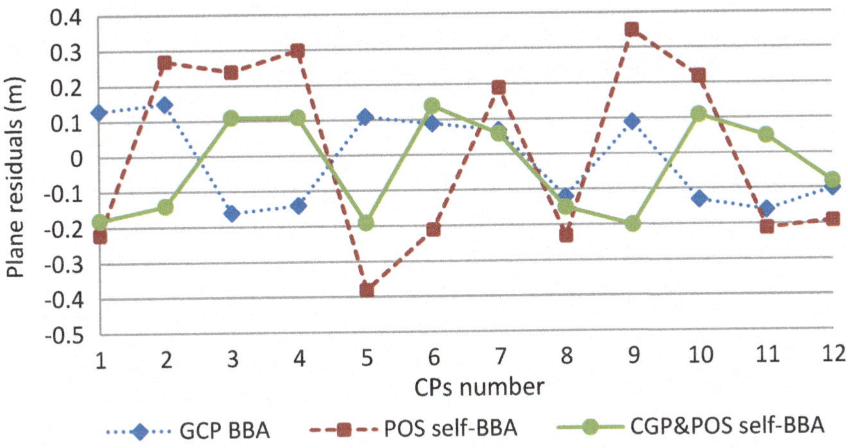

Fig. 16 CPs plane residuals

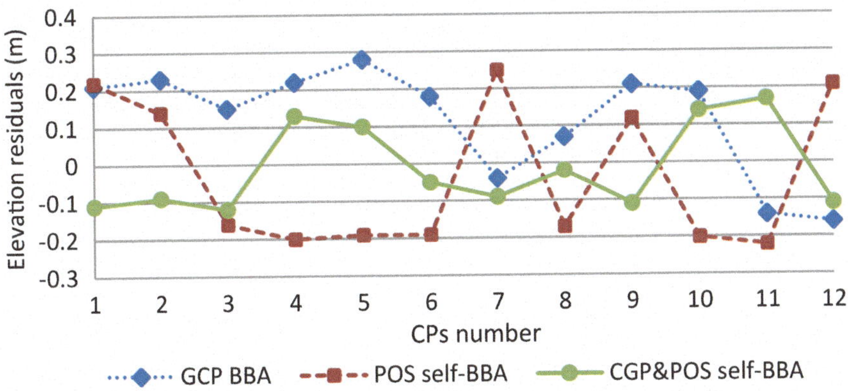

Fig. 17 CPs elevation residuals

Table 4 Error statistics in three methods

Method	Plane median errors (m)	Elevation median errors (m)
GCP BBA	0.124	0.185
POS self-BBA	0.258	0.271
CGP&POS self-BBA	0.135	0.110

than 0.2 m, which proves that the scheme gets the accurate image pose and camera distortion parameters of the PPK UAV. Its accuracy can meet the requirements of large-scale mapping. Therefore, it is feasible to apply PPK technology to the UAV system reasonably. The system can not only meet the accuracy requirements of general geographic information results, but also eliminate the complicated processes of laying a large number of GCPs and camera calibration, so that the production efficiency of surveying and mapping results can be improved.

References

1. Xu Y, Ou J, He H et al (2016) Mosaicking of unmanned aerial vehicle imagery in the absence of camera poses. Remote Sens 8(3):204
2. Hemmelder S, Marra W, Markies H et al (2018) Monitoring river morphology & bank erosion using UAV imagery—a case study of the river Buëch, Hautes-Alpes, France. Int J Appl Earth Obs Geoinf 73:428–437
3. Jiang S, Jiang W (2017) On-board GNSS/IMU assisted feature extraction and matching for oblique UAV images. Remote Sens 9(8):813
4. Templin T, Popielarczyk D, Kosecki R (2018) Application of low-cost fixed-wing UAV for inland lakes shoreline investigation. Pure Appl Geophys 175(9):3263–3283
5. Forlani G, Dall Asta E, Diotri F et al (2018) Quality assessment of DSMs produced from UAV flights georeferenced with on-board RTK positioning. Remote Sens 10(2):311
6. Colomina I, Molina P (2014) Unmanned aerial systems for photogrammetry and remote sensing: a review. ISPRS J Photogramm Remote Sens 92:79–97
7. Liang Y, Qu Y, Cui T (2017) A three-dimensional simulation and visualization system for UAV photogrammetry. ISPRS Int Arch Photogramm Remote Sens Spat Inf Sci XLII-2/W6:217–222
8. Dempewolf J, Nagol J, Hein S et al (2017) Measurement of within-season tree height growth in a mixed forest stand using UAV imagery. Forests 8(7):231
9. Khan M, Heurtefeux K, Mohamed A et al (2018) Mobile target coverage and tracking on drone-be-gone UAV cyber-physical testbed. IEEE Syst J 12(4):3485–3496
10. Prošek J, Šímová P (2019) UAV for mapping shrubland vegetation: Does fusion of spectral and vertical information derived from a single sensor increase the classification accuracy? Int J Appl Earth Obs Geoinf 75:151–162
11. Virone G, Paonessa F, Capello E et al (2016) UAV-based antenna and field measurements. IEEE
12. Wang J, Li Y, Zhu H et al (2018) Interpolation method research and precision analysis of gps satellite position. J Syst Sci Inf 6(3):277–288
13. Stöcker C, Nex F, Koeva M et al (2017) Quality assessment of combined IMU/GNSS data for direct georeferencing in the context of UAV-based mapping. ISPRS Int Arch Photogramm Rem Sens Spat Inf Sci XLII-2/W6:355–361

14. Chiang K, Tsai M, Naser E et al (2015) New calibration method using low cost MEM IMUs to verify the performance of UAV-Borne MMS payloads. Sensors 15(3):6560–6585
15. Erenoglu RC, Erenoglu O, Arslan N (2018) Accuracy assessment of low cost UAV based city modelling for urban planning. Tehnicki vjesnik - Technical Gazette 25(6)
16. Stempfhuber W, Buchholz M (20120) A precise, low-cost RTK GNSS system for UAV applications. Int Arch Photogramm 289–293
17. Hedgecock W, Maroti M, Sallai J et al (2013) High-accuracy differential tracking of low-cost GPS receivers. ACM
18. Xue W, Zhang Y, Zhao L et al (2017) Compare the accuracy of incremental SFM with POS-aided bundle adjustment. Acta Geodaetica et Cartographica Sinica 46(2):198–207
19. Pengfei CHEN, Haifeng HU, Xugang LIAN et al (2017) On influence of post-processed difference technique and image control point density on UAV photogrammetry accuracy. Bull Surv Mapp 11:144–147
20. Liang WU (2017) A method for measuring eccentricity component of airborne gps antenna and aerial camera. Geomat Technol Equip 03:67–69
21. Wang X, Cheng Y, Wu S et al (2016) An effective toolkit for the interpolation and gross error detection of GPS time series. Surv Rev 48(348):202–211
22. Saur G, Krüger W (2016) Change detection in UAV video mosaics combining a feature based approach and extended image differencing. ISPRS Int Arch Photogramm Remote Sens Spat Inf Sci XLI-B7:557–562
23. Fazeli H, Samadzadegan F, Dadrasjavan F (2016) Evaluating the potential of RTK-UAV for automatic point cloud generation in 3D rapid mapping. ISPRS Int Arch Photogramm Remote Sens Spat Inf Sci XLI-B6:221–226
24. Zhou Y, Rupnik E, Faure P et al (2018) GNSS-assisted integrated sensor orientation with sensor pre-calibration for accurate corridor mapping. Sensors 18(9):2783
25. Hamidi M, Samadzadegan F (2015) Precise 3D geo-location of UAV images using georeferenced data. ISPRS Int Arch Photogramm Remote Sens Spat Inf Sci XL-1-W5:269–275
26. Grayson B, Penna NT, Mills JP et al (2018) GPS precise point positioning for UAV photogrammetry. Photogramm Rec
27. Wang T (2017) Comparative analysis of GPS satellite precise ephemeris interpolation methods. Geomat Spat Inf Technol 09:44–46
28. Roul P, Thula K (2019) A fourth-order B-spline collocation method and its error analysis for Bratu-type and Lane-Emden problems. Int J Comput Math 96(1):85–104
29. Mlambo R, Woodhouse I, Gerard F et al (2017) Structure from motion (SfM) photogrammetry with drone data: a low cost method for monitoring greenhouse gas emissions from forests in developing countries. Forests 8(3):68

Extraction of Helicopter Rotor Physical Parameters Based on Time-Frequency Image Processing

Chenxiao Lai and Daiying Zhou

Abstract This paper proposed a method for extracting the physical parameters of helicopter rotors through processing micro-Doppler time-frequency spectrum. We applied image filtering and image segmentation to the time-frequency spectrum of narrow-band RCS data so as to reduce background noise, improve the definition of the spectrum and accurately extract the time-frequency signal line. Then, the parameters such as rotation period, blade length and blade count of the helicopter rotor could be derived directly from the time-frequency signal line, which can be employed to identify the type or even the model of helicopter target. This method solved the problem that the physical parameters of helicopter rotor cannot be extracted precisely from narrow-band RCS data. The simulation results verified the effectiveness of the approach.

Keywords Feature extraction of micro-Doppler · Analysis of time-frequency spectrum image processing · Estimation of physical parameters

1 Introduction

The micro-Doppler signature generated due to the additional frequency modulation on the radar echo induced by the micro-motions such as vibration, rotation, or oscillation of a target or structures [1]. Resulting of the difference in rotational speeds and component configurations (such as the blade count and length) of various aircraft target engines, micro-Doppler signatures enable some properties of the target to be determined. It is possible to classify and identify the aircraft target by extracting the JEM (engine modulation) characteristics of the target from a micro-Doppler spectrum.

Chen et al. first introduced micro-Doppler into the field of radar target recognition and verified the feasibility of using micro-Doppler to identify radar targets [2–4]. Since then, more and more attention is paid to feature extraction from micro-Doppler

C. Lai · D. Zhou (✉)
School of Information and Communication Engineering, University of Electronic Science and Technology, Xiyuan Avenue, Chengdu 2006, China
e-mail: daiyingzhou@163.com

L. Wang et al. (eds.), *Proceedings of the 6th China High Resolution Earth Observation Conference (CHREOC 2019)*, Lecture Notes in Electrical Engineering 657,
https://doi.org/10.1007/978-981-15-3947-3_3

signature. S. H. Darwish et al. used FFT and wavelet transform to obtain a variety of time-domain and frequency-domain features from aircraft target echo as the input data of artificial neural network (ANN) to classify targets [5]. Fei Dayong proposed a method based on complex local mean decomposition, which adaptively extracts features reflecting the difference of micro-motion components from the target echo [6]. Du Lan et al. used fractional Fourier transform to acquire the fractional features of aircraft target echoes [7]. Yang Shoufeng et al. extracted the amplitude deviation coefficient, energy ratio and waveform entropy of the target echo frequency domain [8]. However, the features mentioned above neither can characterize the actual physical parameters of the aircraft target, nor can they further identify the aircraft model by extracting the aircraft target physical parameters.

To solve these problems, this paper proposes a method for extracting the physical parameters of helicopter rotor based on time-frequency spectrum image processing. The method extracts the rotation period of rotor, blade length and blade count directly from the helicopter's narrow-band RCS sequence time spectrum to reliably identify the helicopter targets.

2 Image Preprocessing for Time-Frequency Spectrum

A time-frequency spectrogram is obtained through a short-time Fourier transform on the radar echo of the helicopter rotor. Before estimating the rotation period of the helicopter rotor, the time-frequency signal spectrogram needs to be preprocessed including filtering, quantization and image segmentation.

2.1 Image Preprocessing Methods

The median filter works well in eliminating salty noise, of which the formula can be expressed as:

$$b_{(i,j)} = \text{median}\big(a_{(i-r,j-r)} \ldots a_{(i,j)} \ldots a_{(i+r,j+r)}\big) \tag{1}$$

where $a_{(i,j)}$ and $b_{(i,j)}$ represent, respectively, the pixel point of original spectrum and output spectrum, i and j are the coordinates of the pixel, r is the width of the cut window.

Guided filter is much more complicated which takes into account differences in pixel space and intensity. The algorithm strategy is to introduce a guide image so that the output image has the same gradient as the guide image and utilizes linear regression to minimize the gap between the output image and the original image. The specific conversion formula is as follows [9]:

$$q_{(i,j)} = \frac{1}{|\omega|} \sum_{m,n \in \omega_{(i,j)}} \left(a_{(i,j)} I_{(m,n)} + b_{(i,j)} \right) \tag{2}$$

$$a_{(i,j)} = \frac{\frac{1}{|\omega|} \sum_{m,n \in \omega_{(i,j)}} I_{(m,n)} p_{(m,n)} - \mu_{(i,j)} \overline{p_{(i,j)}}}{\sigma_{(i,j)}^2 + \varepsilon} \tag{3}$$

$$b_{(i,j)} = \overline{p_{(i,j)}} - a_{(i,j)} \mu_{(i,j)} \tag{4}$$

where

$q_{(i,j)}$ is the pixel of the output image,
$p_{(i,j)}$ is the pixel of the input image,
$I_{(m,n)}$ is the pixel of the guiding image in the window $\omega_{(i,j)}$,
$u_{(i,j)}$ is the average of guiding image in the window $\omega_{(i,j)}$,
$\sigma_{(i,j)}^2$ is the variance of guiding image in the window $\omega_{(i,j)}$,
$|\omega|$ is the number of pixels in the window $\omega_{(i,j)}$,
$\overline{p_{(i,j)}}$ is the average value of image P to be filtered in the window $\omega_{(i,j)}$,
m, n is the pixel index of the image in the window $\omega_{(i,j)}$,
P is the normalization factor.

When the guiding image is input image and $\varepsilon = 0$, the filter cuts no ice in this situation which means the output image would be equal to input image. If $\varepsilon = 1$, the guided filter would perform as a weighted mean filter in the region where the pixel intensity changes little (or a monochrome region), while in the region with large fluctuation of pixel intensity, the filtering effect on the image is weak, which helps to maintain the edge.

Figure 1 shows the simulation time-frequency spectrogram of the AH-64 helicopter rotor with a signal-to-noise ratio of 25 dB. The yellow vertical line in the figure is the maximum bandwidth line. When the blade is perpendicular to the radar line of sight, the rotor echo produces flicker and the echo energy is largest. Moreover, at this time, the largest relative speed of the blade tip and the radar come up, resulting in the generation of the maximum Doppler frequency. The yellow sinusoidal signal in the spectrogram is the time-frequency signal line of the blade. Under a certain signal-to-noise ratio, the signal of each blade can be clearly observed. The rotation period of the helicopter target rotor can be obtained by estimating the period of the time-frequency signal line. In the time-frequency diagram, the closer the color of the pixel is to yellow, the higher the amplitude is, and the closer to blue, the lower the amplitude is.

As can be seen from Fig. 1, at the signal-to-noise ratio of 25 dB, there are still a lot of noise spots in the target echo time-frequency diagram. If not be processed, these noise spots would seriously affect the estimation of the time-frequency signal period, due to their similar magnitude to the time-frequency signal line.

Figure 2 is the time-domain diagram after median filtering. The comparison of Figs. 1 and 2 illustrates that the median filter is able to reduce the background noise but also affect the clarity of the time-frequency signal line. Figure 3 is the time-domain diagram after median filtering with guided filtering. As can be seen from

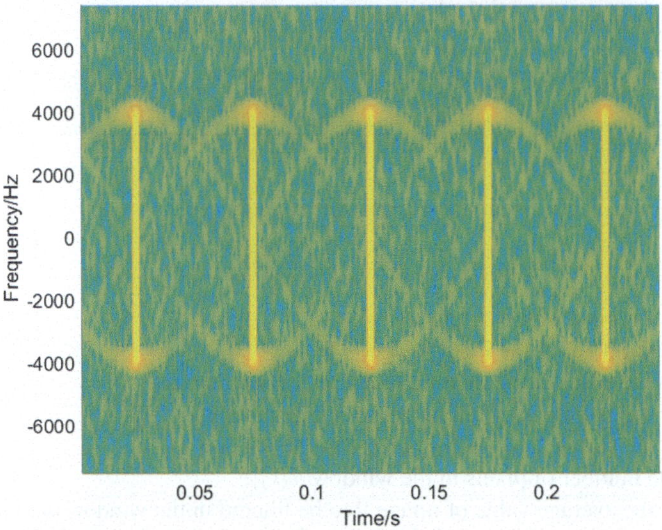

Fig. 1 Radar echo time-frequency diagram of AH-64 rotor

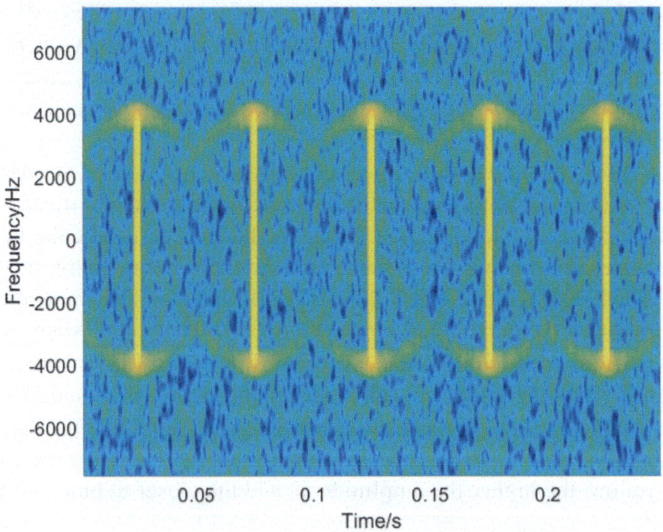

Fig. 2 AH-64 rotor radar echo time-frequency diagram after median filter

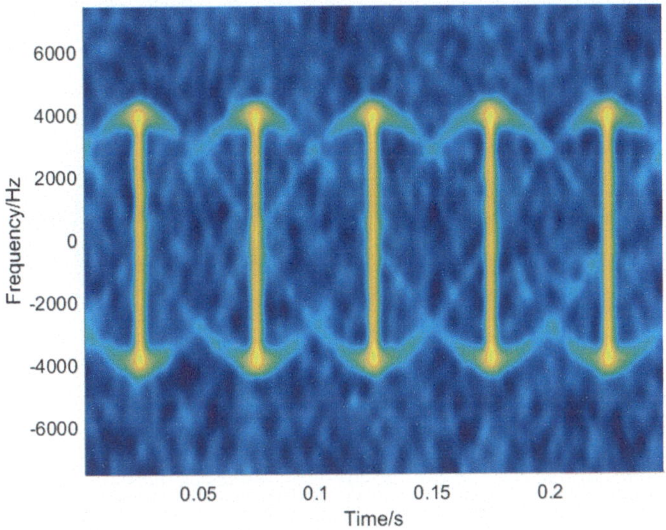

Fig. 3 AH-64 rotor radar echo time-frequency diagram after median filter and guided filter

the comparison of Figs. 2 and 3, the guided filter with the median filter can not only greatly reduce the background noise, but also ensure that the time-frequency signal line has sufficient sharpness, which is beneficial to subsequent image segmentation and extraction of time-frequency signal lines. The same conclusion can be drawn from the comparison of the time–amplitude of the time-frequency signal in Figs. 4 and 5.

2.2 *The Extraction of Time-Frequency Signal Line*

The time-frequency spectrogram is quantized according to the following rules after being filtered:

$$v_{(i,j)} = \begin{cases} 0 & q_{(i,j)} < \text{th2} \\ 0.5 & \text{th2} < q_{(i,j)} < \text{th1} \\ 1 & q_{(i,j)} > \text{th1} \end{cases} \tag{5}$$

where $q_{(i,j)}$ and $v_{(i,j)}$ are the amplitudes of the pixel points on the time-frequency diagram before and after the ternary, and th1 and th2 are two thresholds, which are derived from experimental experience. The time-frequency signal line and the maximum bandwidth line can be segmented after time-frequency diagram being quantized. Figure 6 shows the diagram of the time-frequency spectrogram in Fig. 3 after being quantized.

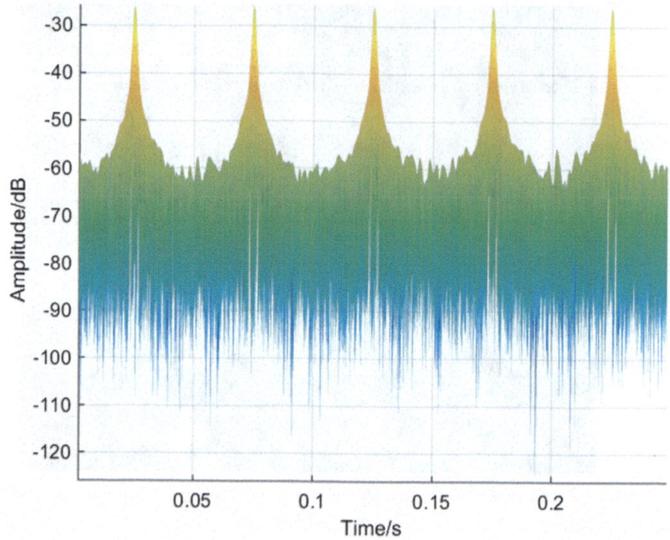

Fig. 4 AH-64 rotor radar echo time-frequency diagram (time–amplitude) before being filtered

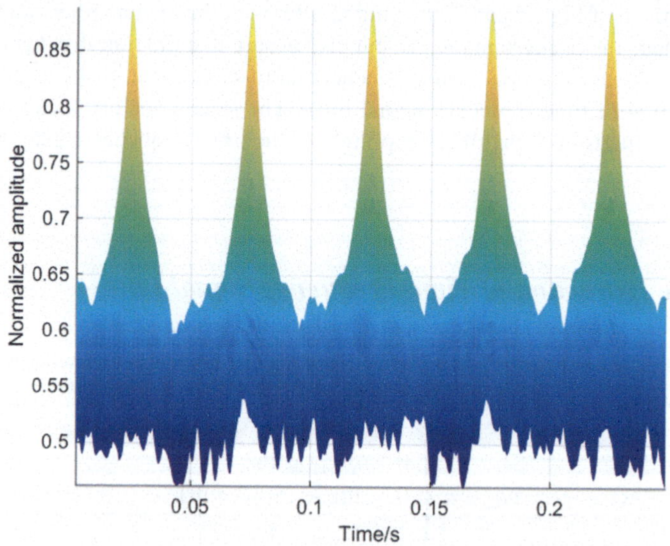

Fig. 5 AH-64 rotor radar echo time-frequency diagram (time–amplitude) after being filtered

Let the set of all pixels satisfying $v_{(i,j)} = 1$ be $\mathbf{W} = [\mathbf{w}_1\mathbf{w}_2 \ldots \mathbf{w}_K]$, and W_k represents the set of pixels belonging to the same maximum pulse width line, where k is an integer and $k \in [1, K]$, and K is the number of maximum pulse width lines. The column coordinates of the kth maximum pulse width line are

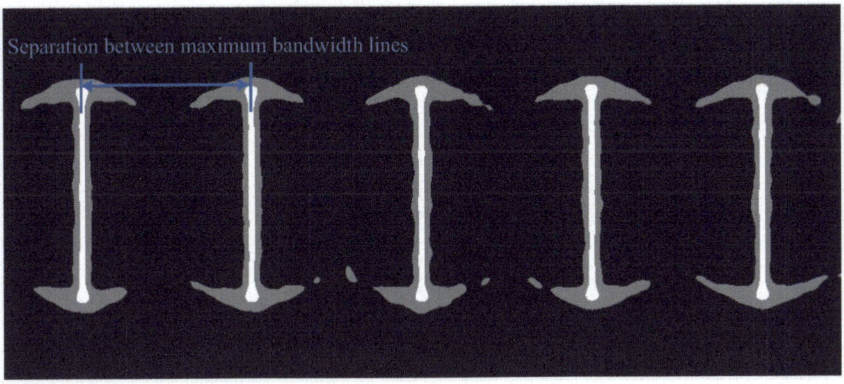

Fig. 6 Diagram of the time-frequency spectrogram in Fig. 3 after being quantized

$$C(k) = \frac{1}{N_k} \sum_{d=1}^{N_k} c_d^k \tag{6}$$

where N_k is the number of pixels in W_k and c_d^k indicate the ordinate of the dth pixel belonging to W_k. Thus, the average separation between maximum bandwidth lines is given by:

$$g_{\text{ave}} = \frac{1}{K_m(K_m - 1)} \sum_{k=1}^{K_m} |C(k) - C(k + K_m)| \tag{7}$$

where $K_m = \lfloor \frac{K}{2} \rfloor$, $\lfloor \bullet \rfloor$ is floor function, $|\bullet|$ represents the sign for absolute value.

The position of the time-frequency signal line can be determined according to $C(k)$, g_{ave} and the minimum abscissa of the pixel in W_k. Figure 7 represents an extracted part of a time-frequency signal line.

3 The Estimation of Helicopter Rotor Physical Parameters

Research shows that the frequency of the time-frequency signal line is the rotation speed of the rotor, so the time-frequency signal line to be estimated can be expressed as

$$\vec{y}(t) = A\cos(2\pi f_r t + \varphi) \tag{8}$$

where f_r is the rotation speed to be estimated of rotor and A is the amplitude of the signal. Translation makes the phase φ zero, so it is unnecessary to spend time on the estimation of φ. The error function can be obtained by applying the least squares

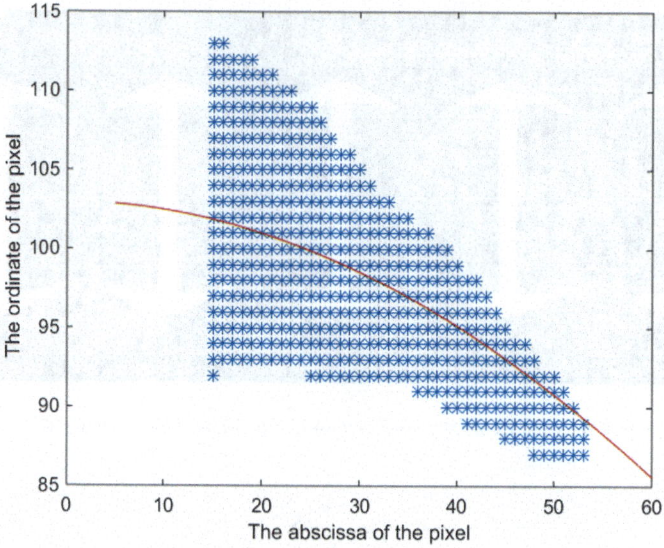

Fig. 7 A fraction of time-frequency signal line

rule, that is

$$E(A, f_r) = \frac{1}{M}\sum_{l=1}^{M}(A\cos 2\pi f_r t - y_l(t))^2 \tag{9}$$

where $y_l(t)$ is the point on the time-frequency signal line extracted from the time-frequency diagram and M is the number of pixels collected. In order to avoid over-fitting, it is vital to add a regular term to the error function:

$$E(A, f_r) = \frac{1}{M}\sum_{i=1}^{M}(A\cos 2\pi f_r t - y_i(t))^2 + \eta f_r^2 \tag{10}$$

where η is the regularization parameter. Partial derivation of A and f_r form Eq. 10 which is as follows:

$$\frac{\mathrm{d}E(A, f_r)}{dA} = \frac{1}{M}\sum_{l=1}^{M} A(\cos 4\pi f_r t + 1) - 2y_l(t)\cos 2\pi f_r t \tag{11}$$

$$\frac{\mathrm{d}E(A, f_r)}{\mathrm{d}f_r} = \frac{1}{M}\sum_{l=1}^{M} 2Ay_l(t)\sin 2\pi f_r t - A^2 t\sin 4\pi f_r t + 2\eta f_r \tag{12}$$

The f_r and A which minimize $E(A, f_r)$ can be found by combining the grid search method and the gradient descent. Besides, averaging multiple estimations from various parts of sinusoidal signal could facilitate the reduction of contingency and the increasing of estimation accuracy. An available way to f_r is shown as Formula 13

$$f_r = \frac{1}{S}\sum_{s=1}^{S} f_{r,s} \tag{13}$$

where $f_{r,s}$ represents the estimated speed from the sth time-frequency signal line and S is the number of time-frequency signal lines. According to the number of pulses N_r in a rotation cycle and the parity factor h, we obtain the blade count N:

$$N = \frac{1}{hN_r f_r} \tag{14}$$

where $h = 1$ or $h = 2$. The blade length L_2 can be derived from the following equation:

$$L_2 = \frac{B\lambda}{A8\pi f_r \cos\theta} \tag{15}$$

where B is the Doppler spectral bandwidth of the echo signal, λ is the wavelength and θ is the pitch angle [10].

4 Experiments

The simulation experiment includes three parts, which are, respectively, to investigate the effects of rotor rotation speed, blade length and blade counts on the simulation results. The simulated parameters of radar are as follows: The radar carrier frequency is 3 GHz that means the electromagnetic wave with a wavelength of 0.1 m; the pulse repetition frequency is determined according to the maximum Doppler frequency generated by different suspense; the pitch angle of the radar is 30°. The azimuth angle is 0, and the initial phase angle of the blade rotation is 45°.

4.1 Experimental Results on Different Rotor Rotation Speed

This experiment is to investigate the influence of helicopter rotor speed on the simulation error. The length of the helicopter blades is 5 m, the number of blades is 4 and the rotor speed is 240, 360, 480, 600, 720 and 1080 r/min. The simulation results are shown in Fig. 8.

From Fig. 8, a conclusion can be drawn that the relative error of estimated rotation speed increases as the rotor rotation speed increases. This is because as the rotational

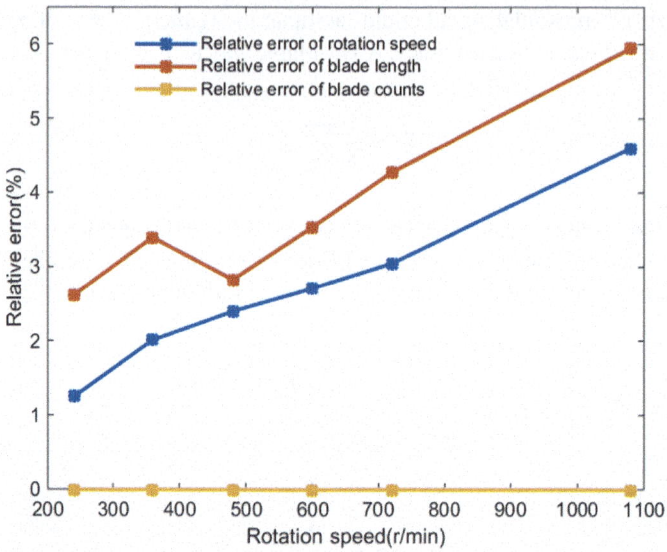

Fig. 8 Simulation results at different rotation speeds

speed increases, the spacing between the maximum bandwidth lines in the time-frequency domain is gradually decreasing at the same pulse repetition frequency which will affect the extraction of the time-frequency signal lines, leading to poor estimation. Since the estimation of the blade length is also related to the bandwidth estimation, there may be cases of error superposition or error cancelation. So, the relative estimation error of the blade length does not show a strong consistent trend with the rotational speed, but they are still positively correlated.

4.2 Experimental Result on Different Blade Lengths

This experiment shows the effects of rotor blade length on relative estimated error. The simulation parameters of helicopter are as follows: The rotor speed is 360 r/min; the number of blades is 4; the length of the helicopter blades is 3, 4, 5, 6, 7, 8 m, and the simulation results are shown below.

Regarding the tendency of broken line in Fig. 9, we have the following observations. As the length of the blades increases, the signal echo of the rotor becomes more intense, and the ratio of the time-frequency signal line to the background noise is higher, which is beneficial to extract a relatively reliable time-frequency signal line in the time-frequency spectrum. Therefore, the larger the blade length, the more accurate the rotor speed estimation.

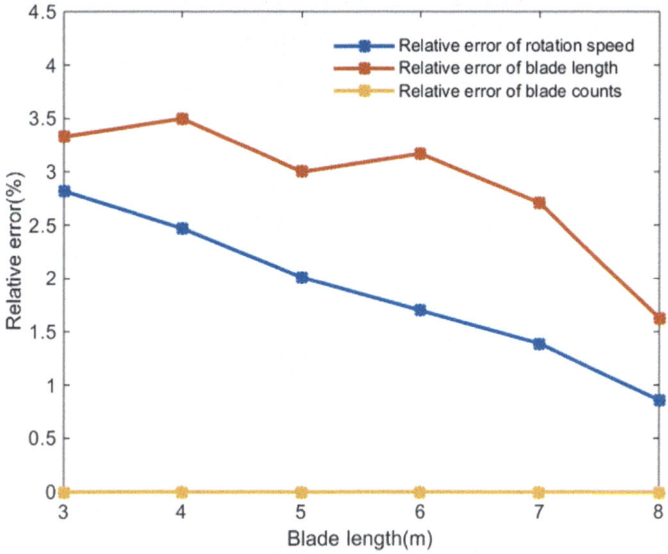

Fig. 9 Simulation results at different lengths of blades

4.3 Experimental Result on Different Blade Counts

In this section, the influence of the number of helicopter rotor blades on the simulation experiment error is conducted. The rotor speed is 360 r/min, the blade length is 5 m and the number of blades is 2, 3, 4, 5, 6 and 8, respectively. The simulation results are as follows.

It can be seen from Fig. 10 that if targets are divided into two portions, the set with odd blades and the another with even blades, the more the number of target blades, the greater the error of the simulation experiment in each set. The simulation results of even blades set are better than that of the odd blades set under the premise that the number of blades is not much different. This is because when the number of blades is even, there will always be two blades perpendicular to the line of radar sight simultaneously, of which the two radar echoes are superimposed on each other, enhancing the energy of the echo. Meanwhile, if the number of blades of the target is odd, there is only one blade perpendicular to the line of radar sight at a time leading to the lower amplitude of target echo signals and the worse signal-to-noise ratio. It is easier to mix in the background noise when extracting the time-frequency signal line under this situation, so that the estimation error of the physical parameters of the aircraft target rotor is increased.

From the comparison of Figs. 11 and 12, we can know that the background noise of the time-frequency spectrogram of the five-blade rotor radar echo is larger than that of the four-blade rotor radar echo. This undoubtedly increases the difficulty of extracting the time-frequency signal line, which makes the parameter estimation effect of the target with odd blades worse than that of the even blades. In addition,

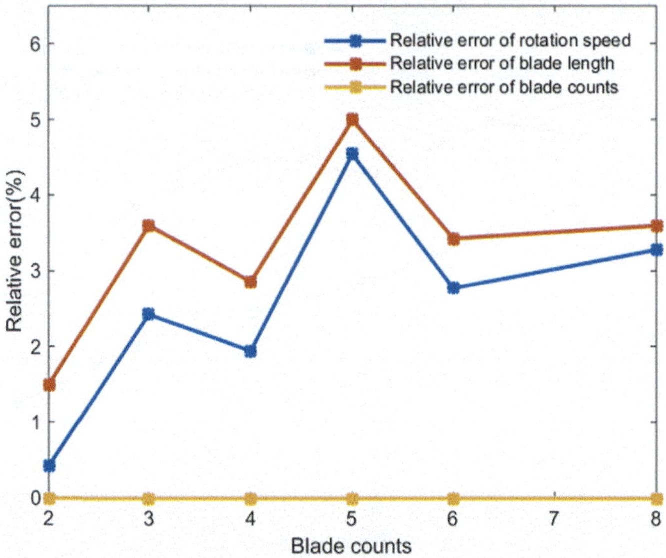

Fig. 10 Simulation results at different blade counts

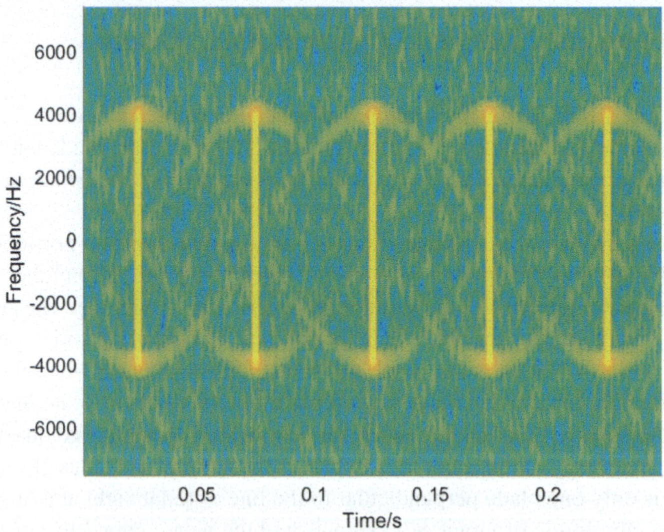

Fig. 11 Time-frequency diagram of radar echo of rotor with four blades

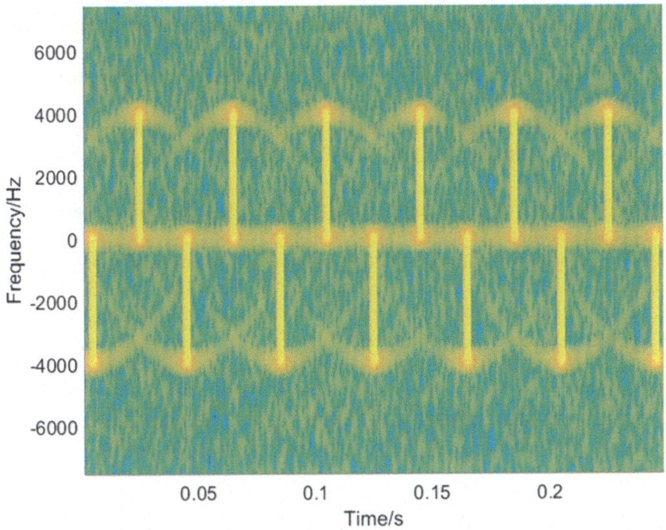

Fig. 12 Time-frequency diagram of radar echo of rotor with five blades

the fault tolerance rate of estimation for the number of blades is very high since the number of blades can only take an integer, and the determination of the parity of blades is relatively easy and accurate. This is also the reason why the estimation error of the number of blades is so small in the simulation experiment.

5 Conclusion

The physical parameter such as the rotational period of the rotor, the number and length of the blades can be extracted from the time-frequency spectrum image by applying image processing method (filtering, quantization, segmentation, etc.) and the least squares parameter estimation under a certain observation time and a sufficiently high signal-to-noise ratio. Simulation experiments indicate that it is effective to extract the physical parameters of the helicopter target through the time-frequency spectrum of the radar echo.

References

1. Zhou W (2011) BMD radar target recognition technology. Publishing House of Electronic Industry
2. Chen VC (2000) Analysis of radar micro-Doppler with time-frequency transform. In: Proceedings of the tenth IEEE workshop on statistical signal and array processing (Cat. No. 00TH8496), Pocono Manor, PA, USA, 2000, pp 463–466

3. Chen VC, Li F, Ho S, Wechsler H (2003) Analysis of micro-Doppler signatures. IEE Proc Radar Sonar Navig 150(4):271
4. Chen VC, Li F, Ho S, Wechsler H (2006) Micro-Doppler effect in radar: phenomenon, model, and simulation study. IEEE Trans Aerosp Electron Syst 42(1):2–21
5. Darwish SH, El-latif MA, Morsy M (2012) Micro-Doppler detection and target identification using artificial neural network. In: 2012 IEEE aerospace conference, Big Sky, MT, 2012, pp 1–5
6. Fei D (2017) Study on feature extraction and classifier and design of airplane targets based on narrowband radar. Xidian University
7. Du L, Shi H, Li L (2016) Feature extraction method of narrow-band radar airplane signature based on fractional fourier transform. J Electron Inf Technol 38(12):3093–3099
8. Yang SF, Wu H (2015) Target feature extraction and recognition based on low-resolution radar. Electron Inf Warfare Technol 30(04):15–20
9. He K, Sun J, Tang X (2013) Guided image filtering. IEEE Trans Pattern Anal Mach Intell 35(6):1397–1409
10. Martin J, Mulgrew B (2002) Analysis of the effects of blade pitch on the radar return signal from rotating aircraft blades. In: Radar 92 international conference. IET

Design of an Airborne DIAL Measurement System for Measuring Concentrations of Atmospheric Pollutants

Yong Chen, Ding-fu Zhou, Ze-hou Yang, Chun-li Chen, Yong-ke Zhang, Xiang-hua Niu, Jing Li, Xiao-feng Li, Guo-juan Zhang, and Guo-hua Jin

Abstract This article presents the research aimed at developing a mobile, remote, fast and accurate measurement system for determining the concentrations of atmospheric pollutants in certain area. The proposed airborne different absorption lidar (DIAL) consists of three non-collocated components. The lidar source component consists of a transmitting antenna, receiving antenna, tunable CO_2 laser emitter, information processor, viewing system turntable and other parts. This device can be installed either on a rotor helicopter or operated from unmanned and manned surface installations, such as vehicles, shipboard or floating surfaces. The lidar electronic cabin has the capacity to provide power to source components. In addition, the display control terminal provides man–machine interface, which enables the completion of functions of mode selection, data analysis, storage, playback, display, communication, among others. This system performs DIAL measurements, and the main technical parameters in the system are calculated. The simulation results obtained by using this system show promising performances with respect to the expected error budget in air pollution conditions. It is concluded that the development of this system can improve emergency management in case of disasters or other emergencies.

Keywords Differential absorption · Lidar · Airborne · Pollutant measurement

1 Introduction

In the last decades, environmental pollution has worsened with the steady development of modern industry and the latest technology. In the cities, levels of air pollution have risen due to factory emissions, vehicular discharges and domestic heating [1].

Y. Chen · D. Zhou · Z. Yang (✉) · C. Chen · Y. Zhang · J. Li · X. Li · G. Zhang · G. Jin
Southwest Institute of Technical Physics, Chengdu 610041, China
e-mail: yangzehou05@163.com

Z. Yang
School of Physics, Beijing Institute of Technology, Beijing 100081, China

X. Niu
State Key Laboratory of Geo-Information Engineering, Xian 710054, China

L. Wang et al. (eds.), *Proceedings of the 6th China High Resolution Earth Observation Conference (CHREOC 2019)*, Lecture Notes in Electrical Engineering 657,
https://doi.org/10.1007/978-981-15-3947-3_4

This scenario requires the use of equipment which can detect and track hazardous substances in the atmosphere quickly and efficiently with a long range of coverage [2]. Rapid and real-time detection of exhaust gases, volatile organic compounds and other harmful chemicals are extremely important [3]. Laser remote sensing systems, for instance, lidar and different absorption lidar (DIAL), have been widely used as fundamental techniques in the research of atmospheric physics and atmosphere optics since the 1960s [1].

Technical advancements in airborne DIAL have drastically increased the measurement capabilities of airborne and ground-based DIAL systems for measuring and monitoring pollution levels [4]. In this paper, we have illustrated the design of an airborne DIAL measurement system. Based on the great potential of DIAL, it is inferred that the proposed system will bring great benefit in the technical advancement of tunable, compact laser emitters, enabling farther portions of the optical spectrum to be developed for accurate measurements of the concentrations of multi-species atmospheric pollutants.

2 Principle

DIAL is a laser remote sensing technique used for measuring the concentrations of atmospheric pollutants [5–7]. The airborne DIAL measurement system is based on ground-based DIAL [8], and it is used in specific regions to quickly quantify the concentrations of a number of atmospheric chemical pollutants [9]. This can help to minimize the damage caused by air pollution to humans.

The developed airborne DIAL is based on the principle of long-range differential absorption of long-wave infrared. Many atmosphere pollutants, such as SF_6, C_2H_4, have typical absorption bands in the infrared band of 9–12 μm; hence, the laser output can be tuned to these bands by using the absorption characteristics of gases. By measuring the absorption intensity of gas with respect to specific wavelength of the laser, we can determine the type of gas, its concentration, distribution and other characteristics. The miniature dual-tunable TEA CO_2 laser is used as the light source to emit shock that matches the characteristic absorption peak or valley of the matter. During the process, it is hung outside the lower abdomen of the rotor in unmanned or manned aerial vehicle. When the aerial vehicle reaches the target area, it begins measurement. Optical pulse pair and differential absorption calculation are carried out to measure the concentration and distance of the measured substance.

A simplified version of the airborne DIAL concept is shown in Fig. 1. When an aircraft with the mounted DIAL system flies to the target area, the long-range differential absorption detection of long-wave infrared laser remotely measures the clouds of atmospheric pollutants in the research area. This way, the species and concentrations of atmospheric pollutants can be determined. When working, two tunable lasers emit two laser pulses with different wavelengths along the same path in a certain time interval (~200 μs) through the transmission antenna. One of these is resonant wavelength, λ_{on}; its wavelength is located at the characteristic peak position

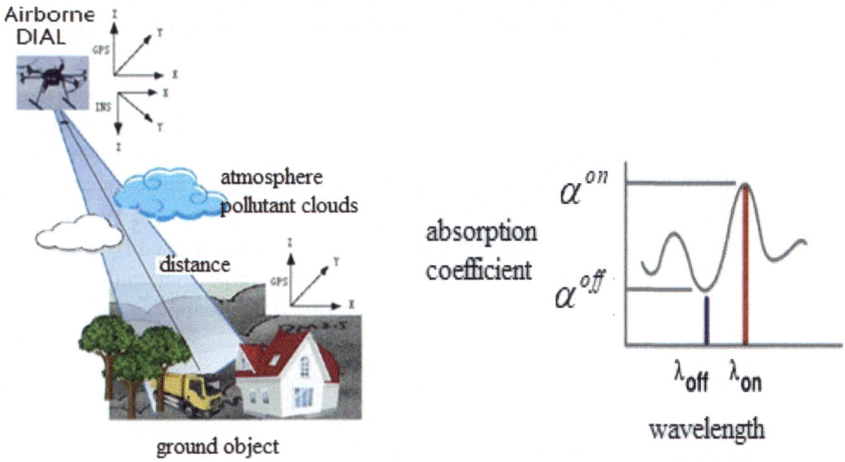

Fig. 1 Diagram of the working principle of airborne DIAL

in the absorption peak of the gas to be measured, and the other is non-resonant wavelength λ_{off} which is located in the absorption valley. Because of the inconsistency in the absorption spectrum characteristics of the two laser beams that occurs after the beams have passed through the affected area of chemical pollutants, differences in the intensity of the echo signals are received by the radar after the beams scatter upon collision from the rear object [10]. By monitoring the relative changes of the scattering echo intensity of the two wavelengths and by comparing the results with the characteristic database of chemical pollutants, we can separate the absorption of the gas to be measured from other factors. Then, we can retrieve the types and concentrations of chemical pollutants in order to completely monitor and study the chemical pollutants.

Based on the lidar equation, the concentration-path length of target gas detected by DIAL can be deduced as follows [11]:

$$\mathrm{CL} = \frac{1}{2\left(\alpha^{\mathrm{on}} - \alpha^{\mathrm{off}}\right)}\left[\ln\left(P^{\mathrm{off}}/P^{\mathrm{on}}\right) + \ln\left(\rho^{\mathrm{on}}/\rho^{\mathrm{off}}\right) + \ln\left(\eta^{\mathrm{on}}/\eta^{\mathrm{off}}\right) - 2L\left(\beta^{\mathrm{on}}/\beta^{\mathrm{off}}\right)\right] \tag{1}$$

Among these parameters, P is the normalized power received by the detector; ρ is the reflection coefficient of ground; η is the optical efficiency of the system; α is the absorption coefficient of the target gas; β is the attenuation coefficient caused by scattering and absorption of other atmospheric and L is detection range. It can be seen from the above formula that the concentration-path length (CL value) of the detected gas is not only related to the ratio of the normalized, received and non-resonant energy to the resonant energy, but also influenced by the characteristic absorption coefficient of the gas to be detected. In addition, the difference of the target reflectivity, optical efficiency of the system and atmospheric attenuation factor for two lasers of different wavelengths is also determined. This process affects the measurement of the concentration of measured gas.

As majority of air pollutants have absorption peaks that fall in the range of 9.2–10.7 μm, CO_2 laser with output branch that matches with the absorption characteristics of atmospheric pollutants is used as a laser irradiation source. This laser can emit multi-branch line pairs to realize measurement of the variety and concentration of air pollutants.

The principle of long-range differential absorption is characterized by eliminating the influence of environmental factors. Thus, it can be used to actualize highly sensitive detection of atmospheric pollutants in various states (vapor, aerosol, small droplet) along the optical path.

3 System Design of Airborne DIAL

The airborne DIAL measurement system is composed of a lidar host, a lidar electronic cabin and a display control terminal (Fig. 2). The lidar host is the equipment used outside the cabin. It completes the detection of the biochemical environment. It is hung on the lower abdomen of the rotorcraft unmanned aerial vehicle (UAV) or the human-operated machine. It uses a two-axis turret scanning structure, and it can conduct multi-mode optical scanning in a downward or sloping-downward motion. The lidar system automatically carries out beam scanning according to the flight parameters information received by the aircraft. The electronic cabin provides power supply, scanning control, temporary data storage and system communication functions for the main engine system. In addition, the display control terminal is a general desktop or notebook computer, and it has an installed system control and

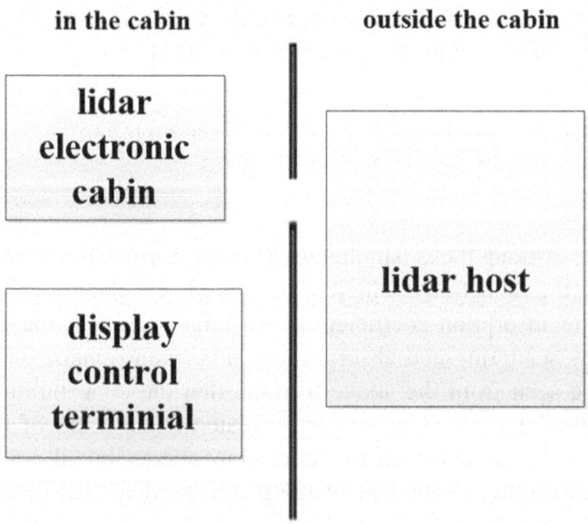

Fig. 2 Airborne DIAL measurement system composition diagram

Table 1 Main parameters of the airborne DIAL measurement system design

Item	Parameter
Working wavelength	9.2–10.7 μm
Detectable gas	Industrial harmful gases such as SF_6, C_2H_4
Maximum distance of pollutant cloud cluster recognition	≥4 km(when visibility is up to 10 km)
Flight speed in dynamic measuring process	≤40 km/h

application software. Moreover, it carries out the functions of mode selection, video monitoring, data acquisition and analysis, storage, playback, display and communication, among others. The main parameters of the airborne DIAL measurement system design are described in Table 1.

Differential absorption measurements are mainly implemented in lidar host which is mainly composed of the following parts: lidar integrated processor, differential laser transmitter, differential detection receiver and differential information processor. The principle framework of the system is shown in Fig. 3. Differential laser transmitter is the core component of the system. It consists of two mini TEA CO_2 lasers, energy detection module and transmitting antenna. Differential detection receiver is composed of long-wave, infrared receiving antenna, in addition to the detector module and linear amplifier. The information processor is a single-board, high-speed information processing and control circuit board that can complete information processing.

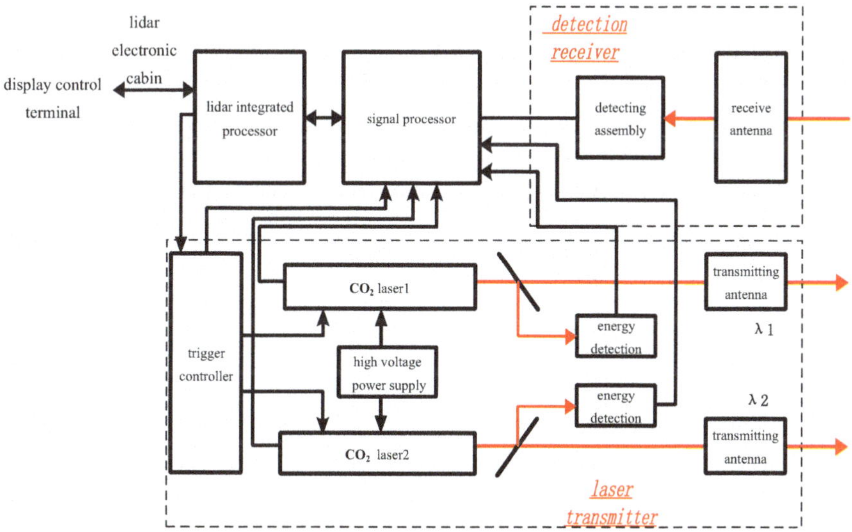

Fig. 3 Principle framework of the system

When the system works, the laser transmitter received instructions, and in turn, the two tunable lasers quickly tune their wavelengths to the specified wavelengths. Then, the two laser pulse pairs are triggered successively; time interval of two lasers is 100 microseconds. Laser pulses are monitored by the energy monitoring module and then emitted to form aerosol cloud clusters. After passing through cloud clusters, the laser pulses scatter back toward the ground, forest, vehicles, buildings and other objects. Subsequently, these pulses pass through the cloud clusters to be measured again and are then collected by the receiving antenna. After photoelectric conversion of detector components and double-passing process of linear amplification, analog electrical signal is sent to signal processor for digital processing. At this stage, normalization of echo intensity and energy monitoring value occurs through filtering, multi-pulse accumulation, recognition and calculation of echo signal in time domain. The normalization of the target normalized intensity information for each wavelength and pollution cloud of different wavelengths export to the lidar integrated processor. After the above-mentioned process has carried out, values for intensity information, ground target distance, location of contaminated clouds and pollutant concentrations can be obtained.

It should be noted that the effect of atmospheric jittering during measurement is not considered because the interval time of laser pulse pair is very short (200 μs), much shorter than the freezing time of atmosphere. In addition, for a moving platform, if the moving speed of the platform does not exceed 300 km/h, the moving distance of the differential laser pulse pair from the platform during launch falls within the millimeter scale. Thus, the platform can be considered to be relatively static. In other words, the motion correction of the beam is not necessary when the differential measurement is performed by a single pulse.

4 Calculation for Main Parameters

To evaluate the feasibility of the airborne DIAL measurement system design, we have made some calculations for the main parameters. Based on the differential lidar equation, the output energy of the CO_2 laser required can be calculated as shown below [12, 13]:

$$P_{\text{out}} = \frac{8\pi R^2(\text{SNR}) \cdot (\text{NEP}) \cdot \tau \cdot \sqrt{\Delta f}}{T \cdot \rho \cdot A \cdot n^{1/2} \cdot \cos\theta} \cdot \frac{\exp[2\alpha_E R]}{[\exp(-2\alpha_{\text{off}} CL) - \exp(-2\alpha_{\text{on}} \text{CL})]} \quad (2)$$

In the formula:

P_{out}—laser output power;
CL—concentration-path length;
R—the target reflector;
T—optical efficiency of the system is calculated;

θ—the sharp angle between laser beam axis and normal line of the distributed target;
τ—the laser pulse duration;
Δf—instrument noise bandwidth;
n—the accumulative number of laser pulses;
A—the light area of receiving antenna;
ρ—the target reflection;
SNR—signal-to-noise ratio of the system, according to system design;
NEP—noise equivalent power.

The spectroscopic parameters for the simulation calculations were obtained from HITRAN2016 [14]. High-resolution transmission molecular absorption database (HITRAN) was developed by the US Air Force Geophysical Laboratory in the late 1960s. It was opened to public in 1973 to study the infrared characteristics of atmosphere for military purposes. Since then, it has been widely used in many fields, such as for studying weak absorption of trace gases in the atmosphere, gas remote sensing measurement, laser transmission research, radar and many others. HITRAN is a compilation of spectral parameters used to predict and simulate the transmission and emission of light in the atmosphere. Figure 4 shows absorption curves of several gases (9–11 μm) [15]. Based on the HITRAN database, we can acquire information about the infrared absorption spectra of common gases in the atmosphere. For characteristic absorption wavelength of some common atmospheric pollutants [15], such as O_3, C_2H_4, SF_6, refer to Table 2.

By calculating the lidar equation based on the above data, the curves of the output energy of CO_2 laser with different detection distances can be obtained under the condition of satisfying the detection sensitivity (CL value).

Figure 5 shows the output energy requirement curve of several typical atmospheric pollutants for CO_2 laser when visibility is in range up to 10 km. It is evident from the figure that the range requirement of not less than 5 km can be met when the laser output is at 15 mJ.

5 Summary

In this paper, we have introduced the design of the developed airborne DIAL measurement system for measuring concentration of atmospheric pollutants. The airborne DIAL measurement system is based on ground-based DIAL, and it is used in specific regions to quickly recognize a number of atmospheric chemical pollutants. This can effectively prevent the damage caused by air pollutants to humans. In the study, the main technical parameters in the system are calculated. Simulation results show promising performances with respect to the expected error budget in conditions of air pollution. It is concluded that the development of this system can improve the management in the face of major disasters and other emergencies.

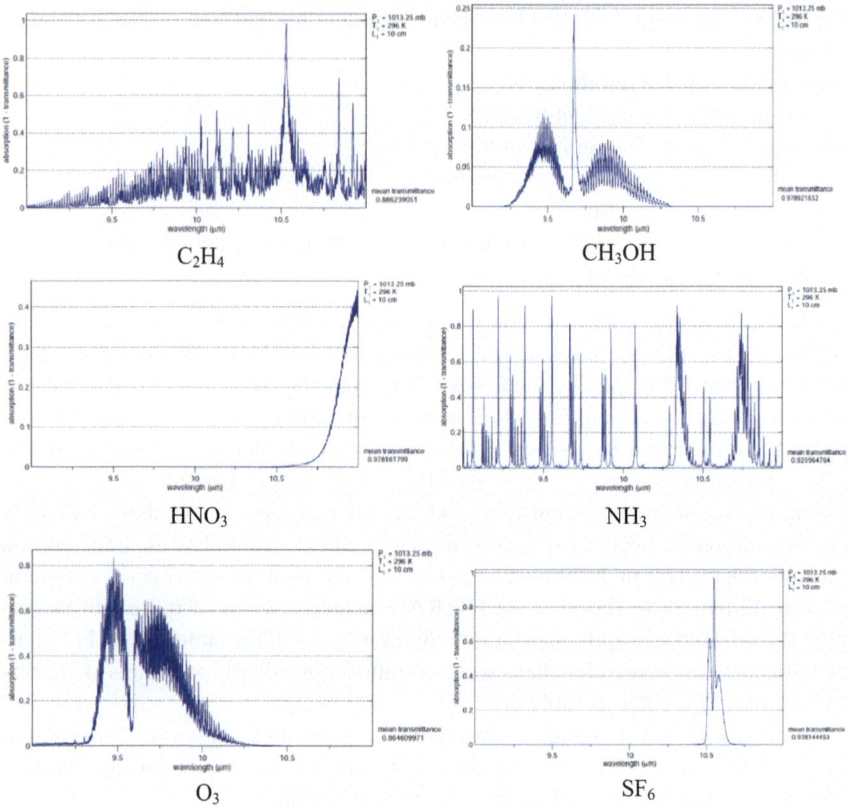

Fig. 4 Absorption curves of several gases (9–11 μm) [15]

Table 2 Characteristic absorption wavelength of some atmospheric pollutants

Atmosphere pollutant	λ_{on} (μm)	λ_{off} (μm)
O_3	9.4581	9.305
C_2H_4	10.513	10.333
SF_6	10.551	10.653

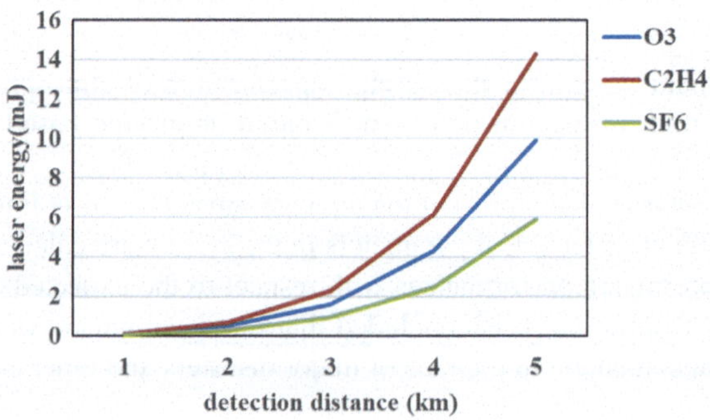

Fig. 5 Curve of laser output energy versus detection distance when visibility is up to 5 km

References

1. Gaudio P (2017) Laser based standoff techniques: a review on old and new perspective for chemical detection and identification. In: Cyber and chemical, biological, radiological, nuclear, explosives challenges. Springer, Cham, pp 155–177
2. Rossi R, Ciparisse JF, Malizia A et al (2018) Multiwavelength differential absorption lidar to improve measurement accuracy: test with ammonia over a traffic area. J Appl Phys B 124:148
3. Xiang C, Ma X, Liang A et al (2016) Feasibility study of multi-wavelength differential absorption LIDAR for CO_2 monitoring. J. Atmos 7:89
4. Gardi A, Sabatini R (2015) Design and development of a novel bistatic dial measurement system for aviation pollutant concentrations. J Int J Sci Eng Invest 41:46–54
5. Browell EV, Ismail S, Grant WB (1998) Differential absorption lidar (DIAL) measurements from air and space. J Appl Phys B 67:399–410
6. Geiko PP, Smirnov SS (2014) Remote sensing of chemical warfare agent by CO_2-lidar. In: 20th international symposium on atmospheric and ocean optics: atmospheric physics, vol 9292. International Society for Optics and Photonics, pp 92922Z
7. Romanovskii OA, Burlakov VD, Dolgii SI et al (2016) A technique for retrieval of ozone vertical distribution from DIAL measurements. In: Lidar remote sensing for environmental monitoring XV, vol 9879. International Society for Optics and Photonics, pp 98791G
8. Abshire J, Ramanathan A, Riris H et al (2014) Airborne measurements of CO_2 column concentration and range using a pulsed direct-detection IPDA lidar. J Remote Sens 443–469
9. Tehrani MK, Mohammad MM, Jaafari E et al (2015) Setting up a mobile Lidar (DIAL) system for detecting chemical warfare agents. J Laser Phys 25:035701
10. Quagliano JR, Stoutland PO, Petrin RR et al (1997) Quantitative chemical identification of four gases in remote infrared (9–11 μm) differential absorption lidar experiments. J Appl Opt 36:1915–1927
11. Warren RE (1985) Detection and discrimination using multiple-wavelength differential absorption lidar. J Appl Opt 24:3541–3545
12. Pilon G (1972) A study of tunable laser techniques for remote mapping of specific gaseous constituents of the atmosphere. J Opto-electron 4:141–153
13. Byer RL, Garbuny M (1973) Pollutant detection by absorption using Mie scattering and topographic targets as retroreflectors. J Appl Opt 12:1496–1505
14. Gordon IE, Rothman LS, Hill C et al (2017) The HITRAN2016 molecular spectroscopic database. J Quant Spectrosc Radiat Transfer 203:3–69
15. Information on http://www.weld.labs.gov.cn

A Multiple Moving Ships Detection Method for GF-4 Satellite Image in Thin-Cloud Environment

Peng Lv, Yuxin Hu, Qianqian Li, Yangshuan Hou, Xiaohui Wang, and Bin Lei

Abstract At present, the GF-4 satellite is the world's highest resolution geostationary orbit optical imaging satellite. The GF-4 satellite has the advantages of wide-swath and high-frequency imaging, so it can provide quasi-real-time images for ship surveillance. In fact, using the GF-4 satellite image to detect moving ships is to detect the gray value of the wake formed by the ship moving at sea. However, because of the geometric features of the ship's wake vary with the speed, the ship detection method based on geometric features is not applicable. Since, not only the ship is moving in the image but also the cloud is moving all the time, the optical flow method is not applicable. Ship detection methods based on machine learning and deep learning are also not applicable due to too few texture features of the wake. In this paper, we present a novel method for ship detection in thin-cloud environments. The real image processing of GF-4 satellite shows that the proposed method can detect multiple moving ships accurately and quickly in a thin-cloud environment.

Keywords GF-4 satellite · Image processing · Multiple ships detection · Thin cloud

1 Introduction

The GF-4 satellite, launched on December 29, 2015, is China's first civilian high-resolution geostationary optical satellite. The successful launch of the GF-4 satellite marked the beginning of an era of the high-temporal, high-spatial resolution, and large-scale earth observation from the geosynchronous orbit [1]. The GF-4 satellite as a geostationary satellite, located approximately 36,000 km over the earth's equator,

P. Lv · Y. Hu (✉) · Q. Li · Y. Hou · X. Wang · B. Lei
Institute of Electronics, Chinese Academy of Sciences, Beijing, China
e-mail: yxhu@mail.ie.ac.cn

Key Laboratory of Spatial Information Processing and Application System Technology, Chinese Academy of Sciences, Beijing, China

Y. Hu · Y. Hou · X. Wang · B. Lei
University of Chinese Academy of Sciences, Beijing, China

L. Wang et al. (eds.), *Proceedings of the 6th China High Resolution Earth Observation Conference (CHREOC 2019)*, Lecture Notes in Electrical Engineering 657,
https://doi.org/10.1007/978-981-15-3947-3_5

has many important advantages. Because, geostationary orbit satellite has a relatively fixed earth observation position, it can have the advantages of a short revisit period and wide coverage at the same time. At present, the GF-4 satellite has the highest resolution from geostationary orbit in the world [2].

The staring camera of the GF-4 satellite can provide 50-m spatial resolution images every 20 s, and the coverage of the images is 500 km × 500 km. As China's first geosynchronous orbit optical remote sensing satellite and the world's highest spatial resolution geosynchronous orbit optical remote sensing satellite, GF-4 satellite can provide fast, reliable, and stable optical remote sensing data for application fields such as disaster reduction, forestry, earthquake and meteorology, supplement brand new technological means for disaster and risk warning and forecasting, forest fire disaster monitoring, seismotectonics information extraction, and weather monitoring and open up a new field of geosynchronous orbit high-resolution earth observation [3–5]. Because of its wide coverage, high resolution, and short revisit period, GF-4 can provide quasi-real-time images for ship surveillance [6].

In recent years, the use of remote sensing satellites for ship surveillance has been a hot topic. Yu et al. [7] used the band-to-band time lags of GF-4 satellite to detect ships. Yang et al. [8] analyzed the characteristics of the sea surface to detect ship in optical images. Liu et al. [9] used the two-parameter constant false alarm rate (CFAR) detector to detect ships using GF-4 images. However, the above methods are only applicable to environments without clouds. If there is cloud interference, the performance of the above methods will decrease rapidly. In this paper, we propose a multiple moving ships detection method for a thin-cloud environment.

The remainder of this paper is organized as follows. Section 2 presents the proposed multiple moving ships detection method and analyzes the computational complexity analysis of the proposed method. Section 3 presents the experimental results and discussions, and Sect. 4 provides the conclusions.

2 Proposed Multiple Moving Ships Detection Method

The resolution of the GF-4 satellite image is 50 m. In theory, a ship, 300 m long and 50 m wide, has a length of 6 pixels and a width of 1 pixel. Therefore, the wake can be seen in the GF-4 image rather than the ship. This is because the ship's gray value is much smaller than the wake's gray value, and the number of pixels in the wake is much larger than the number of pixels in the ship's wake. Therefore, compared with the problem of target detection in high-resolution images, there is little feature information available for ship detection using GF-4 images. In addition, due to satellite motion and attitude control errors, slight jitter occurs between frames. Therefore, using the GF-4 satellite image sequence for moving ship detection is suitable for single-frame detection. In summary, the detection of moving ships in GF-4 images is to detect the wakes of moving ships in a GF-4 satellite image.

2.1 Analysis of Wake Characteristics in GF-4 Satellite Image

Figure 1 illustrates a sketch of a typical ship-generated waveform caused by a moving ship in deep water [10]. A typical ship-generated waveform caused by a moving ship is mainly composed of the turbulent wave, Kelvin wave, divergent waves, transverse waves, and breaking wave region. In the GF-4 image, we can only see the turbulent wave. The turbulent wake appears to be dependent on the ship structure, length, breadth, draft depth, speed, the water depth, the wind speed, direction, etc. Among them, the main factors influencing the turbulent wake are the ship structure, speed, the wind speed, and direction. However, all current analysis of the wake is based on the assumption that the ship is moving in a uniform linear motion. From a mathematical perspective, the problem of calculating the turbulent wave is highly nonlinear. More about the turbulent wake modeling can be found in [11]. However, in real situations, the ship's acceleration and maneuvering make the situation more complicated.

Figure 2 shows GF-4 satellite images of two typical thin-cloud environments. We define Fig. 2a as a simple thin-cloud environment and Fig. 2b as a complex thin-cloud environment. In Fig. 2a, b, the white dots in the red rectangle are the wakes of multiple moving ships. Figure 2c, d shows three-dimensional mesh plots of the original images of Fig. 2a, b. In Fig. 2c, the wakes are located at (330, 818), (663, 889), (838, 818), and the maximum gray value is located at (512, 49). As can be seen from the enlarged view of the wake of (838, 818), there are two peaks with similar gray values near the wake. In Fig. 2d, the wakes are located at (494, 538), (750, 83), and the maximum gray value is located at (245, 448). As can be seen from the enlarged view of the wake of (750, 83), the gray value of the wake is obviously smaller than that of the cloud nearby.

Through the previous analysis, we can draw the following conclusions: (1) the wake may change dramatically with time; (2) in the GF-4 image, the wake is a white dot; (3) the gray value of different wakes may vary greatly; (4) the maximum gray value in the image is often a bad pixel; (5) the gray value of cloud may be much

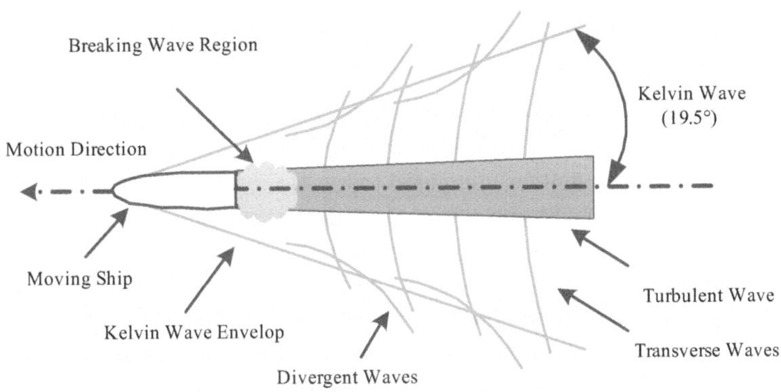

Fig. 1 Definition sketch of wave pattern generated by a moving

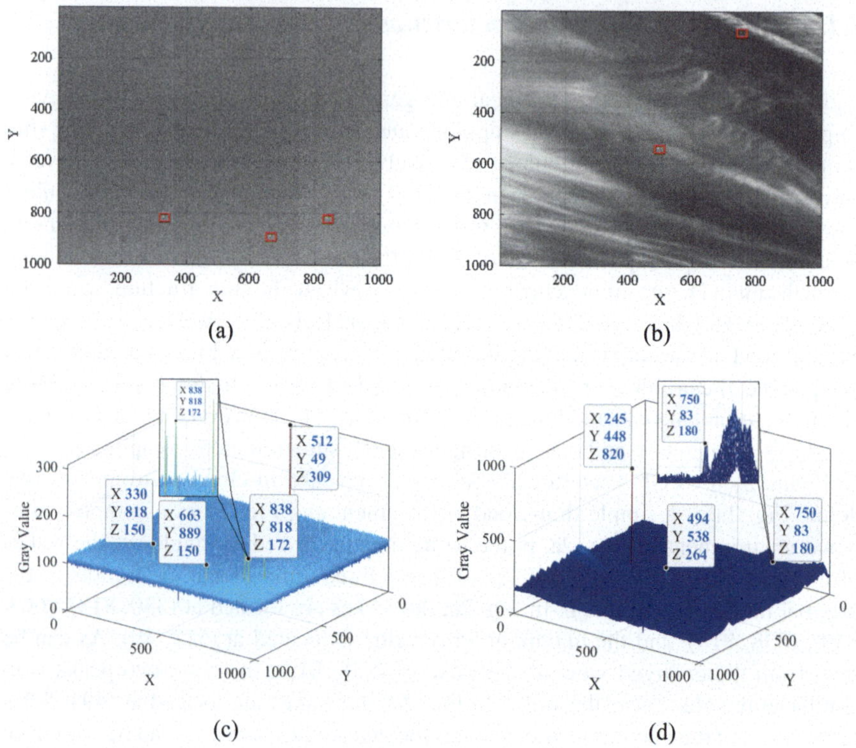

Fig. 2 GF-4 satellite images of two typical thin-cloud environments

larger than the gray value of the wake. Therefore, it is very difficult to detect the wake accurately in the GF-4 image.

2.2 *Proposed Multiple Moving Ships Detection Method*

Since the GF-4 satellite is located in geosynchronous orbit, the thermal environment and radiation environment of the GF-4 satellite are much worse than those of the medium and low-orbit satellites compared to remote sensing satellites operating in low or medium orbits. In addition, due to the influence of multi-mode switching and jitter, the image of GF-4 satellite appears blurred details and unclear edges. This makes the signal-to-noise ratio and edge sharpness of the GF-4 image lower than other satellite images. These reasons make it more difficult to detect ships using GF-4 images. Therefore, we propose a multiple moving ships detection method based on a single GF-4 image. The main steps of our proposed method are as follows.

Firstly, the original image needs image enhancement because the gray value of the wake of a moving ship is much smaller than that of cloud and land. Therefore, it is necessary to increase the ratio of the wake to the gray value of the sea by image enhancement processing, in order to make wake detection easier. Generally speaking, image enhancement can be divided into spatial domain enhancement and frequency domain enhancement [12]. Because there are many changes in the wake's shape, it is impossible to enhance the wake's gray value by enhancing the signal in some frequency bands. At the same time, since the gray value of the wake is always within a certain range in the spatial domain, it is more suitable to enhance the gray value of the wake in the spatial domain.

$$\mathbf{F}(x, y) = \frac{1}{1 + \left(\frac{u}{\mathbf{E}(x, y)}\right)^{v}}, \tag{1}$$

where $\mathbf{E}(x, y)$ is the original image gray value of the image data matrix coordinate (x, y); u is the two-dimensional mean of the original image $\mathbf{E}$; v is the stretching factor.

Secondly, a small neighborhood two-dimensional median filter and a large neighborhood two-dimensional median filter are applied to the image data matrix after image enhancement:

$$\mathbf{G}_{\mathrm{S}}(x, y) = \frac{\sum_{p=-a/2}^{a/2} \sum_{q=-b/2}^{b/2} \mathbf{F}(x + p, y + q)}{ab}, \tag{2}$$

$$\mathbf{G}_{\mathrm{L}}(x, y) = \frac{\sum_{p=-c/2}^{c/2} \sum_{q=-d/2}^{d/2} \mathbf{F}(x + p, y + q)}{cd}, \tag{3}$$

where $\mathbf{F}(x, y)$ is the enhanced image gray value of the image data matrix coordinate (x, y); $\mathbf{G}_{\mathrm{S}}(x, y)$ is the enhanced image gray value of small neighborhood two-dimensional median filter; $\mathbf{G}_{\mathrm{L}}(x, y)$ is the enhanced image gray value of large neighborhood two-dimensional median filter; a is the x-direction kernel radius of small neighborhood two-dimensional median filter, and b is the x-direction kernel radius of small neighborhood two-dimensional median filter, respectively; c is the x-direction kernel radius of small neighborhood two-dimensional median filter, and d is the x-direction kernel radius of large neighborhood two-dimensional median filter, respectively.

Let the small neighborhood two-dimensional median filtering result $\mathbf{G}_{\mathrm{S}}(x, y)$ and the large neighborhood two-dimensional median filtering result $\mathbf{G}_{\mathrm{L}}(x, y)$ be subtracted:

$$\mathbf{G}_{\mathrm{D}}(x, y) = \mathbf{G}_{\mathrm{S}}(x, y) - \mathbf{G}_{\mathrm{L}}(x, y), \tag{4}$$

where $\mathbf{G}_{\mathrm{D}}(x, y)$ is the difference between $\mathbf{G}_{\mathrm{S}}(x, y)$ and $\mathbf{G}_{\mathrm{L}}(x, y)$. The small neighborhood two-dimensional median filtering preserves the outline of the wake, and the large neighborhood two-dimensional median filtering can average the gray value of the wake around it. Therefore, subtract them from each other and get the signal of the

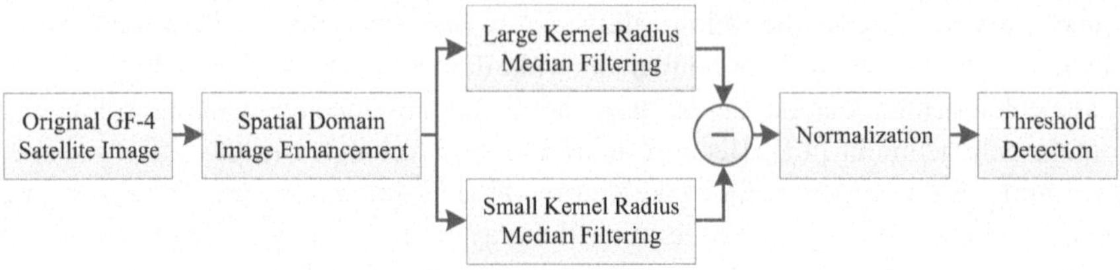

Fig. 3 Procedure of the proposed method

trail. Therefore, the wake signal can be obtained by subtracting them. In addition, the two-dimensional median filter can also reduce the impact of remote sensing image noise.

Thirdly, normalize the two-dimensional median filtering difference data:

$$\mathbf{G}_{\mathrm{N}} = \frac{\mathbf{G}_{\mathrm{D}}}{\max(\mathbf{G}_{\mathrm{D}})}, \tag{5}$$

where $\mathbf{G}_{\mathrm{N}}$ is the normalized result matrix, max(•) represents the maximum value in the matrix. Then, detect the grayscale peak in the normalized image to obtain a grayscale peak detection result matrix $\mathbf{G}_{\mathrm{P}}$. Threshold detection is performed on the grayscale peak detection result matrix to obtain a coordinate matrix of the wake.

$$\mathbf{R}(x, y) = \begin{cases} 0 & \mathbf{G}_{\mathrm{P}}(x, y) < G_{\mathrm{TH}} \\ 1 & \mathbf{G}_{\mathrm{P}}(x, y) \geq G_{\mathrm{TH}} \end{cases}, \tag{6}$$

where $\mathbf{R}(x, y)$ is the normalized grayscale peak detection result matrix, G_{TH} is the gray peak threshold.

The workflow of the proposed method is shown in Fig. 3. Firstly, the contrast stretch transformation is applied to improve the relative gray value of the ship in the image. Secondly, the enhanced image is filtered by a small radius kernel median filter and a large radius kernel median filter, respectively, and the filtering results are subtracted. Finally, normalize the subtraction results and perform threshold detection.

2.3 Computational Complexity Analysis

For evaluating the performance of the proposed method, we compare its computational complexity with that of other popular algorithms. From Fig. 3, we can know that the total time complexity consists of three parts: spatial domain image enhancement, different kernel radius median filtering, normalization, and threshold detection. In theory, any spatial domain image enhancement method can make the gray level of the wake be improved by setting appropriate parameters. The classical image enhancement methods include logarithmic transformation, gamma transformation and contrast stretch transformation [13]. The computational complexity of these three image enhancement methods is O(1). Because the computational complexity of the median filter algorithm for each pixel is O(1) [14], the computational complexity of an image with M pixels $\times$ N pixels is $O(M \times N)$. The computational complexity of normalization and threshold detection is O(1). Hence, the total computational complexity of our proposed method is $O(M \times N)$.

3 Experimental Results

In this section, we present the results of numerical experiments performed to verify the operation of our proposed method. The proposed method is implemented in MATLAB2018b with Intel Core i5-6300 2.3 GHz CPU. The imaging region of panchromatic sensor of GF-4 satellite is approximately 500 km $\times$ 500 km (10,000 pixels $\times$ 10,000 pixels). In order to observe the environment and wake, the original images are subdivided into sub-images with 50 km $\times$ 50 km (1000 pixels $\times$ 1000 pixels) as the region of interest.

3.1 Implementation of the Proposed Method

In order to demonstrate the performance of the proposed algorithm, the key steps in GF-4 satellite image processing are presented and discussed. The image was taken on April 5, 2017, 02:30:00 Universal Time Coordinated (UTC), and the location of the image is the East China Sea. We use the contrast stretch transformation for image enhancement. The stretching factor v is 6. The kernel radius $a = 3$, $b = 3$, $c = 8$, and $d = 8$, respectively.

The experimental results of key steps are presented in Fig. 4. Figure 4a shows the original image of GF-4. The position of the two ship's wakes is marked by a red rectangle. In Fig. 4a, only a few thin clouds can be seen faintly, and no wake can be seen in the red rectangles. Figure 4b shows the result of the contrast stretch transformation of the original image Fig. 4a. The detail of the cloud in the picture becomes very clear, and the wakes in the red rectangle can be seen clearly. Figure 4c shows the three-dimensional grayscale image of the original image. The locations of the two wakes are (494, 538) and (750, 82), respectively. It is interesting to see from this picture that there is a very large gray value of a pixel in the position of (245, 448). This is due to the damage of the photosensitive device of this pixel. Figure 4d shows the three-dimensional image of gray value after contrast stretching transformation. Figure 4e, f show three-dimensional grayscale images filtered by median filters with kernel radius $r = 3$ and $r = 8$, respectively. It can be seen from Fig. 4e that the gray values of the two wakes still exist and the gray value of the bad pixel disappears. However, the gray values of both wakes in Fig. 4f disappear. Figure 4g shows the difference of median filtering result with different kernel radius. It can be seen from Fig. 4g that the gray values of the two wakes' position are significantly higher than in other areas. Figure 4h shows the normalized result of Fig. 4g. It can be seen from Fig. 4h that the locations of the two wakes are clearly detected, and our proposed method has good detection threshold robustness.

3.2 Comparison with Other Methods

To evaluate the performance of the proposed method, we implement Yang's method in literature [8] and Liu's method in literature [9] as a benchmark to detect ships in the simple thin cloud and complex thin-cloud environment using GF-4 images, respectively. The two types of cloud environment datasets consist of 50 GF-4 images. These images have the size of 1000 × 1000 pixels with 50 m resolution. The first frame of the two environments of images is shown as Fig. 2a, b. The parameters of Yang's method are set: the threshold $T_0 = 0.2$; Compactness = [1, 10]; and

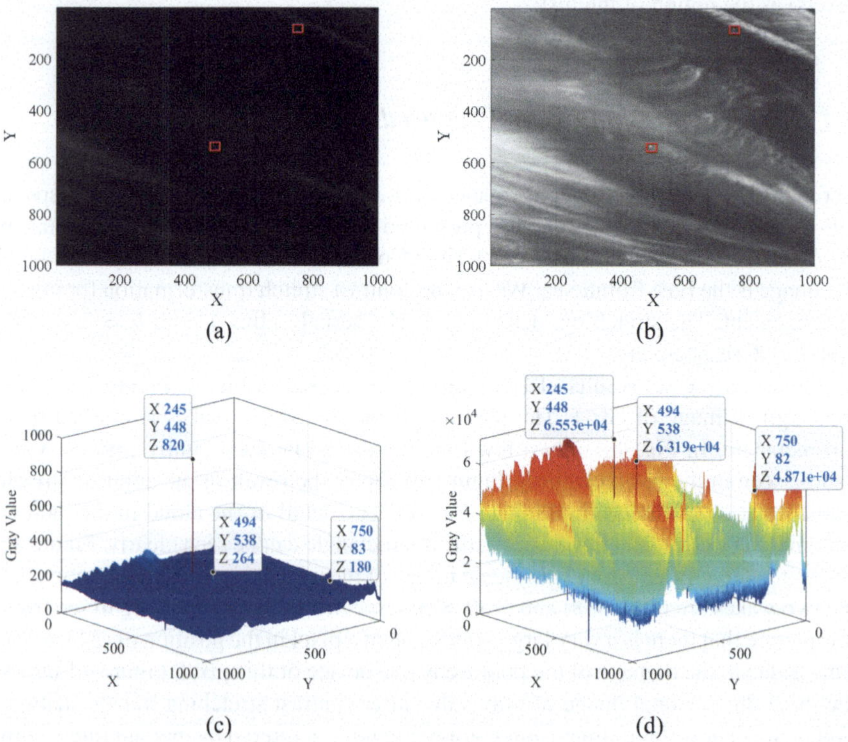

Fig. 4 Experimental results of key steps

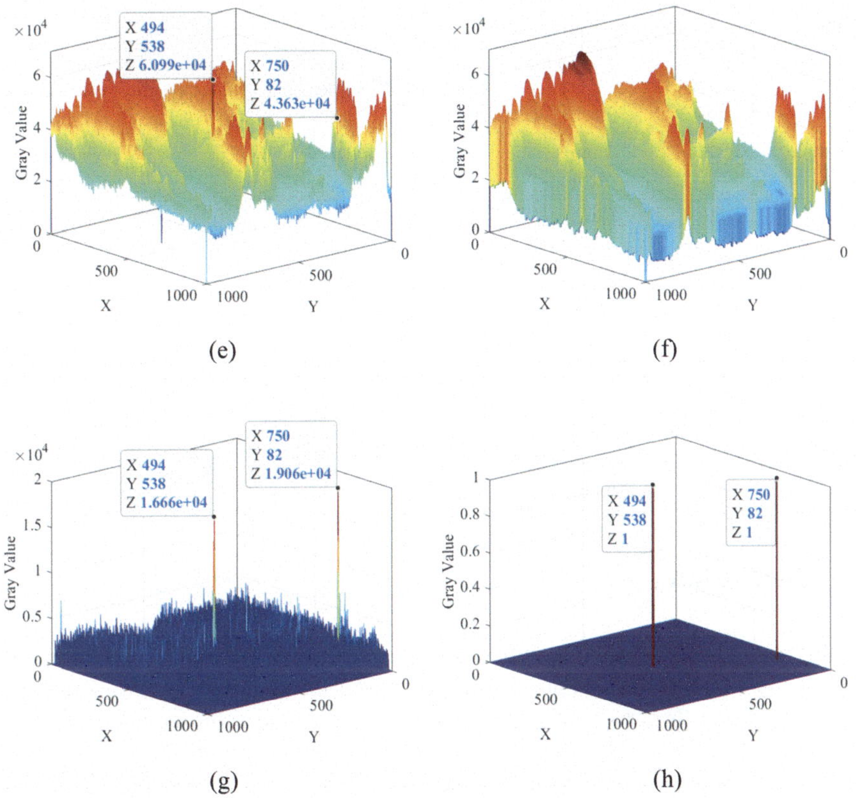

Fig. 4 (continued)

Length/Width = [1, 5]. The parameters of Liu's method are set: the minimum and maximum blob sizes are 2 and 50 pixels. The detection results are shown in Table 1.

The following conclusions can be drawn through the above experiment of GF-4 sequence images: (1) in the simple-cloud environment, the detection results of the three methods are very good, but the running time of CFAR detector is obviously higher than that of the other two methods; (2) in the complex-cloud environment, the detection performance of Liu's method is better than that of Yang's method, and the detection performance of our proposed method is better than that of Liu's method; (3) the running time of the proposed method is similar to that of the Yang's method and obviously less than that of the Liu's method.

Table 1 Ships detection results of different method

Type	Total number of ships	Method	Number of detected ships	Number of undetected ships	Number of false alarm	Running time (s)
Simple	150	Yang's method	152	0	2	9.17
		Liu's method	150	0	0	14.25
		Proposed method	150	0	0	10.29
Complex	67	Yang's method	104	19	56	9.14
		Liu's method	91	7	31	14.26
		Proposed method	75	0	8	10.27

4 Conclusion

In this paper, we proposed a multiple moving ships detection method for GF-4 satellite in the thin-cloud environment. The proposed method uses the GF-4 satellite image to detect the wakes of multiple moving ships, and achieves fast and accurate multiple moving ships detection in the thin-cloud environment. The proposed method is especially suitable for the detection of non-cooperative ships. We believe that the proposed method has broad application prospects.

Acknowledgment The authors would like to thank the China Center for Resources Satellite Data and Application for providing GF-4 satellite images. This project was supported by the China Postdoctoral Science Foundation(Grant No. 2019M650836).

References

1. Wang M, Cheng Y, Tian Y et al (2018) A new on-orbit geometric self-calibration approach for the high-resolution geostationary optical satellite GaoFen4. IEEE J Sel Top Appl Earth Observ Remote Sens 11:1670–1683
2. Yang B, Pi Y, Li X et al (2017) Relative geometric refinement of patch images without use of ground control points for the geostationary optical satellite GaoFen4. IEEE Trans Geosci Remote Sens 56:474–484
3. Information on http://www.cresda.com/EN/satellite/9907.shtml
4. Wu W, Liu W (2018) Remote sensing recognition of residential areas based on GF-4 satellite image. In: 2018 fifth international workshop on earth observation and remote sensing applications (EORSA). IEEE (2018):1–4

5. Sun F, Min M, Qin D et al (2018) Refined typhoon geometric center derived from a high spatiotemporal resolution geostationary satellite imaging system. IEEE Geosci Remote Sens Lett 16:499–503
6. Zhang Z, Shao Y, Tian W et al (2017) Application potential of GF-4 images for dynamic ship monitoring. IEEE Geosci Remote Sens Lett 14:911–915
7. Yu Y, Ai H, He X et al (2018) Ship detection in optical satellite images using Haar-like features and periphery-cropped neural networks. IEEE Access 6:71122–71131
8. Yang G, Li B, Ji S et al (2013) Ship detection from optical satellite images based on sea surface analysis. IEEE Geosci Remote Sens Lett 11:641–645
9. Liu Y, Yao L, Xiong W et al (2018) GF-4 satellite and automatic identification system data fusion for ship tracking. IEEE Geosci Remote Sens Lett 16:281–285
10. Heiselberg H (2016) A direct and fast methodology for ship recognition in Sentinel-2 multispectral imagery. Remote Sens 8:1033
11. Sun RQ, Luo G, Zhang M et al (2011) Electromagnetic scattering model of the Kelvin wake and turbulent wake by a moving ship. Waves Random Complex Media 21:501–514
12. Sundararajan D (2017) Digital image processing. Springer, Singapore, pp 23–142
13. Gonzalez RC, Woods RE (2018) Digital image processing, 4th edn., pp 119–202
14. Perreault S, Hébert P (2007) Median filtering in constant time. IEEE Trans Image Process 16:2389–2394

Identification and Extraction of Nutrient Content in Hyperspectral Black Soil in Frequency Domain

Dong-hui Zhang, Ying-jun Zhao, Kai Qin, Dong-hua Lu, Cheng-kai Pei, Ning-bo Zhao, Yue-chao Yang, and Ming Li

Abstract Hyperspectral remote sensing technology, with its high spectral resolution and high spatial resolution, plays a more and more important role in the quantitative remote sensing monitoring of black soil. In order to extract the information from the current spectrum and the object-oriented method, it is impossible to integrate the spectral domain and the space domain to explore the feasibility of the frequency domain processing method to improve the recognition accuracy. The CASI/SASI aero hyperspectral data are obtained from the Jiansanjiang area of Northeast China, and 60 samples are collected on the ground, and the content of organic matter is tested. The characteristics of the amplitude spectrum and phase spectrum of the typical black land are studied, and an adaptive classifier based on Gauss filter is designed for the hyper spectral space spectrum analysis algorithm, and an air spectrum classification framework based on the optimization of the ground laboratory data is constructed. Compared with the traditional hyperspectral classification algorithm, the frequency domain recognition and extraction algorithm proposed in this paper have described and characterized the hyperspectral data from a new viewpoint, which solve the uncertainty of hyperspectral data. In the future, this method may be a new thought to improve the traditional data processing method.

Keywords Black soil nutrient · Frequency domain characteristics · Adaptive gauss low-pass filter · Aero hyperspectral · Hyperspectral remote sensing

1 Introduction

The spectral data of ground objects are understood as electromagnetic radiation characteristics energy distribution maps in the recognition of nutrient content in black soil based on hyperspectral data. This treatment only describes the spectral distribution characteristics of the energy. It is a single pixel analysis method of black soil, which can extract limited information. Especially, in the extraction of

D. Zhang (✉) · Y. Zhao · K. Qin · D. Lu · C. Pei · N. Zhao · Y. Yang · M. Li
National Key Laboratory of Remote Sensing Information and Imagery Analyzing Technology, Beijing Research Institute of Uranium Geology, Beijing 100029, China
e-mail: donghui222@163.com

L. Wang et al. (eds.), *Proceedings of the 6th China High Resolution Earth Observation Conference (CHREOC 2019)*, Lecture Notes in Electrical Engineering 657,
https://doi.org/10.1007/978-981-15-3947-3_6

black soil information of high-resolution remote sensing data, the same pixel area of land structure, shape, texture, and so on is greatly increased. The efficiency and accuracy of information recognition based on spectral energy are difficult to meet the practical application requirements. In essence, it is a method of mathematical statistics. The result satisfies the optimal value of numerical simulation, but disjoins with the expert experience and background knowledge, and it is difficult to further improve the precision to a certain extent.

The Parseval energy conservation theorem proves that the total energy of the remote sensing objects in the spatial domain E_s and the frequency domain E_f is equal [1]. Under the support of this theory, the research direction at home and abroad is concentrated in two aspects; one is to improve the quality of remote sensing image by exploring the response relationship between the airspace and the frequency domain. In order to improve the resolution of remote sensing images, a super-resolution reconstruction method is established by studying the response relationship between resolution and frequency domain information. It shows that the resolution of 2M and 3M remote sensing images is increased by 1.75 and 1.90 times by the frequency domain aliasing [2]. In order to master the information transmission law of remote sensing optical system, from the frequency domain channel matrix derivation of remote sensing optical system, the frequency domain information transmission parameters of the imaging system are calculated from the frequency domain, and the research results have played a reference role to the design of remote sensing optical system [3]. In view of the problem of image degradation in the detector imaging system, an improved multi-frame image super-resolution enhancement method is proposed in the frame of frequency domain, the joint Gauss distribution model is established, and the interpolation reconstruction results are restored with Bayesian method [4].

The second research field, combined with the characteristics of the identified objects, improves the recognition accuracy through the transformation of airspace and frequency domain. According to the characteristics of the direction and frequency of the linear objects, a Fourier transform method is designed to extract roads from high-resolution images [5]. The image of urban river channel is transformed by Fourier transform, and the spectrum is divided into two parts of edge feature and low-frequency information. The log Butterworth filter and low-pass Butterworth filter are designed to extract the edge features and low-frequency information of the urban river, and the information extraction of the urban river is effectively realized [6]. Based on the characteristics of frequency domain, the principle of high vegetation coverage is extracted. After obtaining the maximum energy direction of the road through the frequency curve analysis, the road center line is extracted by Gabor filter. The method has high calculation efficiency [7]. A study on comprehensive utilization of classical spectral features and texture features of the city land use extraction method, combined with the city spectral changes in frequency domain information on different geographical conditions caused by the sensitive characteristic, in a few large city in the USA carried out tests show that the frequency domain extraction methods with spatial domain, and the accuracy has been greatly improved [8].

With the deepening of hyperspectral remote sensing in digital mapping in black land, the introduction of frequency domain recognition technology can effectively combine the energy expression of the pixel spectrum with the representation of frequency spectrum energy in the frequency domain [9]. The significance is not only to redistribute the original spectral data to another space but also be beneficial to the processing and extraction of information [10]. The variation of energy in the spatial image is reflected, and the spatial features such as color, texture, direction, and boundary of the black soil are involved in the soil quality assessment, and a more comprehensive set of black soil features is obtained to improve the precision of nutrient information extraction [11].

In this paper, an adaptive Gauss low-pass filtering algorithm is designed based on the analysis of the characteristics of the black soil spectrum. By eliminating the data of different energy levels gradually, the purpose of land classification is achieved. The cut-off frequency optimal model is established, and it balances image smoothing with detail preservation. The improved Gauss low-pass filter in this paper not only has a smooth function but also imports black soil nutrient information into the frequency domain. CASI hyperspectral imaging experiments are tested in Jiansanjiang area, Heilongjiang Province, which verified the good performance of the proposed frequency domain recognition and extraction classification method.

2 Black Soil Spectrum Characteristics Analysis

2.1 Black Soil Texture Features Description

There are three main features of the black land texture: one is the close correlation between the texture features and the resolution [12]. Under a certain resolution, the field of view is consistent when it moves in the texture region [13]. The two is the basic graph element, not random arrangement, and the three is that all kinds of textures are homogeneous and uniform in the research area. The place has roughly the same structure size. Therefore, statistical parameters can be used to characterize the distribution characteristics of black soil texture. Three typical black soil texture features are analyzed by selecting five eigenvalues of information entropy, two moments, contrast, synergy, and correlation (Table 1).

The results show that (1) the information entropy of the lattices character black soil is the highest, reflecting the high uncertainty of hyperspectral data under this kind of black soil texture, and the amount of nutrient information contained in black

Table 1 Typical black land texture feature statistics results

Texture name	Information entropy	Two-order moment	Contrast	Synergistic	Correlation
Strip	1.819	0.196	18.689	0.307	0.107
Bright spot	1.994	0.121	66.602	0.206	0.010
Lattices	2.003	0.123	65.564	0.190	0.072

soil is also larger. (2) The two-order moment refers to the discrete degree of the spectral value relative to the mean value. The higher the discrete degree is the more characteristic of the spectral data is, and the stripe texture has the highest degree of dispersion in the three kinds of black land texture. (3) The contrast degree affects the change level of the image and the dark, the higher the contrast, the less number of color order from ash to pure black and pure white, the greater the reflectance contrast. The bright spot black soil is significantly higher than that of the other two types of black soil texture. (4) The synergistic calculation can be used to evaluate the enhancement and coherence of different bands of hyperspectral bands, and the highest mutual gain between bands is strip textures. (5) The correlation reflects the redundancy degree of spectral data in different bands, and the correlation of black soil band in strip is the highest, and there is a lot of redundant information in the data.

2.2 Black Soil Phase Spectrum Characteristic Analysis

The amplitude spectrum of the black land expresses the number of each frequency component in the hyperspectral image, and the phase spectrum determines the position of each frequency component in the image, which is an angle between $-\pi$–π [14]. Although the phase spectrum contains the main structure of the original black soil spectrum, the information of black soil content cannot be obtained directly.

By calculating the phase spectrum and histogram of three typical black land, it is found that there is no close relation with the black land texture, especially the information related to the characteristics of the black land direction is not significant. This phenomenon is consistent with the conclusions of previous studies [15]. Although the information on black land cannot be obtained directly from the phase spectrum, the phase spectrum contains the main structure of the hyperspectral image, and it can play a role in the image segmentation in the detection of the black soil edge based on phase consistency. Therefore, the algorithm design based on frequency domain filtering is mainly based on amplitude spectrum.

3 An Adaptive Gauss Low-Pass Filtering Algorithm

3.1 Basic Principle

In order to make the spatial and spectral information as the reference data of the hyperspectral image classification at the same time, an adaptive Gauss low-pass filtering algorithm is designed to add the information of the nutrient content of the black soil in the frequency domain classifier.

The core idea of the algorithm is that a Gauss low-pass filter model can be extracted (or excluded) to transform the energy of hyperspectral images, and the data of different energy levels are eliminated gradually to achieve the purpose of the black land classification [16]. When smoothing the image with Gauss model, the Gauss function determines the filtering result by calculating variance σ. Combining the interpolated data of the known nutrient data in the geospace of the black soil, the adaptive Gauss filter is designed to select different σ on the basis of preserving the local features of the black soil image to realize the original image classification.

In the frequency domain, the basic filter model is:

$$G(u, v) = H(u, v) \times F(u, v) \tag{1}$$

where $F(u, v)$ is a filtered Fourier transform image; $H(u, v)$ is a filter transform function; $G(u, v)$ is the smoothed image generated after the attenuation of high-frequency information [17].

The Gauss low-pass filter which can transform the spatial domain into the frequency domain is:

$$H(u, v) = \mathrm{e}^{-D^2(u,v)/2d^2} \tag{2}$$

where $D(u, v)$ is the distance from the origin of transformation. d is the expansion degree of the Gauss curve, that is, the cut-off frequency. An adaptive Gauss filter is designed to automatically select different d according to the black soil content characteristic of the smooth image so that the corresponding content of nutrient data is obtained in the processed image.

Therefore, suppose the Gauss smoothing of black soil images is expressed in the following function:

$$I_0(x, y) = I_d(x, y) + e_d(x, y) \tag{3}$$

where for pixel point at (x, y), $I_0(x, y)$ is the gray value of the original hyperspectral data, $I_\mathrm{d}(x, y)$ is the low-pass gray value under the cut-off frequency d, $e_\mathrm{d}(x, y)$ is the residual value under the cut-off frequency d.

A method based on energy function is designed:

$$d_{\mathrm{best}} = \arg\min\left\{c/\sigma^3 + e^3\right\} \tag{4}$$

where, c is a constant term and is determined according to the nutrient test data of the sampling points; σ is the variance; e is the residual value.

It is concluded that under the known constant terms of c, the optimal cut-off frequency d needs variance σ as large as possible, and transforms hyperspectral data to more smooth data. The residual error e must be as small as possible, that is, the smaller the change of reflectivity of the original pixel (x, y) after Gauss filtering, the smaller the better. In this way, a comprehensive Gauss filtering method for nutrient content is established, which achieves the balance between smoothing classification and maintaining details.

3.2 Algorithm Steps

The adaptive Gauss low-pass filtering algorithm is as follows:

(1) The original hyperspectral data of black land are transformed in frequency domain to generate frequency domain data.
(2) The initial Gauss low-pass filter is used for the frequency domain data, and the initial cut-off frequency of nutrient content of the first kind is set d_1. The initial variance σ_1 and residual value e_1 are obtained.
(3) The filtering results are compared with the ground test data. If the extracted information is in accordance with a certain level of nutrient content, it is recorded as c_1, otherwise, d_1 is set up to continue filtering until the filtered information is distributed in this nutrient interval and is recorded as c_1.
(4) In the hyperspectral data, the corresponding pixels in the airspace are removed from the hyperspectral data, and the initial cut-off frequency of second nutrient contents is set at d_2, Repeat step 3 until c_2 is obtained.
(5) Repeat step 4 and step 3 until the five levels of nutrient content are generated by c_3, c_4, and c_5.

4 Black Soil Nutrient Data and Information Extraction

4.1 Data

Data are obtained from the CASI/SASI aviation hyperspectral imaging system. The spectral range is 380–2450 nm, the spatial resolution is 4 m, the continuous spectrum channel number is 137, and the spectral bandwidth is 12.5 nm. The experimental data are collected on April 10, 2017, with a length of 9.27 km and a width of 5.36 km, with an area of about 50 km^2 and a flight altitude of 3 km (Fig. 1). The black and white cloth is laid on the ground. The calibration spectrum is obtained by ASD Field Spec spectrometer. The spectral range is 350–2500 nm, and the spectral resolution is 1 nm.

The sampling point is 60, the coordinates of sample 1 are 132.747E, 47.232N, and the coordinates of sample 60 are 132.857E, 47.272N, and the soil samples are collected at 0.75 km intervals. The surface of the survey area is a black humus layer, thick 30–60 cm, thickest than 1 m, and many cylindrical or granular structures. On the same day, the soil samples of the surface 0–20 cm are collected synchronously, and the large plant residue and stone and other debris are removed. The soil samples are

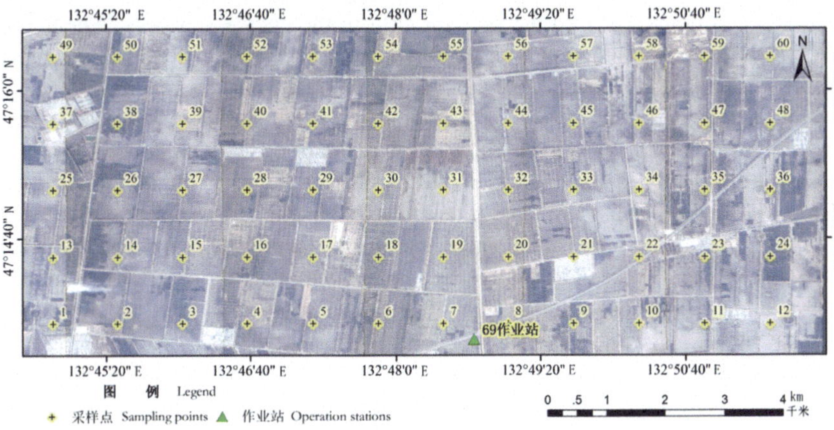

Fig. 1 Aeronautical data acquisition area and sampling point distribution map

Table 2 Information table of black soil organic matter content at different sample points

Sample class		Minimum/(g kg^{-1})	Maximum/(g kg^{-1})	Average/(g kg^{-1})	Standard deviation/(g kg^{-1})
Soil organic matter	Modeling samples	3.39	4.46	3.85	0.23
	Prediction samples	3.30	4.14	3.79	0.24
	All samples	3.30	4.46	3.83	0.23

used in the laboratory to dry and grind, and 0.15 mm screening is used to determine the content of the soil. The organic matter is determined by potassium dichromate volumetric external heating method. In soil nutrient content determination, sample 1–45 is used for training set, and the remaining 15 samples are used for evaluation accuracy (Table 2).

4.2 *Extraction of Nutrient Information from Black Soil*

The values of initial variance σ_1 and e_1 of residual value are 0.1 and 1.5, respectively. The adaptive Gauss low-pass filter is used to calculate the variance value and gradually reduce the residual value. The nutrient test data of the black land in the measured area are brought into Formula 4, and the grade assessment map of the 5 levels of nutrient content is obtained.

The adaptive Gauss low-pass filtering algorithm combines the results of the ground test data. By optimizing the whole energy function, the spatial information of the image is transformed into the space item in the energy function, and the spectral information is converted into the spectral term in the energy function. When constructing the energy function, it is assumed that the adjacent eight neighboring center pixels are identical to the results of the ground test data.

Although the increase of variance σ can improve the classification accuracy of the pixel-level filter in the smooth region, the probability of misclassification of pixels at the edge of the block is also increased. Different from the variation of variance, the algorithm based on Gauss low-pass filter is used to evaluate the difference between the adjacent pixels by the residual value, and the probability optimization is carried out by the residual value. The higher the residual value is, the lower the similarity of gray value of adjacent pixels is, the lower the probability of being labeled as the same category.

Therefore, residuals can play a positive role in the detection of pixels on the edge of the plot. Therefore, on the basis of reasonable initial variance of σ_1 and residual value e_1, the adaptive Gauss low-pass filtering algorithm not only improves the accuracy of the pixel classification accuracy of the smooth region but also makes the edge detection of the soil nutrient content more accurate in the classification results.

5 Black Soil Nutrient Extraction Precision Analysis

5.1 Precision Evaluation Method

The ground test data points are scored by kriging interpolation, and the soil nutrient grading map is obtained as the basis for evaluating the accuracy of information extraction. The total accuracy ($P_c = \Sigma_{k=1}^{n} P_{kk}/P$), mapping precision ($P_{Ai} = P_{jj}/P_{j+}$), leakage error ($1 - P_{ui}$), user accuracy ($P_{ui} = P_{ii}/P_{i+}$), and error ($1 - P_{Ai}$) between the image information extraction results of different initial variance σ_1 and residual e_1 are calculated, and the objective evaluation results are obtained.

5.2 Extraction Precision Analysis Results

The error confusion matrix table is constructed, and the five quantitative indexes of different initial variance σ_1 and residual value e_1 are calculated, respectively.

(1) Overall accuracy evaluation (Fig. 2). The parameters of the overall accuracy and Kappa coefficient are set to $\sigma_1 = 0.5$ and $e_1 = 1.1$, respectively, 92.24% and 0.8933, respectively. The parameters with the lowest extraction accuracy are $\sigma_1 = 0.1$ and $e_1 = 1.5$, 40.51% and 0.2566, respectively.
(2) Determination of the best information extraction parameters for each nutrient level. The best initial parameters for 1–5 level are: $\sigma_1 = 0.1/e_1 = 1.5$ (100%), $\sigma_1 = 0.5/e_1 = 1.1$ (91.10%), $\sigma_1 = 0.5/e_1 = 1.1$ (91.47%), $\sigma_1 = 0.6/e_1 = 1.0$

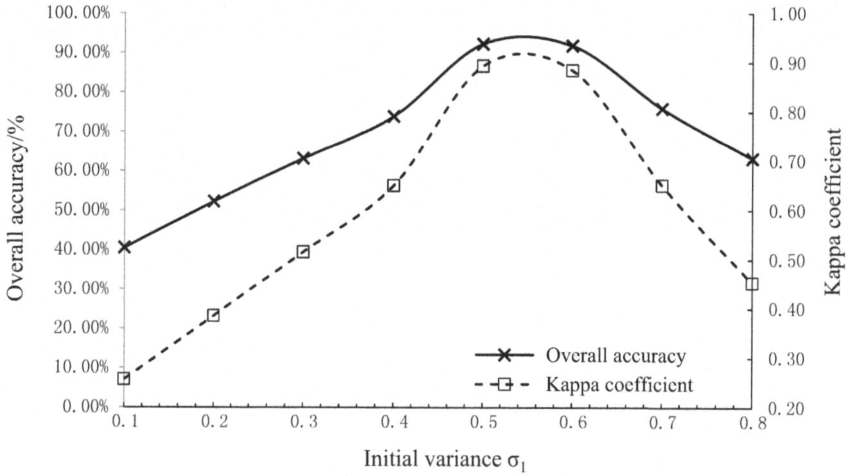

Fig. 2 Comparison of black soil nutrient extraction accuracy with eight initial parameters

(89.24%), and $\sigma_1 = 0.6/e_1 = 1.0$ (99.95%). It is concluded that for the black soil with higher nutrient content, the initial parameter is set to $\sigma_1 = 0.6/e_1 = 1.0$, and the extraction accuracy is higher.

(3) Targeted analysis of nutrient extraction grades of black soil under eight parameter settings. The $\sigma_1 = 0.3/e_1 = 1.3$ in extraction content grade 1, $\sigma_1 = 0.3/e_1 = 1.3$ in extraction content grade 1, $\sigma_1 = 0.4/e_1 = 1.2$ in extraction content grade 1, $\sigma_1 = 0.5/e_1 = 1.1$ in extraction content grade 1, $\sigma_1 = 0.6/e_1 = 1.0$ in extraction content level 5, $\sigma_1 = 0.7/e_1 = 0.9$ in extraction content level 5, $\sigma_1 = 0.8/e_1 = 0.8$ in extraction content level 5, the user accuracy is more higher. It is found that the setting of initial parameters is mainly related to the lowest and two highest nutrient levels.

5.3 *Extraction Results*

The parameters of the highest precision and Kappa coefficient are set up $\sigma_1 = 0.5$, and $e_1 = 1.1$ are applied to extract nutrient in black land. After classification, the nutrient content of each level is given according to the ground analysis data, and the spatial distribution of nutrient content in black soil is obtained (Fig. 3).

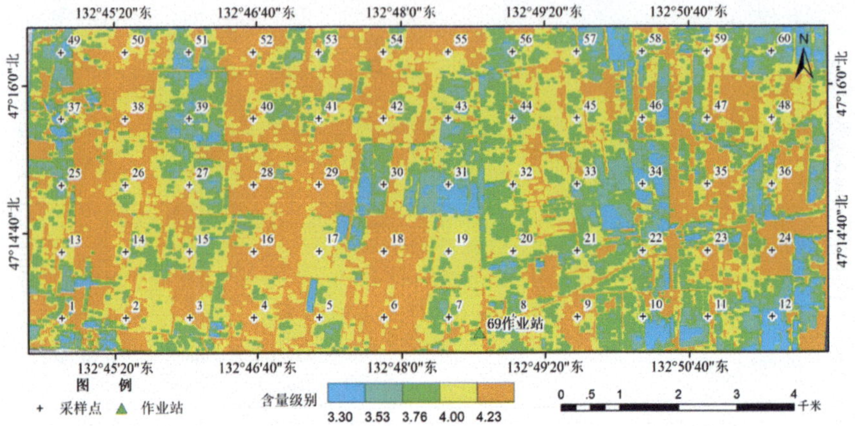

Fig. 3 Spatial distribution of black soil nutrient content by optimal initial parameters

6 Conclusion and Discussion

On the basis of the research of traditional spectral method and object-oriented target recognition method, the feasibility of combining hyperspectral image space and spectral information is explored to improve the accuracy of nutrient recognition in black land [18]. According to the characteristics of black land texture, the amplitude spectrum and phase spectrum characteristics of the typical black land are studied. On this basis, an adaptive classifier based on Gauss filter is designed for the hyperspectral space spectrum joint analysis algorithm, and a space spectrum classification framework based on the ground test data optimization is constructed, and the spatial information of the image is extracted from the frequency domain information. The spatial information and spectral information are scientifically integrated through the smoothing filtering of plots.

Using the data of CASI/SASI aerial hyperspectral imaging system, the proposed adaptive high and low-pass filtering algorithms are tested. The results show that the smooth filtering of the massif can effectively combine the spatial and spectral information of the image. On the basis of effectively removing the salt and pepper noise similar to the traditional classification method, the contour of the black land in the classification result is basically consistent with the real black land plot contour.

Compared with the traditional hyperspectral classification algorithm, the frequency domain recognition and extraction algorithm proposed in this chapter are more efficient. After defining the level of nutrient content, the result data of each level can be quickly calculated. Frequency domain method can partly solve the uncertainty of information extraction, because it expresses hyperspectral data from a new perspective. In the future, this method may be able to break through the traditional atmospheric correction and radiation correction factors and bring new methods.

References

1. Xiao PF, Feng XX, Wang PF et al (2012) High resolution remote sensing image segmentation and information extraction. Science Press, Beijing
2. Yang XF (2011) Remote sensing images in frequency domain and spatial domain super-resolution reconstruction technology. Harbin Institute of Technology, Harbin
3. Ma C (2015) Study on information transfer performance of remote sensing imaging system in spatial and frequency domain. Harbin Institute of Technology, Harbin
4. Tan Z, Xiang LB, Lv QB et al (2017) A super resolution enhancement method for image sequences based on frequency domain. J Actaoptica Sinica 37(7):83–88
5. Zhou LG, Feng XZ, Xiao PF et al (2011) A linear feature detection method for high resolution remote sensing images in frequency domain. J Surv Mapp 40(3):312–317
6. Wang K, Xiao PF, Feng XZ et al (2013) Extraction of urban river information from high-resolution remote sensing images based on frequency domain filtering. J Remote Sens 17(A02):277–285
7. Zhao HH, Feng XZ, Xiao PF (2014) Contour extraction of green cover along urban roads from remote sensing imagery based on frequency domain features. J Remote Sens Inf 29(3):50–56
8. Doustfatemeh I, Baleghi Y (2016) Comprehensive urban area extraction from multispectral medium spatial resolution remote-sensing imagery based on a novel structural feature. Int J Remote Sens 37(18):4225–4242
9. Steinberg A, Chabrillat S, Stevens A et al (2016) Prediction of common surface soil properties based on Vis-NIR airborne and simulated EnMAP imaging spectroscopy data: prediction accuracy and influence of spatial resolution. Remote Sens 8(7):613–627
10. Zhang P, Li Y (2016) Study on the comparisons of the establishment of two mathematical modeling methods for soil organic matter content based on spectral reflectance. Spectrosc Spect Anal 36(3):903–910
11. Žížala D, Zádorová T, Kapička J (2017) Assessment of soil degradation by erosion based on analysis of soil properties using aerial hyperspectral images and ancillary data, Czech Republic. Remote Sens 9(1):28–40
12. Eisele A, Lau I, Hewson R et al (2012) Applicability of the thermal infrared spectral region for the prediction of soil properties across semi-arid agricultural landscapes. Remote Sens 4(11):3265–3286
13. Malec S, Rogge D, Heiden U et al (2015) Capability of spaceborne hyperspectral EnMAP mission for mapping fractional cover for soil erosion modeling. Remote Sens 7(9):11776–11800
14. Niang MA, Nolin M, Bernier M et al (2012) Digital mapping of soil drainage classes using multitemporal RADARSAT-1 and ASTER images and soil survey data. Appl Environ Soil Sci 430347:1–17
15. Liu WD, Baret F, Zhang B et al (2004) Extraction of soil moisture information from hyperspectral remote sensing. Acta Sinica 41(5):700–706
16. Flach P (2012) Machine learning. People's Post and Telecommunications Press, Beijing
17. Yu L, Liu XB, Liu GZ et al (2016) Experiment research and analysis of spectral prediction on soil leaking oil content. Spectrosc Spect Anal 36(4):1116–1120
18. Liu HJ, Pan Y, Dou X et al (2018) Soil organic matter content inversion model with remote sensing image in field scale of black soil area. Trans Chin Soc Agric Eng 34(1):127–133

Evaluation of Geometric Performances of the Gaofen-6 PMS Camera

Liping Zhao and Hongzhou Li

Abstract The Gaofen-6 (GF-6) is an optical satellite consisting of a 2/8 m spatial resolution panchromatic/multispectral (PMS) camera and a 16 m spatial resolution wide angle camera. It was launched in June 2018. The geometric performances are an important basis for many applications, and it was evaluated using multiple revisited data of GF-6 PMS in Northeast China in the paper. According to the GF-6 PMS camera parameters and imaging geometric model, the instantaneous field of view at different side angles was analyzed, and the influence of control data error on image adjustment accuracy was simulated. Based on the high-precision digital ortho-image map and digital surface model data, the direct georeferencing performance and the adjustment accuracy were analyzed, and the root mean square errors (RMSE) of panchromatic and multispectral images with ground control points (GCPs) were approximately 0.5–1.0 pixels. Band registration accuracy between the multispectral bands was also evaluated, and the results showed the average accuracy was approximately 0.3 pixels RMSE. In addition, no significant effects of satellite vibration were detected by GCP residual analysis and band registration residual analysis. In general, this paper quantitatively analyzed the geometric performances of GF-6 PMS camera from the aspects of the direct georeferencing accuracy, adjustment accuracy and band registration accuracy. The results show that the attitude of the GF-6 satellite platform is stable, and the geometric accuracy of the PMS camera is good, meeting the geometric performance requirements of satellite design.

Keywords Gaofen-6 · PMS · Geometric performances · Accuracy · Multi-spectral registration

1 Introduction

In the frame of the Chinese civilian high-resolution remote sensing satellites, the Gaofen-6 (GF-6) and Gaofen-1 (GF-1) form a constellation, are mainly devoted to the operational application services of agriculture and rural affairs, natural resources,

L. Zhao (✉) · H. Li
Land Satellite Remote Sensing Application Center, Ministry of Nature Resource, Beijing, China
e-mail: zhaolpwww@163.com

L. Wang et al. (eds.), *Proceedings of the 6th China High Resolution Earth Observation Conference (CHREOC 2019)*, Lecture Notes in Electrical Engineering 657,
https://doi.org/10.1007/978-981-15-3947-3_7

emergency management, ecology and environment. The constellation is operated in a sun-synchronous and an altitude of 645 km, with a 10:30 AM equator crossing time. The satellite has high resolution, wide coverage, high quality imaging, high performance imaging, etc. and has a design life of eight years. It was launched in June 2018 on a CZ-2D rocket from China's Jiuquan space center.

GF-6 is an optical satellite consisting of a 2/8 m spatial resolution panchromatic/multispectral (PMS) camera with a swath width of 90 km, and a 16 m spatial resolution wide angle camera includes eight-band multispectral imagery with an 800 km swath width. Both cameras cover visible light to near-infrared (NIR) bands. The PMS camera (2/8 m resolution) with large field of view off-axis three-mirror anastigmat telescope uses a zero-distortion optical design, with excellent imaging quality, signal-to-noise ratio and geometric accuracy, and it can provide panchromatic imagery of 2.0 m resolution, and four-band multispectral imagery with blue, green, red and NIR at 8.0 m resolution.

The geometric performances are an important topic for many applications [1–3]. It was evaluated using multiple revisited data of GF-6 PMS in Northeast China in the paper. Using the adjustment schemes based on polynomial compensation model in image space, the accuracy evaluations of panchromatic image and multispectral image were performed separately. Then, the influence of a variety of different precision reference data on the accuracy evaluation was further compared. In addition, the registration accuracy of multispectral images was also analyzed.

2 Data

In order to evaluate the geometric performance of the GF-6 PMS camera, Harbin, Northeast China, with good weather conditions was selected as the experimental area. The main topographic features of the test area are flat, with a height difference of less than 60 m. Flat terrain facilitates the analysis of the geometric properties of the image itself, eliminating terrain effects as much as possible.

During this evaluation, two sets of reference data were available, one set was high-precision data with a planimetric accuracy of 0.5 m RMSE covering only part of the test area, and the other set was low-precision data with the planimetric RMSE of 3 m covering all test ranges. High-precision data was mainly used to analyze the PAN image with 2 m ground sample distance (GSD), and low-precision data was used to evaluate 8 m GSD MS data.

In the region, from July 2018 to April 2019, many revisited data was obtained. Six scenes panchromatic images and 18 scenes multispectral data with good quality are preferred for accuracy assessment, and Figs. 1 and 2 show their distribution ranges. Table 1 summarizes the main characteristics of the GF-6 satellite and PMS cameras.

In this paper, the horizontal datum and ellipsoidal height of WGS84 are used unless otherwise specified.

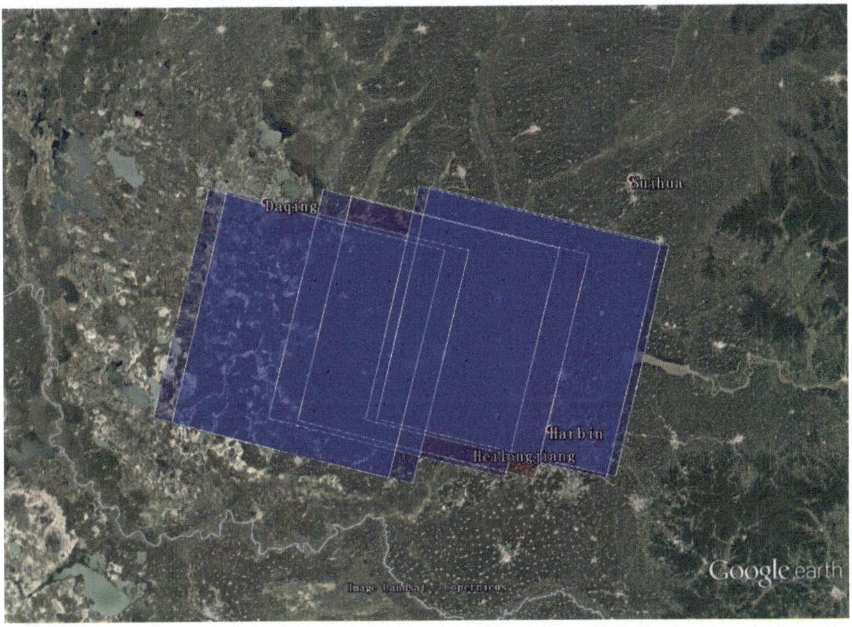

Fig. 1 Test data distribution diagram (PAN)

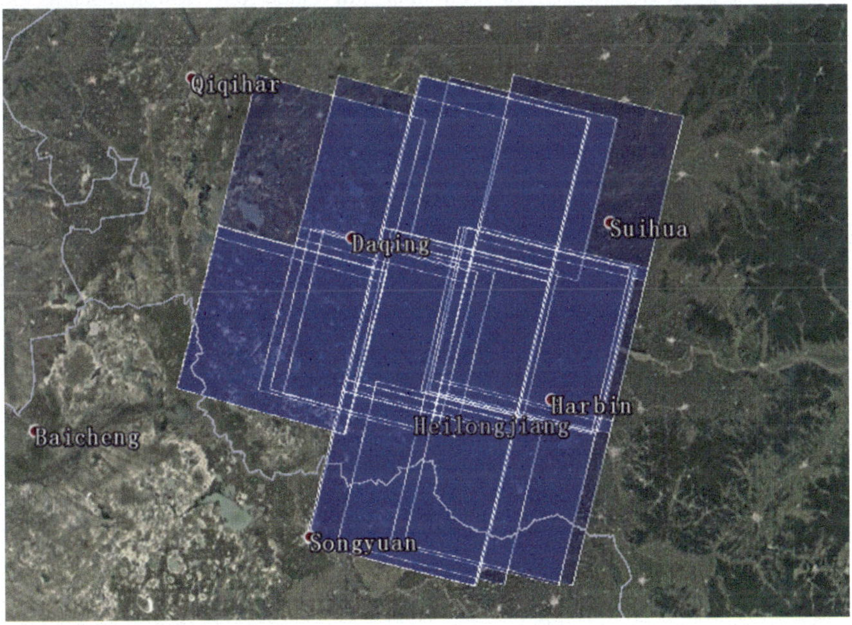

Fig. 2 Test data distribution diagram (MS)

Table 1 Main characteristics of the GF-6 and PMS camera

GF-6		PMS	
Launch periods	June 2, 2018	Field of view	~8°
Design lifetime	8 years	Swath	>95 km at Nadir
Altitude	645 km	Sample	PAN: 48312 pixels MS: 12078 pixels
Cycle	41 days	GSD	PAN: 2 m MS: 8 m

3 Evaluation

Before the evaluation, this paper analyzed the instantaneous field of view (IFOV) changes at different side angles and the influence of ground reference data error on image accuracy. Then, direct georeferencing accuracy was determined using the provided rational polynomial coefficients (RPC) and the ground control points (GCPs) extracted based on ground reference data, and the applicability of various polynomial adjustment models in image space was compared. For panchromatic and multispectral data, affine transformation models in image space were chosen as the formal accuracy assessment model, and the impact of different reference data on accuracy assessment had also been explored. In addition, a preliminary evaluation of the registration accuracy between the four bands of multispectral images was also made.

3.1 Simulation

In order to obtain prior knowledge of GF-6 PMS satellite data processing and analysis, according to the GF-6 PMS camera parameters and imaging geometric model, for PAN and MS data, respectively, the IFOV at different side angles was given in Figs. 3 and 4, and the influence of ground reference data error on image adjustment accuracy was simulated (Figs. 5, 6, 7, 8, 9 and 10).

It could be seen from Figs. 3 and 4 that the small angle side view has little effect on the IFOV of the PAN and MS images. The largest side angle in the experimental data is −2.50°, which is negligible.

Analysis of Figs. 5, 6, 8 and 9 shows that the ground reference data planimetric error has a significant influence on the image accuracy, and the error of one ground sample distance (GSD) roughly corresponds to one pixel on the image. Therefore, for PAN and MS images, if the error in image space is required to be better than 0.3

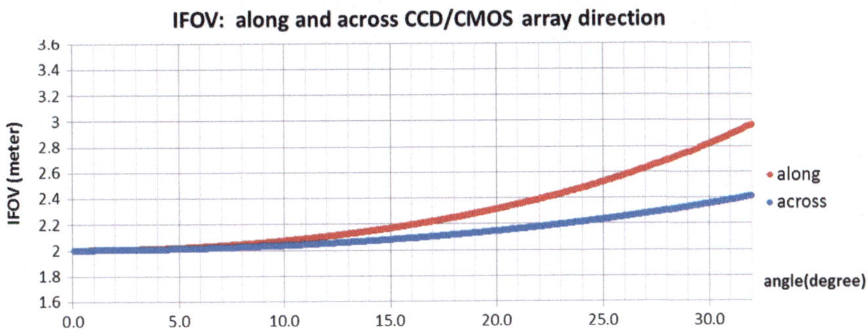

Fig. 3 IFOV with different side views (PAN)

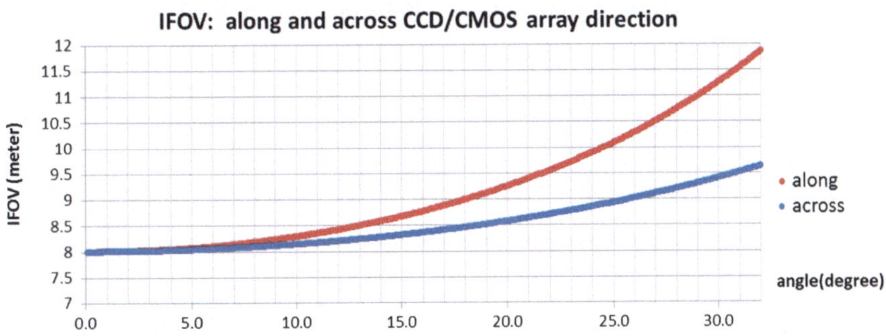

Fig. 4 IFOV with different side views (MS)

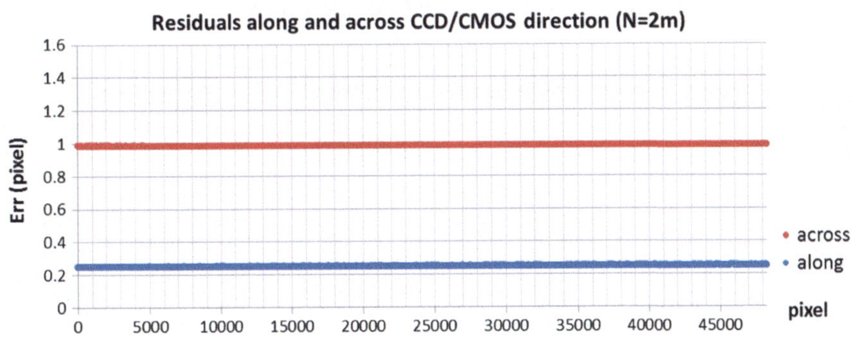

Fig. 5 Influence of ground reference data error on the PAN image (err_N = 2 m)

pixels, the planimetric accuracy of the ground reference data needs to be at least 0.6 and 2.4 m, respectively.

It could be found from Figs. 7 and 10 that the influence of the elevation error of the ground data in the direction of the line array is related to the position of the image point in the field of view, but there is a large system influence in the direction of the

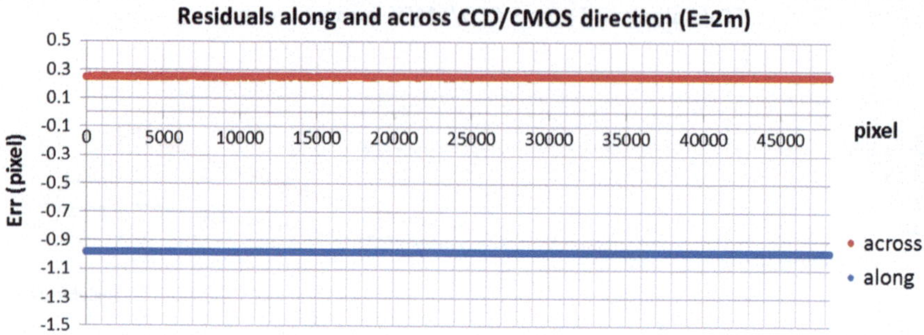

Fig. 6 Influence of ground reference data error on the PAN image (err_E = 2 m)

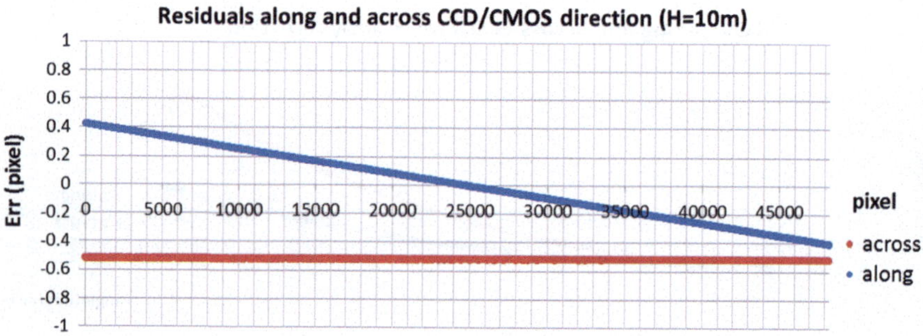

Fig. 7 Influence of ground reference data error on the PAN image (err_H = 10 m)

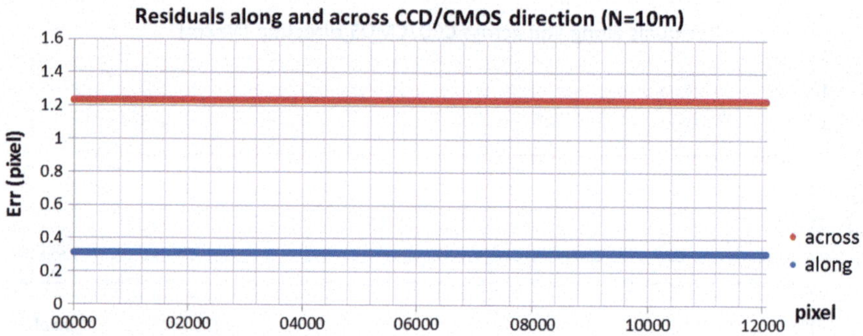

Fig. 8 Influence of ground reference data error on MS image (err_N = 10 m)

vertical line array, which is caused by the camera's boresight direction. The direction does not point to the nadir, which has an angle of about 5° from the vertical. Then, for PAN and MS data, to achieve the accuracy of 0.3 pixels, the ground reference data elevation accuracy needs to be at least 5.0 and 20 m.

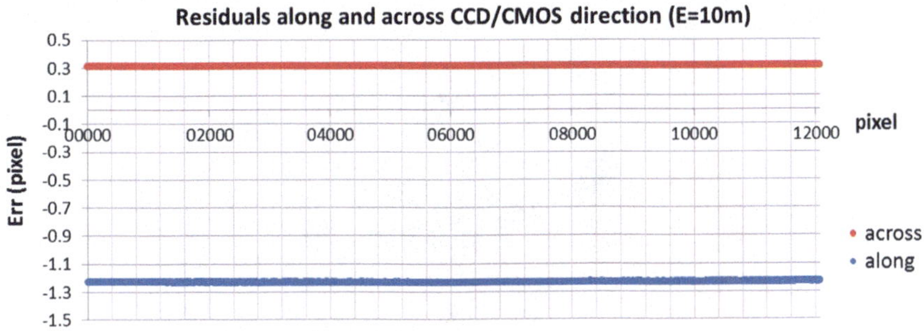

Fig. 9 Influence of ground reference data error on MS image (err_E = 10 m)

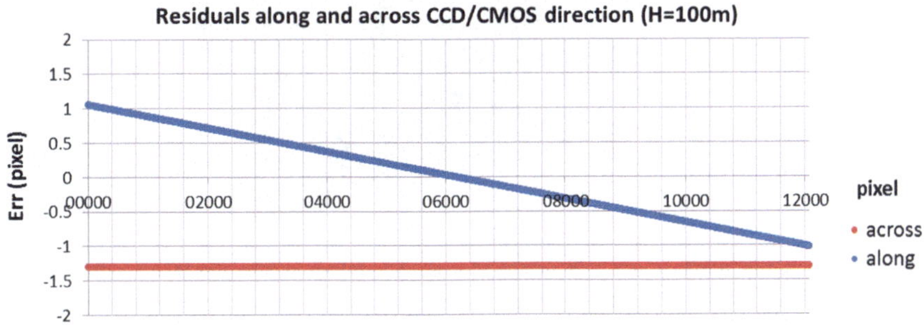

Fig. 10 Influence of ground reference data error on MS image (err_H = 100 m)

3.2 *GCP Collection*

When studying the geometric accuracy of a large amount of data, based on high-precision digital ortho-image map (DOM) and digital surface model (DSM) data, the method of automatically acquiring GCP based on image matching means is fast and reliable. Using this automatic matching method, thousands of GCPs could be extracted on demand on each scene image.

A reference dataset is available over this test area. The dataset is comprised of ortho-image with a GSD of 0.5 m and an accuracy around 0.5 m and digital surface model with a grid of 2 m and an accuracy around 1.0 m. Fast Fourier transform phase matching (FFTP) works in the frequency domain, it pays more attention to the texture information of the image, and the registration of the multi-temporal image is more robust, so FFTP is used to extract tie points between the test data and reference data.

It should be noted that since the range of high-precision reference data is about 120 km × 30 km, and the size of the GF-6 image data reaches 90 km × 90 km, the GCPs extracted based on high-precision data on each scene image could only cover part of the image. Figures 11 and 12 show typical distributions of GCPs extracted from two reference data, respectively.

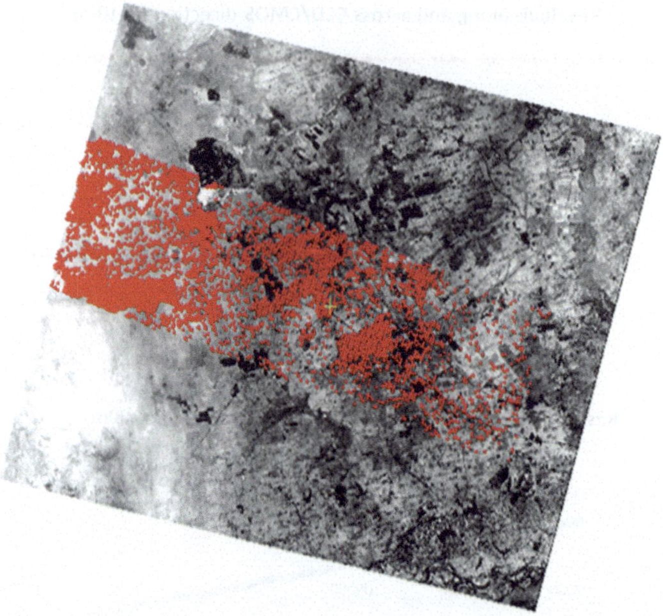

Fig. 11 Distribution diagram of GCPs extracted from the high-precision data (GF-6_20181212)

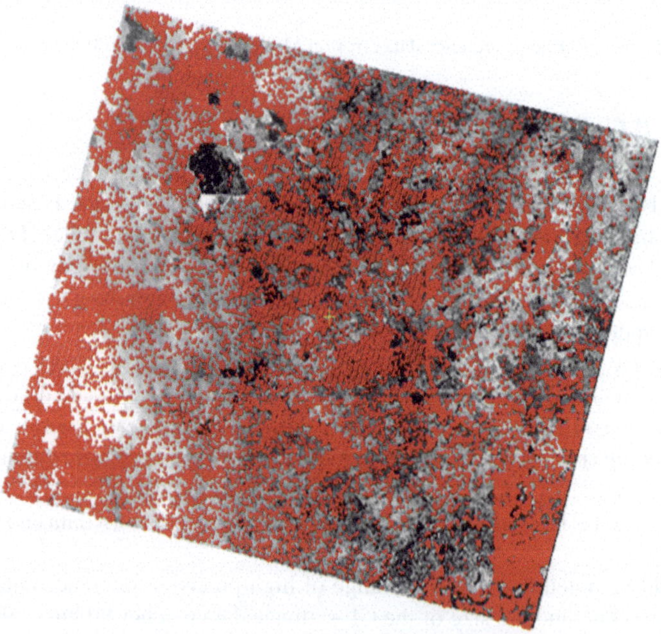

Fig. 12 Distribution diagram of GCPs extracted from the low-precision data (GF-6_20181212)

3.3 Direct Georeference

Rational function model is often used as an imaging geometry model to establish the mapping relationship between image and ground [4, 5]. The direct georeferencing accuracy is an important criterion of satellite geometry performance as it will determine the corrective model used to process imagery. The ground coordinates of the GCPs extracted are mapped to image coordinates using the RPC provided by the supplier, and the direct georeferencing accuracy could be obtained in the image space (see Table 2). Figure 13 shows the accuracy of the GF-6_20181212. The abscissa is the sequence number of the CCD, and the ordinate is the accuracy. As could be seen from the figure, there is mainly an offset, and the drift error is slight. Analysis of the data in Table 2 shows that the absolute geolocation accuracy of GF-6 is high, and the adjustment method with multi-temporal data could effectively improve the direct georeferencing accuracy.

Table 2 Direct georeferencing accuracy of GF-6 PMS (unit: pixel)

No.	Date	Count	mSamp	mLine
1.	20181208	7282	−4.849	3.593
2.	20181212	5918	−7.321	3.498
3.	20190122	4705	−2.319	−3.062
4.	20190228	6552	−4.081	−0.825
5.	20190304	5357	−3.250	−0.577
6.	20190414	6735	3.833	−0.847
Mean		6092	−2.998	0.297

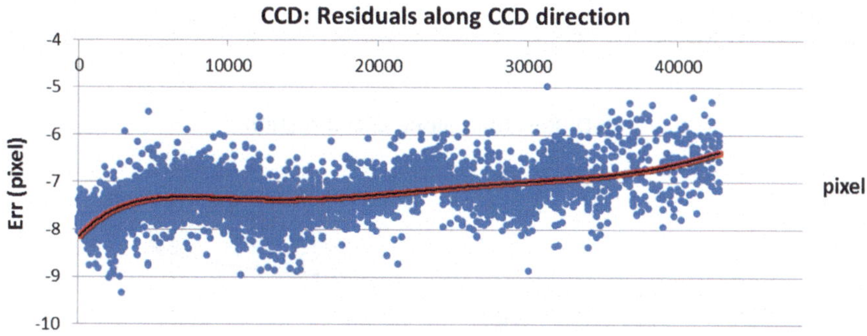

Fig. 13 Direct georeferencing accuracy of CCD direction (GF-6_20181212)

3.4 Adjustment

The bias, affine and second-order polynomial correction models were used to refine the adjustment accuracy, respectively. For typical data such as GF-6_20181208 and GF-6_20181212, Table 3 summarizes the adjustment accuracy using the above models. The accuracy of the different adjustment models of the data GF-6_20181212 is shown in Figs. 14, 15 and 16. As could be seen from Fig. 14, after the data has been corrected by the bias model, there is still a little drift error, which could be corrected using the affine model (Fig. 15). Comparing Fig. 15 with Fig. 16, it is found that the second-order polynomial adjustment model has no significant improvement in accuracy compared with the affine model correction model, indicating that the affine adjustment model could achieve good accuracy in general.

Table 4 gives an overview of the PAN data adjustment accuracy of the six scenes, and the average accuracy of PAN data in the direction of the sample and line could reach 0.380 and 0.515 pixels. Comparing and analyzing the mean error of the data in Table 4, it is found that the adjustment accuracy of the two polynomials models is basically the same, which shows:

- GF-6 PAN camera has good geometric accuracy and small distortion residual
- GF-6 satellite platform has good stability, and jitter is not obvious

Table 3 PAN accuracy with different adjustment models (unit: pixel)

Adjustment model	GF-6_20181208			GF-6_20181212		
	Count	mSamp	mLine	Count	mSamp	mLine
Direct georeference	7282	−4.849	3.593	5918	−7.321	3.498
Bias	7282	0.425	0.707	5918	0.464	0.968
Affine	7282	0.363	0.483	5918	0.392	0.620
Second-order polynomial	7282	0.360	0.459	5918	0.382	0.607

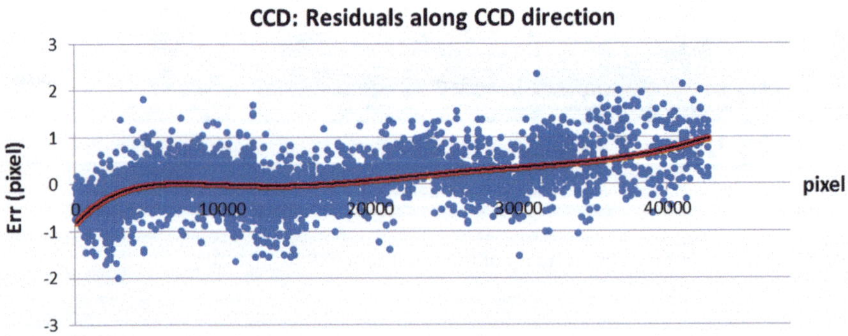

Fig. 14 Residuals of CCD direction with bias adjustment model (GF-6_20181212)

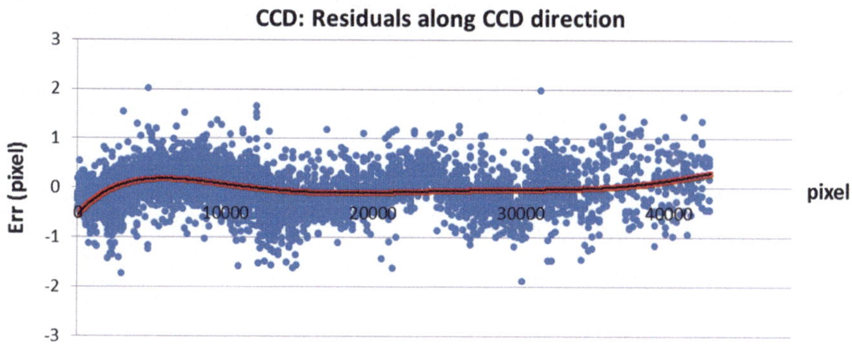

Fig. 15 Residuals of CCD direction with affine adjustment model (GF-6_20181212)

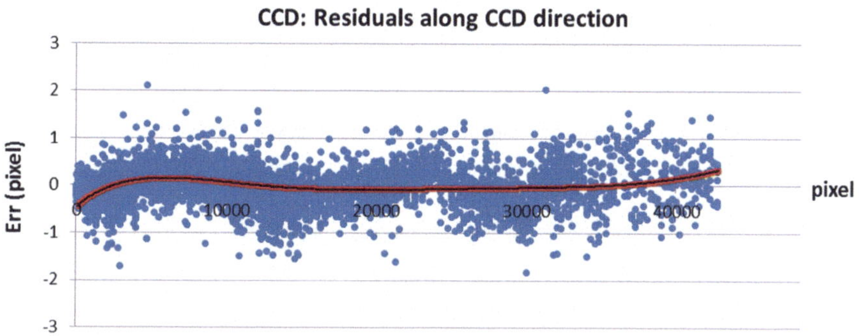

Fig. 16 Residuals of CCD direction with second-order polynomial adjustment model (GF-6_20181212)

Table 4 PAN accuracy comparison with different adjustment models (unit: pixel)

No.	Date	Affine			Second-order polynomial		
		Count	mSamp	mLine	Count	mSamp	mLine
1.	20181208	7282	0.363	0.483	7282	0.360	0.459
2.	20181212	5918	0.392	0.620	5918	0.382	0.607
3.	20190122	4705	0.402	0.563	4705	0.382	0.552
4.	20190228	6552	0.344	0.401	6552	0.336	0.385
5.	20190304	5357	0.412	0.588	5357	0.396	0.578
6.	20190414	6735	0.366	0.437	6735	0.351	0.432
Mean		6092	0.380	0.515	6092	0.368	0.502

Table 5 MS accuracy with affine adjustment models (unit: pixel)

No.	Date	Central location (degree)		RMSE			Count
		Long	Lat	mSamp	mLine	mPlane	
1.	20180701	E125.7	N45.4	0.283	0.326	0.431	41315
2.	20180701	E126.3	N46.8	0.371	0.325	0.493	24812
3.	20180925	E126.2	N45.4	0.309	0.397	0.503	23663
4.	20180925	E126.5	N46.1	0.305	0.387	0.493	55021
5.	20180925	E126.7	N46.8	0.352	0.395	0.529	15112
6.	20181204	E124.6	N46.1	0.326	0.651	0.728	35777
7.	20181204	E124.9	N46.8	0.326	0.568	0.654	14087
8.	20181208	E125.2	N46.1	0.375	0.824	0.905	39201
9.	20181208	E125.5	N46.8	0.389	0.709	0.809	14283
10.	20181212	E125.5	N45.4	0.378	0.647	0.749	11302
11.	20181212	E125.8	N46.1	0.371	0.765	0.850	21899
12.	20181216	E126.4	N46.1	0.357	0.644	0.736	18610
13.	20190114	E124.6	N46.1	0.391	0.547	0.673	12602
14.	20190122	E125.8	N46.1	0.381	0.611	0.720	13512
15.	20190228	E125.3	N46.1	0.355	0.569	0.671	35917
16.	20190304	E125.8	N46.1	0.370	0.524	0.642	33348
17.	20190414	E125.4	N46.1	0.294	0.440	0.529	44873
18.	20190418	E126.3	N46.1	0.345	0.408	0.535	38696
Mean				0.349	0.541	0.647	27446

For the GF-6 MS data, since the GSD is about 8.0 m, even the low-precision image with the GSD of 1.0 m and an accuracy around 3.0 m could be used as the reference data to ensure the reliability of the accuracy evaluation. Moreover, since a very high number (average 27,000) of GCPs are collected, the random error of the reference datasets could be eliminated nearly. For the 18 MS scenes obtained from July 01, 2018 to April 18, 2019, a summary of the accuracy assessment is provided in Table 5, and the average accuracy of MS data in the sample and line direction is 0.349 and 0.541 pixels, respectively.

3.5 *Reference Data Comparison*

In the accuracy analysis, the quality of the reference data will play an important role. In order to compare and analyze the influence of high-precision (Ref_High) and low-precision (Ref_Low) reference data on accuracy evaluation, for the six PAN scenes, the reference data of Ref_High and Ref_Low was used as control points and

Table 6 RMSE of checkpoints with low- precision reference datasets (unit: pixel)

No.	Date	GCPs High-precision			Checkpoints Low-precision		
		Count	mSamp	mLine	Count	mSamp	mLine
1.	20181208	7282	0.363	0.483	34008	0.944	2.159
2.	20181212	5918	0.392	0.620	20286	1.142	1.156
3.	20190122	4705	0.402	0.563	14366	1.002	1.065
4.	20190228	6552	0.344	0.401	34024	1.083	1.721
5.	20190304	5357	0.412	0.588	24043	1.051	1.021
6.	20190414	6735	0.366	0.437	47590	1.087	1.509
Mean		6092	0.380	0.515	29052	1.052	1.439

Table 7 RMSE of checkpoints with high-precision reference datasets (unit: pixel)

No.	Date	GCPs Low-precision			Checkpoints High-precision		
		Count	mSamp	mLine	Count	mSamp	mLine
1.	20181208	34008	0.826	1.513	7282	0.458	0.797
2.	20181212	20286	1.103	1.117	5918	0.437	0.653
3.	20190122	14366	0.988	1.050	4705	0.411	0.566
4.	20190228	34024	0.836	1.190	6552	0.538	0.746
5.	20190304	24043	1.006	0.977	5357	0.455	0.622
6.	20190414	47590	0.790	1.039	6735	0.597	0.687
Mean		29052	0.925	1.148	6092	0.483	0.679

checkpoints of the other party, and the accuracy was recalculated using the affine model (Tables 6 and 7).

Comparing the data of Tables 6 and 7, it could be found that the accuracy of the points extracted by the high-precision reference data is generally close, whether as the control point or the checkpoint, and the same is true for low-precision data, indicating:

- The accuracy evaluation result not only reflects the accuracy of the image itself but also reflects the accuracy of the reference data
- For high-precision reference data, the accuracy evaluation result could correctly represent the accuracy of image data
- For low-precision reference data, the accuracy evaluation result mainly indicates the accuracy of the reference data, not the accuracy of the image data, which is equivalent to the accuracy of the reference data detected using the image data
- Due to the use of a large number of GCPs, whether it is high-precision data or low-precision data, almost eliminate the random error of the reference data

During the evaluation, it is found that the automatic matching control points have the advantages of high efficiency and reliable accuracy, but the cost of obtaining a large range of high-precision data is relatively high. In fact, the high-precision data used in this paper does not completely cover the test data, while the low-precision data is relatively easy to obtain, and the coverage could satisfy all the test data.

3.6 Band Registration

The accuracy of registration between multispectral image bands is a major geometric performance. The GF-6 MS image contains four multispectral bands, which are red, green, blue and near-infrared and are named B1, B2, B3 and B4 in order. Since the band registration accuracy analysis does not require ground reference data, the collected data mainly considers the image quality itself.

Then, fast Fourier transform phase matching is used to extract tie points, which are evenly distributed and highly accurate, and almost all points could be extracted in four bands. The tie points distribution diagram on the image GF-6_20181212 is shown in Fig. 17.

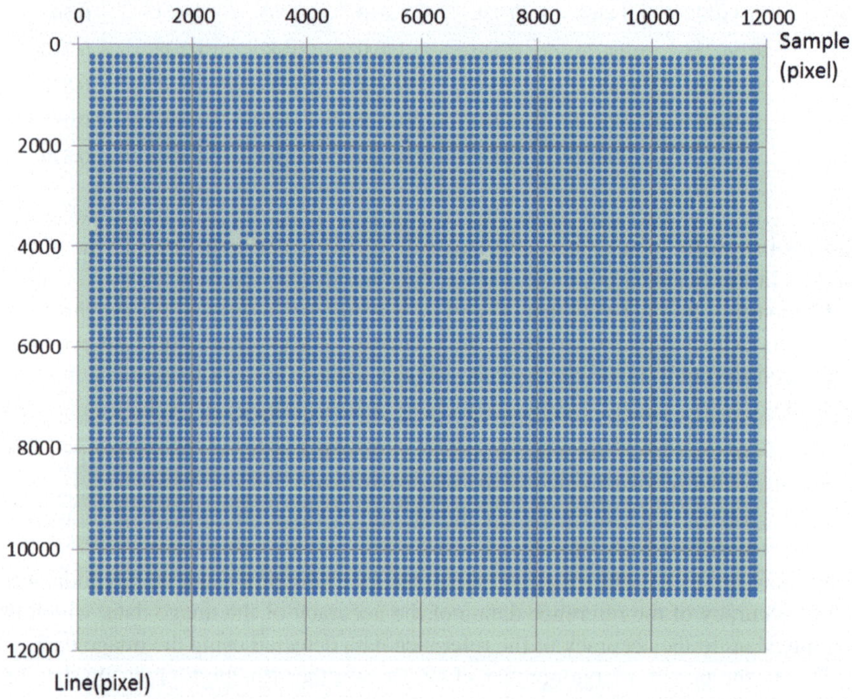

Fig. 17 Tie points distribution diagram on the image (GF-6_20181212)

Taking the data GF-6_20181212 as an example, Table 8 gives detailed accuracy data. Figures 18, 19, 20 and 21 show the distribution of residuals between the bands. Through the residual analysis, no significant effect of satellite vibration is detected. Table 9 summarizes the registration accuracy of the test MS data. In general, the average registration accuracy of GF-6 MS is better than 0.3 pixels RMSE, which meets the design specifications.

4 Conclusion

For the Chinese civil GF-6 PMS camera, this paper selects the data of multiple revisits in Northeast China and uses two sets of reference data and three adjustment models to carry out geometric performance evaluation and analysis. For all test data, the affine model in image space could achieve satisfactory accuracy. The average accuracy of PAN and MS images in the sample and line directions could reach 0.5 pixels, and the average registration accuracy between multispectral bands is better than 0.3 pixels. In general, the GF-6 satellite platform has a stable attitude, and the PMS camera has good accuracy and meets the geometric performance requirements of satellite specifications.

Table 8 Band registration accuracy of MS image (GF-6_20181212) (unit: pixel)

Band	B2			B3			B4		
	Mean	Std.dev	RMSE	Mean	Std.dev	RMSE	Mean	Std.dev	RMSE
B1	0.131	0.129	0.184	0.201	0.169	0.263	0.250	0.182	0.309
B2				0.082	0.060	0.102	0.174	0.129	0.216
B3							0.138	0.098	0.169

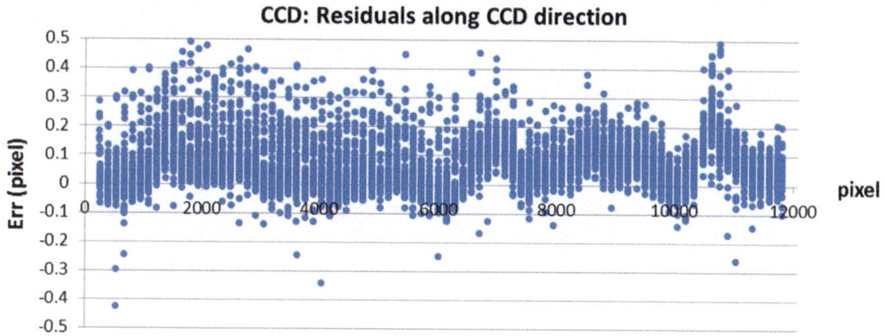

Fig. 18 Residuals distribution diagram of the band registration between B1 and B2 (GF-6_20181212)

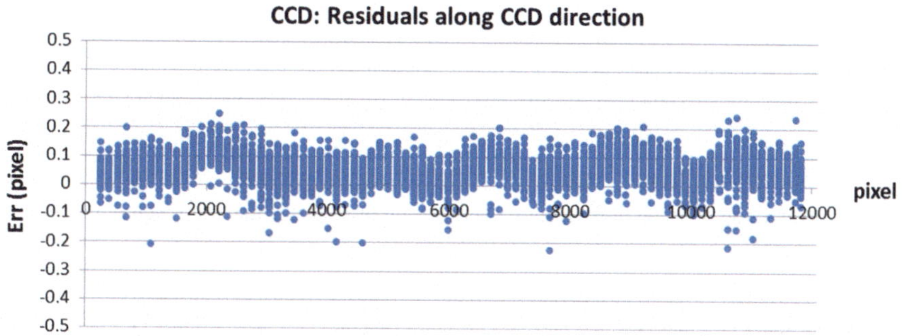

Fig. 19 Residuals distribution diagram of the band registration between B2 and B3 (GF-6_20181212)

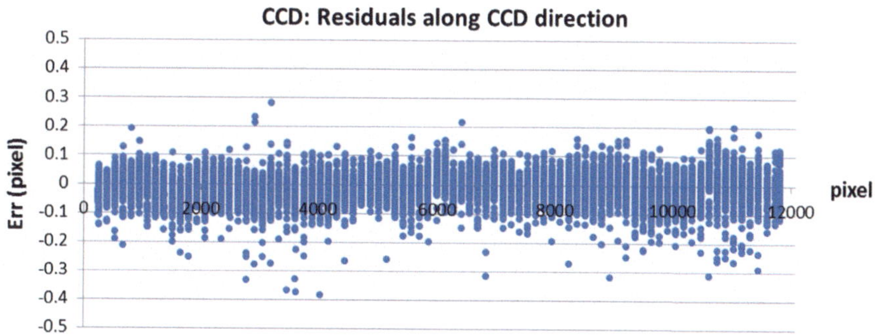

Fig. 20 Residuals distribution diagram of the band registration between B3 and B4 (GF-6_20181212)

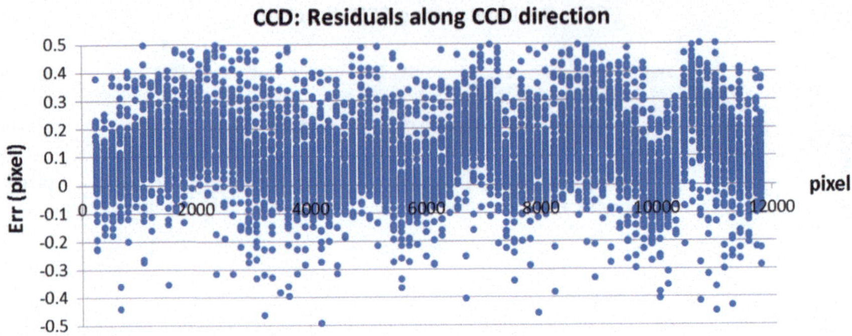

Fig. 21 Residuals distribution diagram of the band registration between B1 and B4 (GF-6_20181212)

Table 9 RMSE of band registration of MS image (unit: pixel)

No.	Data	Count	B1-B2	B1-B3	B2-B3	B1-B4	B2-B4	B3-B4
1.	20180910	3065	0.169	0.229	0.155	0.353	0.295	0.331
2.	20180925	5339	0.107	0.172	0.111	0.470	0.425	0.450
3.	20181029	6430	0.091	0.154	0.087	0.294	0.268	0.300
4.	20181127	6314	0.131	0.185	0.093	0.210	0.150	0.130
5.	20181212	6459	0.184	0.263	0.102	0.309	0.216	0.169
6.	20181213	6362	0.090	0.135	0.075	0.202	0.172	0.141
8.	20190122	6462	0.130	0.190	0.076	0.244	0.180	0.142
Mean		5776	0.129	0.190	0.100	0.297	0.244	0.238

References

1. Jacobsen K (2016) Analysis and correction of systematic height model errors. Int Arch Photogramm Remote Sens Spatial Inf Sci XLI-B1:333–339
2. Jacobsen K (2017) Problems and limitations of satellite image orientation for determination of height models. Int Arch Photogramm Remote Sens Spatial Inf Sci XLII-1/W1:257–264
3. Jacobsen K (2018) Systematic geometric image errors of very high resolution optical satellites. Int Arch Photogramm Remote Sens Spatial Inf Sci XLII-1:233–238
4. Grodecki J (2001) IKONOS stereo feature extraction—RPC approach. In: ASPRS annual conference, St. Louis
5. Grodecki J, Dial G (2003) Block adjustment of high-resolution satellite images described by rational functions. Photogramm Eng Remote Sens 69:59–70

Analysis of Geometric Performances of the Gaofen-6 WFV Camera

Liping Zhao, Xingke Fu, and Xianhui Dou

Abstract The Gaofen-6 (GF-6) is a Chinese civilian optical satellite consisting of a 2/8 m resolution panchromatic/multispectral camera with an image swath of 90 km and a 16 m resolution wide angle camera with an 800 km swath width, and it was launched in June 2018. The geometric performances are an important basis for many applications, imaging geometry simulation, image geometric accuracy, band registration accuracy and elevation extraction accuracy were analyzed in this paper. Evaluation based on digital ortho-image map and digital surface model shows that the direct georeferencing accuracy is about one ground sample distance, and the image geometric accuracy is 0.5–1.0 pixels. According to the analysis of the tie points between eight bands, it is found that the root mean square error (RMSE) of the band registration is about 0.3 pixels. Preliminary tests on the WFV stereo model show that the elevation extraction accuracy of the across-track stereo is between 4 and 10 meters RMSE. In summary, the quantitative test, evaluation and analysis of the above geometrical properties show that the GF-6 WFV camera has excellent geometric performance.

Keywords Gaofen-6 WFV · Geometric performances · Accuracy · Multi-spectral registration · Stereo image-pair

1 Introduction

As part of the Chinese civilian high-resolution remote sensing satellites, the Gaofen-6 (GF-6) mission is mainly devoted to the operational application services of agriculture and rural affairs, natural resources, emergency management, ecology and environment. The satellite has high resolution, wide coverage, high quality imaging, high performance imaging, etc. and has a design life of eight years. It was launched in June 2018 on a CZ-2D rocket from China's Jiuquan space center. GF-6 is an optical satellite consisting of a 2/8 m resolution panchromatic/multispectral (PMS) camera

L. Zhao (✉) · X. Fu · X. Dou
Land Satellite Remote Sensing Application Center, Ministry of Nature Resource, Beijing, China
e-mail: zhaolpwww@163.com

L. Wang et al. (eds.), *Proceedings of the 6th China High Resolution Earth Observation Conference (CHREOC 2019)*, Lecture Notes in Electrical Engineering 657,
https://doi.org/10.1007/978-981-15-3947-3_8

with an image swath of 90 km and a 16 m resolution wide angle camera with an 800 km swath width. The satellite is operated in a sun-synchronous and an altitude of 645 km, with a 10:30 AM equator crossing time and revisit frequency of 41 days.

The wide field of view (WFV) camera carried on the GF-6 is based on a single projection center with large field of view imaging technology, which enhances the image geometry accuracy. The camera with a free-form off-axis four-mirror anastigmat telescope system has an angle of view of more than 60° to achieve 850 km of ultra-wide-width imaging and provides eight-band multispectral imagery with 16 m resolution for efficient data acquisition. In addition to the conventional four bands, such as blue, green, red and near-infrared (NIR), the WFV sensor increases the amount of four bands such as coastal, yellow and double red edge. In particular, the two red edge bands could effectively reflect the unique spectral characteristics of the crop.

The geometric performances are an important topic for many applications [1–3]. In order to evaluate and analyze the geometric performance of the GF-6 WFV camera, three scenes were preferred in northeastern China, and quantitative evaluation and analysis were carried out from three aspects: image geometric accuracy, band registration accuracy and elevation extraction accuracy.

2 Data

Due to the large field of view of the WFV camera, the coverage width of the single scene is about 860 km, which makes it difficult to obtain the whole scene cloudless coverage under normal circumstances. During this evaluation, in order to comprehensively analyze the geometric performance of the whole scene, three scenes cloudless were preferred in northeastern China. However, in these data, there were significant seasonal differences and large changes in land cover. Figure 1 shows the distribution, and Fig. 2 shows the thumbnail image of one of the scenes. The following Tables 1 and 2 provide main characteristics of the GF-6 and WFV camera. Table 3 shows the main parameters of the test data.

For ease of use, the WFV image data is often divided into three files for storage, and the suffixes "*L*", "*M*" and "*R*" are used to distinguish. When stitching into a single image file, the suffix "$L + M + R$" is used to identify the data.

The main terrain feature of the test area is flat, hills and mountains are distributed in a small amount, and the height difference is about 1600 m. A reference dataset was available over this test area. The dataset was comprised of ortho-image with a ground sample distance (GSD) of 5 m and an accuracy around 5 m and digital surface model with 30 m spacing interval and an accuracy around 15 m. In addition, when analyzing the accuracy of across-track stereo of WFV, Ziyuan-3 (ZY-3) stereo data with 2.5 m GSD was used as reference data, and the acquisition time of ZY-3 data was as close as possible to the WFV data. In this paper, the horizontal datum and ellipsoidal height of WGS84 are used unless otherwise specified.

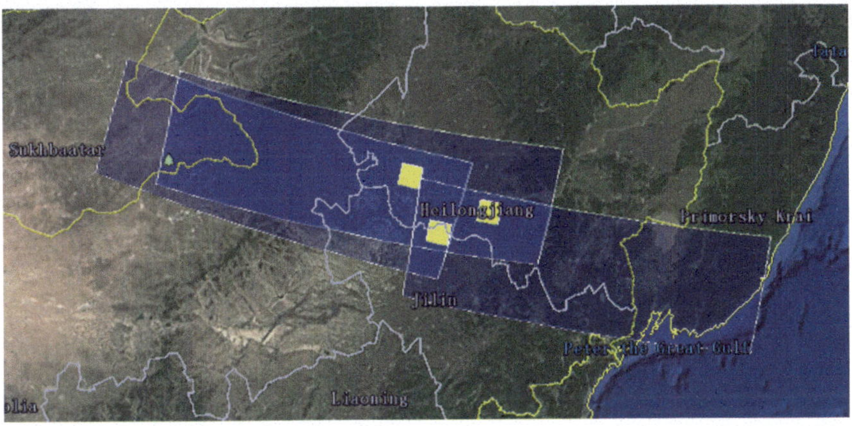

Fig. 1 Test data distribution diagram

Fig. 2 The thumbnail image of test data (GF-6_WFV_20190223)

Table 1 Main characteristics of the GF-6 and WFV camera

GF-6		WFV	
Launch periods	June 2, 2018	Field of view	>60°
Design lifetime	8 years	Swath	>860 km at nadir
Altitude	645 km	Sample	50,715 pixel
Cycle	41 days	GSD	16 m

3 Simulation

Since the GF-6 WFV camera has a large field of view of more than 60°, the complexity of the imaging geometry and the particularity of the geometric performance are caused. In order to obtain prior knowledge of GF-6 PMS satellite data processing and analysis, according to the GF-6 WFV camera parameters and imaging geometric

Table 2 Spectral bands of the WFV camera (specification)

Band	Wavelength	Name
B1	450–520	Blue
B2	520–590	Green
B3	630–690	Red
B4	770–890	NIR
B5	690–730	Red Edge 1
B6	730–770	Red Edge 2
B7	400–450	Coastal Blue
B8	590–630	Yellow

Table 3 Main characteristics of the test data

Date		20180917	20190223	20190309
GSD (m)	Sample	16–27	18–20	18–20
	Line	15	15	15
Image size (pixel)	Width	50,715	50,715	50,715
	Height	17,801	17,801	17,801
Central location (°)	Longitude	123.780	129.927	120.578
	Latitude	46.779	44.624	46.829
Yaw (°)		2.470	2.689	2.584
Pitch (°)		−0.004	0.007	0.000
Roll (°)		−9.998	−0.007	−0.005

model, the instantaneous field of view (IFOV) and ground sample distance (GSD) at different side angles were analyzed (Figs. 3, 4, 5, 6 and 7), and the influence of ground control data error on image adjustment accuracy was simulated (Figs. 8, 9 and 10).

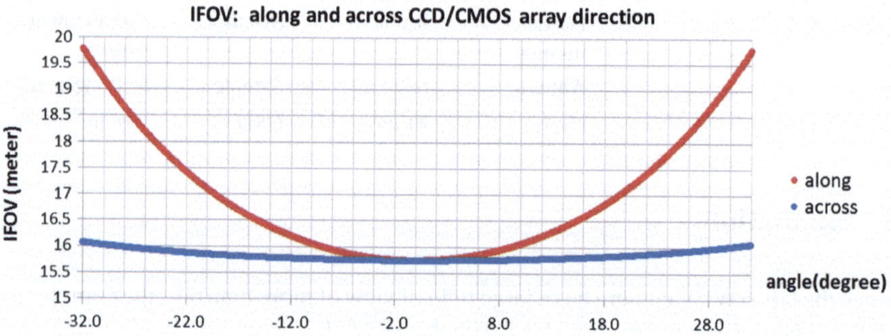

Fig. 3 IFOV of all pixels in the full field of view with 0° roll angle

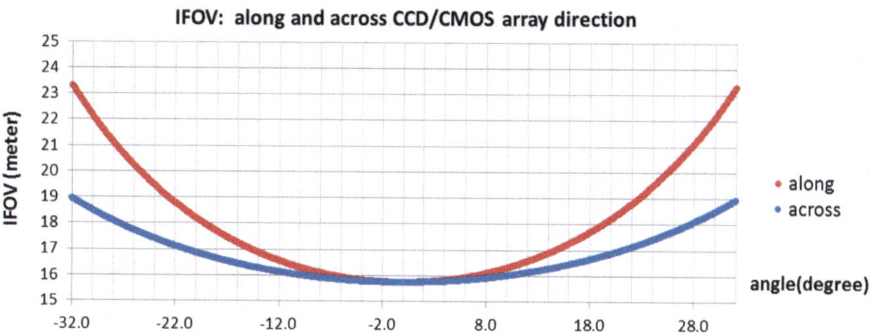

Fig. 4 IFOV of the center pixel with different side views

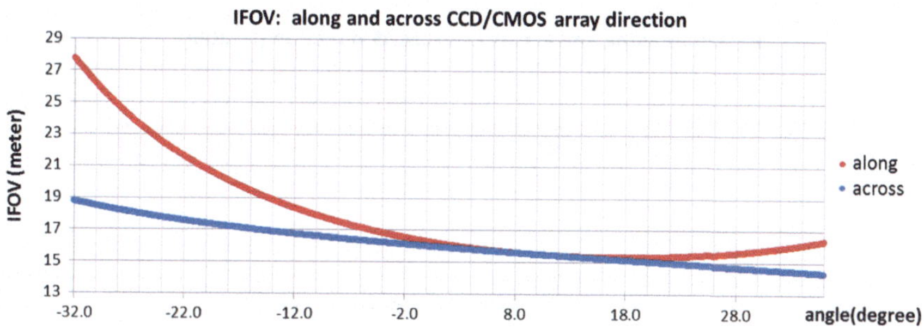

Fig. 5 IFOV of all pixels in the full field of view with $-10°$ roll angle

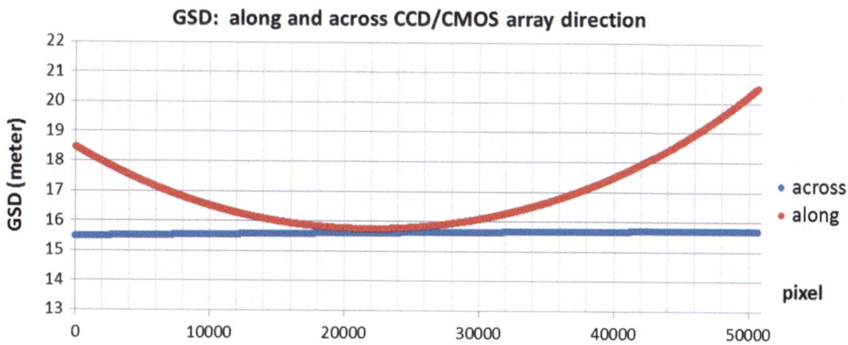

Fig. 6 GSD of all pixels in the full field of view with 0° roll angle

It could be seen from Figs. 3, 4, 5, 6 and 7 that IFOV and GSD change significantly at different fields of view, especially in the case of side view imaging (Figs. 5 and 7, with $-10°$ roll angle), the changes of IFOV and GSD in the entire field of view are particularly dramatic.

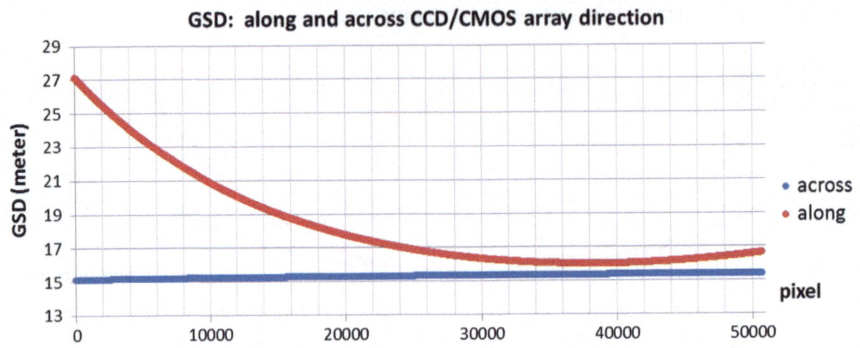

Fig. 7 GSD of all pixels in the full field of view with $-10°$ roll angle

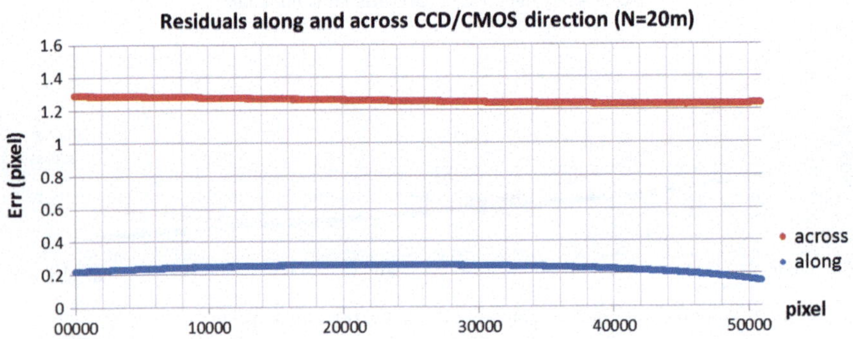

Fig. 8 Influence of ground reference data error on the image (err_N = 20 m)

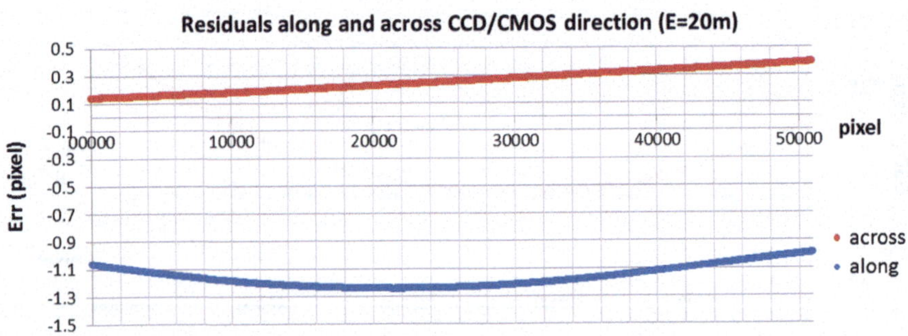

Fig. 9 Influence of ground reference data error on the image (err_E = 20 m)

Analysis of Figs. 8 and 9 shows that the ground reference data planimetric error has a significant impact on image accuracy, and the error of one GSD roughly corresponds to one pixel on the image. Therefore, if the error in image space is required to be better than 0.3 pixels, the planimetric accuracy of the ground reference data needs to

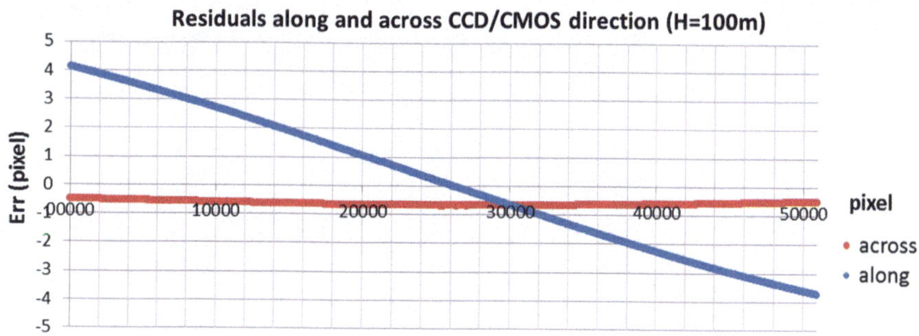

Fig. 10 Influence of ground reference data error on the image (err_H = 100 m)

be at least 4.8 m. As could be seen from Fig. 10, the influence of the elevation error of the ground data in the direction of the vertical line array is negligible. And the effect on the accuracy of the image line array direction is related to the position of the image point in the field of view, especially in the edge field of view. Therefore, for the WFV camera, if the error of the image edge field of view is required to be better than 0.3 pixels, the elevation accuracy of ground reference data needs to be at least 7 m, which is challenging for the reference data. At the same time, this also brings the additional benefits, and the same elevation accuracy could be obtained when the elevation information is extracted by the across-track WFV stereo model.

4 Image Geometry Accuracy

Rational function model is often used as an imaging geometry model to establish the mapping relationship between image and ground [4, 5]. During this evaluation, direct georeferencing accuracy was determined using the provided rational polynomial coefficients (RPC) and reference data, and then, the applicability of various polynomial adjustment models in image space was compared. For WFV multispectral data, the affine transformation method was chosen as the formal accuracy assessment model.

4.1 GCP Collection

For WFV data with a large coverage, a method of automatically acquiring a large number of ground control points (GCPs) based on data of digital ortho-image map (DOM) with 5 m GSD and digital surface model (DSM) with 30 m spacing interval is used, and thousands of GCPs are extracted on each scene image. Figure 11 shows the GCPs distribution of the data GF-6_WFV_20190309.

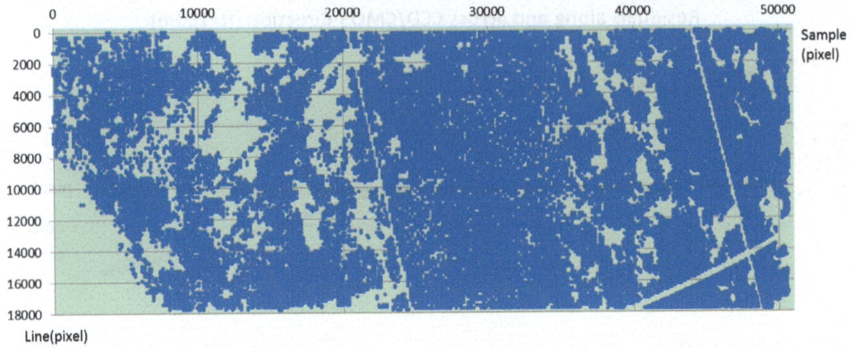

Fig. 11 The GCPs distribution diagram on the image (GF-6_WFV_20190309)

Among them, fast Fourier transform phase matching (FFTP) is used to extract tie points between the WFV image and the reference data, the FFTP works in the frequency domain, it pays more attention to the texture information of the image, and the registration of the multi-temporal image is more robust.

4.2 Direct Georeference

The direct georeferencing accuracy is an important criterion of satellite geometry performance as it will determine the corrective model used to process imagery. The ground coordinates of the GCPs are directly calculated as image coordinates using the RPC provided by the supplier, and as the checkpoints, the direct georeferencing accuracy could be obtained in the image space (see Table 4). Analysis of the data in Table 4, for the 16 m GSD WFV, shows that the absolute geolocation accuracy of GF-6 WFV is higher, better than one GSD.

4.3 Adjustment

Based on results of direct georeferencing accuracy, the affine and second-order polynomial compensation models were used for image adjustment to improve orientation accuracy (Table 5). By analyzing the data in Table 5, it was found that the accuracy of the second-order polynomial adjustment model was not significantly improved compared with the affine model correction model, and the average accuracy of the WFV data calculated by the two models in the sample and line directions was about 0.6 and 0.5 pixels, indicating that the overall WFV geometric accuracy was good.

Analyze the accuracy of each image tile and find that the accuracy of the middle tile is usually better. This could be found from the IFVO (Figs. 3, 4 and 5), GSD (Figs. 6 and 7) and ground error influence diagram (Figs. 8, 9 and 10). In addition,

Table 4 Direct georeferencing accuracy of GF-6 WFV (unit: pixel)

Date	Tile	Count	mSamp	mLine
20180917	*L*	8444	−1.040	0.264
	M	10,086	−0.844	−0.569
	R	10,040	−0.168	−0.749
	L + *M* + *R*	28,570	−0.665	−0.386
20190223	*L*	16,875	−1.881	−1.530
	M	17,597	−0.853	−1.313
	R	4367	−0.359	−0.952
	L + *M* + *R*	38,839	−1.244	−1.367
20190309	*L*	4841	−0.121	−0.596
	M	12,024	−0.385	−0.711
	R	17,287	0.190	−0.802
	L + *M* + *R*	34,152	−0.057	−0.741
Mean			**−0.619**	**−0.788**

Table 5 The WFV accuracy comparison with different adjustment models (unit: pixel)

Date	Tile	Count	Affine		Second-order polynomial		Relief index
			mSamp	mLine	mSamp	mLine	
20180917	*L*	8444	1.087	0.558	1.053	0.556	186
	M	10,086	0.354	0.339	0.347	0.327	158
	R	10,040	0.424	0.332	0.423	0.312	139
	L + *M* + *R*	28,570	0.732	0.483	0.700	0.430	342
20190223	*L*	16,875	0.397	0.383	0.369	0.359	72
	M	17,597	0.396	0.512	0.389	0.501	247
	R	4367	1.092	0.741	1.087	0.709	220
	L + *M* + *R*	38,839	0.602	0.502	0.541	0.498	222
20190309	*L*	4841	0.916	0.717	0.887	0.690	119
	M	12,024	0.565	0.478	0.541	0.450	318
	R	17,287	0.237	0.258	0.223	0.251	31
	L + *M* + *R*	34,152	0.580	0.467	0.545	0.463	342
Mean			0.615	0.481	0.592	0.462	

the image quality in the central portion is often better than the edge image, which is also advantageous for accuracy.

It is found that the accuracy of WFV_20180917_L in the sample direction is low, about twice that of the line direction. From the GSD (Fig. 7) and the ground error influence diagram (Figs. 8, 9 and 10), it could be found that the viewing angle of the data is $-10°$ in roll, causing the GSD along the CMOS array on the left side of the image to degrade significantly, about 18–27 m, and the image edges are more affected by elevation errors. In addition, the WFV_20190223_R has low precision, detecting images and reference data, and found that the terrain of this area is mainly mountainous and the forest is mainly covered, which will reduce the extraction precision of GCP and have a greater impact on the edge of the image. WFV_20190309_L is also low in accuracy and needs further analysis.

However, the accuracy of WFV_20190223_L and WFV_20190309_R at the edge of the field of view is also significantly better than the accuracy of the adjacent central image. By analyzing the topographic relief index of the reference data, it is found that the terrain of the two regions has small fluctuations, which is beneficial to obtain higher precision.

5 Band Registration

The accuracy of registration between multispectral image bands is a major geometric performance and an important indicator of application. Compared with the conventional four-spectral bands camera, the GF-6 WFV contains eight multispectral bands, namely red, green, blue, near-infrared, coastal, yellow and double red edges, which are named B1 to B8 in order. Using the three scenes collected in this article, the eight-band registration accuracy was evaluated (Tables 6, 7 and 8).

The fast Fourier transform phase matching is used to extract tie points between eight bands, which are evenly distributed and highly accurate, and almost all points could be extracted in eight bands. The tie points distribution diagram on the image GF-6_WFV_20190309 is shown in Fig. 12.

The accuracy of the tested three scenes is different. For the data GF-6_WFV_20190309, Figs. 13, 14 and 15 show the residuals distribution of the band registration between bands. Tables 6, 7 and 8, respectively, counts the mean, standard deviation and RMS errors between the eight-band registrations. The overall band registration accuracy of the WFV satisfies the specification of 0.3 pixels RMSE.

6 Elevation Extraction

Although not primarily intended for the generation of DEM, thanks to the field of view of more than 60°, the WFV swaths overlap between orbits could form a good convergence angle and would also allow for the extraction of elevation data.

Table 6 Mean of multispectral band registration (unit: pixel)

Band	B1	B2	B3	B4	B5	B6	B7	B8	Mean
B1		0.048	0.082	0.174	0.176	0.229	0.224	0.258	**0.170**
B2	0.048		0.066	0.164	0.168	0.220	0.224	0.251	**0.163**
B3	0.082	0.066		0.112	0.115	0.171	0.191	0.209	**0.135**
B4	0.174	0.164	0.112		0.083	0.108	0.177	0.173	**0.142**
B5	0.176	0.168	0.115	0.083		0.070	0.129	0.112	**0.122**
B6	0.229	0.220	0.171	0.108	0.070		0.133	0.084	**0.145**
B7	0.224	0.224	0.191	0.177	0.129	0.133		0.121	**0.171**
B8	0.258	0.251	0.209	0.173	0.112	0.084	0.121		**0.172**

Table 7 Standard deviation of multispectral band registration (unit: pixel)

Band	B1	B2	B3	B4	B5	B6	B7	B8	Mean
B1		0.030	0.057	0.124	0.081	0.091	0.123	0.063	**0.081**
B2	0.030		0.055	0.123	0.080	0.088	0.134	0.059	**0.081**
B3	0.057	0.055		0.088	0.039	0.055	0.107	0.041	**0.063**
B4	0.124	0.123	0.088		0.065	0.051	0.123	0.082	**0.094**
B5	0.081	0.080	0.039	0.065		0.037	0.099	0.041	**0.063**
B6	0.091	0.088	0.055	0.051	0.037		0.104	0.052	**0.068**
B7	0.123	0.134	0.107	0.123	0.099	0.104		0.097	**0.112**
B8	0.063	0.059	0.041	0.082	0.041	0.052	0.097		**0.062**

Table 8 RMSE of multispectral band registration (unit: pixel)

Band	B1	B2	B3	B4	B5	B6	B7	B8	Mean
B1		0.056	0.100	0.214	0.194	0.247	0.255	0.265	**0.190**
B2	0.056		0.086	0.205	0.186	0.237	0.261	0.257	**0.184**
B3	0.100	0.086		0.143	0.122	0.179	0.219	0.213	**0.152**
B4	0.214	0.205	0.143		0.106	0.119	0.215	0.191	**0.170**
B5	0.194	0.186	0.122	0.106		0.079	0.162	0.120	**0.138**
B6	0.247	0.237	0.179	0.119	0.079		0.168	0.098	**0.161**
B7	0.255	0.261	0.219	0.215	0.162	0.168		0.155	**0.205**
B8	0.265	0.257	0.213	0.191	0.120	0.098	0.155		**0.186**

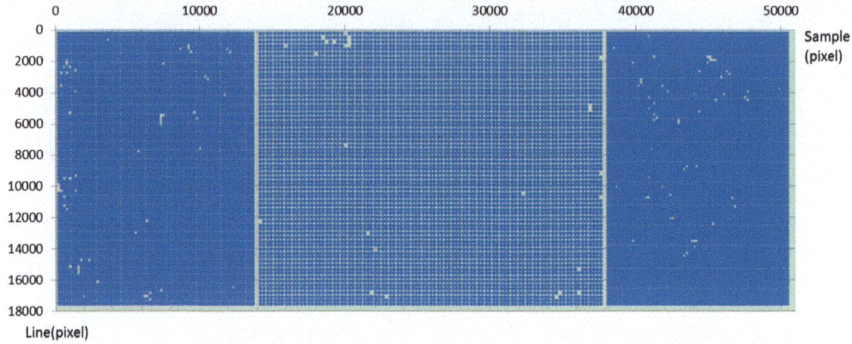

Fig. 12 Tie points distribution diagram on the image (GF-6_WFV_20190309)

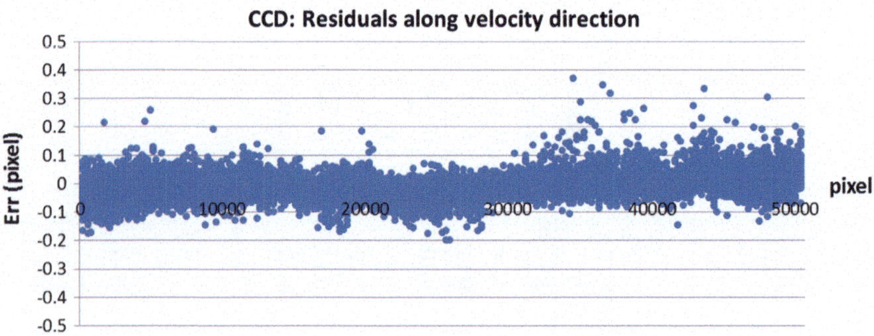

Fig. 13 Residuals distribution diagram of the band registration between B1 and B2 (GF-6_WFV_20190309)

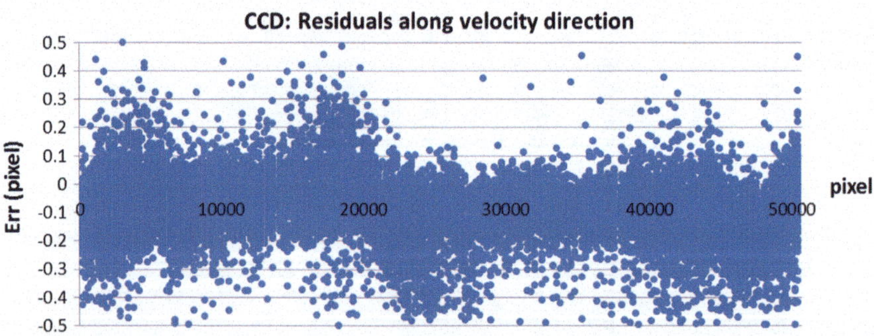

Fig. 14 Residuals distribution diagram of the band registration between B1 and B4 (GF-6_WFV_20190309)

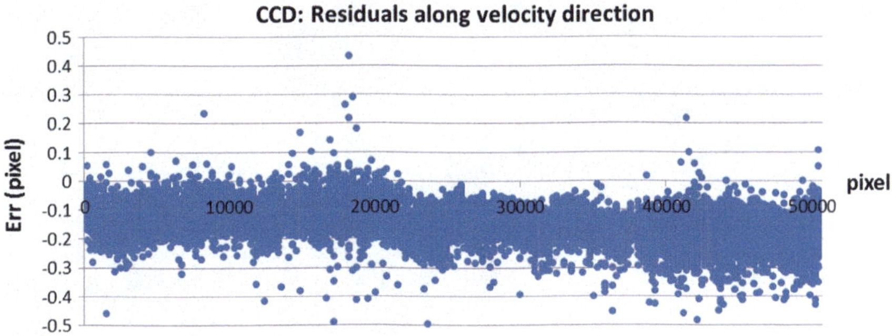

Fig. 15 Residuals distribution diagram of the band registration between B3 and B6 (GF-6_WFV_20190309)

6.1 Stereo Image Pair

Three WFV across-track stereo-pairs with good base-to-height ratio (0.4–1.2) were constructed, and the ZY-3 stereo images in the same region are used as reference data to compare and analyze the elevation extraction performance of the WFV stereo. Table 9 shows the main parameters of these stereo-pairs. For the convenience of subsequent analysis, the average elevation and terrain fluctuation index are also given, which are the data of the test small area, not the data of the whole stereo model. Figures 16, 17 and 18 show schematic diagrams of three WFV stereo-pairs and ZY-3 reference data. It could be seen that the ground coverage of ZY-3 is very small compared to WFV stereo.

Table 9 Main characteristics of the GF-6 WFV stereo

Stereo model	Stereo_A		Stereo_B		Stereo_C	
	Left	Right	Left	Right	Left	Right
Date	20190309	20190223	20180917	20190223	20190309	20180917
GSD ACT[a]	20.1	17.6	16.1	16.5	18.5	16.3
GSD ALT[a]	15.7	15.5	15.4	15.5	15.7	15.3
Base/Height	1.233		0.602		0.441	
Average elevation	164.5		184.7		151.0	
Relief index	19.0		42.8		4.7	
ZY-3 stereo	ZY-3(02)_20190502		ZY-3(02)_20190319		ZY-3(02)_20190217	

[a]GSD ALT: Mean GSD of along track direction; GSD ACT: Mean GSD of across-track direction

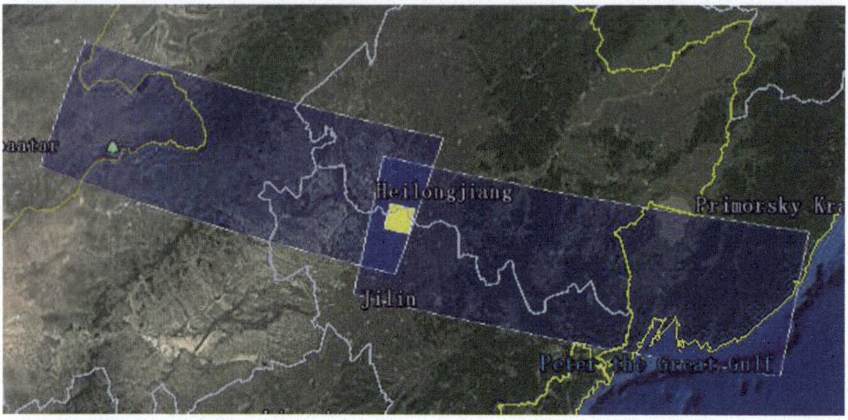

Fig. 16 Distribution diagram of Stereo_A

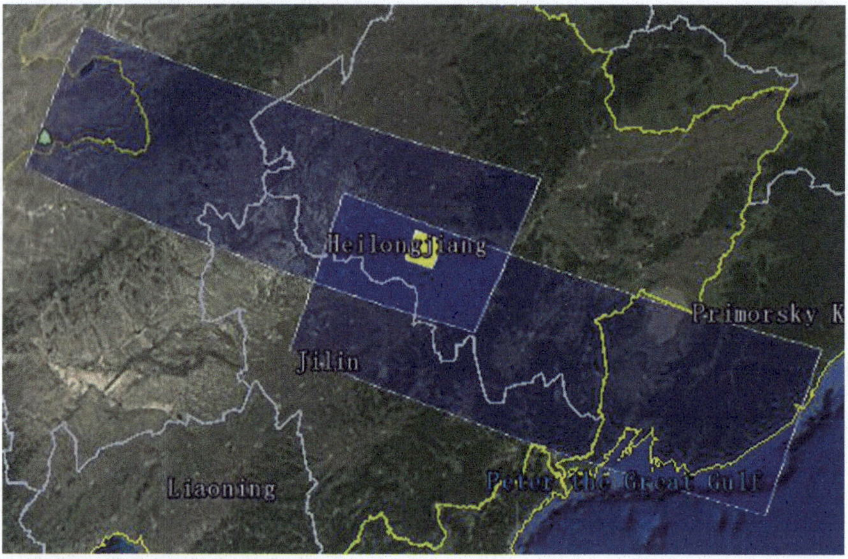

Fig. 17 Distribution diagram of Stereo_B

6.2 Accuracy Analysis

The accuracy of ZY-3 stereo is better than 3 m, which could meet the requirements for evaluating the accuracy of GF-6 WFV stereo. However, since the swath width of the WFV sensor is more than 800 km, and the width of the ZY-3 is only 50 km, only the partial WFV stereo is evaluated by ZY-3. Because of the acquisition time difference between these data, it will cause land cover changes, which makes

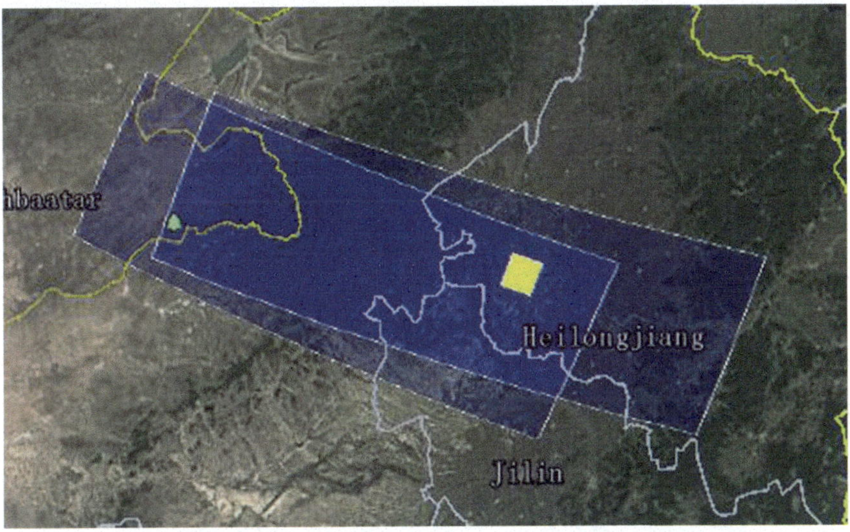

Fig. 18 Distribution diagram of Stereo_C

stereo matching difficult, affecting elevation data extraction and accuracy evaluation. Figure 19 shows the distribution of reference data extracted from the ZY-3 stereo pairs ZY-3(02)_20190319.

The ground coordinates of the reference data are solved by the ZY-3 three stereo images, and the reliability and accuracy are guaranteed, and the GCPs are used to perform the adjustment calculation on the WFV stereo model. Table 10 summarizes the image and ground adjustment accuracy of the three stereo-pairs of GF-6 WFV. It is found that the planimetric accuracy in the image and object space is excellent and amazing. In terms of elevation accuracy, the root mean square error of the three stereo-pairs is about 4–10 m, which is obviously related to the base-to-height ratio and terrain. Among the three stereo models, Stereo_A has the largest baseline, and the terrain is less undulating, and the elevation accuracy is the best. On the other hand, although the base-to-height ratio of Stereo_B is better than that of Stereo_C, the accuracy of elevation extraction is not as good as that of Stereo_C because of its topographic fluctuation much greater than that of Stereo_C.

After preliminary testing, the results show that the WFV across-track stereo based on adjacent-orbit data has a good base–height ratio and could be used to extract elevation information.

7 Conclusion

For the Chinese civil GF-6 WFV camera, the three scenes of Northeast China were optimized, and geometric performance evaluation and analysis were carried out in

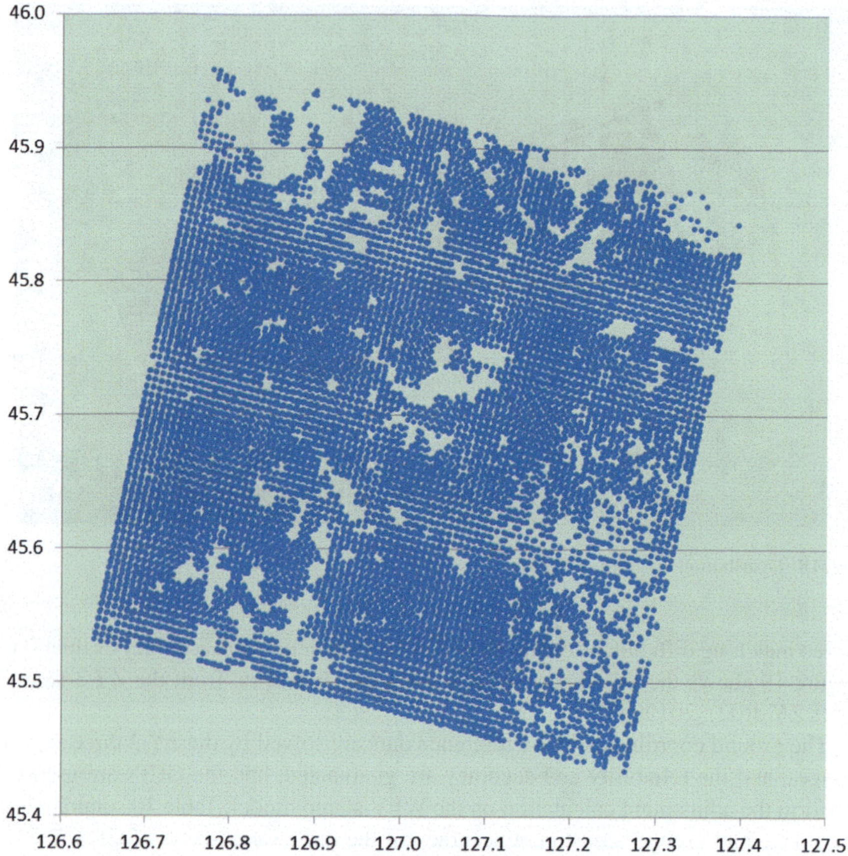

Fig. 19 GCP distribution diagram of Stereo B (ZY-3(02)_20190319)

Table 10 RMSE of the GF-6 WFV stereo

Stereo model		Stereo A	Stereo B	Stereo C
Count		21,379	14,892	29,183
Image (pixel)	mSamp_left	0.224	0.288	0.146
	mSamp_right	0.271	0.235	0.194
	mLine_left	0.183	0.228	0.112
	mLine_right	0.208	0.246	0.176
Object (m)	mN	3.349	3.528	2.582
	mE	2.864	2.323	2.265
	mH	3.449	9.158	5.157

the following four aspects: imaging geometry simulation, image geometric accuracy, band registration accuracy and elevation extraction accuracy. Among them, the direct georeferencing accuracy was better than one GSD, the image adjustment accuracy was about 0.6 and 0.5 pixels in the sample and line direction, and the multispectral registration accuracy between the eight bands could reach 0.3 pixels RMSE. In addition, the elevation accuracy of the WFV across-track stereo could be as high as 4–10 m, which was better than one GSD. In general, in the above geometric performance analysis, no obvious systematic error was found, and the geometric performance of the GF-6 WFV camera was excellent.

References

1. Jacobsen K (2016) Analysis and correction of systematic height model errors. Int Arch Photogramm Remote Sens Spatial Inf Sci XLI-B1:333–339
2. Jacobsen K (2017) Problems and limitations of satellite image orientation for determination of height models. Int Arch Photogramm Remote Sens Spatial Inf Sci XLII-1/W1:257–264
3. Jacobsen K (2018) Systematic geometric image errors of very high resolution optical satellites. Int Arch Photogramm Remote Sens Spatial Inf Sci XLII-1:233–238
4. Grodecki J (2001) IKONOS stereo feature extraction—RPC approach. In: ASPRS annual conference, St. Louis
5. Grodecki J, Dial G (2003) Block adjustment of high-resolution satellite images described by rational functions. Photogramm Eng Remote Sens 69:59–70

Evaluation and Analysis of Geometric Performances of the Gaofen-1 B/C/D Satellite

Liping Zhao and Xingke Fu

Abstract The Gaofen-1 (GF-1) B/C/D is optical satellites with two 2/8 m spatial resolution panchromatic/multispectral cameras, and they were launched in March 2018 on a CZ-4C rocket. The geometric performances were evaluated using multiple revisited data of GF-1 B/C/D in Northeast China in the paper. According to the imaging geometric model, the instantaneous field of view changes and the influence of control data error on image accuracy was simulated. Then, the direct georeferencing accuracy was calculated, and three adjustment models and four reference data were used to evaluate and analyze the geometric performance. In general, with the support of good ground control points, the root mean square error of the panchromatic image based on affine transformation in image space is about 1.0 pixel, which meets the design geometric performances.

Keywords Gaofen-1 B/C/D · Geometric performances · Accuracy

1 Introduction

The Gaofen-1 (GF-1) B/C/D is a Chinese civilian high-resolution optical satellite that uses two identical cameras to achieve a 60 km swath width. Each camera with a coaxial three-mirror anastigmat telescope system could provide panchromatic imagery of 2.0 m resolution and four-band multispectral imagery with blue, green, red and near-infrared (NIR) at 8.0 m resolution. These satellites are similar to Gaofen-1 but have the 16 m camera removed and have a smaller bus that allows three satellites to be launched at a time. They were launched in March 2018 on a CZ-4C rocket from China's Taiyuan space center. The three satellites are operated in a same sun-synchronous but are phased at 120° and an altitude of 645 km, with a 10:30 AM equator crossing time and high revisit frequency of 15 days. With a ±32° lateral sway, the constellation offers about one day revisit capability over any point of the

L. Zhao (✉) · X. Fu
Land Satellite Remote Sensing Application Center, Ministry of Nature Resource, Beijing, China
e-mail: zhaolpwww@163.com

L. Wang et al. (eds.), *Proceedings of the 6th China High Resolution Earth Observation Conference (CHREOC 2019)*, Lecture Notes in Electrical Engineering 657,
https://doi.org/10.1007/978-981-15-3947-3_9

global. The mission is mainly devoted to the operational application services of natural resources, emergency management, ecology and environment, agriculture and rural affairs.

The geometric performances are an important topic for many applications [1–3]. It was evaluated using multiple revisited data of GF-1 B/C/D in Northeast China in the paper. Using the adjustment schemes based on polynomial compensation model in image space, the accuracy assessment was performed. Then, the influence of a variety of different precision reference data on the accuracy evaluation was further analyzed.

2 Data

In order to evaluate and analyze the geometric performance of the GF-1 B/C/D satellites, Harbin, Northeast China, is selected as the experimental area. In the region, weather conditions are good, revisited data is dense, and multiple reference data is available. The main topographic features of the test area are flat, with a height difference of less than 60 m. Flat terrain facilitates the analysis of the geometric properties of the image itself, eliminating terrain effects as much as possible.

In November 2018, GF-1 B and GF-1 C received three valid data from a small angle side view, which is the key data analyzed in this paper. In addition, the other nine scenes of the test area were also acquired at the subsequent time, and they were mainly used for the accuracy comparison analysis of multi-source reference data. The time span of the image dataset is from November 01, 2018 to May 15, 2019, and Fig. 1 shows the distribution range. Table 1 gives the main characteristics of the GF-1 B/C/D satellite, and Table 2 lists the main parameters of the key test data.

Considering that the quality of the reference data will have an important impact on the geometric performance analysis, this paper uses a variety of reference data, which are manually deployed targets, obvious feature points, and two sets of digital orthoimage map (DOM) and digital surface model (DSM) data with different precision. In this paper, the horizontal datum and ellipsoidal height of WGS84 are used unless otherwise specified.

3 Evaluation

Before the evaluation, this paper used the imaging model and satellite design parameters for GF-1 B/C/D panchromatic data to simulate and analyze the influence of ground reference data error on image adjustment accuracy. Then, using ground control points (GCPs) extracted based on reference data, direct georeference accuracy and the refinement methods with polynomials compensation models in image space were analyzed. Finally, the impact of different reference data on accuracy assessment had also been explored.

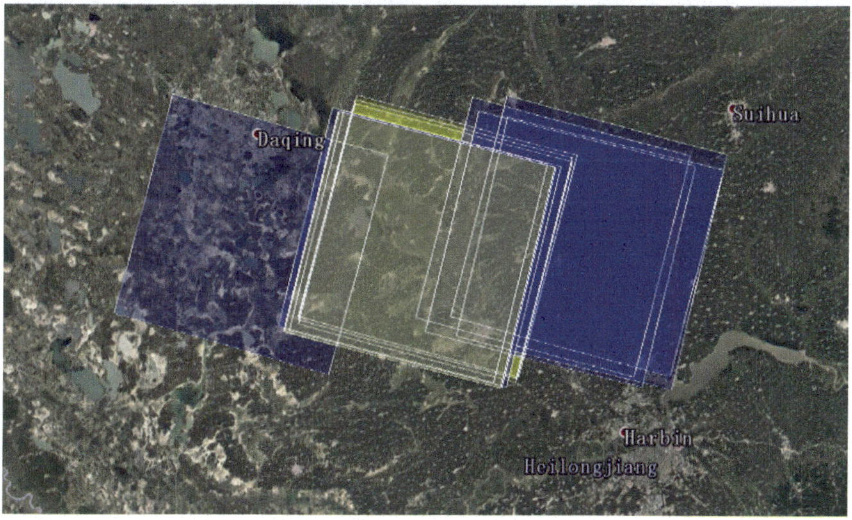

Fig. 1 Test data distribution diagram

Table 1 Main characteristics of the GF-1 B/C/D

GF-1 B/C/D		Sensor	
Launch periods	March, 2018	Swath (km)	60 at nadir
Altitude (km)	645	Sample (pixel)	PAN: 34,160 MS: 8540
Cycle (days)	15	GSD (m)	PAN: 2 MS: 8

Table 2 Main characteristics of the test data

Satellite	Date	Roll	Central location
GF-1B	20181116	3.925	E 125.8°, N 46.3°
GF-1B	20181120	0.033	E 125.7°, N 46.3°
GF-1C	20181101	3.556	E 125.8°, N 46.3°

3.1 Simulation

In order to obtain prior knowledge of the GF-1 B/C/D satellite data processing and analysis, according to the panchromatic (PAN) camera specifications and imaging geometry models, the instantaneous field of view (IFOV) under different lateral viewing angle conditions is presented (Fig. 2), and the simulation analyzes the influence of ground reference data error on the image adjustment accuracy (Figs. 3, 4 and 5).

It could be seen from Fig. 2 that the small side view angle has little effect on the IFOV of the image. The maximum side angle of the test data is 7.270°, which is

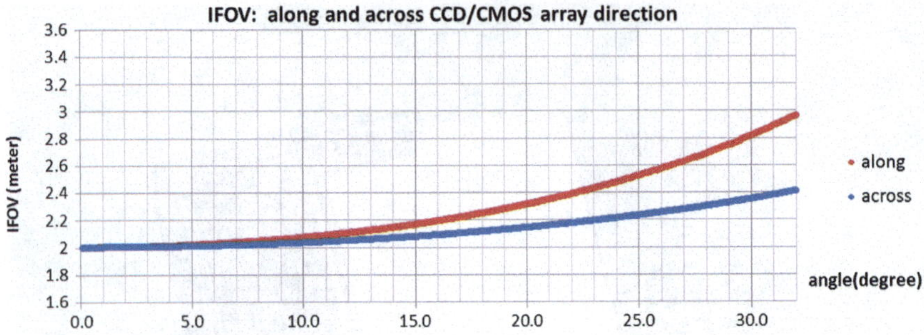

Fig. 2 IFOV with different side views (PAN)

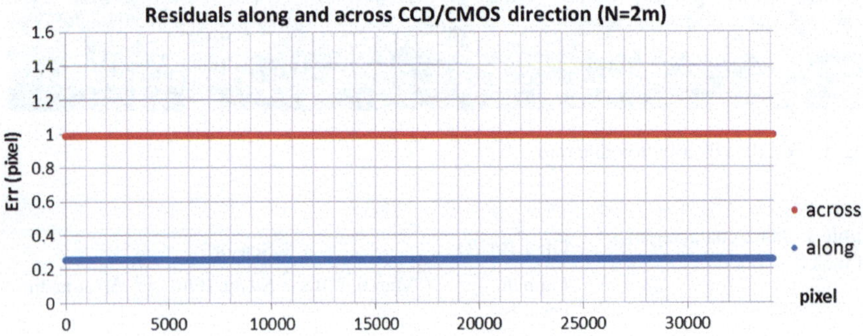

Fig. 3 Influence of ground reference data error on the PAN image (err_N = 2 m)

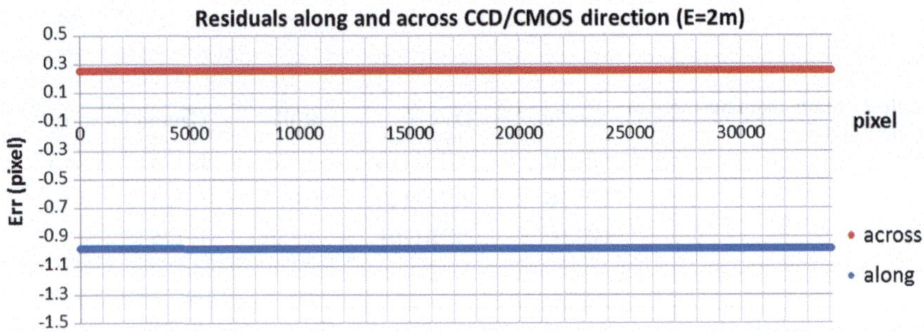

Fig. 4 Influence of ground reference data error on the PAN image (err_E = 2 m)

negligible. Analysis of Figs. 3 and 4 shows that the ground reference data planimetric error has a significant impact on image accuracy, and the error of one ground sample distance (GSD) corresponds to one pixel on the image. Therefore, if the error in image space is required to be better than 0.3 pixels, the planimetric accuracy of the ground reference data needs to be at least 0.6 m. It could be found from Fig. 5 that the influence of the ground data elevation error is related to the position of the image

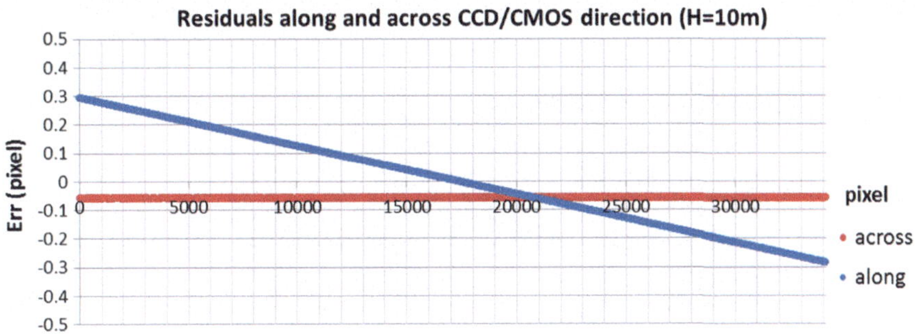

Fig. 5 Influence of ground reference data error on the PAN image (err_H = 10 m)

point in the field of view, and the impact on the adjustment accuracy is small. For the experimental data selected in this paper, the elevation error of GCPs could be ignored.

3.2 GCP Collection

Collecting GCPs manually is a time-consuming and error-prone task when studying the geometrical accuracy of large amounts of data. And based on high-precision DOM and DSM data, automatic acquisition of a large number of GCP is a recommended method. Using this automatic matching method, thousands of GCPs could be extracted on demand in each image.

A reference dataset is available over this test area. The dataset is comprised of ortho-image with a GSD of 0.5 m and an accuracy around 0.5 m and digital surface model with a grid of 2 m and an accuracy around 1.0 m.

Usually, the land cover change between the reference data and the test data changes, which is a challenge to the image matching algorithm. Since the fast Fourier transform phase matching (FFTP) algorithm works in the frequency domain, it pays more attention to the texture information of the image, the registration of the multi-temporal image is more robust, so FFTP is used to extract tie points in this study. Figures 6 and 7 show the GCP distribution of GF-1B_20181116 and GF-1C_20181101, respectively.

3.3 Direct Georeference

Rational function model is often used as an imaging geometry model to establish a mapping relationship between image and ground [4, 5]. The direct georeferencing accuracy is an important criterion of geometry performance as it will determine the corrective model used to process satellite imagery. It is mainly subjected to the

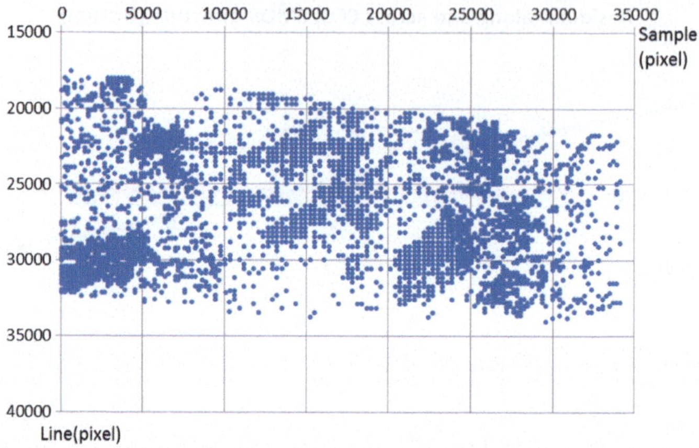

Fig. 6 GCP distribution diagram on the PAN image (GF-1B_20181116)

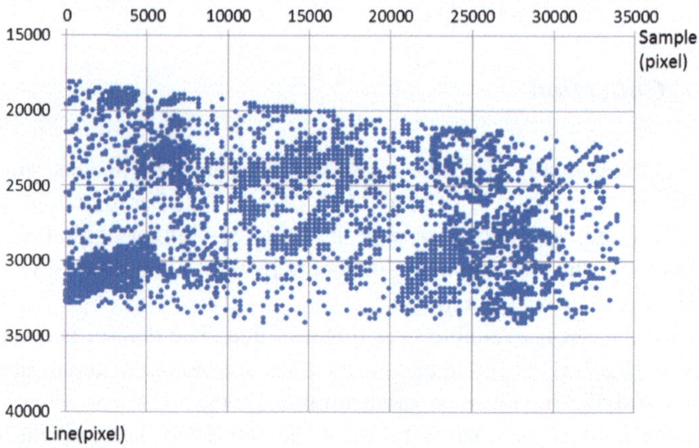

Fig. 7 GCP distribution diagram on the PAN image (GF-1C_20181101)

ephemeris and attitude accuracy of satellite. The ground coordinates of the GCPs extracted are mapped to image coordinates using the rational polynomial coefficients (RPC) provided by the supplier, and the direct georeferencing accuracy could be obtained in the image space (see Table 3).

Figure 8 shows the accuracy of the GF-1C_20181101. The abscissa is the sequence number of the CCD, and the ordinate is the accuracy. As could be seen from the figure, there is not only an offset error but also a drift error. The accuracy of GF-1B_20181116 is given in Fig. 9. The abscissa is the line number of the CCD push sweep, and the ordinate is the accuracy. The figure shows that there is a nonlinear change in camera attitude during dynamic imaging. It should be noted that there is some gross error in the automatically extracted GCP, and the above data has been initially eliminated by the distance from the point to the curve.

Table 3 RMSE with different adjustment models (unit: pixel)

Adjustment Model	GF-1B_20181116			GF-1C_20181101		
	Count	mSamp	mLine	Count	mSamp	mLine
Direct georeference	3149	8.963	28.807	3329	6.277	28.875
Bias	3149	1.934	0.857	3329	1.652	1.027
Affine	3149	0.643	0.600	3329	0.575	0.571
Second-order polynomial	3149	0.451	0.584	3329	0.485	0.567

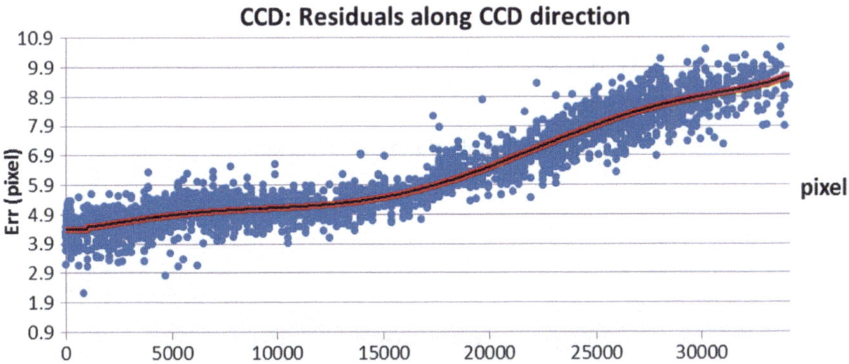

Fig. 8 Residuals of CCD direction with checkpoints (GF-1C_20181101)

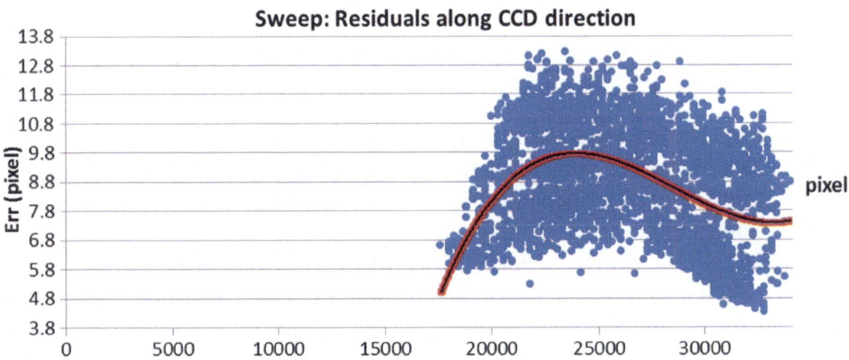

Fig. 9 Residuals of sweep direction with checkpoints (GF-1B_20181116)

3.4 *Adjustment*

The bias, affine and second-order polynomial correction models were used to refine the adjustment accuracy, respectively. For typical data such as GF-1C_20181101 and GF-1B_20181116, Table 3 summarizes the root mean square errors (RMSE) using the above models. The accuracy of the different adjustment models of the data

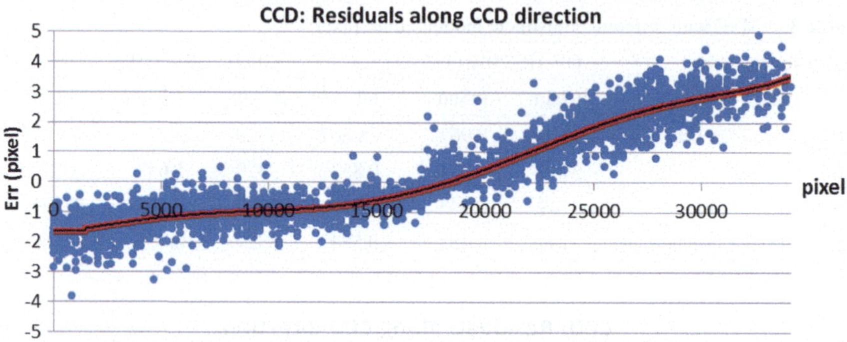

Fig. 10 Residuals of CCD direction with bias adjustment model (GF-1C_20181101)

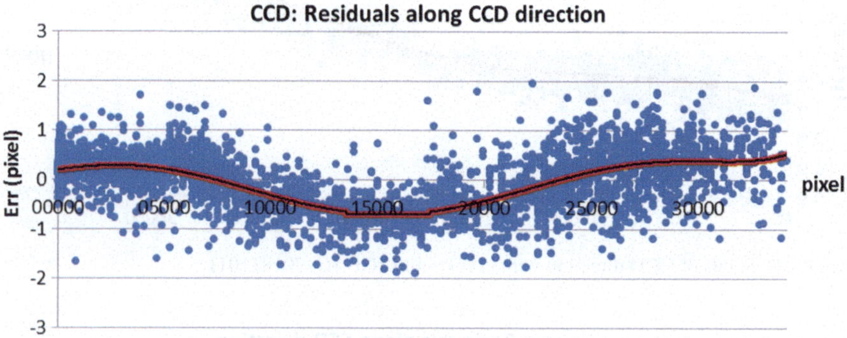

Fig. 11 Residuals of CCD direction with affine adjustment model (GF-1C_20181101)

GF-1C_20181101 is shown in Figs. 10, 11 and 12. As could be seen from Fig. 10, after the data has been corrected by the bias model, there is still drift error, which could be further corrected using the affine model (Fig. 11).

The accuracy of the different adjustment models of the data GF-1B_20181116 is given in Figs. 13, 14 and 15. It could be seen from Fig. 13 that the GF-1B_20181116 data has a quadratic residual in the swept direction. After the affine model correction, the residual is significantly reduced (Fig. 14), and further refined using the second-order polynomial adjustment model, the residual distribution in the sweep direction is no longer systematic (Fig. 15).

From the above analysis, it is found that for GF-1 B/C/D, when the GCP is sufficiently large and uniformly distributed, a good precision could be obtained by using a second-order polynomial in image space. However, with quadratic polynomial adjustment, neither irregular geometric distortion nor high-frequency attitude jitter could be corrected. Moreover, the attitude change of each scene data is uncertain, and the impact on accuracy is also different.

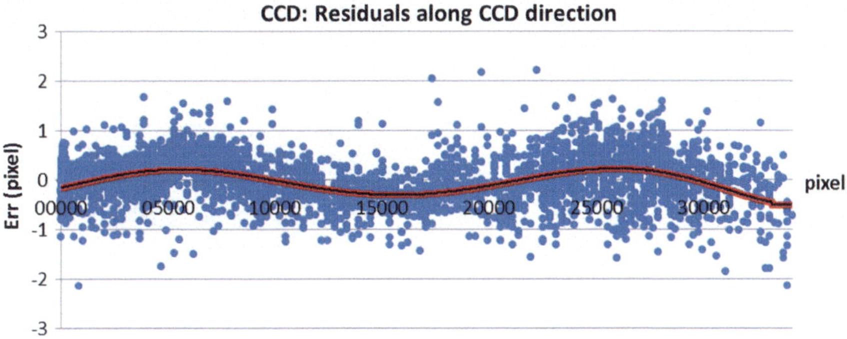

Fig. 12 Residuals of CCD direction with second-order polynomial adjustment model (GF-1C_20181101)

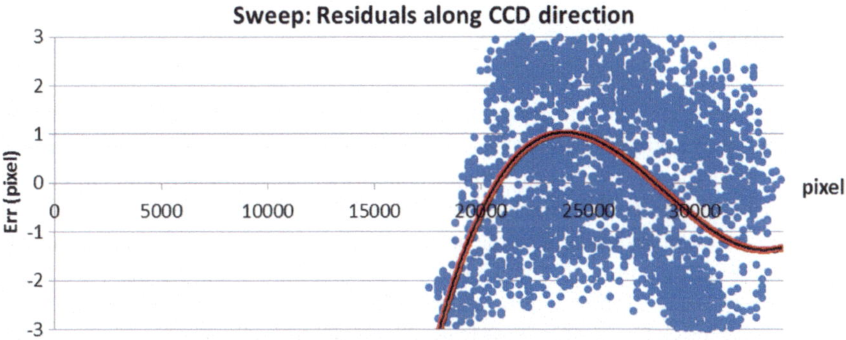

Fig. 13 Residuals of sweep direction with bias adjustment model (GF-1B_20181116)

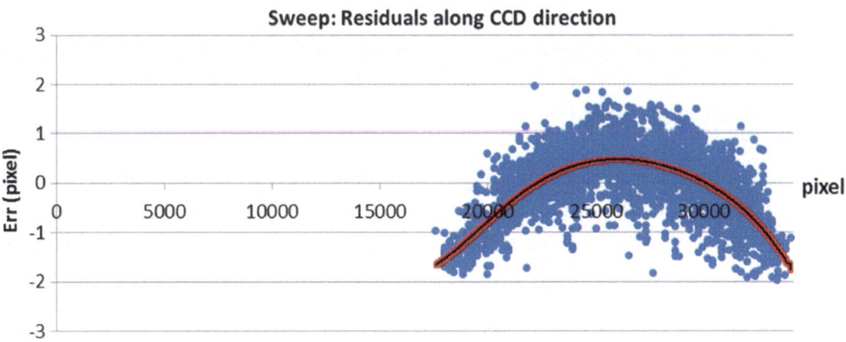

Fig. 14 Residuals of sweep direction with affine adjustment model (GF-1B_20181116)

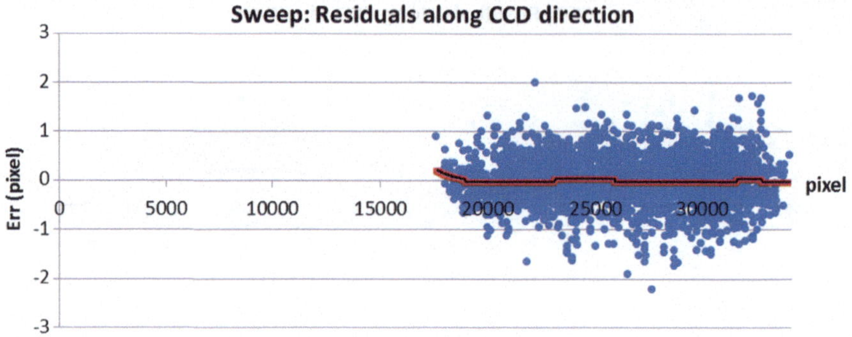

Fig. 15 Residuals of sweep direction with second-order polynomial adjustment model (GF-1B_20181116)

4 Analysis

In the geometric accuracy analysis, the quality of the reference data will play an important role. To compare and analyze the influence of the accuracy of the reference data, this paper uses a variety of reference data, which are artificially laid targets and obvious ground object point, two sets of DOM and DSM data with different precision.

Table 4 gives an assessment of the accuracy based on artificial targets and apparent object points. Figures 16 and 17 show their distribution on the image, respectively. The ground coordinate accuracy of both cases is better than 10 cm. An affine compensation model in image space is used to evaluate the accuracy.

Through experiments, it is found that the artificial target has the characteristics of good recognizability, good reliability and high positioning accuracy. Moreover, the artificial target could overcome the adverse effects of image quality degradation to a certain extent, the measurement accuracy on the image could reach 0.1 pixels, and the image coordinate error could be neglected. In addition, due to the particularity of the artificial target distribution, the accuracy of the entire scene data cannot be represented.

Obvious objects are relatively long-lasting compared with artificial targets, but they are easily affected by factors such as imaging season, image quality, land coverage and operator experience, and the positioning accuracy on images is difficult to

Table 4 Accuracy with affine model based on target and object (unit: pixel)

Scenes	Artificial target			Obvious object		
	Count	mSamp	mLine	Count	mSamp	mLine
GF-1B_20181116	36	0.494	0.228	39	0.672	1.174
GF-1B_20181120	35	0.645	0.274	37	0.872	0.804
GF-1C_20181101	32	0.384	0.406	39	0.926	0.863

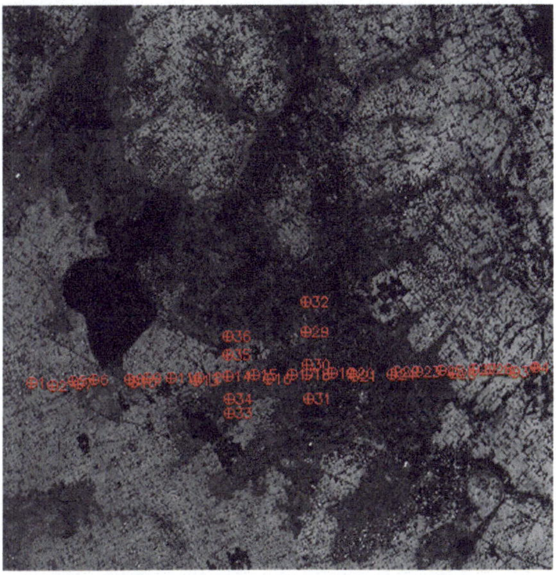

Fig. 16 Artificial target distribution diagram (GF-1C_20181101)

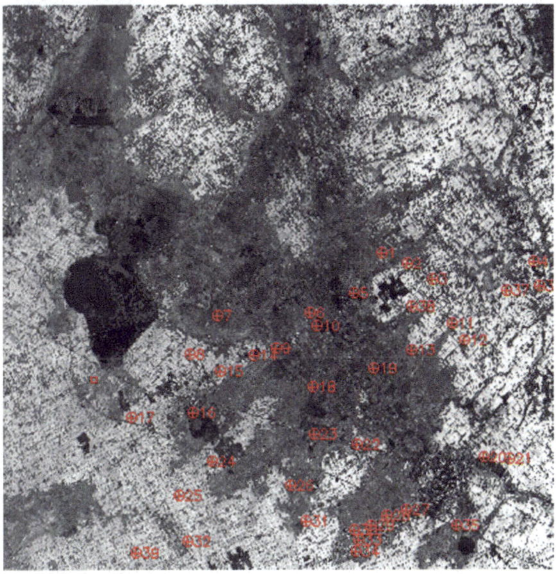

Fig. 17 Obvious object distribution diagram (GF-1C_20181101)

improve. In this experiment, less than 40 of the more than 200 points were identified on the image.

A lower precision DOM and DSM data product for comparison is also available. The accuracy of the ortho-image with the GSD of 1.0 m is approximately 2.0–5.0 m. An affine corrected model in image space is still used for calculation accuracy.

Table 5 provides a summary of the accuracy evaluation results based on high-precision and low-precision data. Figures 18 and 19 show the typical distribution of GCPs on the image based on high-precision and low-precision reference data. By analyzing the data in Table 5, it could be found that the geometric accuracy

Table 5 Accuracy with affine model based on different precision reference datasets (unit: pixel)

No.	Satellite	Date	GCP (High-precision)			GCP (Low-precision)		
			Count	mSamp	mLine	Count	mSamp	mLine
1.	GF-1 B	20181116	9042	0.716	0.677	5450	1.311	1.230
2.	GF-1 B	20181120	9999	0.470	0.703	6208	0.710	1.134
3.	GF-1 B	20190218	7315	0.496	0.676	9892	0.763	1.119
4.	GF-1 B	20190222	2209	0.381	0.503	5920	0.898	1.105
5.	GF-1 B	20190412	2115	0.554	0.625	7772	0.920	1.044
6.	GF-1 B	20190515	2465	0.671	0.766	10,557	1.093	1.039
7.	GF-1 C	20181101	9235	0.696	0.747	5443	0.932	0.964
8.	GF-1 C	20181220	6978	0.433	0.602	7579	0.716	1.038
9.	GF-1 C	20190422	5328	0.364	0.500	15,467	1.010	1.103
10.	GF-1 D	20181127	11838	0.554	0.856	8362	0.706	0.954
11.	GF-1 D	20181205	7129	0.421	0.826	8432	0.719	1.013
12.	GF-1 D	20190415	3080	0.531	0.754	8429	0.848	0.915
Mean			6394	0.524	0.686	8293	0.886	1.055

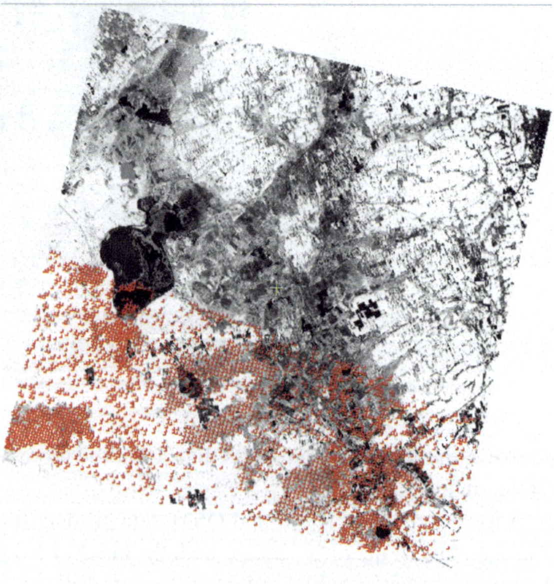

Fig. 18 Distribution diagram of automatic matching GCPs based on high-precision DOM (GF-1C_20181101)

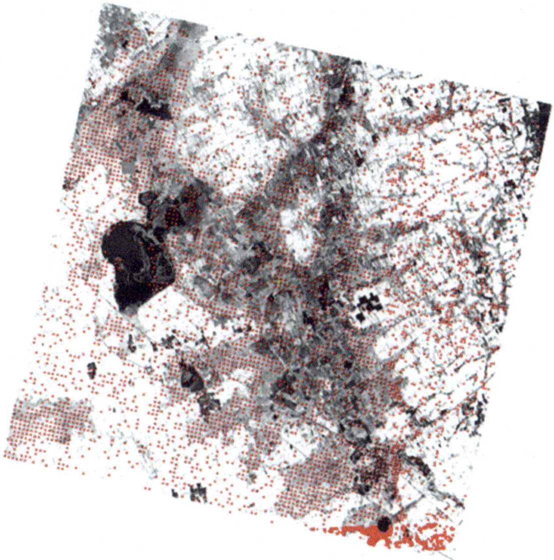

Fig. 19 Distribution diagram of automatic matching GCPs based on low-precision DOM (GF-1C_20181101)

evaluation results based on the low-precision reference data are significantly lower than the "correct" accuracy.

Through the above experiments, it is found that the automatic matching control points have the advantages of high efficiency and good reliability, but the cost of obtaining a large range of high-precision data is relatively high. In fact, the high-precision data used in this paper does not completely cover the test data, while the low-precision data is relatively easy to obtain, and the coverage could satisfy all the test data.

In order to deeply analyze the influence of high-precision (Ref_High) and low-precision (Ref_Low) reference data on accuracy evaluation, the Ref_High and Ref_Low were used as control points and checkpoints of the other party, and the adjustment accuracy was recalculated using the affine model (Tables 6 and 7).

Comparing the data in Tables 6 and 7, regardless of whether it is a control point or a checkpoint, the accuracy of the points extracted by the high-precision reference data is generally close, and the same is true for low-precision data, indicating:

Table 6 Accuracy of checkpoints with low-precision reference datasets (unit: pixel)

Satellite	Date	GCPs (High-precision)			Checkpoints (Low-precision)		
		Count	mSamp	mLine	Count	mSamp	mLine
GF-1 B	20181116	9042	0.716	0.677	5450	1.867	1.283
GF-1 B	20181120	9999	0.470	0.703	6208	0.716	1.164
GF-1 C	20181101	9235	0.696	0.747	5443	1.004	0.977

Table 7 Accuracy of checkpoints with high-precision reference datasets (unit: pixel)

Satellite	Date	GCPs (Low-precision)			Checkpoints (High-precision)		
		Count	mSamp	mLine	Count	mSamp	mLine
GF-1 B	20181116	5450	1.311	1.230	9042	0.935	0.727
GF-1 B	20181120	6208	0.710	1.134	9999	0.477	0.740
GF-1 C	20181101	5443	0.932	0.964	9235	0.749	0.759

Table 8 Summary of accuracy with different reference datasets using affine adjustment models (unit: pixel)

Reference datasets	GF-1B_20181116			GF-1C_20181101		
	Count	mSamp	mLine	Count	mSamp	mLine
Artificial target	36	0.494	0.228	32	0.384	0.406
Obvious object	39	0.672	1.174	39	0.926	0.863
High-precision (1)	3149	0.643	0.600	3329	0.575	0.571
High-precision (2)	9042	0.716	0.677	9235	0.696	0.747
Low-precision	5450	1.311	1.230	5443	0.932	0.964

- The adjustment model calculated from the reference data of two kinds of precision is stable and consistent
- The accuracy evaluation result not only reflects the accuracy of the image itself but also reflects the accuracy of the reference data
- For high-precision reference data, the accuracy evaluation result could correctly represent the accuracy of image data
- For low-precision reference data, the accuracy evaluation result mainly indicates the accuracy of the reference data, not the accuracy of the image data, which is equivalent to the accuracy of the reference data detected using the image data
- Due to the use of a large number of GCPs, whether it is high-precision data or low-precision data, almost eliminate the random error of the reference data.

Finally, Table 8 summarizes the above results for the GF-1B_20181116 and GF-1C_20181101. It could be seen that the reference data has a great influence on the test results, and the evaluation result based on the high-precision reference data is optimal.

5 Conclusion

For the Chinese civil GF-1 B/C/D satellites, using three adjustment models with four kinds of reference data, evaluation and analysis of geometric performance were carried out. Artificial targets, obvious feature points, two sets of DOM and DSM with different precisions were used as reference data to extract GCPs. The bias, affine and second-order polynomial compensation models were used to refine the adjustment accuracy in image space, respectively.

The test results of multiple revisited data using the Harbin experimental area in Northeast China showed that the planimetric accuracy of GF-1 B/C/D image data was about 1.0 pixel RMSE using the affine adjustment model in image space, which satisfied the satellite design performance requirements.

References

1. Jacobsen K (2016) Analysis and correction of systematic height model errors. Int Arch Photogramm Remote Sens Spatial Inf Sci XLI-B1:333–339
2. Jacobsen K (2017) Problems and limitations of satellite image orientation for determination of height models. Int Arch Photogramm Remote Sens Spatial Inf. Sci XLII-1/W1:257–264
3. Jacobsen K (2018) Systematic geometric image errors of very high resolution optical satellites, Int Arch Photogramm Remote Sens Spatial Inf Sci XLII-1:233–238
4. Grodecki J (2001) IKONOS stereo feature extraction—RPC approach. In: ASPRS annual conference, St. Louis
5. Grodecki J, Dial G (2003) Block adjustment of high-resolution satellite images described by rational functions. Photogramm Eng Remote Sens 69:59–70

Preliminary Analysis of Gaofen-1 B/C/D Satellite Stereo Mapping Performance

Liping Zhao and Xianhui Dou

Abstract The Gaofen-1 (GF-1) B/C/D is optical satellite with two 2/8 m spatial resolution panchromatic/multispectral cameras, and they were launched in March 2018 on a CZ-4C rocket. This paper used the GF-1 B/C/D and Ziyuan-3 (ZY-3) data that had been repeatedly visited in Northeast China to carry out mapping performance research. According to the sensor parameters and imaging geometric model, the instantaneous field of view at different side angles was analyzed, and the influence of ground control data error on image adjustment accuracy was simulated. Then, using the GF-1 B/C/D and ZY-3 data acquired simultaneously, multiple stereo models were constructed. Based on high-precision ground reference dataset, a large number of ground control points were automatically extracted for each stereo model, and then the affine model in image space and the provided rational polynomial coefficients were used for the adjustment. The results show that the root-mean-square error of the elevation of the GF-1 B/C/D stereo models is about 2–3 m. Subsequently, using these stereo models, the digital surface model (DSM) was extracted and analyzed. In general, whether it is the accuracy of stereo model or the performance of DSM extracted, it could meet the 1:50,000 scale mapping specification requirements. However, it is found that some inter-chip stitching errors between CCDs cannot be ignored in the GF-1 B/C/D image data, "elevation fracture" and abnormal elevation values are obvious in the DSM, and these defects need continuous improvement.

Keywords Gaofen-1 B/C/D · ZY-3 · Stereoscopic · Accuracy · DSM

1 Introduction

The Gaofen-1 (GF-1) B/C/D is a Chinese civilian optical satellite with two 2/8 m spatial resolution panchromatic/multispectral cameras. They were launched at a time in March 2018 on a CZ-4C carrier rocket from China's Taiyuan Satellite Launch Center. The three satellites are operated in a same sun-synchronous orbit, but are phased

L. Zhao (✉) · X. Dou
Land Satellite Remote Sensing Application Center, Ministry of Nature Resource, Beijing, China
e-mail: zhaolpwww@163.com

L. Wang et al. (eds.), *Proceedings of the 6th China High Resolution Earth Observation Conference (CHREOC 2019)*, Lecture Notes in Electrical Engineering 657,
https://doi.org/10.1007/978-981-15-3947-3_10

at 120° and an altitude of 645 km, with a 10:30 AM equator crossing time and high revisit frequency of 15 days. The satellites have a ±32° side-swing capability, which could combine multi-sensor data to construct the along- or across-track stereo image pair with different base–height ratios for digital surface model (DSM) extraction.

The Ziyuan-3 (ZY-3) satellite is mainly devoted to the operational services of 1:50,000 scale stereo mapping. The triplet stereoscopic instrument on the ZY-3 is dedicated to simultaneous stereo pair acquisition along the track. Among them, two cameras point to the forward (FWD) and backward (BWD), respectively, at a tilted angle of 22°, and the third points to the nadir (NAD). The satellite also has a lateral viewing capability around the roll axis.

The geometric performances are an important topic for many applications [1–3]. Although stereo mapping was not the design purpose of GF-1 B/C/D, satellites with side view imaging capability could be used for stereo mapping. In this study, the GF-1 B/C/D panchromatic (PAN) data was used in conjunction with ZY-3 image to construct the stereo model. The adjustment scheme based on the affine transformation in the image was used for accuracy analysis. Further, digital surface model (DSM) extraction and comparative analysis were performed using each stereo model.

2 Data

In order to carry out the GF-1 B/C/D stereo mapping performance analysis, this paper uses GF-1 B/C/D and ZY-3 stereo data to construct the multi-source stereo pairs. Table 1 gives the main characteristics of the GF-1 B/C/D and ZY-3. Considering that optical data acquisition is greatly affected by seasonal and weather factors, in order to obtain the best possible data in the same experimental area, the Harbin experimental area in Northeast China is selected for research in this paper. In the region, the main topographic features are flat, with a height difference of less than 60 m. Such flat terrain facilitates the extraction of tie points between images, which could eliminate the influence of terrain as much as possible, and is more conducive to the geometric performance analysis of the image itself.

Between November 1 and 28, 2018, GF-1 B/C/D and ZY-3 01/02 passed through the test area several times with small-angle swaying and successfully obtained data.

Table 1 Main characteristics of the GF-1 B/C/D and ZY-3 (specification)

Satellite	GF-1 B/C/D	ZY-3(01)	ZY-3(02)
Launch periods	March 2018	January 2012	May 2016
Altitude (km)	645	505	505
Cycle (days)	41	59	59
Swath (km)	60	50	50
Mean GSD (m)	PAN: 2.0 m	NAD: 2.1 BWD: 3.5 FWD: 3.5	NAD: 2.1 BWD: 2.5 FWD: 2.5

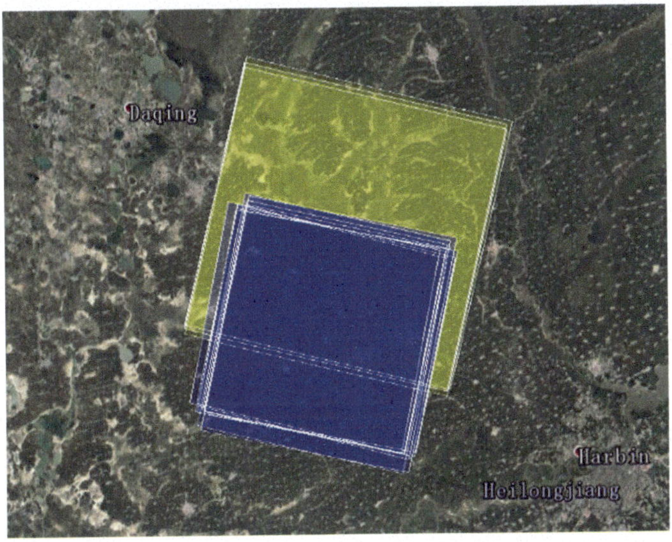

Fig. 1 Test data distribution diagram

Six times of the data were selected according to the test requirements, and the coverage was shown in Fig. 1. Among them, on November 1, 2018, GF-1C and ZY-3 (02) obtained quasi-synchronized data. On November 16, 2018, GF-1B and ZY-3 (02) also transited the test area almost simultaneously. On November 27 and 28, 2018, respectively, GF-1D and ZY-3 (01) successfully obtained data again, and the time difference was one day.

The triplet stereoscopic data of the ZY-3 has two functions. One is to construct the stereo pair with GF-1B/C/D to analyze the mapping performance, and the other is as comparative stereo model data.

As shown in Tables 2 and 3, according to the imaging date, all test data is divided into 3 groups, and each group of data constitutes 5 stereo pairs. Since the above data is obtained by the small-angle side-swing method, in addition to the ZY-3 stereo composed of BWD and FWD, the other stereo pairs constructed have a base-to-height ratio of about 0.44–0.50.

In order to analyze GF-1B/C/D stereo mapping performance, this paper uses two kinds of reference data: One is the traditional high-precision digital ortho-image map (DOM) and digital surface model (DSM) data (referred to as reference DOM and DSM), and the other reference data is the ground three-dimensional (3D) data extracted by the free network adjustment method based on ZY-3 triplet stereo data (referred to as reference ZY-3 3D). In this paper, the horizontal datum and ellipsoidal height of WGS84 are used unless otherwise specified.

Table 2 Main characteristics of the test data

Group	Satellite	Sensor	Date	Time	Roll
Group_B	GF-1B	PAN	20181116	10:50:32	3.925
	ZY-3(02)	NAD	20181116	10:47:40	−5.587
		BWD	20181116	10:48:10	−5.587
		FWD	20181116	10:47:11	−5.587
Group_C	GF-1C	PAN	20181101	10:50:07	3.556
	ZY-3(02)	NAD	20181101	10:52:44	4.165
		BWD	20181101	10:53:13	4.165
		FWD	20181101	10:52:15	4.165
Group_D	GF-1D	PAN	20181127	10:53:14	7.270
	ZY-3(01)	NAD	20181128	10:30:07	−0.007
		BWD	20181128	10:30:36	−0.007
		FWD	20181128	10:29:38	−0.007

Table 3 Stereo models based on GF-1 B/C/D and ZY-3

Group	Satellite	Left image	Right image	Abbreviation
Group_B	GF-1B ZY-3(02)	GF-1B PAN	ZY-3(02) BWD	GF-1B_PB
		GF-1B PAN	ZY-3(02) FWD	GF-1B_PF
		ZY-3(02) NAD	ZY-3(02) BWD	ZY-3(02)_NB_B
		ZY-3(02) NAD	ZY-3(02) FWD	ZY-3(02)_NF_B
		ZY-3(02) BWD	ZY-3(02) FWD	ZY-3(02)_BF_B
Group_C	GF-1C ZY-3(02)	GF-1C PAN	ZY-3(02) BWD	GF-1C_PB
		GF-1C PAN	ZY-3(02) FWD	GF-1C_PF
		ZY-3(02) NAD	ZY-3(02) BWD	ZY-3(02)_NB_C
		ZY-3(02) NAD	ZY-3(02) FWD	ZY-3(02)_NF_C
		ZY-3(02) BWD	ZY-3(02) FWD	ZY-3(02)_BF_C
Group_D	GF-1D ZY-3(01)	GF-1D PAN	ZY-3(01) BWD	GF-1D_PB
		GF-1D PAN	ZY-3(01) FWD	GF-1D_PF
		ZY-3(01) NAD	ZY-3(01) BWD	ZY-3(01)_NB_D
		ZY-3(01) NAD	ZY-3(01) FWD	ZY-3(01)_NF_D
		ZY-3(01) BWD	ZY-3(01) FWD	ZY-3(01)_BF_D

3 Experiments

Before the evaluation, the simulation analyzed the influence of the ground reference data error on the adjustment accuracy. Then, using the reference data, a large number of ground control points (GCPs) were extracted, and image accuracy and stereo model accuracy were analyzed. Finally, DSM extraction and analysis were performed.

3.1 Simulation

In order to obtain prior knowledge of the GF-1 B/C/D and ZY-3 satellite data processing and analysis, based on the specifications of satellite and imaging geometry models, the instantaneous field of view (IFOV) under different lateral viewing angle conditions is presented (Fig. 2), and the simulation analyzes the impact of ground control data error on the adjustment accuracy (Figs. 3, 4, 5, 6, 7 and 8).

It could be seen from Fig. 2 that the small-angle side view has little effect on the IFOV of satellite images such as GF-1 B/C/D. The maximum side angle of the experimental data in this paper is 7.270°, which is negligible. A similar situation is also true for ZY-3.

Analysis of Figs. 3 and 4 shows that the ground reference data planimetric error has a significant impact on the accuracy of the GF-1 B/C/D image. The error of one ground sample distance (GSD) corresponds to one pixel on the image. Therefore, if the error in image space is required to be better than 0.3 pixels, the planimetric accuracy of the ground reference data needs to be at least 0.6 m. It could be seen from Fig. 5 that the influence of the ground data elevation error is related to the position

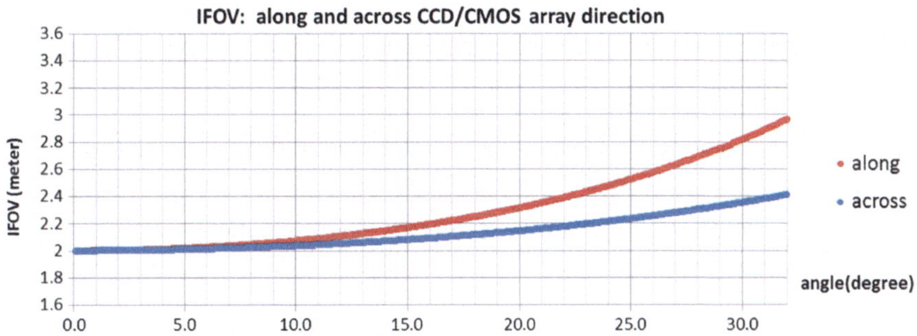

Fig. 2 IFOV with different side views (GF-1 B/C/D PAN)

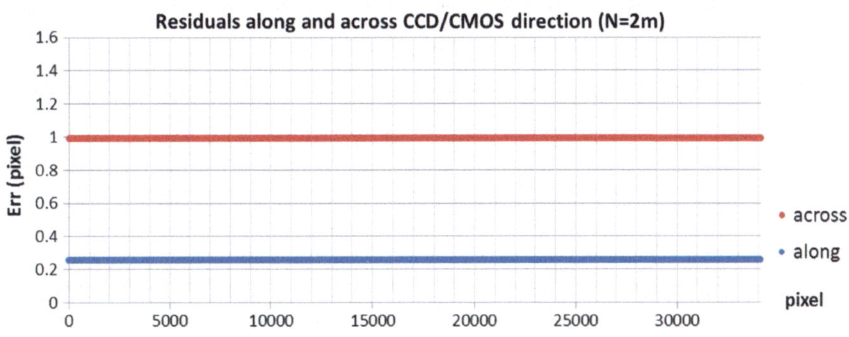

Fig. 3 Influence of ground reference data error on the GF-1 B/C/D PAN image (err_N = 2 m)

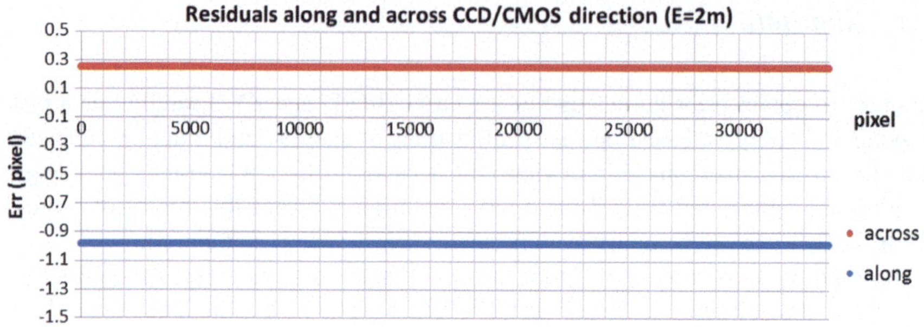

Fig. 4 Influence of ground reference data error on the GF-1 B/C/D PAN image (err_E = 2 m)

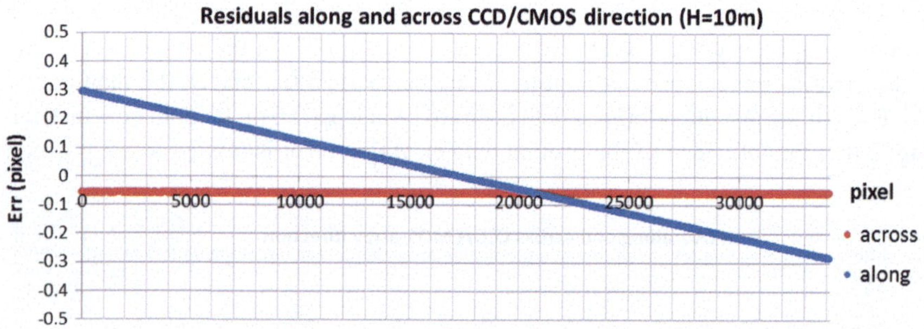

Fig. 5 Influence of ground reference data error on the GF-1 B/C/D PAN image (err_H = 10 m)

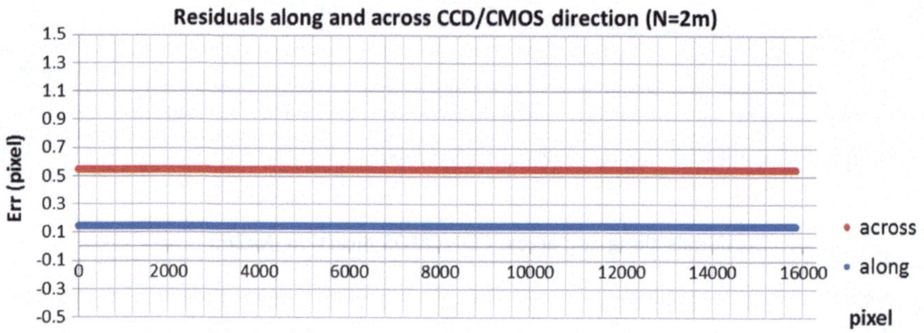

Fig. 6 Influence of ground reference data error on the ZY-3 BWD image (err_N = 2 m)

of the GF-1 B/C/D image in the field of view, and the impact on the adjustment accuracy is small. The effect on the ZY-3 NAD camera is similar.

The ZY-3 BWD and FWD cameras differ from the GF-1 B/C/D PAN cameras in that they have a large angle of inclination along the orbit direction. Taking the ZY-3(01) BWD camera as an example, Figs. 6 and 7 show the influence of the planimetric error and Fig. 8 shows the influence of the elevation error. As could be seen from Figs. 6 and 7, the planimetric error effect is similar to that of the GF-1 B/C/D PAN

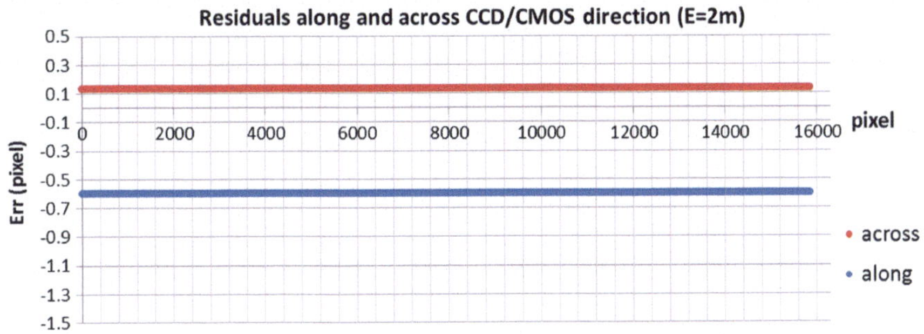

Fig. 7 Influence of ground reference data error on the ZY-3 BWD image (err_N = 2 m)

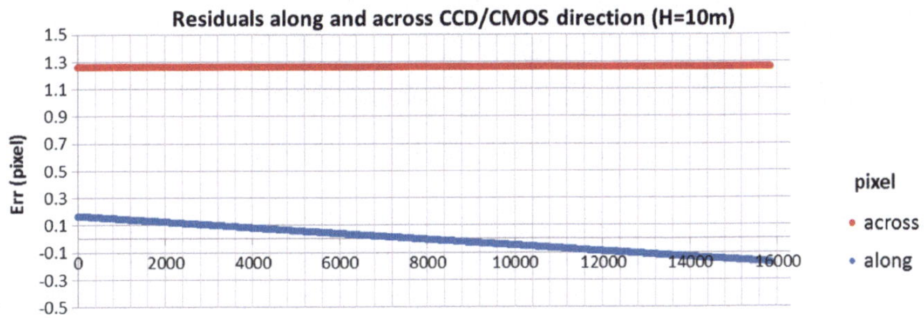

Fig. 8 Influence of ground reference data error on the ZY-3 BWD image (err_H = 10 m)

camera. It could be seen from Fig. 8 that the influence of the elevation error of the ground data in the direction of the line array is negligible, and the influence on the accuracy of the vertical line array direction is very important. Therefore, for the ZY-3 BWD and FWD camera, if the required error in image space is better than 0.3 pixels, the ground reference data planimetric accuracy needs to be better than 1.0 m, and the elevation accuracy needs to be at least 2.5 m, which requires the reference data to have good elevation accuracy.

3.2 GCP Collection

Based on high-precision DOM and DSM data, a large number of GCPs could be automatically acquired by image matching, which is a recommended method. Using this method, on each stereo model, tens of thousands of GCPs could be extracted as needed.

A reference dataset is available over this test area. The dataset is comprised of ortho-image with a GSD of 0.5 m and an accuracy around 0.5 m and digital surface model with a grid of 2 m and an accuracy around 1.0 m.

Table 4 Number of points based on two sets of reference data

Group	Count (reference DOM + DSM)	Count (reference ZY-3 3D)
Group_B	11,982	12,888
Group_C	9862	12,026
Group_D	11,558	14,313

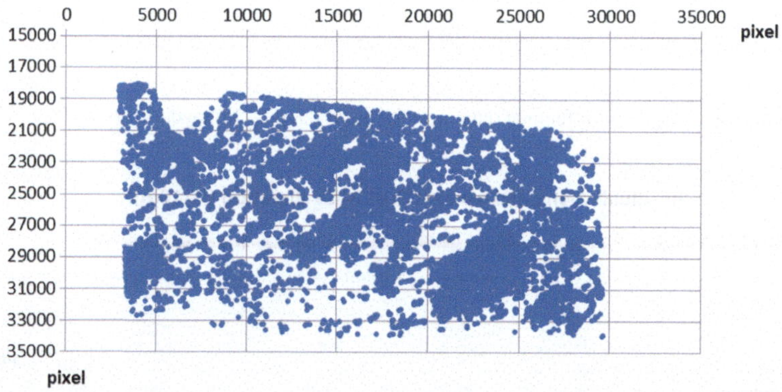

Fig. 9 GCP distribution diagram based on the reference DOM and DSM (Group_B)

Fast Fourier transform phase (FFTP) matching is used to extract tie points due to seasonal differences between the selected test data and the reference data. FFTP works in the frequency domain, it pays more attention to the texture information of the image, and the registration of the multi-temporal image is more robust.

In addition, considering that the ZY-3 stereo model has stable and good precision, the extracted ground planimetric and vertical data based on ZY-3 triplet stereo data is also used as reference data for comparison and analysis. The acquisition time of each group of image data from GF-1 B/C/D and ZY-3 01/02 in this test area is very close, and the land coverage has almost no change. It is very beneficial for extracting tie points between images and has better reliability for analyzing mapping performance.

Table 4 shows the number of GCPs obtained by the two schemes. As for the Group_B data, Figs. 9 and 10, respectively, show the distribution. It could be clearly seen that the GCP distribution of the reference ZY-3 3D is better.

3.3 Adjustment

Rational function model is often used as an imaging geometry model to establish the mapping relationship between image and ground [4, 5]. In the experiments, the vendor-supplied rational polynomial coefficients (RPCs) and affine transformation compensation models in image space are used for accuracy analysis. The ground

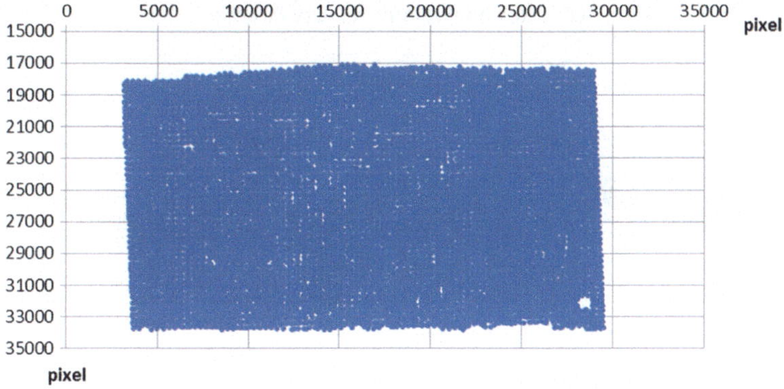

Fig. 10 GCP distribution diagram based on the reference ZY-3 3D (Group_B)

coordinates of the stereo model are calculated by the tie points between the left and right images. The accuracy of the two schemes based on the reference DOM and DSM and reference ZY-3 3D is separately analyzed.

3.3.1 Adjustment with Reference DOM and DSM

According to the conventional method, the adjustment accuracy of the GCPs extracted based on the high-precision reference DOM and DSM dataset is analyzed. Table 5 shows the image accuracy of the GF-1 B/C/D and ZY-3. Taking the data Group_B as an example, the residuals of velocity direction with affine adjustment model on the image of GF-1B and ZY-3 NAD, BWD and FWD are shown in Figs. 11,

Table 5 Image accuracy of the stereo model (unit: pixel)

Group	Sat	Sensor	mSamp	mLine	mPlane
Group_B	GF-1B	PAN	0.611	0.543	0.817
	ZY-3(02)	NAD	0.318	0.318	0.450
		BWD	0.348	0.331	0.481
		FWD	0.309	0.416	0.518
Group_C	GF-1C	PAN	0.569	0.531	0.779
	ZY-3(02)	NAD	0.402	0.499	0.641
		BWD	0.501	0.510	0.715
		FWD	0.373	0.508	0.630
Group_D	GF-1D	PAN	0.458	0.652	0.797
	ZY-3(01)	NAD	0.348	0.364	0.503
		BWD	0.342	0.294	0.451
		FWD	0.286	0.354	0.455

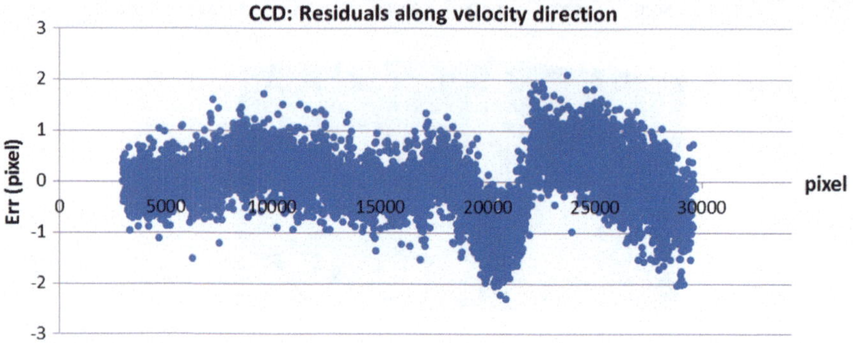

Fig. 11 Residuals of velocity direction with affine adjustment model on the image (GF-1B_20181116)

12, 13 and 14, respectively. In the figure, the abscissa is the sequence number of the pixel, and the ordinate is the residuals. As could be seen from Fig. 11, there are still some minor systematic errors in GF-1B, while the residuals of ZY-3 NAD, BWD and FWD are much smaller (Figs. 12, 13 and 14).

Comparing the data in Table 5, and analyzing Figs. 11, 12, 13 and 14, the following conclusions could be drawn:

- ZY-3 image is more accurate.
- Using ZY-3 data as a reference and comparison data, it is appropriate to perform GF-1 B/C/D accuracy analysis.

Tables 6, 7 and 8 provide a summary of the accuracy of the three sets of data such as Group_B, Group_C and Group_D. It could be seen that the accuracy of each group of data meets the requirements of the 1:50,000 scale stereo mapping specification.

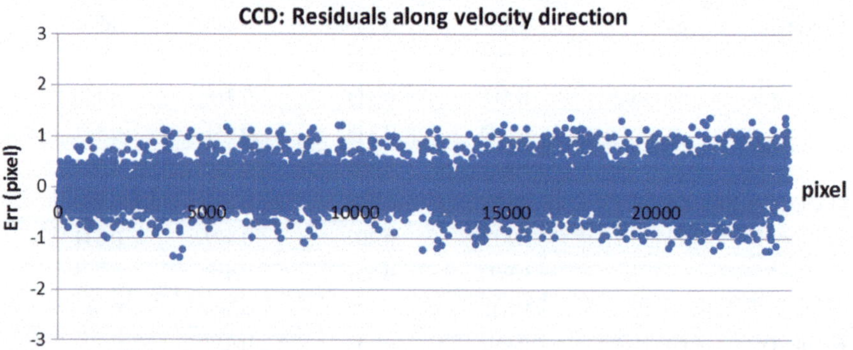

Fig. 12 Residuals of velocity direction with affine adjustment model on the image (ZY-3(02)_NAD_20181116)

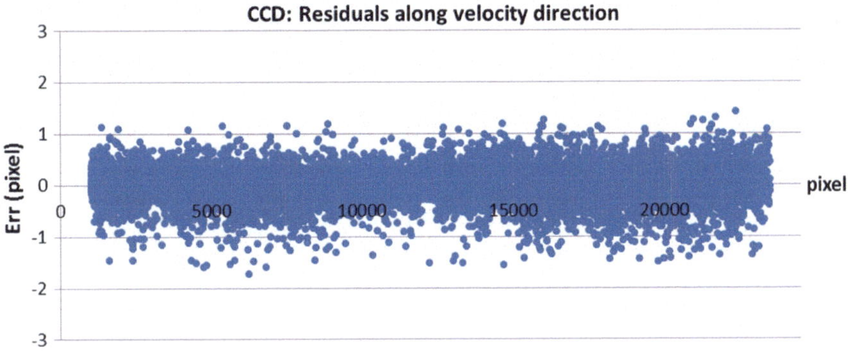

Fig. 13 Residuals of velocity direction with affine adjustment model on the image (ZY-3(02)_BWD_20181116)

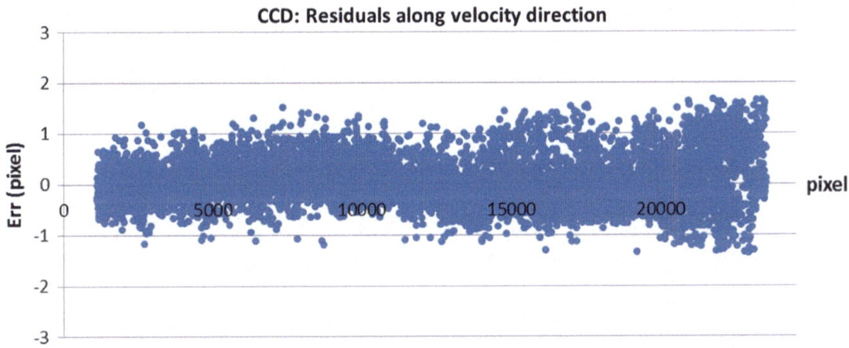

Fig. 14 Residuals of velocity direction with affine adjustment model on the image (ZY-3(02)_FWD_20181116)

Table 6 Ground accuracy of the stereo model (Group_B) (unit: meter)

Stereo model	mN	mE	mH	mH/GSD
GF-1B_PB	1.074	0.863	2.352	0.941
GF-1B_PF	1.020	0.860	2.335	0.934
ZY-3(02)_NB_B	0.691	0.649	1.409	0.564
ZY-3(02)_NF_B	0.678	0.626	1.648	0.659
ZY-3(02)_BF_B	0.808	0.663	1.174	0.470

3.3.2 Adjustment with Reference ZY-3 3D

The GF-1 B/C/D adjustment accuracy analysis based on ZY-3 stereo data is also a good method. Table 9 shows the accuracy in the image space of GF-1 B/C/D, and Table 10 summarizes the accuracy in the object space of Group_B, Group_C and

Table 7 Ground accuracy of the stereo model (Group_C) (unit: meter)

Stereo model	mN	mE	mH	mH/GSD
GF-1C_PB	1.094	0.932	2.305	0.922
GF-1C_PF	1.116	0.833	2.564	1.026
ZY-3(02)_NB_C	1.093	0.918	2.526	1.010
ZY-3(02)_NF_C	1.091	0.733	2.312	0.925
ZY-3(02)_BF_C	1.160	0.855	1.372	0.549

Table 8 Ground accuracy of the stereo model (Group_D) (unit: meter)

Stereo model	mN	mE	mH	mH/GSD
GF-1D_PB	1.236	0.874	2.877	0.822
GF-1D_PF	1.343	0.752	2.516	0.719
ZY-3(02)_NB_D	0.816	0.720	2.011	0.575
ZY-3(02)_NF_D	0.801	0.681	2.056	0.587
ZY-3(02)_BF_D	1.037	0.883	1.426	0.407

Table 9 Image accuracy of the stereo model based on reference ZY-3 3D (unit: pixel)

Group	Sat	Sensor	mSamp	mLine	mPlane
Group_B	GF-1B	PAN	0.717	0.465	0.855
Group_C	GF-1C	PAN	0.451	0.415	0.613
Group_D	GF-1D	PAN	0.354	0.546	0.650

Table 10 Ground accuracy of the stereo model based on reference ZY-3 3D (unit: meter)

Group	Stereo model	mN	mE	mH	mH/GSD
Group_B	GF-1B_PB	0.994	0.751	2.490	0.996
	GF-1B_PF	0.848	0.823	1.845	0.738
Group_C	GF-1C_PB	0.822	0.449	1.846	0.738
	GF-1C_PF	0.916	0.495	2.010	0.804
Group_D	GF-1D_PB	0.982	0.596	2.562	0.732
	GF-1D_PF	1.100	0.495	2.285	0.653

Group_D. Taking GF-1B data as an example, Fig. 15 shows the residuals of velocity direction on the image using affine adjustment model based on reference ZY-3 3D. Overall, the accuracy obtained by this method is similar to that of the method based on the high-precision reference DOM and DSM.

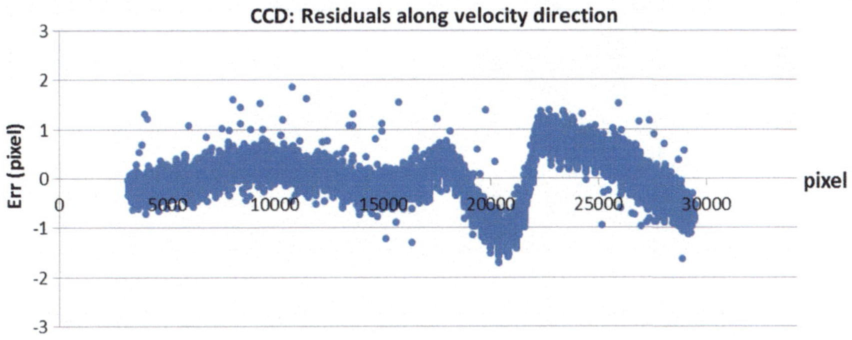

Fig. 15 Residuals of velocity direction on the image using affine adjustment model based on reference ZY-3 3D (GF-1B_20181116)

3.4 DSM Generation and Comparison

After adjustment with the GCPs, the DSM had been generated with a grid spacing of 5 m. For comparative analysis, all DSMs were extracted using the same methods and parameters, without any manual interaction editing. For the stereo image model GF-1B_PB, which is constructed by GF-1B_PAN_20181116 and ZY-3(02)_BWD_20181116 images, Fig. 16 shows the overall shading of the DSM, and more DSM details could be found in the partial shading map of sites B1, B2 and B3 (Figs. 18, 19 and 20). As a comparative stereo model ZY-3(02)_NB_B, which consists of ZY-3(02)_NAD_20181116 and ZY-3(02)_BWD_20181116 images, the corresponding shading maps are given in Figs. 17, 18, 19 and 20, respectively.

Analyzing Figs. 18, 19 and 20, compared with the DSM extracted based on the ZY-3 stereo model, it could be clearly seen that the DSM based on model GF-1B_PB has the following problems:

- Poor detail performance.
- There is an abnormal "elevation fracture" at about 1/3 of the right side, which corresponds to the image residual distribution map shown in Figs. 11 and 15.
- There are more abnormal elevation values in some residential areas and road areas.

In addition, the DSM extracted by other stereo models constructed by GF-1 B/C/D is similar to the DSM from GF-1B_PB, and details are not discussed herein. In terms of DSM extraction, although there are some problems with the automatically extracted DSM, after manual processing, in most cases, the accuracy and quality could meet the 1:50,000 scale mapping specification requirements.

Fig. 16 DSM shaded map from the stereo GF-1B_PB

4 Conclusion

For the Chinese civil GF-1 B/C/D satellites, this paper selected the data of repeated visits in Northeast China and initially carried out the evaluation and analysis of the stereo mapping performance. The ZY-3 mapping satellite data acquired simultaneously was used for comparative analysis.

The affine model in image space and the provided RPC were used for the adjustment calculation. The experimental results show that the root-mean-square error of the elevation of the GF-1 B/C/D stereo model is about 2–3 m, compared with ZY-3, and its accuracy and the performance of DSM extracted by it are slightly worse, but they all meet the 1:50,000 scale mapping specification.

Through analysis of image adjustment residuals and the DSM extracted by the stereo model, it was found that some inter-chip stitching errors between CCDs were obvious in the GF-1B/C/D image data, and it was necessary to continue the refined geometric calibration to reduce the stitching errors.

As far as DSM extraction was concerned, it was found that DSMs automatically extracted by the GF-1 B/C/D stereo model had obvious "elevation fracture" and abnormal elevation values. Especially for artificial buildings and roads, the DSM had many defects and requires manual interactive editing to improve quality.

Fig. 17 DSM shaded map from the stereo ZY-3(02)_NB_B

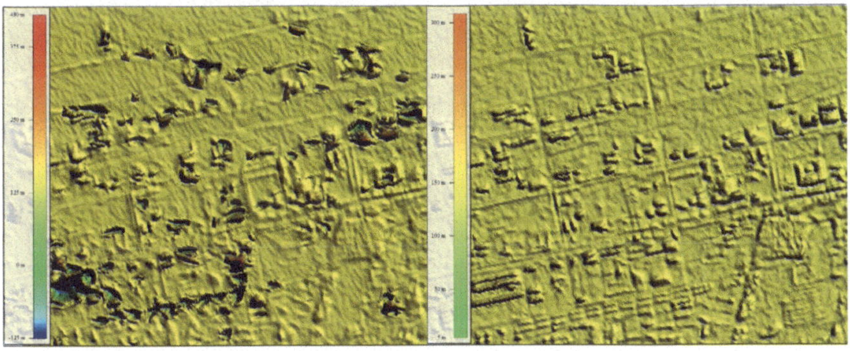

Fig. 18 DSM shaded map of site B1 from the stereo GF-1B_PB and ZY-3(02)_NB_B (left: GF-1B_PB, right: ZY-3(02)_NB_B)

In general, although the GF-1 B/C/D satellite was not designed for mapping, the data could still be used for stereo mapping. Continuous improvement was necessary for some of the shortcomings that currently exist.

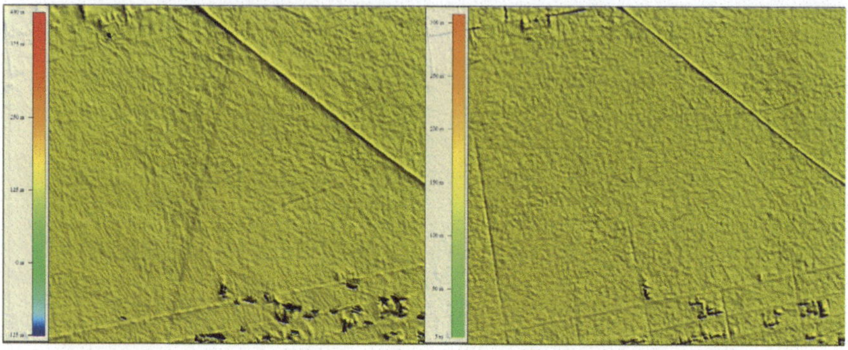

Fig. 19 DSM shaded map of site B2 from the stereo GF-1B_PB and ZY-3(02)_NB_B (left: GF-1B_PB, right: ZY-3(02)_NB_B)

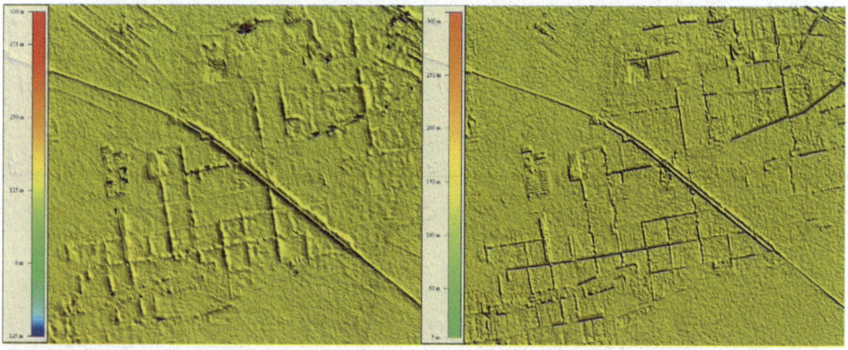

Fig. 20 DSM shaded map of site B3 from the stereo GF-1B_PB and ZY-3(02)_NB_B (left: GF-1B_PB, right: ZY-3(02)_NB_B)

References

1. Jacobsen K (2016) Analysis and correction of systematic height model errors. Int Arch Photogramm Remote Sens Spatial Inf Sci XLI-B1:333–339
2. Jacobsen K (2017) Problems and limitations of satellite image orientation for determination of height models. Int Arch Photogramm Remote Sens Spatial Inf Sci XLII-1/W1:257–264
3. Jacobsen K (2018) Systematic geometric image errors of very high resolution optical satellites. Int Arch Photogramm Remote Sens Spatial Inf Sci XLII-1:233–238
4. Grodecki, J (2001) IKONOS stereo feature extraction—RPC approach. In: ASPRS annual conference, St. Louis
5. Grodecki J, Dial G (2003) Block adjustment of high-resolution satellite images described by rational functions. Photogramm Eng Remote Sens 69:59–70

Ecological Vulnerability Assessment and Cause Analysis of the Farming–Pastoral Zone in Northern China—Taking Yulin City as an Example

Yanmei Zhong, Jianguo Cheng, Lingkui Meng, and Wen Zhang

Abstract The farming–pastoral zone in China is located in the monsoonal region. It has large fluctuations in precipitation and land use. It is one of the most obvious areas of ecological vulnerability. Taking Yulin City of Shanxi Province as an example, this paper builds an indicator system and then uses principal component analysis (PCA) and analytic hierarchy process (AHP) to filter indicators and determine weights. Finally, this paper establishes an ecological vulnerability index (EVI) model. Through the classification of EVI, the spatial distribution map of ecological vulnerability in Yulin City is obtained. The results show that the ecological vulnerability in Yulin City tends to decrease from north to south. Among counties, Fugu, Shenmu, Yuyang and Dingbian are extremely fragile, and the main causes of ecological vulnerability are economics, landscape diversity, annual precipitation and vegetation coverage.

Keywords Remote sensing · Ecological vulnerability · Principal component analysis · AHP · Cause analysis

1 Introduction

When ecosystem is intervened by outside, it will make sensitive reactions and self-recovery reactions, and this is ecological vulnerability [1]. China's farming–pastoral zone is located in the monsoon climate zone of Southeast Asia, and with large precipitation fluctuations, has characteristic of fragile ecology. Yulin City is an area where ecology is particularly fragile. The ecological environment research of Yulin and the whole farming–pastoral zone has attracted extensive attention from scholars. However, the main method for ecological assessment of Yulin is single-factor evaluation method [2–6]. And a few methods for ecological vulnerability evaluation still have

Y. Zhong · L. Meng · W. Zhang (✉)
School of Remote Sensing and Information Engineering, Wuhan University, Wuhan 430079, China
e-mail: wen_zhang@whu.edu.cn

J. Cheng
Information Center of Ministry of Water Resources, Beijing 100053, China

L. Wang et al. (eds.), *Proceedings of the 6th China High Resolution Earth Observation Conference (CHREOC 2019)*, Lecture Notes in Electrical Engineering 657,
https://doi.org/10.1007/978-981-15-3947-3_11

problems of large correlation between indicators and strong subjectivity of weight determination [7–9]. These years, remote sensing is used in ecological research, and it makes data acquisition simpler and data types more abundant [10–12]. This paper uses remote sensing, meteorological and statistical yearbook data, principal component analysis (PCA) and analytic hierarchy process (AHP) to obtain ecological vulnerability index and analyzes the overall status, spatial distribution and causes of ecological vulnerability of Yulin.

2 Study Area

Yulin is located in the northernmost part of Shanxi Province, between N 36°57′ – N 39°35′ and E 107°28′–E 111°15′. It has typical semiarid and arid climate, and the average annual precipitation is close to 400 mm. There are three landforms: windblown grass beach, loess hilly and gully, and beam-shaped low hills. The mineral resource is diverse and rich in total. It is an important energy export city in China. Special climate, topography, soil texture and high-intensity human activities make it a typical ecologically fragile area in the northern farming–pastoral zone.

3 Methods and Data

3.1 Construction of Ecological Vulnerability Indicator System

This paper constructs an evaluation indicator system from three aspects: ecological sensitivity, ecological restoration and ecological stress [13, 14]. Taking into account the availability of data, 13 indicators were chosen to constitute the indicator system. The indicator system is shown in Fig. 1.

3.2 Evaluation Indicator Selection and Weight Determination

3.2.1 Standardization of Primary Selection Indicators

We standardize the primary selection indicators to make them be unified to 0–1, so that the values are closer to 1 indicating that the ecology is more fragile [4]. Eight indicators of temperature, elevation, slope, aspect, soil erodibility factor, soil erosion intensity, GDP density and population density are positively correlated with ecological vulnerability; landscape diversity index, vegetation coverage, precipitation, relative humidity and GPP are negatively correlated with ecological vulnerability.

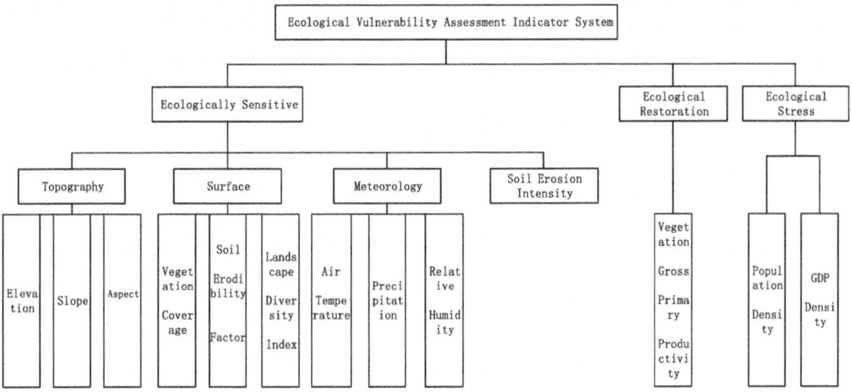

Fig. 1 Ecological vulnerability assessment indicator system

The standardized formulas are shown in Eqs. 1 and 2, respectively.

$$\text{posively correlated:} \quad P^{*i} = (P_i - P_{\min})/(P_{\max} - P_{\min}) \tag{1}$$

$$\text{negatively correlated:} \quad P^{*i} = 1 - (P_i - P_{\min})/(P_{\max} - P_{\min}) \tag{2}$$

P^{*i} represents the value after the normalization of the *i*th evaluation indicator, P^i represents the value before the *i*th evaluation indicator is normalized, $P_{\max}$ represents the maximum value of the *i*th evaluation indicator, and $P_{\min}$ represents the minimum value of the *i*th evaluation indicator [4].

3.2.2 Selection of Evaluation Indicators

We use the PCA to analyze 13 standardized primary evaluation indicators [15–19]. Table 1 shows the results of sorting the principal components according to the contribution value. The cumulative contribution rate of the first five is 86.81%, which is more than 85%, indicating that the first five principal components can cover most information of the primary selection indicators, so the first five principal components are selected as evaluation indicators. Table 2 shows eigenvectors corresponding to each principal component.

3.2.3 Determination of Indicator Weights

We use AHP to determine indicator weights. The assignment principle of the pairwise comparison matrix is shown in Table 3. The larger the value, the greater the influence of the ith element on the previous indicator layer compared to the *j*th element.

Table 1 Principal component contribution rate and cumulative contribution rate

Principal component	Eigenvalues	Contribution rate (%)	Cumulative contribution rate (%)
1	0.10317	32.7911	32.7911
2	0.08283	26.3266	59.1177
3	0.03872	12.3056	71.4233
4	0.03104	9.8670	81.2902
5	0.01737	5.5220	86.8123
6	0.01226	3.8962	90.7085
7	0.00934	2.9697	93.6782
8	0.00720	2.2894	95.9676
9	0.00390	1.2381	97.2057
10	0.00341	1.0833	98.2890
11	0.00252	0.8021	99.0910
12	0.00156	0.4959	99.5869
13	0.00130	0.4131	100.00

Table 2 Eigenvectors corresponding to each principal component

Eigenvector	Principal component				
	1	2	3	4	5
α_1	0.64236	−0.27545	0.32387	−0.03409	0.04721
α_2	−0.11891	−0.25524	−0.16227	0.06069	0.08317
α_3	−0.00889	0.01013	0.01042	−0.01733	0.09875
α_4	−0.01328	0.09345	0.16424	0.97881	0.06298
α_5	0.02509	0.27287	0.09066	−0.02346	−0.06305
α_6	0.08523	0.12261	−0.12268	−0.00493	0.16803
α_7	0.44726	0.28017	0.46675	−0.07925	−0.11404
α_8	0.16351	0.28050	−0.05602	−0.00666	0.08573
α_9	−0.32953	−0.19795	0.55427	−0.11678	0.70066
α_{10}	−0.32737	0.49174	0.36467	−0.09927	−0.25902
α_{11}	−0.07814	−0.00431	0.15554	−0.05549	−0.03140
α_{12}	−0.28965	0.15145	0.14463	−0.04015	−0.15021
α_{13}	0.18040	0.54730	−0.33484	−0.04125	0.58566

α_1–α_{13}: GDP density, population density, soil erodibility factor, aspect, DEM, GPP, landscape diversity index, relative humidity, soil erodibility factor, annual precipitation, slope, annual average temperature, vegetation coverage

Table 3 Assignment of pairwise comparison matrix

Scaling	Meaning
1	Element i and element j have the same importance to the previous-level factor
3	Element i is slightly more important than element j
5	Element i is more important than element j
7	Element i is much more important than element j
9	Element i is extremely more important than element j
2, 4, 6, 8	Indicates the intermediate value of the above adjacent judgment
Reciprocal	$A_{ji} = 1/A_{ij}$

This paper constructs a pairwise comparison matrix based on the contribution rates of the first five principal components. Equation 3 is proposed to calculate relative importance scale, A_{ij} represents the relative importance scale of elements i and j, and α_i and α_j represent the contribution rates of elements i and j. A is the pairwise comparison matrix constructed. After calculation, the weights of the evaluation indicators are: 0.5350, 0.2868, 0.0853, 0.0555, 0.0374.

$$A_{ij} = \frac{(\alpha_i - \alpha_j)}{3.4086} + 1, (i < j) \tag{3}$$

$$A = \begin{vmatrix} 1 & 3 & 7 & 8 & 9 \\ 1/3 & 1 & 5 & 6 & 7 \\ 1/7 & 1/5 & 1 & 2 & 3 \\ 1/8 & 1/6 & 1/2 & 1 & 2 \\ 1/9 & 1/7 & 1/3 & 1/2 & 1 \end{vmatrix}$$

3.3 Evaluation Model

The five evaluation indicators are graded by the natural discontinuity grading method shown in Table 4. The higher the level, the higher the ecological vulnerability.

In Eq. 4, EVI represents the ecological vulnerability index; w_i represents the weight of the ith evaluation indicator; f_i represents the level of the ith evaluation indicator [4].

$$\text{EVI} = \sum_{i=1}^{n} w_i f_i \tag{4}$$

Table 4 Indicators' grading results

Grade	Principal component				
	1	2	3	4	5
1	0–0.47	0.09–0.46	0–0.71	0–0.36	0–0.52
2	0.47–0.79	0.46–0.74	0.71–0.93	0.36–0.56	0.52–0.71
3	0.79–1.18	0.74–1.03	0.93–1.16	0.56–0.76	0.71–0.88
4	1.18–1.58	1.03–1.42	0.16–1.40	0.76–0.97	0.88–1.08
5	1.58–1.94	1.42–2.03	1.40–1.82	0.97–1.33	1.08–1.62

3.4 Data Sources and Preprocessing

The soil erosion intensity is calculated according to RUSLE model (modified general soil loss model) based on the above data [20–22]. Thirteen indicators are unified into 1 km resolution, UTM_Zone_49N (Table 5).

Table 5 Data sources

Data type	Data resource	Indicator
Soil	HWSD	Soil erodibility factor
Vegetation type	MCD12Q1	Landscape diversity index
Statistical data	Shanxi provincial bureau of statistics	Population and GDP density
Topographic	SRTM	Elevation, slope and aspect
GPP	MOD17A2	Annual average GPP
Vegetation coverage	MOD13A2	Vegetation coverage index
Meteorological	China meteorological data network	Temperature, humidity, precipitation

Table 6 Division of ecological vulnerability levels in Yulin

Vulnerability level	EVI value	Area ($\times 10^3$ km^2)	Proportion of total area (%)
Micro	<1.9	9.17	21.8
Mild	1.9–2.7	4.26	10.1
Moderate	2.7–3.2	9.59	22.8
Severe	3.2–3.6	10.83	25.7
Extreme	>3.6	8.26	19.6

4 Results and Analysis

4.1 Overall State of Ecological Vulnerability

The EVI is classified as shown in Table 6. Yulin has the largest area with severe vulnerability, accounting for 25.7% of the city's area. Moderate and above vulnerability account for nearly 70% of the city's area, and only 20% of the city's area is micro-vulnerability.

4.2 Spatial Distribution of Ecological Vulnerability

The distribution of ecological vulnerability levels is shown in Fig. 2. The proportion of vulnerability levels in various districts and counties is shown in Fig. 3. The ecological vulnerability level of Yulin is characterized by high north and low south. Fugu, Shenmu, Yuyang and Dingbian are most fragile. These four counties have 20–45% extremely vulnerable areas. Especially in Shenmu, extremely vulnerable areas account for 45% of the county's area.

4.3 Analysis of the Causes of Ecological Vulnerability

Figure 4 depicts the standardized evaluation indicators of various districts and counties in Yulin. The closer the evaluation indicator is to 1, the greater the vulnerability of the county. The main causes of ecological vulnerability in Dingbian are single landscape, low vegetation coverage and low precipitation. The reason for the ecological vulnerability of Fugu and Shenmu is human production activities and single landscape types. These two counties develop coal industry vigorously, and this leads to a decline in landscape diversity and increase in ecological vulnerability. The main reason for the ecological vulnerability of Yuyang is the landscape type is not rich enough and coupled with the pressure of human activities on the environment, and vegetation coverage is low. The main reason for the ecological fragility in Jingbian

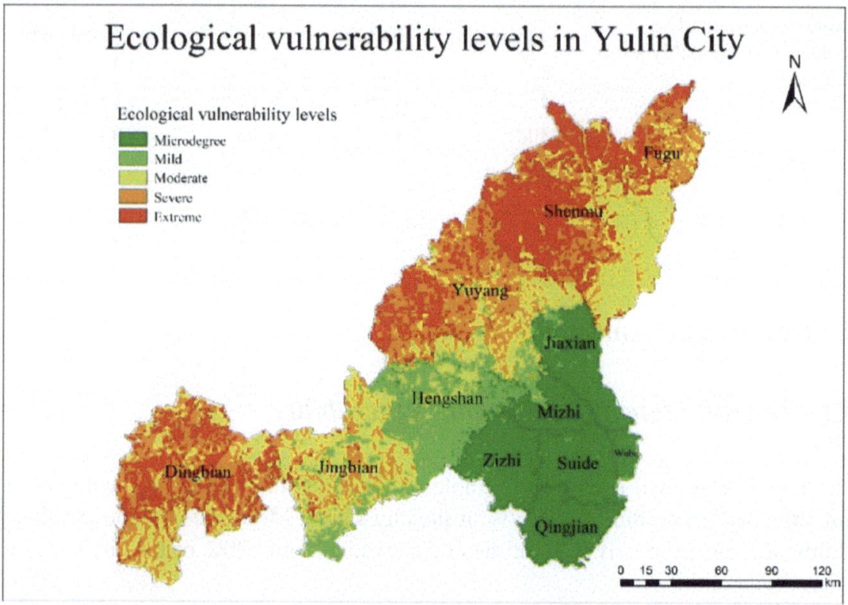

Fig. 2 Spatial distribution of ecological vulnerability levels in Yulin City

Fig. 3 Statistics on ecological vulnerability levels of various districts and counties in Yulin City

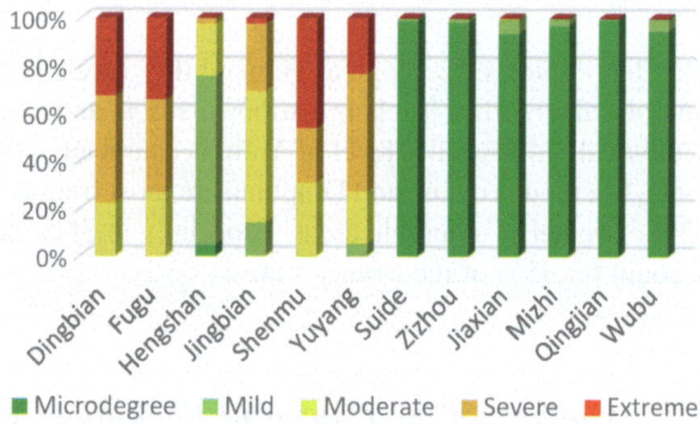

is that the landscape type is not rich and the precipitation is small. The 5/6 population in Jingbian is agricultural population, and less precipitation in arid areas will restrict the growth of crops and reduce the stability of the agricultural and livestock ecosystem. Vegetation coverage and aspect are the main reasons for the ecological fragility of Hengshan. There is a large area of half-sunny slope in Hengshan, and water evaporation on sunny slope is fast, which is not conducive to vegetation growth. One of the most obvious features of the six southern counties is the large K value and high vegetation coverage, and the vegetation has a tightening effect on the soil thus decreasing EVI.

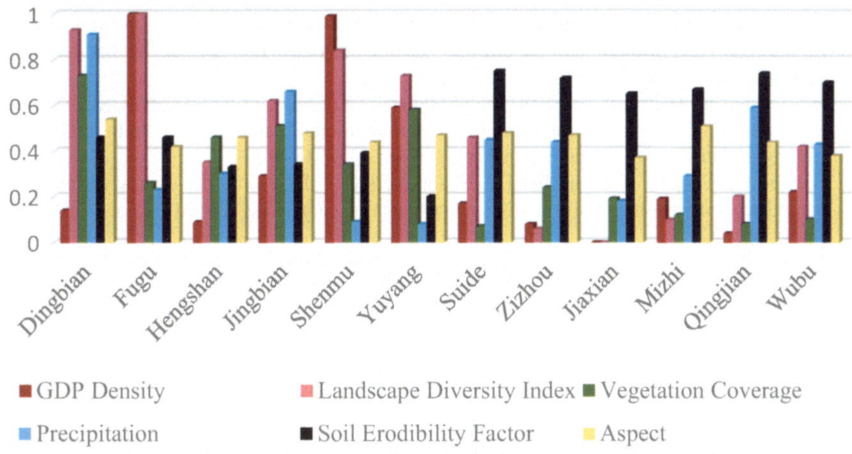

Fig. 4 Evaluation indicators after standardization in various districts and counties of Yulin City

5 Summary

This paper processes remote sensing, and meteorological and socioeconomic data, and uses PCA and AHP to calculate EVI. The main research results were as follows:

1. The method of combining PCA and AHP is used to construct the EVI model, which reduces the correlation between indicators and improves the objectivity of weight determination.
2. Obtain the distribution map of vulnerability level. It is found that the ecological of six northern counties is more fragile than six southern counties. The extremely fragile areas are Fugu, Shenmu, Yuyang and Dingbian.
3. The main causes for the ecological vulnerability of the farming–pastoral zone in Yulin were economics, landscape diversity, annual precipitation and vegetation coverage.

Acknowledgements **Fund projects** National Key Research and Development Program of China (2017YFC0405806).

References

1. Zhengjia L, Xingxiu Y, Lei L, Mei H (2011) Vulnerability assessment of eco-environment in Yimeng mountainous area of Shandong Province based on SRP conceptual mode. Chin J Appl Ecol 8:2084–2090
2. Qiufang S, Peihao P, Jie H, Liu Z, Xiaofei S, Huaiyong S (2016) Monitoring eco-environmental vulnerability in Anning River Basin in the upper reaches of the Yangtze River using remote sensing techniques. Remote Sens Land Resour 28(2):175–181

3. Furong (2008) Remote sensing monitoring of ecological environment in the typical region of fragile ecological restoration in the farming-pastoral ecotone—Taking Zhenglan banner as an example. Inner Mongolia Normal University
4. Pan L, Deyong H, Wenji Z (2010) The monitoring of changes of land vegetation covers by remote sensing in farming-pastoral mixed zones of North Chiana—A case study in Guyuan County, Hebei Provinc. Remote Sens Land Resour 2:113–117
5. Jiangbo G, Wenjuan H, Dongsheng Z, Shaohong. W (2016) Comprehensive assessment of natural ecosystem vulnerability in Tibetan Plateau based on satellite-derived datasets. Sci Geogr Sin 4:580–587
6. Jingli L (2009) Quantitative evaluation of Gansu Wen County soil erosion based on RUSLE model. Lanzhou University
7. Jing Z (2016) Based on RS and GIS the Erguna Wetlands ecological vulnerability assessment. Southwest Jiaotong University
8. Wei D, Xingzhong Y, Rong S, Yuewei Z (2016) Eco-vulnerability Assessment based on remote sensing in the argo-pastoral ecotone of North China. Environ Sci Technol 39(11):174–181
9. Shao H, Liu M, Shao Q, Sun X, Wu J (2014) Research on eco-environmental vulnerability evaluation of the Anning River Basin in the Upper Reaches of the Yangtze River. Environ Earth Sci 72(5):1555–1568
10. Shuang X, Runping S, Xiaoyue Y (2012) A comparative study of different vegetation indices for estimating vegetation coverage based on the dimidiate pixel model. Remote Sens Land Resour 4:95–100
11. Scanlon TM, Albertson JD, Caylor KK, Williams CA (2002) Determining land surface fractional cover from NDVI and rainfall time series for a savanna ecosystem. Remote Sens Environ 82(2):376–388
12. Carlson TN, Ripley DA (1997) On the relation between NDVI, fractional vegetation cover, and leaf area index. Remote Sens Environ 62(3):241–252
13. Tingting X (2016) Assessment of soil erosion in karst area Based on RUSLE model—A case study in the Southwest of Songzi. Huazhong University of Science and Technology
14. Yonghua L, Qiang F, Xue W et al (2015) Spatial and temporal differentiation of ecological vulnerability under the frequency of natural hazard based on SRP model: A case study in Chaoyang county. Sci Geogr Sin 35(11):1452–1459
15. Jian S, Jun C (2012) Principal component analysis and its application (A). Light Ind Technol. 9:12–13
16. Vaughan RA (1990) A review of: "Introduction to Remote Sensing". In: Campbell JB (ed) The Guilford Press, Guilford, 1987. p 551. Price£ 34.95. Remote Sen 11(10):1932
17. Jensen JR Lulla K (2007) Introductory digital image processing: a remote sensing perspective. Science Press
18. Lillesand T, Kiefer RW, Chipman J (2000) Remote sensing and image interpretation, 6th edn. Remote sensing and image interpretation. Wiley, pp 3035
19. Elkington MD (1987) A review of: "Remote Sensing Digital Image Analysis: An Introduction". In: Richards JA (ed) Springer, Berlin, Heidelberg. New York, 1986, p 281. Price DM (1987) 138. Int J Remote Sen 8(7):1075–1075
20. Bin W, Jinbai H, Xinglong G (2015) Grid soil moisture constants estimation based on HWSD over basin. Hydrology 35(2):8–11
21. Bin W, Xingchen D, Jinbai H et al (2017) Grid runoff parameters estimation and adjustment of GSAC model based on HWSD. Trans Chin Soc Agric Mach 48(9):250–256
22. Nachtergaele FO, Velthuizen H, Verelst L et al (2012) Harmonized world soil database (version 1.2)

Airport and Ship Target Detection on Satellite Images Based on YOLO V3 Network

Ren Ying

Abstract Airplane and ship play a very important role in both civil life and military operations. It is a meaningful to detect airplane and ship around the world through remote sensing images. Target recognition algorithms based on deep learning technology are proven to be effective and gradually replacing traditional algorithms. This paper builds a target detection system based on NVIDIA TX2 development platform and YOLO V3 algorithm and focuses on both the ship and airplane targets. The training data comes from image fragments generated by satellites such as Jilin No. 1, DigitalGlobe, and Planet. The label of each target includes a bounding box and category information. The image processing method such as rotation and noise is added to increase the robustness of the trained YOLO V3 network for different sensors and atmosphere. The training of the input image takes about 3 days on an NVIDIA Titan X GPU. At test time, we partition testing images of arbitrary size into cutouts with a fixed size of 1 k × 1 k and run each cutout through our trained model to find ships and airplanes. The experimental results show that the F_1-score values of the airport and the ship are 91.48% and 93.89%, respectively, and the detection speed of one cutout on NVIDIA TX2 development platform is about 0.56 s.

Keywords Target detection · Deep learning · YOLO V3 · Satellite images · Training data

1 Introduction

In recent years, with the development of related technologies such as aerospace, remote sensing, and sensors, the data and information contained in remote sensing images are becoming more and more abundant, which also facilitates the analysis of remote sensing images by researchers. Airplane and ship play a very important role in both civil life and military operations. It is an important technology to identify airplane and ship around the world through remote sensing images.

R. Ying (✉)
Chang Guang Satellite Technology Co., Ltd., Changchun, China
e-mail: renying1009@163.com

L. Wang et al. (eds.), *Proceedings of the 6th China High Resolution Earth Observation Conference (CHREOC 2019)*, Lecture Notes in Electrical Engineering 657,
https://doi.org/10.1007/978-981-15-3947-3_12

Remote sensing image target recognition based on traditional algorithms mainly relies on manual interpretation, which is low in efficiency, high in cost, and poor in timeliness. The traditional model can no longer meet the current operational needs. Fortunately, a variety of target recognition algorithms based on deep learning technology are gradually replacing traditional algorithms. With the introduction of networks such as deep residual networks and deep dense networks, the number of layers of deep convolutional neural networks is getting deeper, and the over-fitting phenomenon brought about by the deepening of the network is greatly reduced, and the recognition effect is also increasing accurate.

At present, the methods of target recognition of convolutional neural networks can be divided into two categories. The first category is region-based target recognition methods such as Faster R-CNN [1] and Mask R-CNN [2]. This type of method works well for small targets, but the detection speed is slow. The other type is regression-based target recognition methods, such as SSD [3] and YOLO [4]. The regression-based target recognition method uses end-to-end target recognition, and the speed is much faster than the region-based target recognition method.

In this paper, an on-board target detection system based on NVIDIA TX2 development platform and YOLO V3 [5] algorithm is built, which is focused on the detection of airplanes and ships. Satellite images with the two targets are collected and labeled. The experimental results show that the F_1-score values of the well-trained network on airports and ships are 91.48% and 93.89%, respectively, and the detect speed of a signal image with a size of 1 k $\times$ 1 k on NVIDIA TX2 development platform is about 0.56 s.

2 Annotation of Dataset

2.1 *Motivation*

Datasets play an important role in data-driven technology, such as deep learning. There are many conventional target detection datasets, and the front-end target detection algorithms (such as Faster R-CNN, Mask R-CNN, SSD, and YOLO) are basically experimented on these regular datasets. But the classification training based on conventional datasets performs poorly on satellite images, because the satellite images have its particularity, such as scale diversity, particular perspective, and small target problem.

Satellite images are of different orbital altitudes from a few hundred kilometers to thousands of kilometers, and the ground targets are even different in size, such as the aircraft carrier is more than 300 m, and the small ship is only a few 3 m. The perspective of the satellite image is basically a high-altitude view, but the conventional datasets are mostly the horizontal perspective, so the same target mode is different. Even a good detector trained on the regular dataset may have a poor performance on detecting targets of satellite images. Many targets of satellite images are with small size (tens or even a few pixels), which leads to a small amount of target information. The CNN-based target detection method has a good performance on the conventional target detection dataset, but for small targets, CNN's pooling layer will further reduce the amount of information. A 24 × 24 target has only about 1 pixel after 4 layers of pooling, making the dimension too low to distinguish.

Based on the above reasons, for the target detection task of satellite images, the conventional dataset is often difficult to train the ideal target detector, and a special satellite database is needed.

2.2 *Image Collection and Annotation*

As mentioned before, two categories are selected and annotated in our dataset, i.e., airplane and ship. In target detection of satellite images, in addition to resolutions of satellite images, the variety of sensors are also effective to the performance of classifier. To ensure that the classifier of targets on satellite images is robust to different resolutions and sensors, and satellite images in our dataset are collected from multiple satellites with multiple resolutions. Training data is collected from small chips of large images mainly from three satellites, i.e., Jilin No. 1, DigitalGlobe, and Planet. To increase the diversity of training data, the satellite images are from different cities. In addition, the dataset is expanded by some image processing operations, such as image noise, rotation, and contrast change.

As for dataset labeling, many visual concepts such as region descriptions, and objects, can be annotated with bounding boxes, as shown in Fig. 1. A common description of bounding boxes is (x_c, y_c, w, h), where (x_c, y_c) is the center location, w and h are the width and height of the bounding box, respectively. That is to say, each label includes a bounding box and category identifier for each object.

Compared with those in regular images dataset (e.g., PASCALVOC and MSCOCO), these satellite images are mostly very large in size. The original size of images in our dataset ranges from about 800 × 800 to about 4 k × 4 k. We make annotations on the original full image without partitioning it into pieces to avoid the cases where a single instance is partitioned into different pieces. And then, the original images are partitioned into pieces with sizes smaller than 1 k × 1 k. The cross-dataset generalization is an evaluation for the generalization ability of a dataset. We randomly select 14,662 for training and 3258 for testing. Our final dataset and the train/test split for each category are shown in Table 1.

3 Network Architecture

You Only Look Once (YOLO) algorithm is a regression-based target recognition method, which is proposed in 2015 [6]. The third generation of YOLO V3 has been developed in 2018. Just like its name, it only needs to do a forward calculation to make

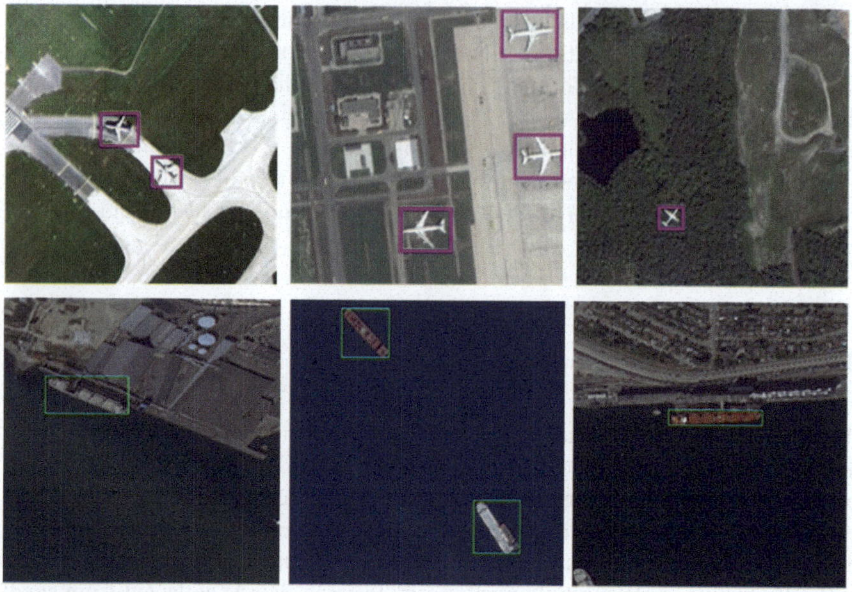

Fig. 1 Samples of annotated images in our dataset

Table 1 Train/test split

Object class	Training examples	Test examples
Airplanes	6686	1486
Ships	7976	1772
All	14,662	3258

a variety of objects are detected, so the YOLO series algorithm detects quickly. The network still maintains the advantages of the fast detection of the YOLO V2 network, and the correct rate of recognition is greatly improved. Especially in the detection and identification of small targets, the accuracy is greatly improved. YOLO V3 network draws on the idea of residual neural network, introducing multiple residual network modules and using multi-scale prediction improves the defect of YOLO V2 network in small target recognition. Because of the high accuracy and timeliness of detection, this algorithm is one of the best algorithms in the field of target detection. The model uses a number of well-formed 3×3 and 1×1 convolutional layer and later uses multi-scale predictions to structure some residual networks. Finally, it has 53 convolutional layers and is called as Darknet-53.

The YOLO V3network introduces the idea of using anchor boxes in Faster R-CNN. Three scales are used for COCO datasets and VOC datasets. Each scale has 3 anchor boxes, and the scale features are large. A small a priori box, so you can select the appropriate a priori box anchors according to the target you want to identify and modify the network structure according to the scale of the prediction.

The YOLO V3 network uses anchor boxes as a priori boxes to detect targets in the image. In the Faster R-CNN and SSD, you need to manually set the a priori box, which will make the selection subjective. If you can choose a suitable a priori box, the deep convolutional neural network will be easier to learn. Hence, the K-means algorithm is used in the YOLO V3 network to cluster the target frame size of our dataset.

4 Evaluations

4.1 Network Training

We train with stochastic gradient descent. Besides the anchor boxes, the main parameters in the training are set as follows: each batch of 32 images, number of iterations of 50,200, an initial learning rate of 10^{-3}, a weight decay of 0.0005, and a momentum of 0.9. Meanwhile, rotating training dataset is used during training to increase the contrast and exposure of the image. Each iteration trains 1 batch and performs 1 scale transformation to realize the expansion of the dataset. In addition, by doing k-means clustering on our dataset, the size of the corresponding prediction box is set to the center of the nine clusters, which are (44, 65), (63, 37), (135, 189), (124, 85), (76, 67), (217, 103), (76, 130), (115, 50), (243, 219). Training takes 3 days on a single NVIDIA Titan X GPU.

4.2 Test Results

If the number of iterations of the training network exceeds a certain number, the phenomenon of over-fitting may occur. In order to select the best weight file, the network weight file is saved every 10,000 iterations during training. The first 500 images of the test images are selected to test all the weights using the network and calculate the recall rate, which is expressed by

$$R = T_{\mathrm{P}}/(T_{\mathrm{P}} + F_{\mathrm{N}}) \tag{1}$$

where T_{P} denotes the number of targets that are correctly detected and F_{N} denotes the number of targets that are not detected.

Part of the test results is shown in Table 2. It can be seen from Table 2, the optimal iteration time for the YOLO V3 network is around 50,000.

We select the weight file with 50,000 iteration times to evaluate the well-trained network on our dataset and calculate the accuracy rate, which is expressed by

$$P = T_{\mathrm{P}}/(T_{\mathrm{P}} + F_{\mathrm{P}}) \tag{2}$$

Table 2 Relationship between recalling rate and iteration times

Iteration times	Average recalling rate (%)
10,000	88.13
20,000	89.61
30,000	90.19
40,000	91.62
50,000	**92.02**
50,200	91.89

Bold represents the maximun value at the iteration number.

Table 3 Relationship between recalling rate and iteration times

Object class	Accuracy rate (%)	F_1-score
Airplane	92.12	91.48
Ship	94.57	93.89

where T_P denotes the number of targets that are correctly detected and F_P denotes number of targets that are miss detected.

In this paper, F_1-score, i.e., balanced F score, is used to evaluate the target detection effect of our well-trained network on satellite images, which can be expressed by

$$F_1 = 2^* \frac{P * R}{P + R} \tag{3}$$

As shown in Table 3, the F_1-score of the well-trained network on airplane and ship is 91.48% and 93.89%, respectively.

Then the well-trained network is transplanted to NVIDIA TX2 development platform to evaluate the detection speed. At test time, we partition testing images of arbitrary size into manageable cutouts and run each cutout through our trained model. Partitioning takes place via a sliding window with user-designed bin sizes and overlap (30% by default). As shown in Table 3, the average detection time of a single image with a size of 1 k × 1 k on NVIDIA TX2 development platform is 0.56 s. The visualization results are shown in Figs. 2 and 3.

5 Conclusion

In this paper, an on-board target detection system based on NVIDIA TX2 development platform and YOLO V3 algorithm is built, which is focused on the detection of airplanes and ships. Satellite images with the two targets are collected and labeled. The experimental results show that the F_1-score values of the well-trained network on airports and ships are 91.48% and 93.89%, respectively, and the detect speed of

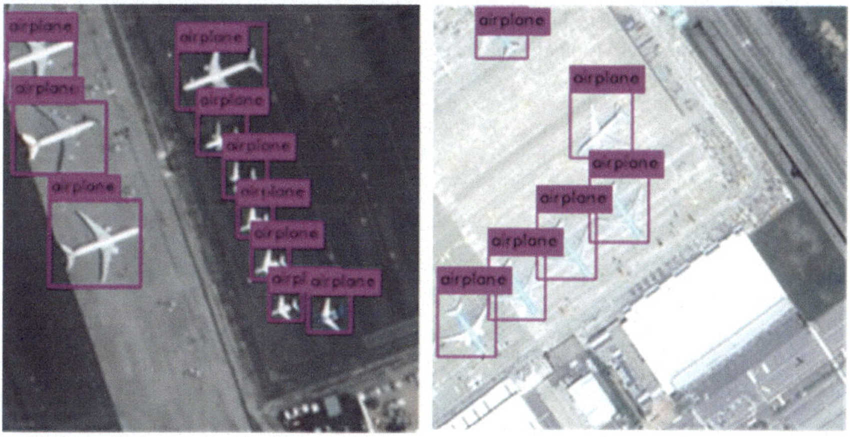

Fig. 2 Visualization results of airplane detection

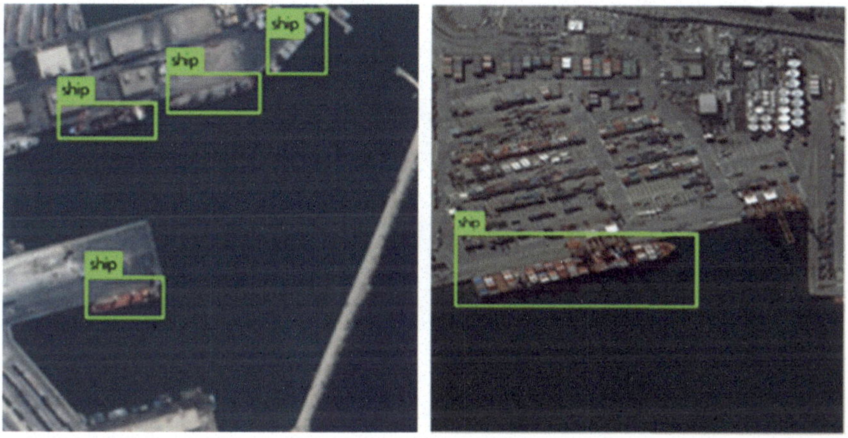

Fig. 3 Visualization results of ship detection

a signal image with a size of 1 k $\times$ 1 k on NVIDIA TX2 development platform is about 0.56 s.

References

1. Girshick R (2015) Fast R-CNN. In: IEEE conference on computer vision and pattern recognition, pp 1440–1448
2. He KM, Gkioxari G, Dollar P et al (2018) Mask R-CNN. In: IEEE conference on computer vision and pattern recognition, pp 1–12

3. Liu W, Anguelov D, Erhan D, et al (2016) SSD: signal shot multibox detector. In: IEEE conference on computer vision and pattern recognition, pp 21–37
4. Redmon J, Farhadi A (2018) YOLO V3: an incremental improvement. In: IEEE conference on computer vision and pattern recognition, pp 1–6
5. Redmon J, Farhadi A (2017) YOLO9000: better, faster, stronger. In: IEEE conference on computer vision and pattern recognition, pp 6517–6525
6. Redmon J, Divvala S, Girshick R, Farhadi A (2015) You only look once: unified, real-time object detection. arXiv:1506.02640

River Segmentation Based on Visual Salience Calculation of Spectral Residual and Region Growing in SAR Images

Guiming Zhang, Gong Zhang, Liyan Luo, Jiantao Wang, and Qing Ding

Abstract In the problem of the SAR image river segmentation, the threshold method and the region growing method are two widely used segmentation methods based on water pixel information. In view of the fact that the precision of the segmentation result for the threshold method is low and the segmentation result of the region growing method has voids which lead to the problem of the high missing alarm; for the first time, this paper introduces the visual salience detection theories into the SAR image river segmentation and presents a method that combines visual salience calculation of spectral residual and region growing. This method first binarizes the preprocessed SAR image, then extracts the saliency map of the binary image by utilizing the spectral residual model, and finally segments the river region by using the region growing method on the saliency map. Compared with the threshold method, the region growing method and the method that combines the threshold method and the region growing method, the experiments demonstrate that the method proposed by this paper has much better precision and comprehensive segmentation performance. Apart from that, it also effectively solves the problem of the voids existing in the segmentation results for the traditional region growing method. Therefore, the method presented by this paper can be applied to the SAR image river segmentation in the applications such as water resources planning and flood disaster prevention.

Keywords Synthetic-aperture radar · Spectral residual · Region growing · River · Binarization

1 Introduction

Segmentation of rivers is of great importance in water resources assessment, soil and water conservation, transportation, and flood disaster prevention. Synthetic-aperture radar (SAR), as an imaging radar working in microwave band, has attracted more and more attention due to its advantages of all-time, all-weather, strong penetration

G. Zhang (✉) · G. Zhang · L. Luo · J. Wang · Q. Ding
Science and Technology on Electronic Information Control Laboratory, Chengdu, China
e-mail: zguiming@126.com

L. Wang et al. (eds.), *Proceedings of the 6th China High Resolution Earth Observation Conference (CHREOC 2019)*, Lecture Notes in Electrical Engineering 657,
https://doi.org/10.1007/978-981-15-3947-3_13

ability, and wide monitoring range. Generally speaking, the backscattering coefficient of the river surface is relatively small and the echo signal is weak compared with the adjacent objects, which results in the lower gray values for the river surface in SAR images, showing obvious differences from the gray values of the land surface in the image.

Among the SAR image river detection methods, a common method is to segment and draw the river region from the image by hand; but this method is time-consuming and labor-consuming. In order to improve the efficiency and precision of river segmentation in SAR images, many researchers have done a lot of in-depth work on it. The threshold method has become the most widely used one for river segmentation in SAR images due to its advantages of computational efficiency, but its anti-noise ability is poor [1–3]. In addition to that, it cannot solve the problem of the different objects with the same spectra characteristics, either. Region growing method is another commonly used method for river segmentation in SAR images [4–6]. It has the characteristics of the simplicity and high efficiency, but the noise and non-uniformness of the gray level in SAR images may lead to holes and over-segmentation in segmentation results.

The research shows that in the face of complex scenes, human visual system can quickly screen out important targets and ignore insignificant information. This selective and active physiological and psychological activity is called visual attention mechanism [7, 8]. In SAR images, the river region and background show obvious grayscale differences, which make the river region significant relative to the background. The application of visual salience model that helps to reduce the computational load of river segmentation and improve the efficiency of information processing is used to segment the river region quickly. At present, the commonly used visual attention calculation models include visual attention calculation model of ITTI, visual attention calculation model of AIM, and visual attention calculation model of spectral residual (SR) [9]. Compared with the first two visual attention calculation models, the spectral residual model is based on frequency domain processing [10], which is relatively simple and fast, and is conducive to extracting saliency maps of wide SAR images and building real-time system.

Based on the theory of human visual salience attention and the saliency of river region in SAR images, a river segmentation method for SAR images based on salience calculation model of spectral residual and region growing is proposed for the first time in this paper. Firstly, the SAR images are binarized after preprocessing, and then the binarized images are used to extract the saliency maps by taking the SR model. Finally, the salient maps are used to segment the river regions by region growing method on the salient maps.

2 River Segmentation Based on Visual Salience Calculation of Spectral Residual and Region Growing

2.1 *Rough Segmentation Processing by Binarization*

Rough segmentation by binarization for the SAR images is used to form the binary images for further processing. Concretely, the pixels the values of that are below the threshold are divided into the river region, and the pixels the values of that are above the threshold are considered as background. It is obvious that the determination of the segmentation threshold is the key. The commonly used methods of the threshold determination are empirical method, experiment method, two-peak method, and mathematical statistics method, etc. [11]. Among them, the experiment method needs iterations between human and machine, which is time-consuming and laborious. The two-peak method is often affected by noise, and it is easy to get false threshold. Meanwhile, the method of maximum square error, that is the most widely used one belong to the mathematical statistics method, often misclassifies some land areas into river because the pixel values of that are close to the pixel values of the river, such as grassland. Particularly, when the ratio of the river area to the whole image is less than 30%, the segmentation performance of the method of maximum square error will decrease significantly [12]. For the reason that the backscattering coefficient of rivers is relatively small and the gray values of rivers in SAR images are obviously different from those of adjacent riverbanks; the empirical method is adopted to define appropriate thresholds and separate the river regions from the riverbanks roughly in this paper.

2.2 *Visual Salience Calculation Model of Spectral Residual*

After binarization, the river region is more significant when compared with the background and is more conducive to extracting the saliency map accurately. The basic idea of the spectral residual model is that the logarithmic amplitude spectrum of each natural image has an approximate trend [10]. Therefore, the change part of the image that is the saliency area is caused by the spectral residual. Given an image $I(x)$, after Fourier transform the image is decomposed into two parts which are the amplitude spectrum $A(f)$ and the phase spectrum $P(f)$, as shown in Eqs. 1 and 2.

$$A(f) = \mathrm{Amp}\{\mathrm{FFT}[I(x)]\}. \tag{1}$$

$$P(f) = \mathrm{Pha}\{\mathrm{FFT}[I(x)]\}. \tag{2}$$

Then the logarithmic transformation of the amplitude spectrum is shown in Eq. 3. Then the logarithmic amplitude spectrum is subtracted by the convolution between the logarithmic amplitude spectrum and the low-pass filter, and the spectral residual can be obtained, as shown in Eq. 4.

$$L(f) = \log[A(f)]. \quad (3)$$

$$R(f) = L(f) - h_n(f) * L(f). \quad (4)$$

Finally, the saliency map of the original image can be obtained through the inverse Fourier transform of the spectral residual and the initial phase spectrum. In addition, a Gauss filter can be taken to smooth the image for better visual effect, as shown in Eq. 5.

$$S(x) = g(x) * \left|\mathrm{FFT}^{-1}\{\exp[R(f) + i \cdot P(f)]\}\right|^2. \quad (5)$$

In the equations above, FFT and FFT^{-1} denote Fourier transform and its inverse Fourier transform, respectively. $A(f)$, $P(f)$, $L(f)$, $R(f)$, $S(x)$ denote the amplitude spectrum, phase spectrum, logarithmic amplitude spectrum, spectral residual, and saliency map of the original image. $h_n(f)$ denotes the low-pass filter in the frequency domain. The mean filter is chosen in this work. $g(x)$ denotes the low-pass filter in the spatial domain. The Gauss filter is taken in this paper.

2.3 *Region Growing Method*

Region growing is the process of joining neighboring points (or collections of points) into larger regions, subject to the predetermined growing criteria [13]. Its basic idea is to determine whether the adjacent pixels satisfy the conditions according to the growing rules when starting from the seed points. If the conditions are satisfied, the area is continuously growing by merging the adjacent pixels with the seed points, and then the above qualified pixels are taken as the new seed points and the procedures mentioned before are repeated again. This process stops until no pixels can be merged, and the region growing segmentation is completed. Traditional region growing method is vulnerable to noise, so how to select seed points and determine the conditions of region growing become difficult problems. They both affect the segmentation results. If the growing conditions are harsh, there will be voids in the segmentation results. If the growing conditions are loose, the segmentation precision may be reduced.

In this paper, seed points are selected manually on the saliency maps. When the segmentation stops due to gray difference or encountering bridges, seed points will be selected in the undivided region until the river segmentation in the whole image is completed. In this paper, the threshold method is carried out after the preprocessing of the original images, which makes the river region pixel values smooth and convergent.

And then the influence of the river region pixel values' difference on the following segmentation is greatly solved.

2.4 Evaluation Method for the River Segmentation Performance

Comparing the results of river segmentation by various methods and the manual mark, four performance indicators which are precision, missing alarm, *F*-score, and operation time, respectively, as shown in Eqs. 6–9, can be used to quantitatively evaluate the segmentation performance.

$$P = \frac{A(W_{\mathrm{r}}) \cap A(W_{\mathrm{e}})}{A(W_{\mathrm{e}})}. \tag{6}$$

$$R = \frac{A(W_{\mathrm{r}}) \cap A(W_{\mathrm{e}})}{A(W_{\mathrm{r}})}. \tag{7}$$

$$\mathrm{MA} = 1 - R. \tag{8}$$

$$F = \frac{2PR}{P + R}. \tag{9}$$

Among these equations, $A(W_{\mathrm{r}})$ represents the set of true river pixels in the image, which is plotted by manual mark in the experiment. $A(W_{\mathrm{e}})$ represents the set of river pixels in the image obtained by the river segmentation method. The value of *P* represents the precision of the method, and the higher the value, the more precise the result of river segmentation. The values of *R* and MA are recall ratio and missing alarm of the method, respectively, representing the integrity of the river segmented by the method. The higher the value of *R*, the lower the value of MA will be, and the closer the river area segmented is to the true river range. In practice, the values of *P* and *R* sometimes appear contradictory, so *F*-score that is the weighted harmonic average of precision and recall ratio is introduced to evaluate the comprehensive segmentation performance for the method. In the same environment, the shorter the operation time is, the more practical the project will be.

3 River Segmentation Procedures in SAR Images

In this section, the detailed procedures of SAR image river segmentation method based on the salience calculation model of spectral residual and region growing are given. The flowchart of the proposed method is shown in Fig. 1.

The specific processing steps are as follows:

Fig. 1 Flowchart of the proposed method for the SAR image river segmentation

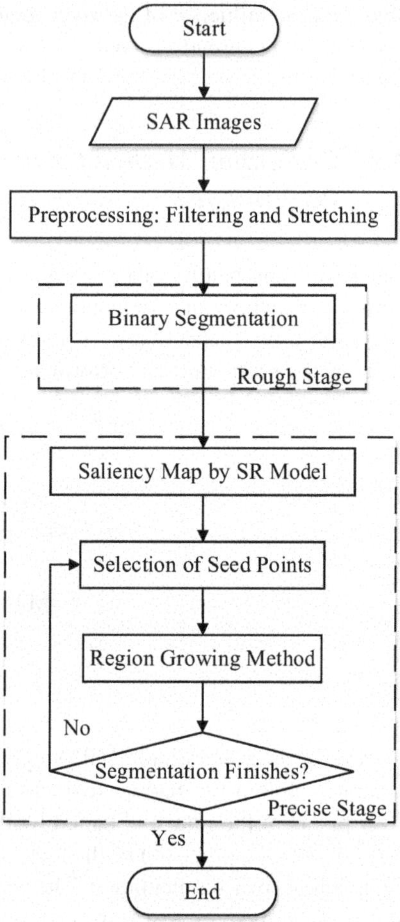

1. Preprocessing. First, the original image is smoothed by using an enhanced Lee filter which effectively removes the speckles and retains the fine information of the image. Secondly, after filtering the image is stretched by piecewise linear transformation to improve the contrast between the river area and the background. The pixel value of 2% the lower gray value in the image is set to 0, and that of the 2% higher gray value is set to 255. The rest of the pixel values are extended to the range of 0–255 by linear transformation.
2. Binary segmentation. After the preprocessing is completed, a custom threshold is used to segment the image. When the pixel value in the image is less than or equal to the threshold, the pixel value is set to 0. Meanwhile, when the pixel value in the image is greater than the threshold, the pixel value is set to 255. The threshold here is set to 60.

3. Using the SR model to extract the saliency map. For binary segmentation images, the SR model is used to extract the saliency map. The uniformly connected region on the saliency map will be a closed area with boundary.
4. Selection of seed points. Manual interaction is used to select seed points in river area on the saliency map.
5. Using the region growing method to segment river. The method of region growing is used to segment and grow river region. If the river region is not segmented completely, the seed points are selected in the area to be segmented to grow until the whole river area is fully divided.

4 Experimental Results and Analysis

In order to verify the validity of the purposed method in this paper, the experiments of river segmentation are carried out on two randomly selected SAR images by MATLAB R2017b, and the performance comparison results of the widely used threshold method, region growing method, improved region growing method based on threshold (threshold + region growing method for short), and the purposed method are given in this section.

Figure 2 is the two original SAR images (named as Image 1 and Image 2, respectively) selected for river segmentation experiments and the images generated in the process of using the purposed method (including the images after preprocessing, the binary segmentation images, and the SR saliency images). The pixel size of Image 1

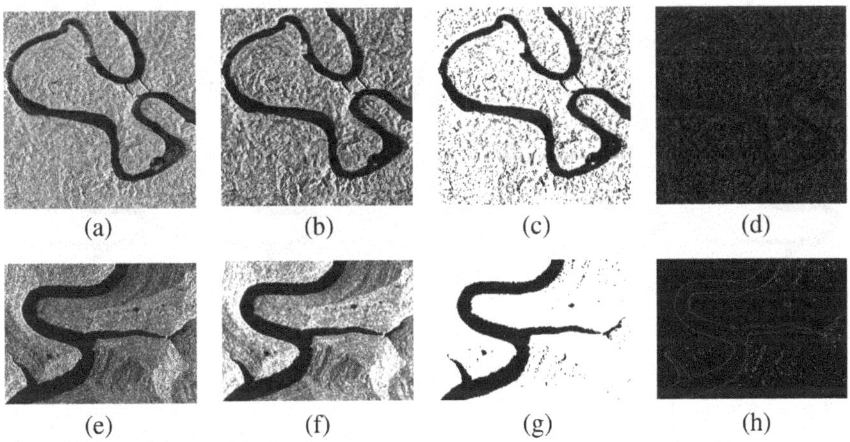

Fig. 2 Original images and images produced in the process of using the proposed method. **a** the original image for Image 1; **b** the image after preprocessing for Image 1; **c** the binary segmentation image for Image 1; **d** the saliency map for Image 1; **e** the original image for Image 2; **f** the image after preprocessing for Image 2; **g** the binary segmentation image for Image 2; **h** the saliency map for Image 2

is 920 × 919, and that of Image 2 is 480 × 341. From Fig. 2a, e, it is not difficult to see that the river area is significant relatively to the background. Because of the inherent reasons, there are still many speckles in the image, which will affect the implementation of the subsequent binary segmentation. Therefore, an enhanced Lee filter is used to suppress speckles in the image and a 2% linear transformation is taken to enhance the contrast of the image. The preprocessed image is shown in Fig. 2b, f, respectively. It can be seen that the speckles are effectively suppressed, and the contrast between the river area and the background is enhanced after preprocessing. Then the rough binary segmentations of Fig. 2b, f are carried out by using the custom threshold. The segmentation results are shown in Fig. 2c, g. It can be seen from the figures that the threshold method is easy to mistake the objects whose gray values are close to the river, resulting in a serious decline in the precision of river segmentation. Finally, the SR model is used to extract the saliency maps of the binary segmentation images, as shown in Fig. 2d, h, respectively. Combining with the binary segmentation image, we can also see that the uniformly connected region is a closed area with boundary on the saliency map, which is suitable for further precise river segmentation by region growing method.

Figure 3 is the river segmentation results of Image 1 and Image 2 using the manual mark, the threshold method, the region growing method, the threshold + region growing method, and the purposed method, respectively. It can be seen from the graphs that the threshold method cannot solve the problem of the different objects with the same spectra characteristics, which leads to the low precision of river segmentation. It should be pointed out that the region growing method, the threshold

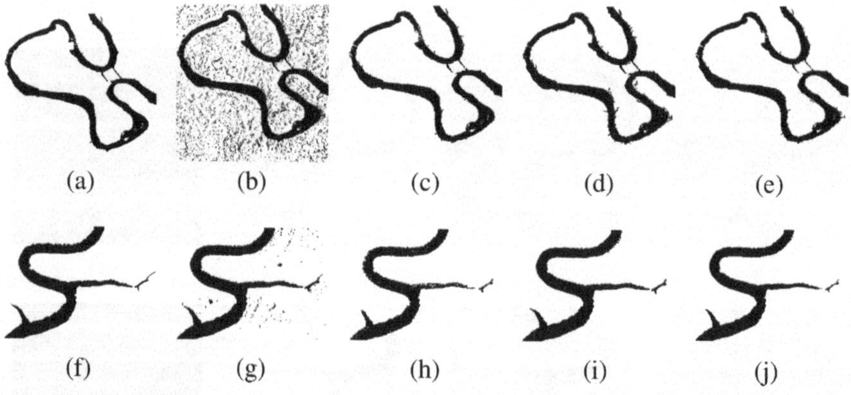

Fig. 3 Comparison result of the five methods. **a** the manual segmentation result for Image 1; **b** the segmentation result of threshold method for Image 1; **c** the segmentation result of region growing method for Image 1; **d** the segmentation result of combining threshold method and region growing method for Image 1; **e** the segmentation result of the proposed method for Image 1; **f** the manual segmentation result for Image 2; **g** the segmentation result of threshold method for Image 2; **h** the segmentation result of region growing method for Image 2; **i** the segmentation result of combining threshold method and region growing method for Image 2; **j** the segmentation result of the proposed method for Image 2

+ region growing method, and the region growing part of the purposed method all follow the same growing rules and termination conditions. It can be seen from the figures that compared with the traditional region growing method, both the threshold + region growing method and the method presented in this paper can solve the problem of voids existing in the segmentation results. This is mainly due to the smoothing of river pixel gray values by using the threshold method in advance, so as to avoid the problem of insufficient region growing due to the difference of pixel gray level. Of course, the problem of voids can also be eliminated by morphological filtering (such as expansion and corrosion methods), but at the same time the void disappears while the details of the river edge change. Therefore, in order to precisely compare the segmentation performance of these methods, the results of river segmentation by each method in this paper have not been carried out by morphological filtering.

For the images to be segmented in this paper, the advantages and disadvantages of the four river segmentation methods are quantitatively analyzed by calculating the precision, the missing alarm, *F*-score, and the operation time, respectively. The performance comparison results are shown in Table 1. As can be seen from Table 1, the precision of the threshold method is the lowest, which agrees with the above analysis. Meanwhile, the threshold method and the threshold + region growing method have the lowest missing alarm. However, compared with the former, the threshold + region growing method has a significant improvement in precision, especially in the situation that the gray values of background and river gray values are close to each other. Compared with the threshold method and the threshold + region growing method, the traditional region growing method has a higher precision as well as a higher missing alarm. In this paper, the purposed method is the one that the SR salience calculation model is introduced on the basis of the threshold + region growing method. From the performance comparison results in Table 1, it can be seen that the introduction of SR model reduces the amount of river segmentation calculation and improves the efficiency of information processing. The purposed method has the highest precision in four methods. The missing alarm of the purposed method is

Table 1 Performance comparison of four methods

Method name	Precision (%)		Missing alarm (%)		*F*-score (%)		Operation time (s)	
	Image 1	Image 2	Image 1	Image 2	Image 1	Image 2	Image 1	Image 2
Threshold method	58.0	90.9	0.5	1.0	73.3	94.8	0.24	0.21
Region growing method	92.4	98.8	1.7	8.2	95.3	95.2	14.47	5.31
Threshold + region growing method	87.4	94.4	0.4	1.0	93.1	96.6	24.58	6.33
The purposed method	94.8	99.6	1.7	4.4	96.5	97.6	14.81	5.32

higher than the threshold method and the threshold + region growing method, but lower than the region growing method. In terms of *F*-score, the purposed method has the best comprehensive segmentation performance, followed by region growing method and the threshold + region growing method, and finally the threshold method. In terms of the segmentation efficiency, the threshold method has the shortest operation time, followed by the regional growing method. The threshold + region growing method has the longest segmentation time. In this paper, because that the purposed method is the one that the SR model is introduced on the basis of the threshold + region growing method, the operation time of the purposed method is greatly shortened, which makes the operation efficiency of this method equal to that of the region growing method. Further, it has the best comprehensive segmentation performance of these four methods.

5 Summary

In this paper, a river segmentation method in SAR images based on SR salience calculation model and region growing method is proposed for the first time. Firstly, the preprocessed SAR image is binarized. Then the saliency map of the binary image is extracted by SR model, and finally, the river region is segmented by region growing method on the saliency map. By comparing the river segmentation results of two randomly selected SAR images, we can see that the proposed method has higher accuracy and better comprehensive segmentation performance than the widely used threshold method, region growing method, and threshold + region growing method. At the same time, it also solves the problem of the voids existing in the segmentation results by traditional region growing method. Finally, in terms of operational efficiency, the proposed method is comparable to that of the region growing method, but significantly better than that of the threshold + region growing method. The reason is that the introduction of SR model reduces the amount of river segmentation calculation and improves the efficiency of information processing.

References

1. Lingyan Chen, Zhi Liu, Hong Zhang (2014) SAR image water extraction based on scattering characteristic. Remote Sens Technol Appl 29:963–969
2. Niu S, Guo Z, Li N et al (2018) Research progress and trend analysis of water extraction by spaceborne SAR. J Liaocheng Univ (Nat Sci) 31:72–86
3. An C, Niu Z, Li Z et al (2010) Otsu threshold comparison and SAR water segmentation result analysis. J Electron Inf Technol 32:2215–2219
4. T Su, S Zhang, H Li (2017) Segmentation algorithm based on texture feature and region growing for high-resolution remote sensing image. Remote Sen Land Resour 29:72–81
5. He F, He X, Ding X et al (2018) Extracting water bodies from synthetic aperture radar images. Electron Opt Control 25:21–24 + 61

6. Ling X, Hou X (2015) River extracted from remote sensing images based on region growing. Geomat Spat Inf Technol 38:198–200
7. Guo C, Zhang L (2010) A novel multiresolution spatiotemporal saliency detection model and its applications in image and video compression. IEEE Trans Image Process 19:185–198
8. Rodríguez-Sánchez R, Fdez-Valdivia J, Toet et al A (2004) The relationship between information prioritization and visual distinctness in two progressive image transmission schemes. Pattern Recogn 37:281–297
9. Gao D, Han S, Vasconcelos N (2009) Discriminant saliency, the detection of suspicious coincidences, and application to visual recognition. IEEE Trans Pattern Anal Mach Intell 31:989–1005
10. Hou X, Zhang L (2007) Saliency detection: a spectral residual approach. In: IEEE computer society conference on computer vision and pattern recognition, Minneapolis, Minnesota, pp 1–8
11. Li J, Huang S, Li J (2010) Research on extraction of water body from ENVISAT SAR images: a modified Otsu threshold method. J Nat Disast 19:139–145
12. Lee SU, Chung SY (1990) A comparative performance study of several global thresholding techniques for segmentation. Comput Vis Graph Image Process 52:171–190
13. Zucker SW (1976) Region growing: childhood and adolescence. Computer Vis Graph Image Process 5:382–399

Remote Sensing Road Extraction by Refining Road Topology

Huiqin Gao, Yuan Yuan, and Xiangtao Zheng

Abstract Remote sensing road extraction is one of the research hotspots in high-resolution remote sensing images. However, many road extraction methods cannot hold the edge interference, including shadows of sheltered trees and vehicles. In this paper, a novel remote sensing road extraction (RSRE) method based on deep learning is proposed, which considers the road topology information refinement in high-resolution image. Firstly, two parallel operations, which named dilation module (DM) and message module (MM) in this paper, are embedded in the center of semantic segmentation network to tackle the issue of incoherent edges. DM containing dilated convolutions is used to capture more context information in remote sensing images. MM consisting of slice-by-slice convolutions is used to learn the spatial relations and the continuous prior of the road efficiently. Secondly, a new loss function is designed by combining dice coefficient term and binary cross-entropy term, which can leverage the effects of different loss. Finally, extensive experimental results demonstrate that the RSRE outperforms the state-of-the-art methods in two public datasets.

Keywords High resolution · Road extraction · Deep learning · Feature fusion

1 Introduction

Road extraction in high-resolution remote sensing image aims at detecting and segmenting road pixels in images. It refers to judging pixels as road or non-road, usually regarded as a binary classification problem. Road is an integral part of vehicle navigation, city planning, geographic and information updating, and so on. At present,

H. Gao · X. Zheng (✉)
Key Laboratory of Spectral Imaging Technology CAS, Xi'an Institute of Optics and Precision Mechanics, Chinese Academy of Sciences, Xi'an 710119, China
e-mail: zhengxiangtao@opt.cn

H. Gao
University of Chinese Academy of Sciences, Beijing 100049, China

Y. Yuan
School of Computer Science and Center for OPTical IMagery Analysis and Learning (OPTIMAL), Northwestern Polytechnical University, Xi'an 710072, China

L. Wang et al. (eds.), *Proceedings of the 6th China High Resolution Earth Observation Conference (CHREOC 2019)*, Lecture Notes in Electrical Engineering 657,
https://doi.org/10.1007/978-981-15-3947-3_14

the task of road extraction mainly contains road surface detection [1] and road centerline extraction [2, 3]. The former extracts all road pixels out, while the latter only label the skeletons of roads, which used to provide directions. Some methods also extract surface and centerline of road simultaneously [4].

In the field of high-resolution remote sensing road extraction, numerous methods have been proposed in recent years. Early methods extract low-level features (*e.g.*, edge, corner, gradients) and define heuristic rules (*e.g.*, geometrical shape) to classify pixels into road or non-road. S. Hinz and A. Baumgartner combined road and context information to extract road, including radiation measurements and geometry information [5]. In [6, 7], uniform areas with shape or geometric characteristics in the image were detected firstly and then used region growth technique to generate road map. The problem of these methods is that the features and rules used are only for simple scenes, while roads in the high-resolution remote sensing images are complex and irregular.

Several methods applied deep learning to road extraction in high-resolution remote sensing image. Y. Zhang et al. used multi-source data and multi feature to improve accuracy of road extraction [8]. Y. Wei et al. designed a road structural loss function to constrain road edge [9]. G. Máttyus et al. inferred the correct road in the result of initial segmentation [10]. F. Bastani et al. designed an iterative graph construction method to output road map [11]. Recent methods adopted the idea of semantic segmentation [12–15] and take roads as foreground and non-roads as background. Z. Zhang et al. combined the strengths of residual learning and U-Net [16] to extract road [17]. Y. Xu et al. fused attention mechanisms in DenseNet to capture local and global road information simultaneously [18]. L. Zhou et al. used larger receptive filed to preserve detailed information, thus obtained a better result of road extraction [19]. Although methods based on deep learning made some progress, it still had incoherent issues of road edge. The issues mainly caused by edge interference, including shadows of roadside trees or buildings and vehicles on the roads, which can be observed in high-resolution remote sensing images. To solve the issues, a novel high-resolution remote sensing road extraction (RSRE) method is proposed to refine road topology information. In addition to increasing the receptive field to keep context information, RSRE also considers the spatial relations of road pixels of an image. The spatial relations of pixels in image contribute to learning topology information of road with weak coherence in high-resolution remote sensing image. Therefore, RSRE can alleviate the incoherent issues which occurred in many existed methods based on deep learning.

In this paper, RSRE focuses on road topology information refinement in high-resolution remote sensing image. Topology information refinement means the maintenance of the shape, structure, or connection of the road throughout the whole image. Based on the encoder–decoder architecture which often used in semantic segmentation network, RSRE adopts dilation module (DM) and message module (MM) between encoder and decoder to enhance connectivity of road edge. Dilated convolutions in DM can increase the receptive field to keep the detailed context information in image, while slice-by-slice convolutions in MM enable messages passing across rows and columns in image to capture spatial relations of pixels. After extracting the features of image by encoder, DM and MM are used to reprocess the features between encoder and decoder. Finally, RSRE uses sigmoid layer and threshold value to output road maps. Furthermore, a new loss function is proposed to make RSRE not favor the non-road which has most of pixels in image. Experimental results show that RSRE is excellent both on DeepGlobe Road dataset [20] and Massachusetts Road dataset [21].

2 Method

2.1 RSRE Architecture

Due to the initial high-resolution remote sensing image has large size, and road always span the whole image with natural properties like topology and connectivity. Therefore, RSRE receives 1024 × 1024 high-resolution image as input to reduce the loss of detail caused by cropping images and generates road map with road topology information refinement and better road connectivity recovery. As shown in Fig. 1, RSRE has an encoder–decoder structure and combines low-level detail information and high-level semantic information to extract road in high-resolution image.

There are three parts in RSRE: encoder, center, and decoder parts. Like the architecture of D-LinkNet, the encoder part extracts feature maps of input high-resolution remote sensing image and uses ResNet34 [22] pretrained on ImageNet [23] dataset. The center of RSRE fuses the results of feature reprocessing by DM and MM to keep topology information of road. The decoder part uses transposed convolution layers [24] to do up-sampling and restores the resolution from 32 × 32 to 1024 × 1024. Finally, RSRE uses sigmoid layer and threshold value to output road maps. Pixels which probability of sigmoid layer output larger than 0.5 are considered roads, while others are considered as non-roads.

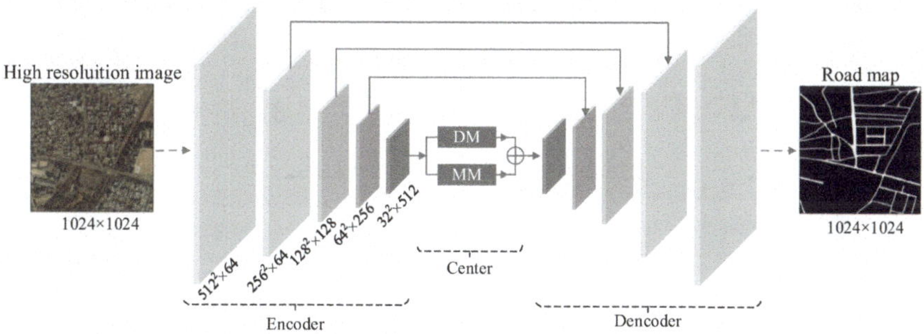

Fig. 1 RSRE architecture. Symmetrical blocks represent features with the same size and channels. The below expression $n^2 \times c$ means that the size is $n \times n$, and the number of channels is c. RSRE has an encoder–decoder structure. Center part is the core of RSRE, including DM and MM

2.2 DM and MM

The center part can refine the topology structure of road in the high-resolution remote sensing image. The high-dimensional hidden layer features are selected as the input of center part because of rich information. This part is composed of two parallel operations: DM and MM. DM increases receptive field without reducing the image resolution through dilated convolution layers [25] with series and parallel connections. As shown in Fig. 2, DM stacks result of each dilation rate, which contributes to capturing multi-scale context.

Although DM contributes to obtain multi-scale context information by increasing receptive field, it has an issue of lacking correlation information between pixels. Due to the values of dilated convolution is obtained from the pixels of mutually independent rows and columns, and these pixels lack correlation on each other. The issue causes loss of local information might be not relevant and thus bring poor continuity but is critical to the road, which has long-distance continuous, and strong spatial relation while weak appearance clue in the high-resolution remote sensing image.

In order to solve the potential issue of DM, MM uses the module of spatial CNN [26] in the field of computer vision to enhance spatial relations of pixels in high-resolution remote sensing image. Though slice-by-slice convolutions within feature maps, it can better propagate spatial information of pixels on rows and columns and thus can effectively preserve the topology information of road with long thin structure in high-resolution image.

As shown in Fig. 3, MM also applies on high-dimensional hidden layer features. The height, width, and the number of channel of the input feature are 32, 32, 512. MM has four directions to slice: upward, down, left, and right. Only the down and upward directions are shown in figure, and the left and right directions are similar. In each direction, the feature is sliced along height (upward and down) or width (left and right) of feature. The first slice goes through convolution and rectified linear unit (ReLU) and then adds next slice to formulate a new slice. The new slice repeats the same processing sequentially until the last slice is updated and thus obtains a new

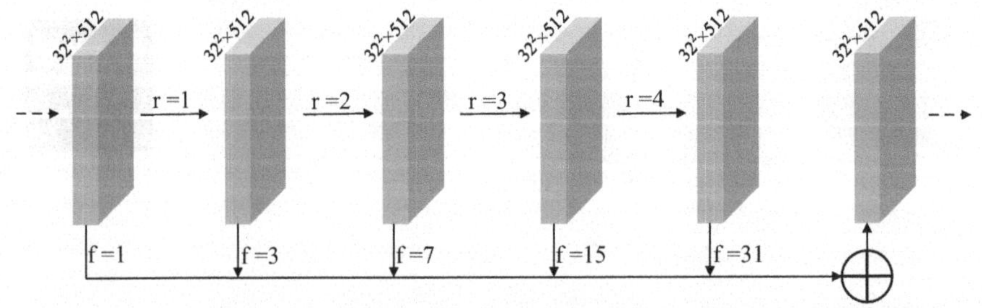

Fig. 2 DM architecture. DM contains dilated convolution with series and parallel connections. The expression $n^2 \times c$ on the feature block means that the size is $n \times n$, and the number of channels is c. The parameters r represents the dilation rate, and f means the receptive field

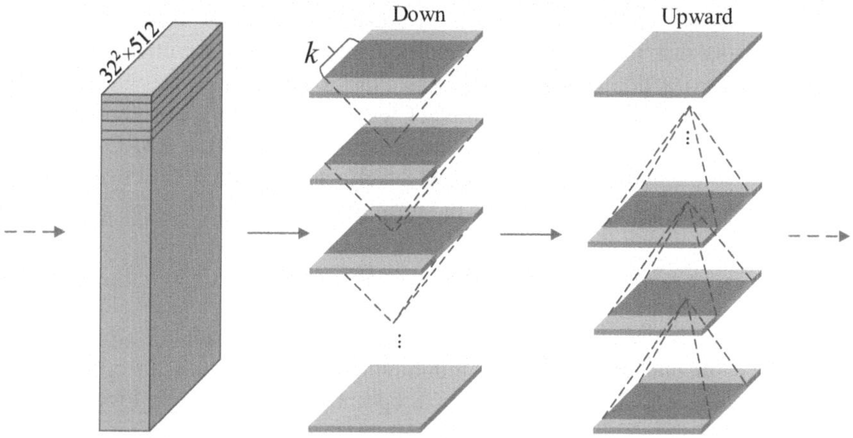

Fig. 3 MM Architecture. MM contains slice-by-slice convolutions within features. Expression $n^2 \times c$ on the feature means that the size is $n \times n$, and the number of channels is c. Parameter k is kernel width used in MM

feature of size $32 \times 32 \times 512$. Weights of the slice-by-slice convolutions are shared with the same direction and are initialized randomly different with spatial CNN.

2.3 Loss Function

Although roads are distributed in the entire high-resolution image, the imbalance in road pixels and non-road pixels in an image has the potential to tilt the training result toward non-road class with more pixels. Therefore, a new loss function is used in this paper, which uses dice coefficient item (1) and binary cross-entropy (2) simultaneously.

$$l_{\text{dice}} = 1 - \frac{\sum_{n=1}^{N} y_n f_{\text{w}}(x_n) + m}{\sum_{n=1}^{N} y_n + f_{\text{w}}(x_n) + m}. \tag{1}$$

$$l_{\text{bce}} = -\frac{1}{N} \sum_{n=1}^{N} \left[y_n \cdot \log f_{\text{w}}(x_n) + (1 - y_n) \cdot \log(1 - f_{\text{w}}(x_n)) \right]. \tag{2}$$

x_n means nth high-resolution remote sensing image, where $n = 1, 2, 3, \ldots, N$ and N is the mini-batch size. y_n indicates the ground truth (GT) of image x_n, and GT is a binary map. Expression $f_{\text{w}}(x_n)$ denotes the output of RSRE, where w represents the weight of RSRE that needs to be optimized. Parameter m is an adjustable parameter between numerator and denominator in l_{dice}.

Loss l_{dice} can be regarded as the degree of similarity of road contours between GT and predicted $f_{\text{w}}(x_n)$. Loss l_{bce} is often used in the training of semantic segmentation network. However, in the case of extremely unbalanced data, the cross-entropy loss is much smaller than the dice loss after multiple iterations, and the effect of l_{bce} will be lost. Road extraction can be regarded as a task of pixel-level recognition, only pixels of road can view as positive sample. Clearly, there is great imbalance in road pixels and non-roads pixels. To release the question of imbalance, RSRE combines l_{dice} and l_{bce} with λ and adjusts m to control the effects of different losses on training results. Thus, the final loss function is the form in (3):

$$\mathrm{loss}_{\mathrm{w}} = l_{\mathrm{dice}} + \lambda l_{\mathrm{bce}}. \tag{3}$$

Parameter w is the weight of RSRE that needs to be updated. λ is a constant coefficient by manual setting. Though minimizing the loss function, the optimal w is obtained gradually. RSRE chooses Adam as optimizer to optimize the loss function.

3 Experiment

3.1 Datasets

The method is tested on two big datasets. The first is DeepGlobe Road dataset. The resolution of each image is 1024 × 1024. Image scenes include urban, rural, wilderness, seaside, tropical rainforest, and others. On account of only training images have labels, for the convenience of measuring the accuracy of road extraction, the experiment divides the labeled 6226 training image into 4358 for training and 1868 for test.

The second dataset is Massachusetts Road dataset. The size is 1500 × 1500 with a resolution of 1.2 m per pixel. In the original training data, there are some images that do not match labels; because in the work of the original thesis, the dataset is used to study the effectiveness of robustness. In this paper, first delete the mismatch image and GT pairs, then crop to size 1024 × 1024 from center, and train RSRE on the rest of 737 images and test in 49 images.

3.2 Implementation Details

In the experiments, PyTorch [27] is used as the deep learning framework. In the training phase, the min-batch size is 16 and uses 2 GPUs. The learning rate was initially set to 2e−4 and reduced by a factor of 0.1 in every 20 epochs. RSRE adopts data augmentation to avoid the problem of over-fitting without cross-validation, including flip, vertical flip, diagonal flip, color jittering, image shifting, and scaling. In the predicting phase, each image has operations of horizontal flip, vertical flip, and

Table 1 Results of RSRE compared with other methods in two datasets, and the best values of precision (*P*), recall (*R*), and F_1-score have been highlighted in bold

Method	DeepGlobe Road dataset			Massachusetts Road dataset		
	P	*R*	F_1-score	*P*	*R*	F_1-score
U-Net	0.730	0.768	0.748	0.613	0.760	0.679
LinkNet	0.753	0.734	0.743	0.681	0.750	0.714
D-LinkNet	0.776	**0.829**	0.802	0.737	**0.816**	0.774
RSRE	**0.793**	0.821	**0.807**	**0.755**	0.810	**0.781**

diagonal flip. Hence, each image is predicted 8 times and averaged the probability of each prediction.

3.3 Result and Analysis

To assess the effectiveness of RSRE in road extraction with high-resolution remote sensing image, the precision (*P*), recall (*R*) [28], and F_1-score are introduced as follows:

$$P = \frac{\text{TP}}{\text{TP} + \text{FP}},\ R = \frac{\text{TP}}{\text{TP} + \text{FN}},\ F_1 = 2 \times \frac{P \times R}{P + R}. \tag{4}$$

where TP, FP, TN, FN represent the number of true positives, false positives, false negatives, and false negatives. *P* means the percentage of all road pixels in image that is predicted correctly. *R* represents the proportion of all pixel predicted correctly that detected as road. F_1-score is an evaluation metric for the harmonic mean between *P* and *R*.

RSRE is compared with U-Net, LinkNet, D-LinkNet in two datasets and evaluated *P*, *R*, F_1-score of road in Table 1 The architecture of U-Net and LinkNet is modified to fit the high-resolution remote sensing image input of 1024×1024, and they use cross-entropy as loss function. As baseline of RSRE, D-LinkNet only uses DM, but has no MM and constrained loss. The best result of RSRE is obtained when $m = 0.5$, $\lambda = 0.01$. The results show that after the center feature fusion and loss constrained, the *P* and F_1-score of road extraction have increased. The *R* drops a little, because precision and recall are a pair of contradictory measures, when precision is high, the recall is often low.

The effect of different weight combinations tested in Massachusetts Road dataset is shown in Table 2. As λ decreases, the influence of l_{bce} decreases gradually, and both *P* and F_1-score improve. It shows that the constraint binary cross-entropy can alleviate the poor performance caused by sample imbalance. *m* has the best adjustment effect on l_{dice} at 0.5. Therefore, RSRE gets the best result at $m = 0.5$, $\lambda = 0.01$.

Table 2 Results of RSRN with different combinations of m, λ in Massachusetts Road dataset

m, λ	P	R	F_1-score
0.0, 1.00	0.739	**0.815**	0.777
0.0, 0.50	0.741	0.816	0.776
0.5, 0.01	**0.755**	0.810	**0.781**
1.0, 0.01	0.753	0.808	0.780

In order to observe ability of RSRE proposed in this paper, some typical images tested in DeepGlobe Road dataset are shown in Fig. 4. The test results of RSRE at different background including river, building roof, urban, and rural. The road maps of comparison demonstrate that RSRE can maintain road topology information effectively in the presence of disturbances. The results in D-LinkNet are superior to U-Net and LinkNet because of the use of DM, which proves that it is necessary to increase the receptive field. Though river in image has approximate linear edge, building roof may block part of the road edge or pavement and thus destroy the continuity of the road. The connectivity of roads in a red circle of figure can be well kept through RSRE than other methods.

Results of RSRE compared with other methods in Massachusetts Road dataset are shown in Fig. 5. Compare with the results of U-Net, LinkNet, and D-LinkNet, RSRE can detect most of the roads correctly with the road topology information refinement. Though increase the receptive field and enhance spatial relations by adding DM and

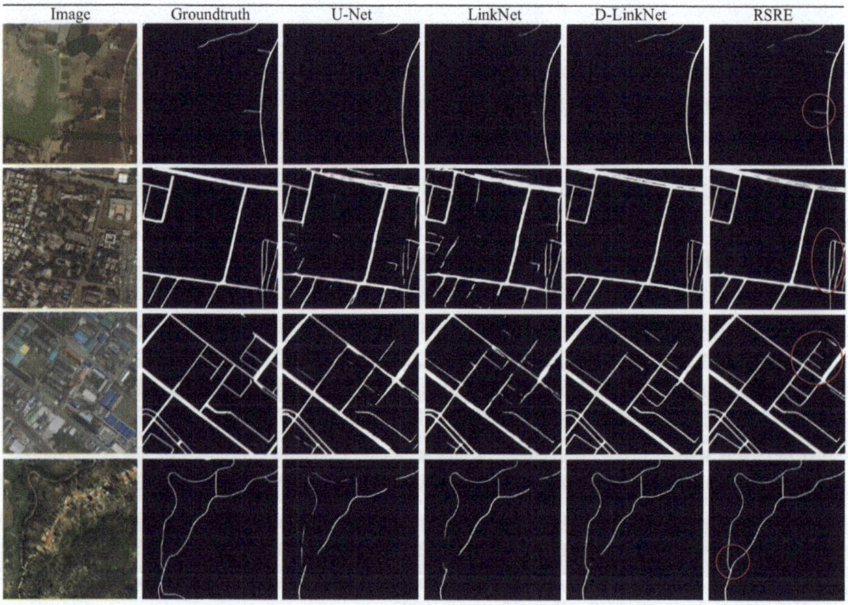

Fig. 4 Example results of RSRE and other methods tested in DeepGlobe Road dataset. From top to bottom, the contained background is river, building roof, urban, and rural

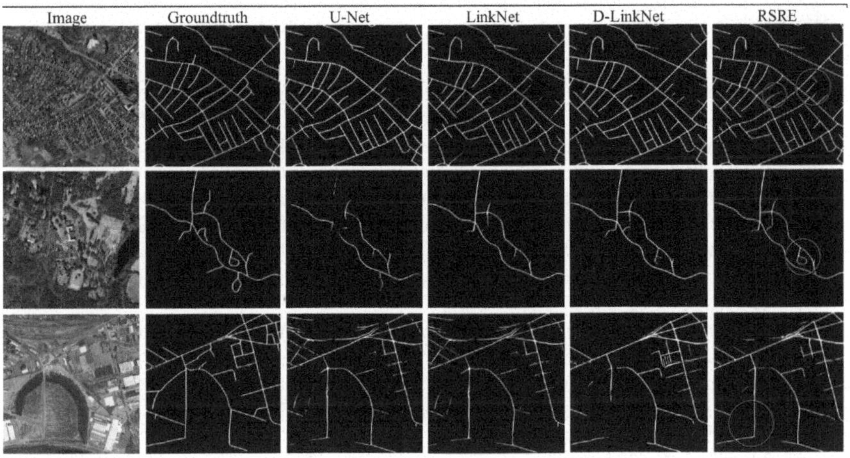

Fig. 5 Example results of RSRE and other methods tested in Massachusetts Road dataset

MM in the center of deep learning network, RSRE keeps rich information of road to refine topology in high-resolution remote sensing image. Therefore, in the phase of generating road map, the context and spatial relations information of road obtained in RSRE alleviate effectively the incoherent issues caused by shadows of sheltered trees and vehicles.

4 Summary

In this paper, RSRE extracts road from high-resolution remote sensing image, which pays more attention to refine road topology information. By the feature fusion processing of DM and MM, it can refine topology information of road and thus effectively preserve the continuity of long thin structure of road. The new combined loss function can solve the imbalance of road and non-road pixels. The results in two datasets show that RSRE could alleviate the discontinuity and incoherence of road which come from edge interference. It also has a good performance in different backgrounds. However, the test results of all methods are not very accurate for very short and thin sections of road, and RSRE has the wrong recognition in a heavily sheltered area, Therefore, the future work will focus on how to solve the above issues.

References

1. Mnih V, Hinton GE (2010) Learning to detect roads in high-resolution aerial images. In: Proceedings of the 11th European conference on computer vision, pp 210–223
2. Gao L, Shi W, Miao Z, Lv Z (2018) Method based on edge constraint and fast marching for road centerline extraction from very high-resolution remote sensing images. Remote Sens 10:900
3. Zhang Z, Zhang X, Sun Y, Zhang P (2018) Road centerline extraction from very-high-resolution aerial image and LiDAR data based on road connectivity. Remote Sens. 10:1284
4. Cheng G, Wang Y, Xu S, Wang H, Xiang S, Pan C (2017) Automatic road detection and centerline extraction via cascaded end-to-end convolutional neural network. IEEE Trans Geosci Remote Sens 55:3322–3337
5. Hinz S, Baumgartner A (2003) Automatic extraction of urban road networks from multi-view aerial imager. ISPRS-J Photogramm Remote Sens 58:83–98
6. Hu J, Razdan A, Femiani JC, Cui M, Wonka P (2007) Road network extraction and intersection detection from aerial images by tracking road footprints. IEEE Trans Geosci Remote Sens 45:4144–4157
7. Song M, Civco D (2004) Road extraction using SVM and image segmentation. Photogramm Eng Remote Sens 70:1365–1371
8. Zhang Y, Xia W, Zhang Y, Sun S, Sang L (2018) Road extraction from multi-source high-resolution remote sensing image using convolutional neural network. In: Proceedings of IEEE 2018 international conference on audio, language and image processing (ICALIP), pp 201–204
9. Wei Y, Wang Z, Xu M (2017) Road structure refined CNN for road extraction in aerial image. IEEE Geosci Remote Sens Lett 14:709–713
10. Máttyus G, Luo W, Urtasun R (2017) Deeproadmapper: extracting road topology from aerial images. In: Proceedings of IEEE international conference on computer vision, pp 3438–3446
11. Bastani F, He S, Abbar S, Alizadeh M, Balakrishnan H, Chawla S, Madden S, DeWitt D (2018) RoadTracer: automatic extraction of road networks from aerial images. In: IEEE conference on computer vision and pattern recognition, pp 4720–4728
12. Ronneberger O, Fischer P, Brox T (2015) Unet: convolutional networks for biomedical image segmentation. In: Proceedings of international conference on medical imaging and computer-assisted intervention, pp 234–241
13. Badrinarayanan V, Kendall A, Cipolla R (2017) Segnet: a deep convolutional encoder-decoder architecture for image segmentation. IEEE Trans Pattern Anal Mach Intell 39:2481–2495
14. Chaurasia A, Culurciello E (2017) Linknet: exploiting encoder representations for efficient semantic segmentation. In: Proceedings of IEEE visual communications and image processing, pp 1–4
15. Chen L, Zhu Y, Papandreou G, Schroff F, Adam H (2018) Encoder-decoder with atrous separable convolution for semantic image segmentation. In: Proceedings of the European conference on computer vision, pp 801–818
16. Ronneberger O, Fischer P, Brox T (2015) U-Net: convolutional networks for biomedical image segmentation. In: Proceedings of international conference on medical imaging and computer-assisted intervention, pp 234–241
17. Zhang Z, Liu Q, Wang Y (2018) Road extraction by deep residual u-net. IEEE Geosci Remote Sens Lett 15:749–753
18. Xu Y, Xie Z, Feng Y, Chen Z (2018) Road extraction from high-resolution remote sensing imagery using deep learning. Remote Sens 10:1461
19. Zhou L, Zhang C, Wu M (2018) D-Linknet: linknet with pretrained encoder and dilated convolution for high resolution satellite imagery road extraction. In: IEEE conference on computer vision and pattern recognition. Workshops, pp 182–186
20. Demir I, Koperski K, Lindenbaum D, Pang G, Huang J, Basu S, Hughes F, Tuia D, Raskar R (2018) Deepglobe 2018: a challenge to parse the earth through satellite images. In: IEEE/CVF conference on computer vision and pattern recognition. Workshops, pp 172–179
21. Mnih V (2013) Machine learning for aerial image labeling. University of Toronto, Canada

22. He K, Zhang X, Ren S, Sun J (2016) Deep residual learning for image recognition. In: Proceedings of IEEE conference on computer vision and pattern recognition, pp 770–778
23. Deng J, Dong W, Socher R, Li L, Li K, Li F (2009) Imagenet: a large-scale hierarchical image database. In: Proceedings of IEEE conference on computer vision and pattern recognition, pp 248–255
24. Zeiler MD, Taylor GW, Fergus R (2011) Adaptive deconvolutional networks for mid and high level feature learning. In: Proceedings of the IEEE conference on computer vision, pp 2018–2025
25. Yu F, Koltun V (2016) Multi-scale context aggregation by dilated convolutions. In: Proceedings of international conference on learning representations
26. Pan X, Shi J, Luo P (2018) Spatial as deep: spatial CNN for traffic scene understanding. In: Proceedings of AAAI conference on artificial intelligence, pp 7276–7283
27. Paszke A, Gross S, Chintala S, Chanan G, Yang E, DeVito Z, Lin Z, Desmaison A, Antiga L, Lerer A (2017) Automatic differentiation in PyTorch
28. Martin DR, Fowlkes CC, Malik J (2004) Learning to detect natural image boundaries using local brightness, color, and texture cues. IEEE Trans Pattern Anal Mach Intell 26:530–549

Cloud Detection Method in GaoFen-2 Multi-spectral Imagery

Zhaocong Wu, Lin He, Yi Zhang, and Jun Li

Abstract Cloud cover is one of the major factors which affect the application of GaoFen-2 imagery. Cloud detection in GaoFen-2 imagery is fairly difficult due to the lack of enough infrared bands. This paper presents a cloud detection method for GaoFen-2 multi-spectral imagery based on the radiation transmission model. The scattering coefficient of remote sensing image is estimated by using radiation transmission, and then the cloud mask was obtained by combining the geometric and texture features in high-resolution remote sensing images. Experiments on GaoFen-2 multi-spectral images show that the accuracy of cloud detection is above 94.70%. The method proposed in this paper can effectively reduce the influence of highlighted buildings during cloud detection and achieve a high accuracy for GaoFen-2 imagery cloud detection with less bands. In addition, this paper provides an alternative distinction method for the quantitative researches of thick and thin clouds in optical satellite imagery.

Keywords Cloud detection · GaoFen-2 · Radiation transmission · Thick cloud · Thin cloud

1 Introduction

Optical sensing satellites are usually subject to clouds. Thin clouds affect the true land surface brightness, and thick clouds even completely obscure the ground which cause problems for applications of the optical remote sensing imagery, such as land cover classification [1] and land-use change [2], especially for quantitative studies, such as vegetation monitoring [3] and water detection [4]. The distribution, thickness, and amount of cloud all restrict the use of satellite data, the accuracy of experiments, and the reliability of conclusions of researches in earth observation.

With the development of satellite sensors, various cloud detection methods have been developed for optical remote sensing images, such as cloud detection algorithms

Z. Wu · L. He (✉) · Y. Zhang · J. Li
School of Remote Sensing Information Engineering, Wuhan, Hubei, China
e-mail: helin19950830@gmail.com

L. Wang et al. (eds.), *Proceedings of the 6th China High Resolution Earth Observation Conference (CHREOC 2019)*, Lecture Notes in Electrical Engineering 657,
https://doi.org/10.1007/978-981-15-3947-3_15

for NOAA advanced very high-resolution radiometer (AVHRR) satellite imagery [5]; for Terra/Aqua MODIS images [6–8]; for Landsat Thematic Mapper (TM)/Enhanced Thematic Mapper (ETM)+ images [9–11]; for SPOT-5 high-resolution geometrical (HRG) imagery [12]; and for GaoFen-1 wide field of view (WFV) imagery [13].

Cloud detection methods can be divided into two categories: single-scene-based methods and multi-scene-based methods [14]. Traditional thresholding methods are generally employed in single-scene-based cloud detection methods [15]. The traditional thresholding approach can yield pleasing cloud detection results [16]. However, it is an empirical cloud detection algorithm, and many coefficients need to be given. Therefore, low adaptability and universality are inevitable. In recent years, cloud detection schemes based on machine learning methods performed well, such as Markov random field (MRF) framework [17], support vector machine (SVM) [18], and deep learning [19]. However, it is inevitable that computational complexity and time-consuming are high for machine learning approaches.

Multi-scene-based cloud detection methods usually utilize time-series images for comparison in short periods of time [20]. Based on the hypothesis of invariance of surface features, change detection is used to determine whether it is a cloud or not, according to pixel-to-pixel contrast of spectral values and correlation parameters between multi-temporal remote sensing images. Multi-scene-based cloud detection method can also achieve great performance [21], but it is closely related to the reference image, as a ground truth.

The existing cloud detection approaches take cloud as the detection object, and clouds are extracted by using spectral parameters and texture features, such as the top-of-atmosphere (TOA) reflectance of Landsat OLI cirrus band [22] and the local binary pattern (LBP) [13]. In this letter, the proposed method considers radiation features of the full scene, regards that clouds obscure the true ground and thus reduce the radiation transmission value of surface features. Based on the principle and methodology of image dehazing, the radiation transmission and scattering coefficient values are first estimated. Then, clustering method is used to extract the initial cloud pixels. The morphology method and texture feature analysis are combined to remove noises, and the final cloud mask is achieved.

The remaining of this paper is organized as follows: The proposed approach is presented in Sect. 2. The cloud detection process is introduced in Sect. 2.1. Error removal procedure is given in Sect. 2.2. The experimental results, analysis, and evaluation are presented in Sect. 3. Finally, the conclusion is given in Sect. 4.

2 Methodology

The radiation transmission on the remote sensing image can directly highlight the difference between the cloud and the surface features. The proposed method first estimates the radiation transmission and scattering coefficient value of each pixel; the potential cloud mask is further acquired by clustering of the scattering coefficient map. The dark channel prior method is first introduced, which is used to estimate radiation transmission value; then, the k-means clustering method clusters the scattering coefficient map converted by radiation transmission; some categories are finally selected as the rough cloud mask. Spectral features combined with geometric and texture features are employed to filter misclassified pixels (Fig. 1).

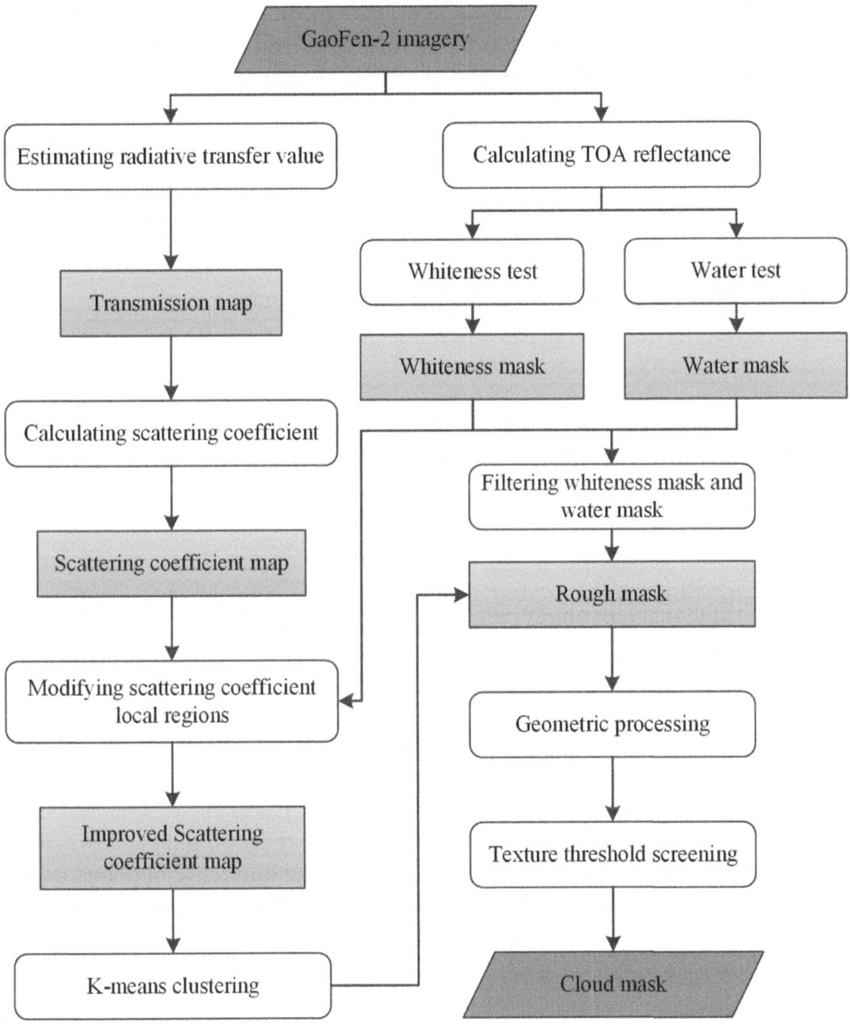

Fig. 1 Flow diagram of cloud detection approach

2.1 Cloud Detection

2.1.1 Estimating Radiation Transmission Map Using Dark Channel Prior Method

Due to the influence of absorption and scattering of medium, electromagnetic waves are attenuated on the path and the survived light is described as *the transmission* [23]. The observed intensity $I(\lambda)$, which is band related, measured by the sensor contains the reflectance of the object and the scattering air light on the path from a surface point to sensor, which are signed as $I_o(\lambda)$ and $I_P(\lambda)$, respectively.

$$I(\lambda) = I_o(\lambda) + I_p(\lambda) \tag{1}$$

The radiation transmission $t(\lambda)$, namely the *transmission*, expresses visibility in the daytime and is defined via Koschmieder's law as follows:

$$t(\lambda) = \mathrm{e}^{-\beta(\lambda)d} \tag{2}$$

where $\beta(\lambda)$ is the scattering coefficient of atmosphere and d is the scene depth. The reflectance intensity of the object on the path from a surface point to sensor is given by

$$I_o(\lambda) = J(\lambda)t(\lambda) \tag{3}$$

where $J(\lambda)$ is the inherent object radiance. The scattering airlight is integrated over the path of length d:

$$I_p(\lambda) = \int_0^d J(\lambda)B(\lambda, \theta)t(\lambda) \tag{4}$$

where θ is the angle between the incident light and the reflected light and $B(\lambda, \theta)$ is the angular scattering function to describe the scattering difference of points on the path. Based on four basic assumptions, labeling the global atmospheric light as A, the simpler observed intensity model is given by [24]:

$$I(\lambda) = J(\lambda)t(\lambda) + A(1 - t(\lambda)) \tag{5}$$

For images with three bands, the estimation of position-related radiation transmission $\tilde{t}(x)$ is written as [25]:

$$\tilde{t}(x) = \tilde{t}(x) \min_{y\in\Omega(x)}\left(\min_{c\in\{r,g,b\}} \frac{J^c(y)}{A^c}\right) + 1 - \min_{y\in\Omega(x)}\left(\min_{c\in\{r,g,b\}} \frac{I^c(y)}{A^c}\right) \tag{6}$$

where J^c is a color channel of J and $\Omega(x)$ is a local patch centered at x. The dark channel prior method makes a basic assumption that in most of the non-sky patches, at least one color channel has some pixels whose intensity is close to zero. Thus, the minimum intensity is as zero in a local patch. For an arbitrary image J, the dark channel J^{dark} is given by:

$$J^{\text{dark}}(x) = \min_{y \in \Omega(x)} \left(\min_{c \in \{r,g,b\}} J^c(y) \right) \tag{7}$$

In non-sky patches, the dark channel J^{dark} is low and tends to be zero. The estimation formula of radiation transmission $\tilde{t}(x)$ can be simplified as follows:

$$\tilde{t}(x) = 1 - \min_{y \in \Omega(x)} \left(\min_{c \in \{r,g,b\}} \frac{I^c(y)}{A^c} \right) \tag{8}$$

In studies of image dehazing, the transmission is an important feature. The scene depth expresses the distance from an object to the camera. It is generally believed that the haze in the scene is uniformly covered, so the scattering coefficient can be assumed to be a constant. The greater the scene depth is, the thicker the haze is, and vice versa. Therefore, the estimation of scene depth can be used to restore haze-free images.

Unlike close range photogrammetry, the scene depth of remote sensing images can be considered as a constant, since the distance between the sensor and the ground is much larger than the height difference between the surface objects. Due to the presence of clouds, the scattering coefficient of light on the path is changed. Thinner cloud and haze correspond to lower scattering coefficients and vice versa. Through Koschmieder's law, we have the scattering coefficient's calculation formulation:

$$\beta(x) = \frac{-\ln\left(\tilde{t}(x)\right)}{d} \tag{9}$$

A parameter which is linear with the scattering coefficient $\beta(\lambda)$ is employed and signed as $\beta(\lambda)'$:

$$\beta(x)' = -\ln\left(\tilde{t}(\lambda)\right) \tag{10}$$

2.1.2 Generating a Rough Cloud Mask Using K-means Clustering Algorithm

This paper uses k-means clustering algorithm instead of thresholding approach to classify pixels on the scattering coefficient map to extract clouds. Clustering methods can be divided into hierarchical algorithms and partitional algorithms. K-means clustering is one of the most classical partitional algorithms which were proposed independently in different scientific fields. The advantages of simplicity and efficiency make k-means clustering algorithm widely utilized over 60 years [26]. The points and centroids of each cluster are updating based on Euclidean distance matrix till converged to local optimal in processing [27]. K-means algorithm allocates each point of the given data set to a corresponding cluster to satisfy the following condition:

$$\sum_{i=1}^{c}\sum_{k\in A_i}||x_k - v_i||_2 \rightarrow \min \tag{11}$$

where A_i and v_i represent to the data set and the mean of the points over cluster i, separately. $|| \, ||_2$ denotes the square of distance norms. Initial clustering center is random selected from the data set, and the other points are clustered according to the minimal distances.

The classified image generates multiple categories A_1, A_2, ... A_n, and will be further determined into two categories: cloud and non-cloud. It is very important to discuss the attribution of a category A_i which is generated as a "*boundary category*." We signed the mean of A_i as v_i. The condition, that A_i and A_{i-1} are classified into one class rather than A_{i+1}, is that v_i is closer to v_{i-1}, or v_i is close to v_{i+1} but not significantly close to $v_i \cdot p_i$ is the parameter which are decided by empirical approaches.

$$A_i = \left(\text{Diff}_{v_{i+1},v_i} - \text{Diff}_{v_i,v_{i-1}}\right) > p_1 \text{ or } 0 < \left(\text{Diff}_{v_i,v_{i-1}} - \text{Diff}_{v_{i+1},v_i}\right) < p_2 \tag{12}$$

2.2 *Error Removal*

2.2.1 Scattering Coefficient Modification Using Spectral Features

The TOA reflectance ρ_λ is calculated, with parameters from the extraterrestrial band solar radiation of GaoFen-2 satellite and the absolute radiation calibration coefficient of domestic land observation satellite in 2016 from China Center for Resources Satellite Data and Application,

$$\rho_\lambda = \pi * L_\lambda * d^2/\text{ESUN}_\lambda * \cos\theta_s \tag{13}$$

where d is the astronomical unit (the distance between sun and earth), ESUN_λ is solar irradiance, θ_s is solar zenith angle, and L_λ is the TOA radiance.

Water and whiteness masks are generated by using spectral features. Spectral test for water and whiteness is widely implemented and proved to be necessary [9, 11, 13]. Water test is estimated to remove the error pixels caused by water. This paper uses the normalized difference water index (NDWI) to extract water.

$$\text{Water Test} = \text{NDWI} > i_2 \tag{14}$$

NDWI is a valid parameter based on the spectral analysis of water and can help to delineate the features of water rapidly and efficiently [28].

$$\text{NDWI} = \left(\rho_{\text{g}} - \rho_{\text{nir}}\right)/\left(\rho_{\text{g}} + \rho_{\text{nir}}\right) \tag{15}$$

Whiteness test is calculated to detect bright pixels with high TOA reflectance on single visible band or on the "MeanWhite" value [9] which is defined for bright white artificial structures. The TOA reflectance of visible bands designed as ρ_{r}, ρ_{g}, ρ_{b} for red, green, and blue bands, respectively. The purpose is to detect surface objects with high brightness in a single band.

$$\text{Whiteness Test} = \text{MeanWhite} > p_3 \text{ or } \rho_r > p_4 \text{ or } \rho_g > p_4 \ or\ \rho_b > p_4 \tag{16}$$

where

$$\text{MeanWhite} = \left(\rho_{\text{r}} + \rho_{\text{g}} + \rho_{\text{b}}\right)/3 \tag{17}$$

The radiant transmission of bright surface objects should be similar to that of other landscapes and larger than that of clouds. While the transmission values of bright surface objects are calculated wrongly, because of the similarity of spectral values in visible bands. In this paper, whiteness test is used to modify radiation transmission and scattering coefficient values before clustering, and reduce the influence of bright surface objects, to a certain extent.

$$\text{Modified Transmission} = \text{Transmission}/(1 - \text{Whiteness Test} * i_5) \tag{18}$$

After these steps, a rough cloud mask by using spectral information is achieved. The condition for a cloud pixel is that it belongs to clustering results, and it is not water.

2.2.2 Error Removal Using Geometric and Texture Features

Moderate-to-high spatial resolution remote sensing images provide a great amount of details of land surface [29]. Rich geographic objects increase the complexity of texture feature and thus generate many trivial errors. GaoFen-2 imagery lacks sufficient spectral information to use, and these errors are distinguishable in geometric and texture features from cloud patches. It is necessary to consider the texture and geometric morphometric information.

After spectral testing, the rough cloud mask is further processed by geometric approaches. Assuming that there are cloud regions with the number of n, the detection errors caused by small ground objects can be removed via judging the area of each cloud region and non-cloud region.

$$\text{Area Test} = \text{RoughMask}_{1\sim n} > p_6 \tag{19}$$

The strategy of geometric detection before texture detection can help to improve the operation efficiency since it is time-consuming for object-oriented texture features calculation, and geometric detection helps to significantly reduce the number of patches.

There are great differences on texture features between the cloudy region and the non-cloudy region in moderate to high-resolution remote sensing image. This paper applies graylevel and gradient information. Graylevel gradient co-occurrence matrix (GLGCM) is utilized to extract texture features comprehensively in this paper.

It is proved by experiments that the characteristics of the cloud on inhomogeneity of graylevel distribution $G_{\text{graylevel}}$ and gradient distribution G_{gradient} are more significant, compared with others statistical vectors. Because the differences on both of statistical vectors between non-cloudy regions and cloudy regions are in magnitude, the results can be easily separated by using thresholding method. The patches should be tested using texture information, named as "Texture Test." In order to prevent the generation of the pseudo "cloud region" which is relatively bright compared with other regions, due to the clustering of non-cloudy images by the method. The pseudo "cloud region" has high texture complexity and can be distinguished via setting the upper bound.

$$\text{Texture Test} = p_8 < G_{\text{gray-level}} < p_9 \text{ or } p_{10} < G_{\text{gradient}} < p_{11} \tag{20}$$

3 Experiments and Discussion

3.1 Data Set and Experimental Setup

To quantitatively evaluate the performance of the proposed method, ten GaoFen-2 high-resolution images were selected as experimental data, which were produced after relative radiometric correction and systematic geometric correction. The selected images, acquired from May 2015 to October 2018, have 7411 * 7025 pixels with multi-spectral (4 m/pixel) bands and four spectral channels (blue, green, red, and near-infrared).

Experiments on the GaoFen-2 image are designed to demonstrate the feasibility and advantages of the method for cloud detection. Reference objects were identified manually, and the precision rate (PR), recall rate (RR), and *F*-measure were taken as metrics. The formulas are as follows:

$$\text{PR} = \frac{\text{CTC}}{\text{TC}} \tag{21}$$

$$\text{RR} = \frac{\text{TC}}{\text{RC}} \tag{22}$$

$$F = \frac{\left(1 + \beta^2\right) \cdot (\mathrm{PR} \cdot \mathrm{RR})}{\left(\beta^2 \cdot \mathrm{PR} + \mathrm{RR}\right)} \tag{23}$$

where TC and CTC denote the number of cloud pixels which are tested and correctly tested, respectively. RC is the number of cloud pixels in reference map. The weights of precision and recall can be adjusted via regulating β index which is taken as 0.5 in this paper; the range of F-measure is [0, 1]; and the high value reflects high similarity.

3.2 Experiments and Evaluation

Test and parameter selections of the algorithm were carried out on local images. To test the effectiveness of the algorithm, we selected GaoFen-2 imageries which cover urban areas and rural areas and include thin and thick clouds with obvious or fuzzy boundaries. The cloud morphology includes planar thin clouds, massive thin clouds, planar thick clouds, massive thick clouds, and others.

Clustering method can help to divide modified scattering coefficient maps into different layers. Through choosing different layers, cloud regions can be determined. After geometry and texture test, the proposed method can effectively eliminate highlight surface objects, such as roads (the first line in Fig. 2), buildings (the third line in Fig. 2), and so on. Through clustering analysis, we can distinguish the thin cloud from the bright homogeneous surface (the second line and the fifth line in Fig. 2). Due to the consideration on bright areas, errors caused by snow can also be removed (the fourth line in Fig. 2).

Ten GaoFen-2 images were used to evaluate the effect of cloud detection on panoramic high-resolution imagery. Reference cloud regions were identified and labeled manually combined with spectral feature extraction. Original images, reference images, and cloud detection results are shown in Fig. 3. The cloud detection algorithm can detect both thin and thick clouds on GaoFen-2 images. Thin clouds (a, b) can be easily extracted via the proposed method (the average F-measure is 96.94%), even if the brightness characteristics of thin clouds are not notable (c). For small thick cloud (d–f) and large bright thick cloud (g–j), the algorithm can also obtain satisfying results (the average F-measure is 92.86 and 96.59%).

In this paper, the size of the detected region was used as one of the methods to remove the highlighted surface objects which cause limitations. If the geometric threshold is large, the algorithm will eliminate more highlighted surface areas; but it will also eliminate some small bright clouds. Conversely, the algorithm cannot eliminate some highlighted artificial buildings (c in Fig. 3), which leads to an increase of errors (F-measure is 88.16%). The precision, recall, F-measure, and cloud coverage are shown in the Table 1.

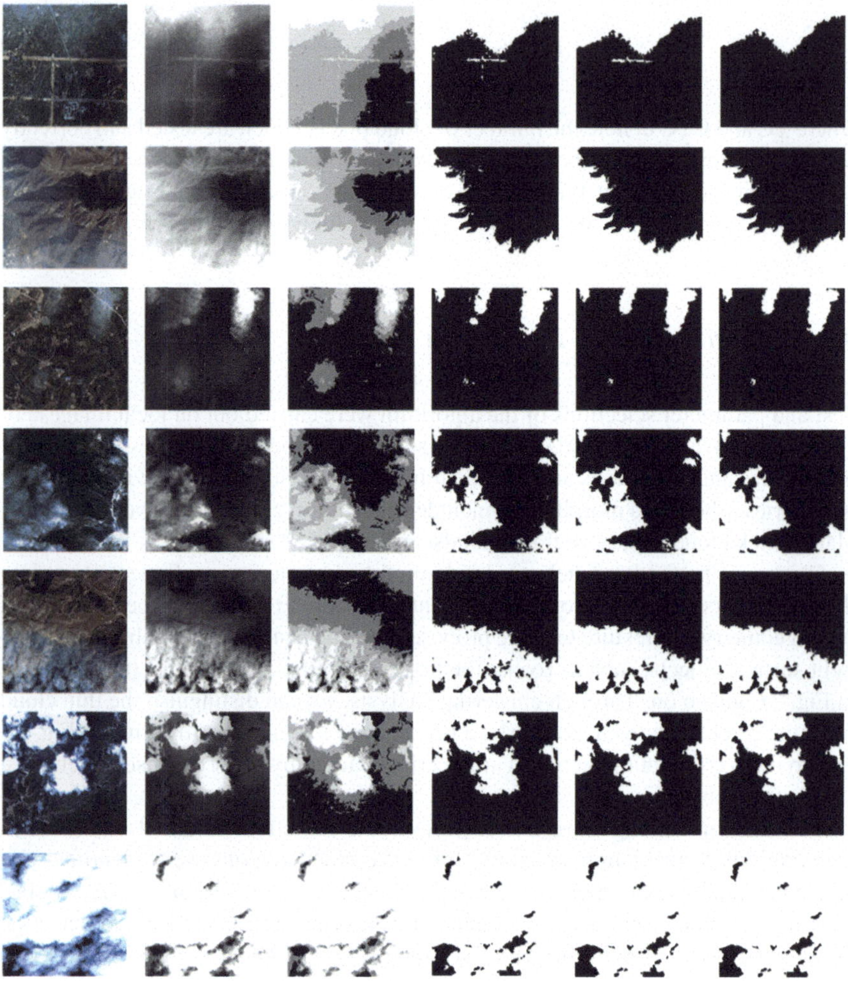

Fig. 2 The results of each step. Columns 1–6 are GaoFen-2 images, scattering coefficient maps, clustering maps, rough masks, results after geometric processing, and cloud masks obtained after texture test

4 Summary

In this paper, the presented cloud detection method for GF-2 multi-spectral imagery estimates the radiation transmission and scattering coefficient values, based on the methodology of image dehazing, and combines clustering method, morphology method, and texture feature analysis to achieve a cloud mask. The effectiveness of the algorithm and its superiority in the detection of thin clouds are proved by

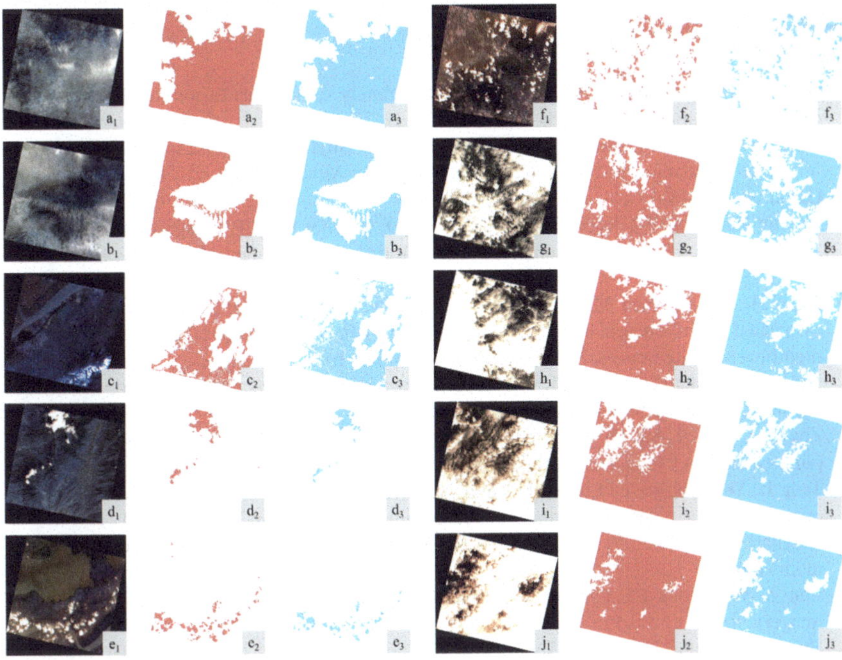

Fig. 3 GF-2images (a_1–j_1), reference masks (a_2–j_2), and cloud masks (a_3–j_3)

Table 1 Precision, recall, *F*-measure, and cloud coverages of ten cloud masks

	Precision	Recall	*F*-measure	Cloud cover
a	0.9998	0.9208	0.9829	0.5397
b	0.9457	0.9989	0.9559	0.4775
c	0.8627	0.9661	0.8816	0.3433
d	1.0000	0.6695	0.9102	0.0315
e	0.9964	0.7270	0.9277	0.0413
f	0.9674	0.8778	0.9481	0.8120
g	1.0000	0.7129	0.9255	0.5714
h	0.9995	0.8907	0.9757	0.7631
i	0.9993	0.9309	0.9848	0.7655
j	0.9988	0.9018	0.9778	0.9306

experiments and evaluation. How to overcome the accuracy of the algorithm in identifying small bright areas and distinguish different cloud thickness quantitatively will be our future researches.

References

1. Chen B, Huang B, Xu B (2017) Multi-source remotely sensed data fusion for improving land cover classification. ISPRS J Photogramm 124:27–39
2. Pellikka PKE, Heikinheimo V, Hietanen J, Schäfer E, Siljander M, Heiskanen J (2018) Impact of land cover change on aboveground carbon stocks in Afromontane landscape in Kenya. Appl Geogr 94:178–189
3. Romijn E, Herold M, Kooistra L, Murdiyarso D, Verchot L (2012) Assessing capacities of non-annex I countries for national forest monitoring in the context of REDD+. Environ Sci Policy 19–20:33–48
4. Duan H, Cao Z, Shen M, Liu D, Xiao Q (2019) Detection of illicit sand mining and the associated environmental effects in China's fourth largest freshwater lake using daytime and nighttime satellite images. Sci Total Environ 647:606–618
5. Turner J, Marshall GJ, Ladkin RS (2001) An operational, real-time cloud detection scheme for use in the Antarctic based on AVHRR data
6. Zhang X, Tan S, Shi G, Wang H (2019) Improvement of MODIS cloud mask over severe polluted eastern China. Sci Total Environ 654:345–355
7. Tang H, Yu K, Hagolle O, Jiang K, Geng X, Zhao Y (2013) A cloud detection method based on a time series of MODIS surface reflectance images. Int J Digit Earth 6:157–171
8. Jouybari Moghaddam Y, Aghamohamadnia M (2013) A novel method for cloud detection in modis imagery
9. Zhu Z, Woodcock CE (2012) Object-based cloud and cloud shadow detection in landsat imagery. Remote Sens Environ 118:83–94
10. Huang C, Thomas N, Goward SN, Masek JG, Zhu Z, Townshend JRG et al (2010) Automated masking of cloud and cloud shadow for forest change analysis using landsat images. Int J Remote Sens 31:5449–5464
11. Richard RI, John LB, Samuel NG, Terry A (2006) Characterization of the Landsat-7 ETM automated cloud-cover assessment (ACCA) algorithm
12. Fisher A (2014) Cloud and cloud-shadow detection in SPOT5 HRG imagery with automated morphological feature extraction. Remote Sens-Basel 6:776–800
13. Li Z, Shen H, Li H, Xia G, Gamba P, Zhang L (2017) Multi-feature combined cloud and cloud shadow detection in GaoFen-1 wide field of view imagery. Remote Sens Environ 191:342–358
14. Zhai H, Zhang H, Zhang L, Li P (2018) Cloud/shadow detection based on spectral indices for multi/hyperspectral optical remote sensing imagery. ISPRS J Photogramm 144:235–253
15. Zhu X, Helmer EH (2018) An automatic method for screening clouds and cloud shadows in optical satellite image time series in cloudy regions. Remote Sens Environ 214:135–153
16. Foga S, Scaramuzza PL, Guo S, Zhu Z, Dilley RD, Beckmann T et al (2017) Cloud detection algorithm comparison and validation for operational landsat data products. Remote Sens Environ 194:379–390
17. Le Hégarat-Mascle S, André C (2009) Use of Markov random fields for automatic cloud/shadow detection on high resolution optical images. ISPRS J Photogramm 64:351–366
18. Ishida H, Oishi Y, Morita K, Moriwaki K, Nakajima TY (2018) Development of a support vector machine based cloud detection method for MODIS with the adjustability to various conditions. Remote Sens Environ 205:390–407
19. Xie F, Shi M, Shi Z, Yin J, Zhao D (2017) Multilevel cloud detection in remote sensing images based on deep learning. IEEE J-Stars 10:3631–3640
20. Champion N (2012) Automatic cloud detection from multi-temporal satellite images: towards the use of Pléiades time series
21. Lin C, Lin B, Lee K, Chen Y (2015) Radiometric normalization and cloud detection of optical satellite images using invariant pixels. ISPRS J Photogramm 106:107–117
22. Zhu Z, Wang S, Woodcock CE (2015) Improvement and expansion of the Fmask algorithm: cloud, cloud shadow, and snow detection for Landsats 4–7, 8, and Sentinel 2 images. Remote Sens Environ 159:269–277

23. Fattal R (2008) Single image dehazing. ACM Trans Graph (TOG) 27:1–9
24. Fabio C, Eric K (1997) Depth from scattering
25. He K, Sun J, Tang X (2011) Single image haze removal using dark channel prior. IEEE Trans Pattern Anal Mach Intell 33:2341–2353
26. Jain AK (2010) Data clustering: 50 years beyond K-means. Pattern Recogn Lett 31:651–666
27. Zahra S, Ghazanfar MA, Khalid A, Azam MA, Naeem U, Prugel-Bennett A (2015) Novel centroid selection approaches for K means-clustering based recommender systems. Inform Sci 320:156–189
28. McFEETERS SK (1996) The use of the normalized difference water index (NDWI) in the delineation of open water features. Int J Remote Sens 17:1425–1432
29. Zhang X, Xiao P, Feng X, Yuan M (2017) Separate segmentation of multi-temporal high-resolution remote sensing images for object-based change detection in urban area. Remote Sens Environ 201:243–255

High-Resolution Land-Use Mapping in Beijing-Tianjin-Hebei Region Based on Convolutional Neural Network

Pan Chen, Zhengchao Chen, Xuan Yang, Baipeng Li, and Bing Zhang

Abstract The land-use mapping of Beijing-Tianjin-Hebei region (Jingjinji) covers an area of over 200,000 km^2. Faced with a large number of objects in tiny size, which is quite difficult for land-use mapping, the land-use mapping of Jingjinji requires a certain design of convolutional neural network (CNN) [1] (Lecun et al. Neural Comput 1(4):541–551, [2]. In this paper, we propose a feature pyramid fusion network with attention mechanism. The module is designed based on the pyramid scene parsing network (PSPNet) (Zhao et al. [3]), and two improvements are made for the large-scale high-resolution land-use mapping task: (1) Since PSPNet concatenate features in a non-selective way at the center of the module, we propose the attention feature pyramid fusion (AFPF) block, which can selectively fuse features with different scales; (2) In order to make the border of results more precise and the tiny objects more accurate, we use an encode–decode structure to merge low-level features and high-level features at the upsample stage. During each fusion operation, the attention mechanism is employed. The final experiment proves that the network designed in this paper performs better in the accuracy of small objects and precise border with respect to the original PSPNet in the land-use mapping task of Jingjinji.

Keywords High-resolution · Land-use mapping · Beijing-Tianjin-Hebei region · Convolutional neural network

P. Chen · X. Yang · B. Zhang
Key Laboratory of Digital Earth Science, Aerospace Information Research Institute, Chinese Academy of Sciences, Beijing 100094, China

Z. Chen (✉) · B. Li
Airborne Remote Sensing Center, Aerospace Information Research Institute, Chinese Academy of Sciences, Beijing 100094, China
e-mail: chenzc@aircas.ac.cn

P. Chen · X. Yang · B. Zhang
University of Chinese Academy of Sciences, Beijing 100049, China

L. Wang et al. (eds.), *Proceedings of the 6th China High Resolution Earth Observation Conference (CHREOC 2019)*, Lecture Notes in Electrical Engineering 657,
https://doi.org/10.1007/978-981-15-3947-3_16

1 Introduction

Large-scale high-resolution land-use mapping algorithms are important methods to achieve accurate resource and environmental monitoring and to understand development trends. The mapping results are quite important for the study of regional land-use condition, the capture of land-use dynamic evolution and the monitoring of key resources. Therefore, it is of great significance to study the land-use mapping algorithm for large-scale high-resolution remote sensing data, which can achieve both accurate and automated land-use mapping.

Traditional remote sensing physics algorithms and machine learning algorithms often do not fit the data well in the face of large-scale high-resolution image data, for the limitation of parameter quantities. Since deep learning was proposed in 2006 [4], the powerful ability to fit complex problems by convolutional neural network (CNN) has made great progress in the past 10 years, bringing new method guidance to remote sensing land-use mapping tasks.

In recent years, deep learning has many applications in land-use mapping, but most of the research is based on small-scale or low-resolution image data. Few people get good performance on large-scale high-resolution land-use mapping tasks [5–8]. The difficulties can be summarized as follows: (1) The label of remote sensing dataset is difficult to be obtained, and the internal difference in dataset is huge; (2) the difference in the proportion of target objects causes a large sample imbalance problem; (3) the sources of remote sensing data are diverse and not uniform; (4) existing CNN cannot comprehensively utilize the spatial and spectral features in remote sensing image; (5) the current pre-trained models of various CNN are based on the field of computer vision. There is no pre-trained model for remote sensing dataset. (6) There is a lack of module structure specifically for remote sensing land-use mapping; (7) insufficient skills for training and optimizing network models; (8) difficulty in organizing and managing massive remote sensing data.

Under the premise of solving some of the above problems, this paper studies the CNN for large-scale high-resolution remote sensing land-use mapping with precise border. The network is mainly based on PSPNet. As for the feature pyramid fusion stage, we have designed a block named attention feature pyramid fusion module (AFPF). At the same time, the encode–decode structure is used and the attention decoder is designed to make the border more precise. We finally use the proposed method to carry out the land-use classification of Jingjinji, and it performs good.

2 Research Area Status and Research Data

2.1 Research Area Status

Jingjinji is located in the North China region, consisting of Beijing, Tianjing and Hebei provinces, covering an area of 218,000 km^2. It accounts for 8.5% of the national population and 9.8% of national Gross Domestic Product (GDP). With the development of social economy, the ecology of the Jingjinji has suffered a certain degree of damage, and there have been frequent occurrences of sandstorms, haze and other bad weather. In recent years, Jingjinji has become a national key area of concern, and it takes a lot of manpower, material resources and financial resources to monitor the environment and resources. Therefore, it is urgent to find the method of high-resolution land-use mapping in Jingjinji.

2.2 Research Data

In this paper, we use GF-1 satellite data as the research data. Launched in 2013, the GF-1 is a multi-spectral high-resolution wide-format imaging Earth observation satellite. The GF-1 payload consists of 2 m resolution full color and 8 m resolution multi-spectral panchromatic and multi-spectral (PMS) high-resolution cameras and four 16 m medium resolution wide multi-spectral wide field-of-View (WFVs) camera and supporting high-speed data transmission system [9]. As a result, the quality and coverage of GF-1 data is sufficient for high-resolution land-use mapping in Jingjinji. In this task, the data of GF-1 will be used as the research data source, and the image quality will be improved by image enhancement, orthorectification, image fusion and other pre-process. Finally, RGB band's 2 m resolution image data captured in 2014 is selected as the research data. The distribution of GF-1 images in Jingjinji is shown in Fig. 1a.

The label for this study was derived from the second national agricultural census in 2014. The classification system we use includes nine categories, namely cultivated land, garden, forest, grass, water, residential area, road, bare land and agricultural facilities. The label is also sampled at a resolution of 2 m. The label we use distribute in some areas of Beijing and Tianjin, with a total data area of 24,100 km^2. The data distribution is shown in Fig. 1b.

3 Methodology

The model used in this paper is modified based on PSPNet. PSPNet was proposed by Zhao et al. [3]. in 2016, using dilated ResNet [10] as the basic network, and added pyramid pooling module at the end of the basic network for multi-scale feature fusion. Finally, the output of pyramid pooling module concatenates with the output of dilated ResNet, and then upsample to the original image size. PSPNet has achieved state-of-the-art on multiple datasets and has become a classic semantic segmentation network.

However, PSPNet directly upsamples feature map by eight times without feature fusion and feature selection, which is hard to map high-resolution remote sensing data. Therefore, based on the PSPNet, we have made some improvements according

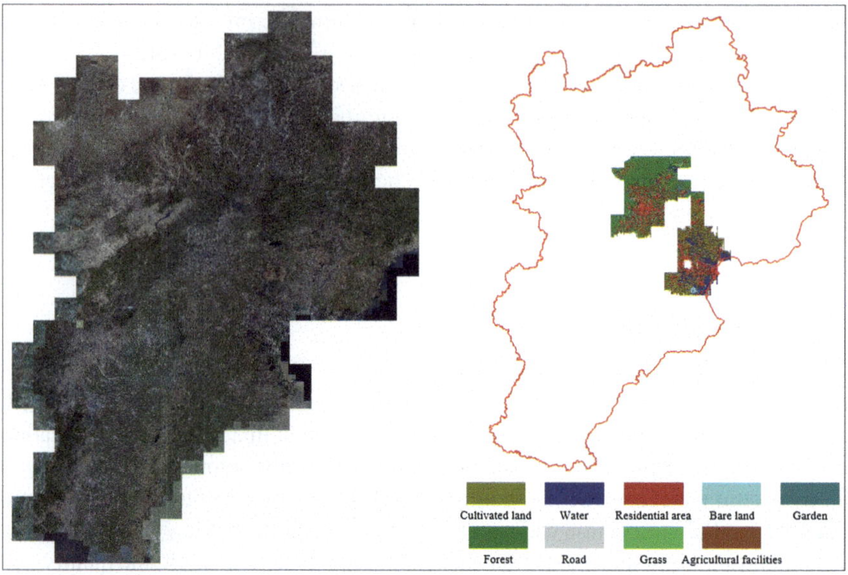

(a) (b)

Fig. 1 Research data of Jingjinji. **a** The GF-1 images' location distribution; **b** the labels' location distribution

to the characteristics of remote sensing objects. The improvement mainly consists of designing the structure of attention feature pyramid fusion (AFPF) and using attention decoder.

3.1 Attention Feature Pyramid Fusion

The PSPNet applies pyramid pooling module, which combines the characteristics of four different pyramid scales. The pyramid pooling module can realize the aggregation of multi-scale features, so that the network can have a good expression for different scales. The structure of pyramid pooling module is shown in Fig. 2a. The input of the module is the last layer of dilated ResNet. The input feature map is downsampled into four different sizes, which are 1×1, 2×2, 3×3 and 6×6. Convolution operations are performed in parallel on four scales. Upsampling and concatenation are followed at the end of the module to fuse last layer of dilated ResNet and multi-scale features.

The AFPF proposed in this paper is shown in Fig. 2b. Different from the original pyramid pooling module, this module has four different branches. Three of sub-branches are a series of convolutions and pooling operations, followed by concatenation in the backend. The other branch is downsampled by global pooling (GP)

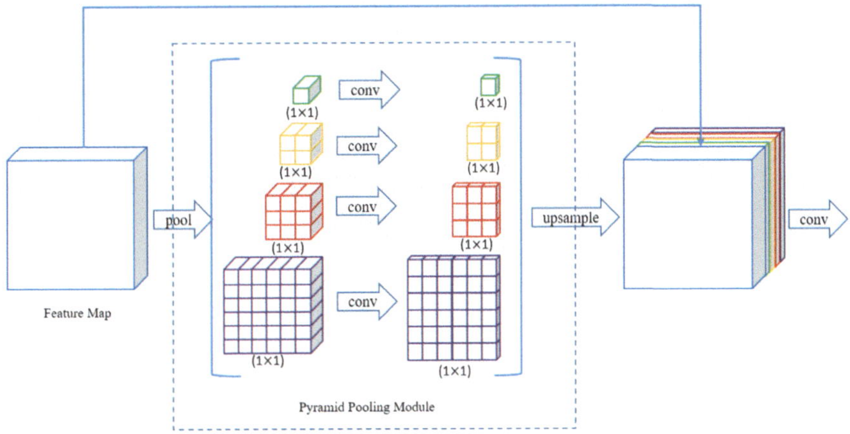

(a) Pyramid Pooling Module in PSPNet

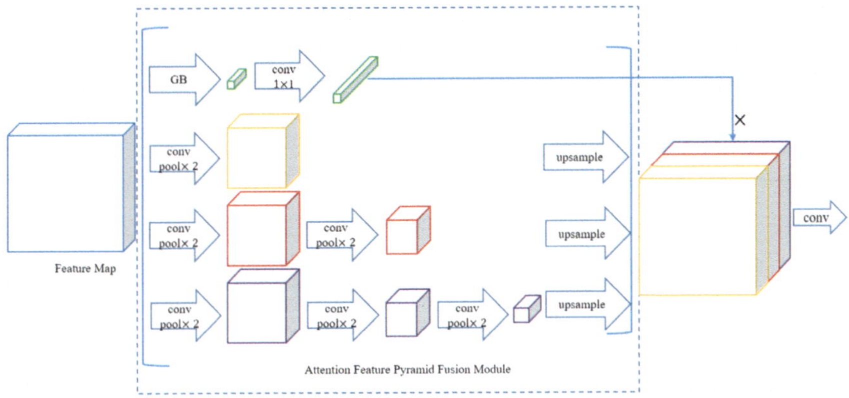

(b) Attention Feature Pyramid Fusion Module

Fig. 2 Two kinds of feature pyramid fusion module. **a** Pyramid pooling module in PSPNet; **b** attention feature pyramid fusion module. GP in this means global pooling

layer, followed by 1 × 1 convolution and upsampling to the original feature map size. In the end, the two parts mentioned above are added together to finish this module.

3.2 Attention Decoder

Since PSPNet directly upsamples the feature map by eight times as the final result, too much detail feature is lost, which is not conducive to the extraction of tiny objects and precise borders. Based on PSPNet, we have changed the simple upsampling operation to the encode–decode structure (Fig. 3) to use low-level features to enrich the details. At the same time, in the process of fusion between low-level features and high-level features, the attention mechanism is used, which can reduce the impact of unwanted information.

In this paper, we use global attention upsample (GAU) [11] as the attention decoder. The structure of GAU is shown in Fig. 4. This block takes low-level features

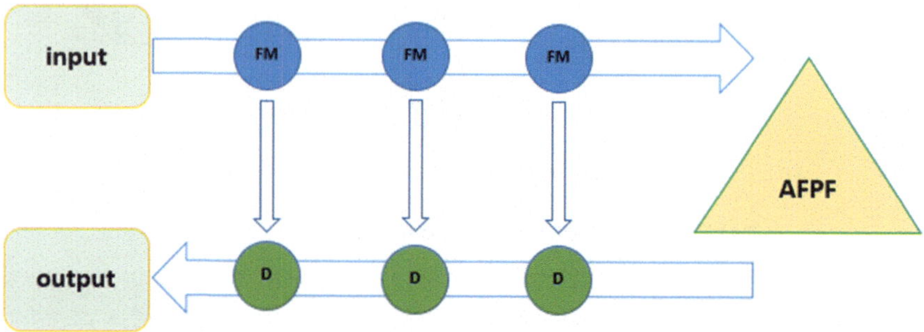

Fig. 3 The encode–decode structure used in this paper. FM refers to lower-level feature maps in basic net. AFPF refers to attention feature pyramid fusion module. D refers to attention decoder

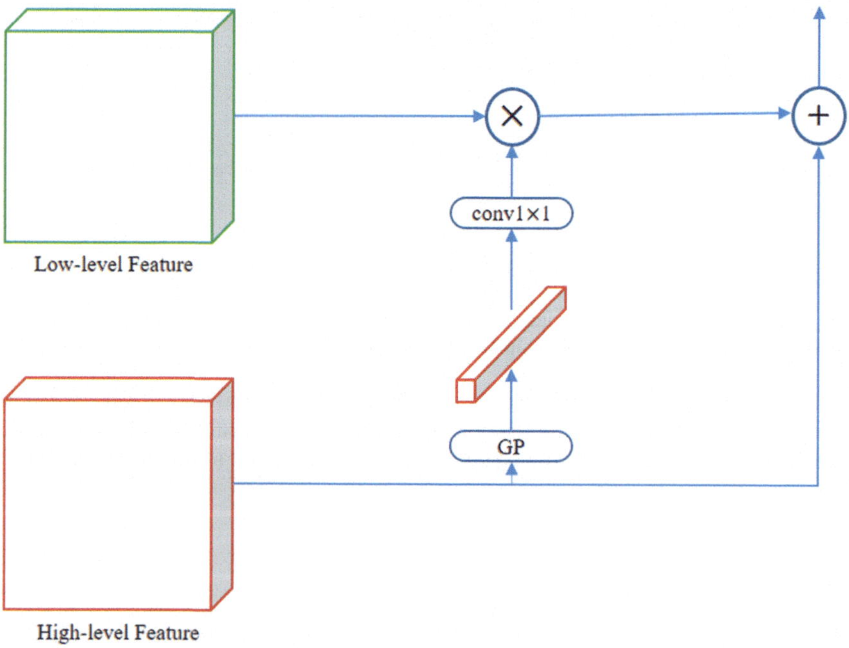

Fig. 4 Structure of GAU, where GP means global pooling

from dilated ResNet and high-level features after AFPF as inputs. The high-level features are first downsampled to a size of 1 × 1 by global pooling, and further abstracted by 1 × 1 convolution. At this time, the high-level features become a one-dimensional vector, which can be multiplied by the low-level features as weights. After weighting the features, the low-level features are concatenate with the high-level features to achieve feature fusion.

4 Experiment and Result

4.1 Experiment

In this paper, the training set and the validation set are randomly selected in a ratio of 10:1. All the data were sliced into the size of 512 × 512. The number of images in training set is around 70,000 images, and the number of images in validation set is about 7000 images. At the same time, in order to avoid over-fitting, we have applied four kinds of data enhancement processing on training set, such as randomly flip, randomly rotation, randomly scaling and randomly color change. As for the network training parameter settings, we set the mini-batch size to 8, select the Adam optimizer, set the learning rate to 1e-3 and reduce learning rate by a factor of 0.1 in every 10 epochs. The training loss function is Lovasz Softmax [12]. The network is trained with 2 NVIDIA TITANXP and converges at around 40 epochs.

In the inference stage, this study slices all 290 GF-1 images of 16,384 × 16,384 in the Jingjinji into 1024 × 1024 size slices, with 512 overlap between each sliced images. After predicting each small sliced image through the network, we remove the overlap part and stitch small slices into a 290 mapping results. The inference process uses two GPUs, mini-batch size is 20, and the entire process takes 16 h.

4.2 Result

This paper aims to study the CNN for high-resolution remote sensing land-use mapping in Jingjinji. We optimize the task based on the original PSPNet to get more accurate results, especially for tiny objects and precise borders. As shown in Fig. 5, through the optimization of the network, the land-use mapping results are significantly better than the original PSPNet in the precise border and small objects, indicating that the network improvement strategy adopted in this study is effective. As for the accuracy of the validation set, the original PSPNet can only achieve a maximum precision of 81.3%, and the improved network that we proposed can achieve an accuracy of 84.5% in the validation.

The result of land-use mapping in Jingjinji is shown in Fig. 6. From the mapping results, the overall land-use of the Jingjinji is as follows: the upper left area is

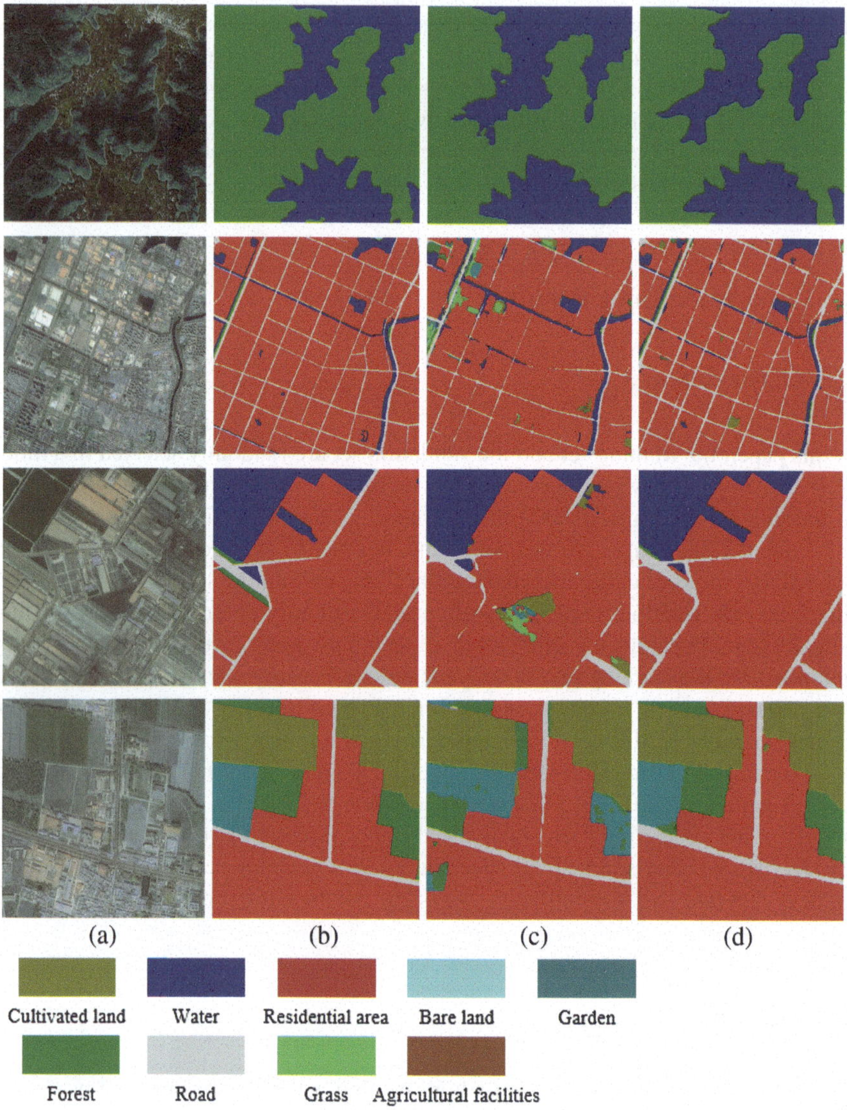

Fig. 5 Details in mapping results. **a** GF-1 images; **b** labels; **c** results from PSPNet; **d** results by the network proposed in this paper

dominated by forest, the lower right part is dominated by cultivated land, and the middle part is mainly residential area. From the statistics of the mapping result in Table 1, we can find that among all classes, forest has the highest proportion, which is 39.00%, and cultivated land accounts for 35.83%. The least class is road, accounting for 0.89%. The statistical results are basically consistent with the 30 m land-use mapping results [13].

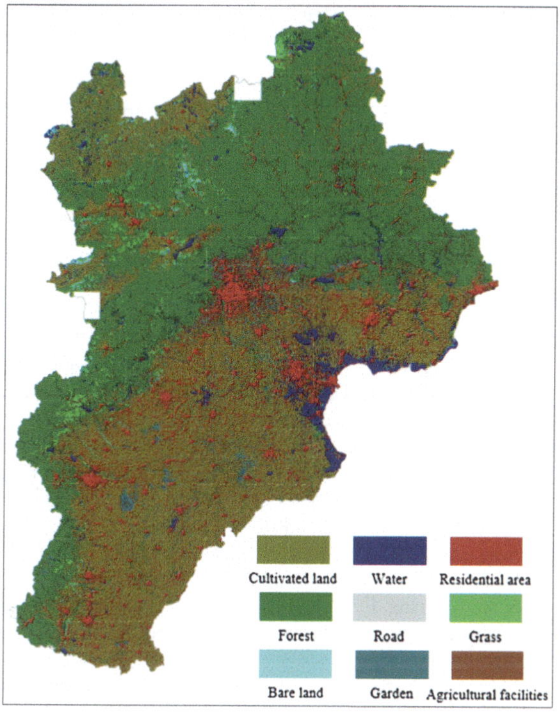

Fig. 6 2 m land-use mapping of Jingjinji

Table 1 Class proportion in Jingjinji

Class name	Proportion (%)
Cultivated land	35.83
Water	5.45
Residential area	12.46
Bare land	1.06
Garden	2.47
Forest	39.00
Road	0.89
Grass	1.93
Agricultural facilities	0.90

5 Summary

In this paper, a new CNN structure is designed for the Jingjinji land-use mapping task. The network is based on the PSPNet architecture. On this basis, the attention feature pyramid fusion block and the attention decoder are designed to make the neural network perform better on small target objects, which makes the mapping results more precise. We have tested our module by mapping GF-1 data of Jingjinji and confirm that the network structure proposed in this paper is indeed superior to the original PSPNet.

Acknowledgements This work was supported by the Strategic Priority Research Program of the Chinese Academy of Sciences (Grant No. XDA 23100304), and is granted by Qin Xin Talents Cultivation Program, Beijing Information Science and Technology University (QXTCP B201903).

References

1. Lecun Y, Bengio Y, Hinton G (2015) Deep learning. Nature 521(7553):436
2. Lecun Y, Boser B, Denker JS et al (2016) Backpropagation applied to handwritten zip code recognition. Neural Comput 1(4):541–551
3. Zhao H , Shi J , Qi X , et al (2017) Pyramid Scene Parsing Network[C]// 2017. IEEE Conference on Computer Vision and Pattern Recognition (CVPR). IEEE, 2017
4. Hinton GE, Salakhutdinov RR (2006) Reducing the dimensionality of data with neural networks. Science 313(5786)
5. Konstantinos K, Grigorios T, Michalis Z, Panagiotis T (2015) Deep learning for multi-label land cover classification. In: SPIE remote sensing
6. Xiao T, Zhang J, Yang K et al (2014) Error-driven incremental learning in deep convolutional neural network for large-scale image classification. In: Acm international conference on multimedia
7. Marmanis D, Datcu M, Esch T et al (2015) Deep learning earth observation classification using ImageNet pretrained networks. IEEE Geosci Remote Sens Lett 13(1):1–5
8. Kussul N, Lavreniuk M, Skakun S et al (2017) Deep learning classification of land cover and crop types using remote sensing data. IEEE Geosci Remote Sens Lett 14(5):778–782
9. Jin W, Hai-Bo Z, Guo-Bin L (2015) The application of GF-1 remote sensing image in the land use change monitoring. J Anhui Agric Sci
10. Yu F, Koltun V, Funkhouser T (2017) Dilated residual networks
11. Li H, Xiong P, An J et al (2018) Pyramid attention network for semantic segmentation
12. Berman M, Triki AR, Blaschko MB (2017) The Lovasz-Softmax loss: a tractable surrogate for the optimization of the intersection-over-union measure in neural networks
13. Chen-Xi LI, Ke-Ning WU, Li-Si Z (2016) Research on land use change characteristics and driving forces in Beijing, Tianjin and Hebei Region. China Popul Resour Environ

Dichotomy: Trajectory Planning Algorithm Based on Point Group Distribution

Naiting Xu, Fan Yang, HaiMing Lian, and Yi Wang

Abstract The historical navigation trajectory carries rich information about the spatiotemporal patterns of a moving target, as well as its transboundary potentials. To obtain the moving track of a ship or a plane, a large amount of point-to-point data is required via the measurements of satellites or aviation equipments. On the basis of time, we can easily draw a track. However, the distribution of the point-to-point data information provided by each detection unit is irregular due to the inconsistency of the standards of the data units. Some of the regional point groups are densely distributed, while the others may be very sparse. The point density can mess up the trajectory detection and make it not ideal to observe the behavior habits of a target unit. In this study, a dichotomy trajectory planning algorithm was proposed to fix the problem mentioned above. The target area was set as a grid with the points in the grid dichotomized. In that case, the distribution of the targeted point group after dichotomization can perform as similar as possible, and an approximate trajectory line was formed. Such a method considers the change of point density. Results were tested against the performance of the clustering, the random, and the greedy methods, showing that the former had a better adaptability in dealing with different data distribution. In this paper, a novel approximate estimation algorithm based on the density change rate is proposed. Finally, we will compare the performance of our dichotomous method with clustering method, random method, and greedy algorithm through experiments and show the trajectory fitting effect through different distribution data.

Keywords Random algorithm model · Greedy algorithm model · Clustering algorithm model · Dichotomy algorithm model · Point group distribution · Trajectory · Trajectory planning

N. Xu (✉) · H. Lian · Y. Wang
Institute of Electronics, Chinese Academy of Sciences, Suzhou 215123, Jiangsu, China
e-mail: sa615237@mail.ustc.edu.cn

Key Laboratory of Intelligent Aerospace Big Data Application Technology, Suzhou 215123, China

F. Yang
Beijing University of Science and Technology, Beijing, China

L. Wang et al. (eds.), *Proceedings of the 6th China High Resolution Earth Observation Conference (CHREOC 2019)*, Lecture Notes in Electrical Engineering 657,
https://doi.org/10.1007/978-981-15-3947-3_17

1 Introduction

With the rapid development of wireless communication technology, mobile terminals, and mobile positioning, the trajectories of a large number of moving targets (aircraft, ships, etc.) through mobile terminals such as smart phones and navigators can be obtained. This allows people to pay attention to spatiotemporal evolution patterns for moving targets. The goal of spatiotemporal data mining is to acquire trajectory knowledge of moving objects and provide effective decision support for modern aviation, navigation, intelligent transportation, and other directions [1–3]. However, due to the accuracy of equipment, weather conditions, human factors, and so on, the derived trajectory would be redundant and chaotic, which is inconvenient for geographic information system technicians to analyze whether navigation behaviors of a ship and aircraft violate relevant laws or regulations.

Generally, the point-scale trajectory analysis methods mainly include clustering analysis [4, 5], correlation analysis [6, 7], and anomaly detection [8, 9]. The purpose of clustering is to detect out the similar objects from different classes. At present, the widely used clustering algorithms are k-means, hierarchical clustering, and density clustering. For trajectory data, clustering is indeed a method of normal moving object motion rules and behavior habits; but in different scenarios, they have their own advantages and disadvantages. For example, k-means clustering is simple, fast deal with large data sets. This algorithm maintains scalability and efficiency. When the cluster is close to the Gauss distribution, its performance is better. However, if there are outliers, it will lead to bias and not suitable for non-convex clustering. DBSCAN is a density-based clustering algorithm, which assumes that categories can be determined by the compactness of sample distribution. It can be applied to a convex or non-convex sample set; however, its parameters are hard to set to make satisfactory clustering results.

In some special cases, the two methods mentioned above cannot be applicable. In our studied case, the historical track was focused on. As we will discuss in this paper, if we simply use clustering algorithm to solve the problem of historical track simplification, the final effect will be very unsatisfactory. We will ignore the important trajectory points and the moving direction of the moving target, which is not conducive to the analysis of the moving law and behavior pattern of the moving target. In paper [10], the idea of VQ-based path analysis is proposed. It is assumed that all trajectory points in a region are covered by a grid, and the point sets in each grid are clustered separately. This method solves the problem of trajectory deviation to a large extent, but the problem of trajectory deviation in each grid is not considered. This paper perfects this point. In order to express the problem more visually, take an example to illustrate. We simulate the trajectory points as desert trees. Generally speaking, the more symmetrical the distribution of trees on both sides, and the road goes along with the forest, not across it. We will feel that the road is integrated into the forest. So this road is more representative of the direction of the forest, so our problem can be expressed as to find a path that can represent the whole region and make the surrounding forest as far as possible road extension.

1.1 Summary of Key Contributions

The following is a list of our main contributions.

- We propose a trajectory analysis model, namely dichotomy algorithm method (DAM), based on symmetric distribution. Assuming that the trajectory point set $\mathbb{P} = \{p_1, \ldots, p_N\}$ of a region has N elements and is covered by a grid, the trajectory points in $\mathbb{P}$ have time attributes, and the corresponding time set is $\mathbb{T} = \{t_1, \ldots, t_N\}$. For each grid, two representative trajectory points are selected. Finally, a new trajectory line is formed by connecting the selected point sets according to time. The trajectory line is concise and reflects the trajectory law to the greatest extent in the context of studying moving targets.
- For each mesh area, find a trajectory, so that the point set in the area can be distributed as far as possible on both sides of the trajectory and can be closely around the trajectory. Then we think that this trajectory is representative and can accurately express the law of trajectory points in this region. To this end, we propose an algorithm based on the symmetric distribution (DAM), which can find this trajectory.
- We also propose random algorithm model, greedy algorithm model, and clustering algorithm model to study trajectory. These models have their own advantages and disadvantages. In this paper, we show the trajectory fitting effect of these models through experiments.

1.2 Article Structure

The remainder of this paper is organized as follows. Section Two provides a review of the state-of-the-art research on trajectory analysis; Section Three presents system overview, random algorithm model, clustering greedy algorithm model, clustering algorithm model, and dichotomy algorithm method; Section Four introduces the mechanisms of four trajectory methods including the random algorithm model, greedy algorithm model, clustering algorithm model, and dichotomy algorithm method; Section Five evaluates performances of dichotomy algorithm method against the other three; Section Six delivers our conclusions for this paper.

2 Related Work

Song and Liu [11] framed a data compression algorithm which is developed on the basis of track data curve fitting idea. Because the core technology of the algorithm is fixed partition fitting, the error is large. Karimbergki [12] proposed a method of trajectory fitting including curvature, direction, and position parameters of circle. The method is based on the explicit solution of nonlinear least squares problem, and the error estimation of trajectory fitting parameters is reliable. Liu et al. [13] proposed a trajectory information fitting algorithm with adaptive selection of step size. Although the algorithm improves the fitting accuracy, the fitting curve is discontinuous and the inflection point is obvious. Liang and Li [14] proposed an adaptive fitting algorithm for trajectory information based on the optimization of least squares method and constrained quadratic programming method, so as to automatically select the optimal fitting interval and generate the key points and coefficients of the fitting interval. Peng and Xiguo [15] deduced a trajectory fitting method based on genetic programming (GP) and ant colony algorithm. A large number of calculations show that the fitting method has clear physical relationship, high accuracy, and fast calculation speed. Lee and Xu [16] proposed a model which is very suitable for finding the best-fitting trajectory from many examples. The model is a spline smoother considering local velocity information. The existing smoothing design only considers the position information, but does not consider the local velocity information; so it is difficult to apply to the dynamic system with time smooth trajectory. Zhang and Liu [17] proposed a mathematical model based on polynomial fitting of sliding window. The model processes the historical data sequence of target position by polynomial fitting. The main idea is that every time the target's future position is calculated, the target data is updated with time, so as to achieve real-time prediction. This method overcomes the problem that the prediction error of general trajectory fitting increases with time. The method of trajectory fitting proposed by Ge and Chen [18] is based on Fourier series function to fit discrete trajectory points. Genetic algorithm is used to optimize the trajectory fitting model. The simulation results verify the correctness of the trajectory planning algorithm.

3 Preliminaries

The study of historical trajectory is of great significance. Generally, the original trajectory has various phenomena, such as aggregation, abnormality, round trip, and so on, which is helpful for judging and researching moving targets. It is inconvenient to judge a target (e.g., plane, ship) cross the border or break the law. So it is meaningful to select representative trajectory points and form simple and effective trajectory lines. In addition, for the sake of narrative convenience, we abbreviate "trajectory planning" as TP. In this section, we will summarize *system overview*, *random algorithm model*, *greedy algorithm model*, *clustering algorithm model*, and *dichotomy algorithm method* for trajectory planning.

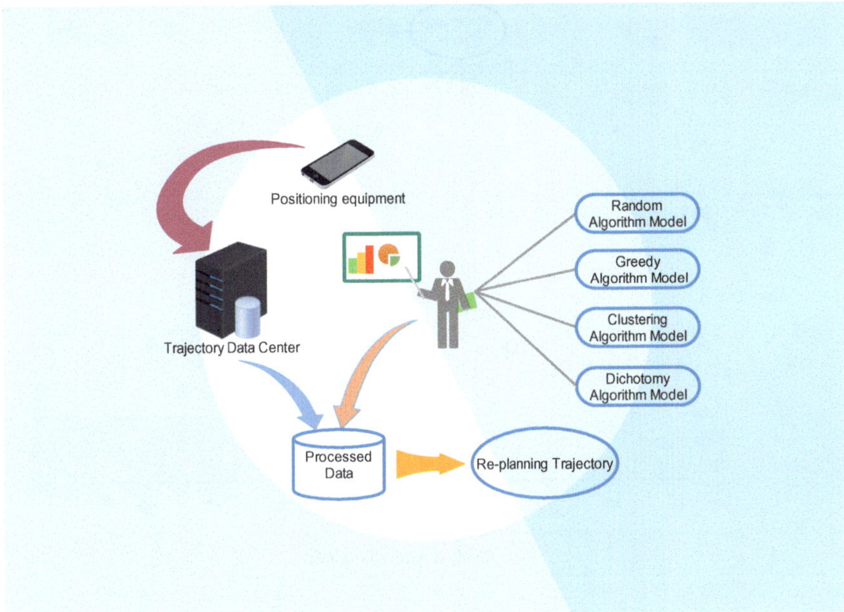

Fig. 1 System overview

3.1 System Overview

In this section, we will briefly introduce the framework of trajectory planning as shown in Fig. 1. The specific implementation process is as follows:

- Firstly, the trajectory information of specified time period in specified area is collected by positioning device or smart mobile device, and the data is provided to data analysis center.
- After analyzing and extracting the data, representative trajectory points can be selected, such as random algorithm model, clustering algorithm model, greedy algorithm model, and VQ-based dichotomy model.
- A new trajectory is formed by linking the selected trajectory point data with the corresponding time information. This simple and effective trajectory is convenient for studying the behavior pattern of moving targets.

3.2 Random Algorithm Model

The idea of random algorithm model is to randomly select k points from trajectory point set $\mathbb{P}$ and link them up according to the corresponding time information $\mathbb{T}$ of K points. The advantage of this method is that it is simple and fast, but because of its

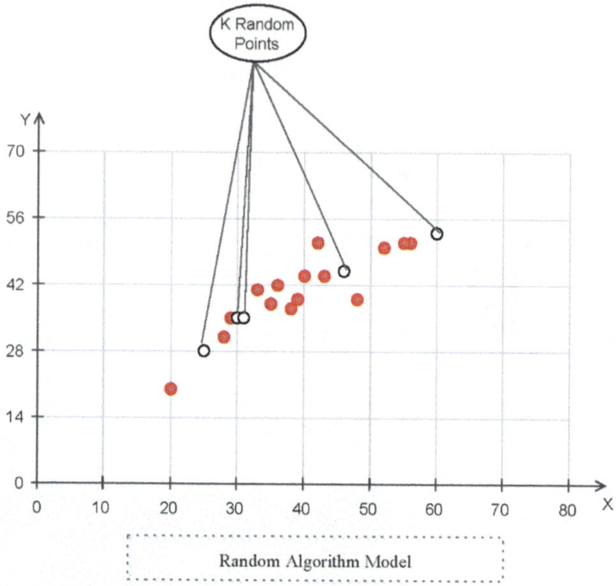

Fig. 2 Random algorithm model

randomness, it is difficult to capture key point information. Figure 2 shows a simple illustration of random algorithm model.

3.3 *Greedy Algorithm Model*

Greedy algorithm means that when solving a problem, always make what seems to be the best choice at present. That is to say, without considering the global optimum, the local optimal strategy will be selected. Figure 3 shows the computational process of the greedy model.

The aim of the greedy algorithm model is to continuously find the point with the highest density. Firstly, we assume that the scanning radius of each point as the center of the circle is r, and find the point with the highest density from the point set scanned and sequentially behind the center of the circle. The density of the point of p_i is the ratio of the number of points within the radius to the area of the circle, with p_i as the origin. Definition 1 gives the definition of distance.

Definition 1 (*Distance*). Suppose there are two trajectory points p_i, p_j in two-dimensional space. We define the coordinates of two points as (x_i, y_i), (x_j, y_j), and then the distance between these two trajectory points is

$$\mathrm{dis}(p_i, p_j) = \sqrt{(x_i - x_j)^2 + (y_i - y_j)^2} \tag{1}$$

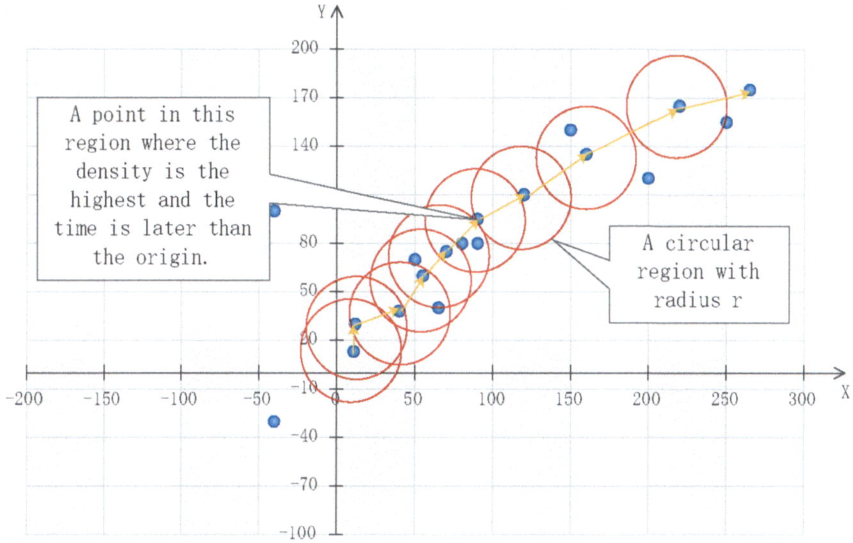

Fig. 3 Greedy algorithm model

In order to facilitate calculation, we put forward the concept of density weight. The specific definition is shown in Definition 2.

Definition 2 (*Density Widget*). Assuming that a point p_j is in a circular region with origin p_i and radius r, we define the weight of this point as 1. Otherwise, we think that the weight of this point is 0.

$$\omega(p_j)_{p_i,r} = \begin{cases} 1, & \text{dis}(p_i, p_j) \le r \\ 0, & \text{otherwise.} \end{cases} \quad (2)$$

According to Definition 2, we can generalize the lemma of Point Density in Lemma 1.

Lemma 1 (Point Density). *With point p_i as the origin and r as the radius, $\mathcal{P}_{-i}$ is a set of all trajectory points without p_i; in this circular area, we define that the density of this point p_i based on parameter r is the ratio of the number of elements in the region to the area of the region, that is*

$$\rho(p_i) = \frac{\sum_{j:p_j \in \mathcal{P}_{-i}} \omega(p_j)_{p_i,r}}{\pi r^2} \quad (3)$$

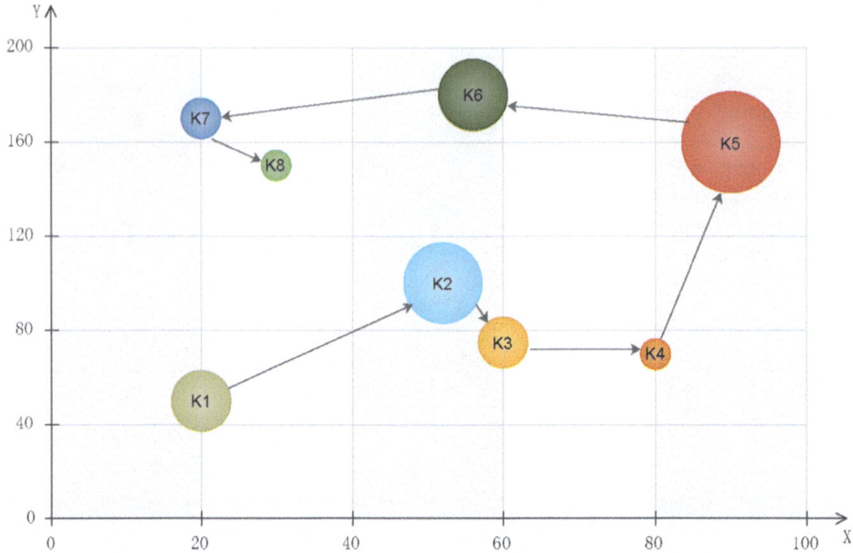

Fig. 4 Clustering algorithm model

3.4 Clustering Algorithm Model

Clustering algorithm is a clustering algorithm, so-called clustering; that is, according to the principle of similarity, data objects with high similarity are divided into the same cluster, and data objects with high similarity are divided into different clusters. The biggest difference between clustering and classification is that the clustering process is unsupervised; that is, the data objects to be processed do not have any prior knowledge; while the classification process is supervised, that is, there is a training data set with prior knowledge.

K-means algorithm is a partition-based clustering algorithm, which takes distance as the criterion of similarity measurement between data objects. That is, the smaller the distance between data objects, the higher their similarity, and the more likely they are in the same cluster. There are many ways to calculate the distance between data objects. K-means algorithm usually uses Euclidean distance to calculate the distance between data objects. The clustering details are shown in Fig. 4.

3.5 Dichotomy Algorithm Model

Dichotomy is a new algorithm proposed in this paper, which is mainly an extension of the idea of symmetric distribution. Paper [10] proposed a trajectory analysis algorithm based on vector quantization. This algorithm needs to assume that a grid covers the area to be analyzed, and then a simple clustering algorithm is used for each grid element to calculate the clustering center as the representative of the grid element area. In this paper, the proposed algorithm (dichotomy algorithm) also assumes that the trajectory point distribution area is covered by meshes; but for each mesh element, we use the idea of density distribution symmetry to find the best partitioning line. If the distribution of trajectory points is mainly concentrated around the partitioning line, then assuming that the intersection point between the splitting line and the element boundary of the mesh is a, b, and then the two closest trajectory points p_α and p_β will be the approximate optimal splitting line. The trajectory points p_α and p_β can be used as the representative points of the trajectory points set in the mesh area. If the distribution of the trajectory points is mainly concentrated in the normal direction of the partition line, then we will take the clustering center of the two partitioned points set as the representative points. The representative points obtained by this scheme cannot only describe the direction of the trajectory, but also be more representative.

The ideological framework of the model is shown in Fig. 5. Firstly, the central point of each grid element is determined to be the origin of the coordinate system, and the grid element region is divided into two equal parts A_1, B_1 by the linear function $y = kx$. With the change of slope k, we can find the most similar critical point for the distribution of two trajectory point sets P_α, P_β. The judgment of similar distribution of trajectory point set can be made by using the change rate of trajectory point set density. This algorithm is also a new algorithm proposed in this paper, which calculates the average change rate of trajectory point set density by continuously reducing the area. As shown in Fig. 5, flags 1, 2, and 3 represent three states. For area A_1, assuming that state 1 changes to state 2, the area reduction rate is ε, the area of state 1 is defined as $S(A_1)$, and the number of point sets of state 1 is $\mathcal{N}(A_1)$. Define the edge length of the grid element to be l. Then $S(A_1) = l^2/2$. It is easy to get that the area of state 2 and state 3 is $S(A_2) = (1-\varepsilon)S(A_1)$ and $S(A_3) = (1-\varepsilon)^2 S(A_1)$, respectively. Densities at states 1, 2, and 3 are $\rho(A_1) = \frac{\mathcal{N}(A_1)}{S(A_1)}$, $\rho(A_2) = \frac{\mathcal{N}(A_2)}{S(A_2)}$ and $\rho(A_3) = \frac{\mathcal{N}(A_3)}{S(A_3)}$, respectively.

Then the definition of the density (defined in Definition 3) used to judge the distribution similarity of the trajectory point set is as follows.

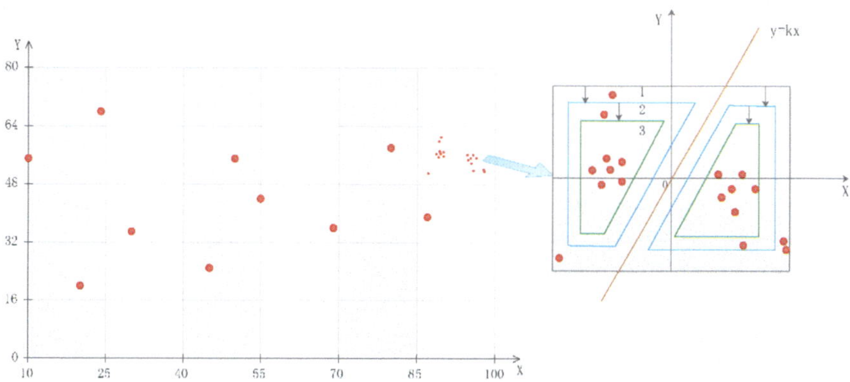

Fig. 5 Dichotomy algorithm model

Definition 3 (*Density Base on Area*). Suppose that a grid element region is divided into two parts, A_1 and B_1, by a linear function $y = kx$. For region A_1, if the initial state area is $S(A_1)$ and the number of trajectory points is $\mathcal{N}(A_1)$, then the area of A_1 is $S(Ar + 1) = (1 - \varepsilon)^m S(A_1)$ with the reduction rate ε changed by m times, and the number of current trajectory points is $\mathcal{N}(A_{m+1})$, the area name is A_{m+1}. The density base on area A_{m+1} can be expressed as the following formula:

$$\rho(A_{m+1}) = \frac{\mathcal{N}(A_{m+1})}{S(A_{m+1})} = \frac{2\mathcal{N}(A_{m+1})}{l^2(1-\in)^m} \tag{4}$$

In this section, we design a trajectory planning algorithm based on dichotomy. Assuming that the trajectory points $\mathbb{P} = \{p_1, \ldots, p_N\}$ in a region are covered by a grid, the corresponding time series is $\mathbb{T} = \{t_1, \ldots, t_N\}$, and the mesh size of the grid element is l. For trajectory points in each grid element, we want to draw a partition line and divide the trajectory points in each grid into two parts according to the distribution, so that the distribution of the two parts should be as similar as possible. Assuming that the center of the mesh is the origin of the Cartesian coordinate system, the partition line can be expressed as a linear function $y = kx$ with slope k, where $k = \tan\theta$ (θ is the angle to the X-axis). Details are shown in Fig. 1. Then, the intersection points of the linear function $y = kx$ and the meshes are taken as the basic points of trajectory analysis, and the two points p_i, p_j closest to the intersection point in the mesh are selected to represent the group of points in the mesh. How to judge whether the distribution of two parts of point groups is similar after the linear function $y = kx$ segmentation is the key to solve the problem. We measure the similarity of distribution by area reduction.

Define a set of trajectory points in a grid element as $\mathcal{P} = \{p_1, \ldots, p_n\}$, the corresponding time series is $\mathcal{T} = \{t_1, \ldots, t_n\}$. We measure the similarity of distribution by area reduction. Assume that the linear function $y = kx$ divides a grid element into two regions a and b, and their corresponding areas are $S(a)$ and $S(b)$, and we can get the lemma (Lemma 2) of point group distribution similarity.

Lemma 2 (PGD Similarity). *Suppose that the reduction rate of edge length in a and b regions is defined as ε, and that the density of point groups in a and b regions can be defined as ρ(a) and ρ(b). Assuming that the number of area reductions is defined as m, for region a, after m times of area iteration reduction, the total change of point-group density is* $\Delta\rho(a, m) = |\rho(a)_1 - \rho(a)_2| + \cdots + |\rho(a)_\iota - \rho(a)_{\iota+1}| + \cdots + |\rho(a)_m - \rho(a)_{m+1}|$, *where* $\rho(a)_1, \ldots, \rho(a)_{m+1}$ *is the density sequence corresponding to all changes* ($m > 1$). *That is* $\Delta\rho(a, m) = \sum_{j=1}^{m} |\rho(a)_j - \rho(a)_{j+1}|$. *Then the average density change rate about a is* $\overline{\Delta\rho}(a, m) = \frac{\Delta\rho(a,m)}{m}$. *Similarly, the average density change rate about b is* $\overline{\Delta\rho}(b, m) = \frac{\Delta\rho(b,m)}{m}$. *The difference of point group distribution similarity between a and b is* $\xi = \left|\overline{\Delta\rho}(a, m) - \overline{\Delta\rho}(b, m)\right|$, *namely*

$$\xi = |\sum_{j=1}^{m} |\rho(a)_j - \rho(a)_{j+1}| - |\rho(b)_j - \rho(b)_{j+1}|| \tag{5}$$

Table 1 Notations of follow algorithms

Notation	Description
$\mathbb{P}$	Set of all trajectory points in a region
$\mathbb{T}$	Time set of all trajectory points in a region
$\mathcal{P}$	A set of trajectory points in a grid element region
$\mathcal{T}$	A time set of trajectory points in a grid element region
N	Number of elements in P set
$\mathbb{P}'$	New set of trajectory points after processing
k, $\tan\theta$	Slope of mesh element segmentation line

4 Algorithm Analysis

In this paper, four kinds of trajectory analysis models are involved, which are random algorithm model (RAM), greedy algorithm model (GAM), clustering algorithm model (CAM), and dichotomy algorithm model (DAM). In this chapter, the implementation algorithms of these four models will be described in detail. Table 1 lists frequently used notations.

4.1 Random Algorithm

Random algorithm model, as its name implies, randomly extracts k trajectory points from the set of trajectories as representative points and links them up according to their time characteristics. Algorithm 1 (RAFTP) shows the algorithm process.

Algorithm 1: RAFTP

Input: $\mathbb{P}$, $\mathbb{T}$, k

Output: $\mathbb{P}'$

1. random k elements from trajectory point set $\mathbb{P}$;
2. drawing trajectory line based on time series $\mathbb{T}$;
3. return $\mathbb{P}'$;

4.2 Greedy Algorithm

Greedy algorithm cannot get the overall optimal solution for all problems, and the key is the choice of greedy strategy. The greedy strategy must have no aftereffect; that is, the process before a certain state will not affect the later state, only related to the current state.

In short, the main idea of the greedy algorithm for trajectory analysis is to select the starting point of p_j, scan the surrounding area with p_j as the starting point and r as the radius to find the maximum density point. For the definition of density, see Lemma 1. After each scan, the trajectory points in the scanning area that are later than the origin time will be removed from the trajectory set to avoid repeated calculation. Through continuous iteration, until the time does not meet the conditions; that is, no longer can find the trajectory point of later time. Algorithm 2 (GAFTP) shows the detailed algorithm steps of greedy algorithm model.

Algorithm 2: GAFTP

Input: $\mathbb{P}, \mathbb{T}, r$

Output: $\mathbb{P}'$

1. $\mathbb{Q} \leftarrow \mathbb{P}$

2. $t_{max} \leftarrow t_N$

3. Determine the starting point as $p_j \in \mathbb{Q}$

4. $\mathbb{P}' \leftarrow p_j$

5. While $t_j \leq t_N and \mathbb{Q} \neq \emptyset$ **do**

6. the set of trajectory points scanned from $\mathbb{Q}$ **with** p_j **as the center and** r **radius is as the radius is** $\mathcal{P}(p_j)$

7. $\mathbb{Q} \leftarrow \mathbb{Q} \ominus \backslash \mathcal{P}(p_j)$

8. filter out the trajectory-point set with times are later than t_j**, and get the set** $\mathcal{P}'(p_j)$

9. for each $p_t' \in \mathcal{P}(p_j)$ **do**

10. if $t(p_t') \geq t_j$ **then**

11. $\mathcal{P}'(p_j) \leftarrow p'$

12. $p_0 \leftarrow \max_{p_j \in \mathcal{P}'(p_j)} \rho\left(p_j'\right)$

13. $\mathbb{P}' \leftarrow p_0$

14. return $\mathbb{P}'$

4.3 Clustering Algorithm

Clustering algorithm aggregates the set of trajectory points $\mathbb{P}$ into k classes and calculates the central points of k clusters. Based on the time information of the central points, the selected trajectory points are joined together to form a new trajectory line. Algorithm 3 (CAFTP) shows the steps of clustering.

Algorithm 3: CAFTP

Input: $\mathbb{P}, \mathbb{T}, k$

Output: P'

1. Using the idea of K-means algorithm, the set $\mathbb{P}$ of trajectory points is aggregated into k classes

2. $\mathbb{P}' \leftarrow$ calculating k cluster centers

3. connect trajectory points to form a new trajectory based on the time information of clustering center

4. return $\mathbb{P}'$

4.4 Dichotomy Algorithm

Dichotomy algorithm model divides all trajectory points into grid elements according to the idea of vector quantization and then analyzes the set of trajectory points in each grid element. DAM detailed steps in Algorithm 4 (GMFTP). Assuming that the set of elements in a grid element is $\mathcal{P}$, the corresponding time information sequence is $\mathcal{T}$, and the slope of the point set function $y = kx$ is θ. Because we need to find the best dividing line by changing the slope, we need to determine the magnitude of the angle change, denoted as $\Delta\theta$. We need to use the concept of density change rate to compare the distribution of trajectory points in the two regions after segmentation. See Lemma 2 for the specific definition. This involves the concept of the number of regional area reductions used to calculate the rate of density change. Here we define the number of reductions as m. In Algorithm 4 (GMFTP), 2–6 rows are searched for the best segmentation line. Assuming that the segmented region is a and b, then 8–12 rows determine the region where the trajectory points belong. Lines 13–22 determine the points within the grid area that can represent the set of trajectory points in the region. All grid elements are calculated according to Algorithm 4.

Finally, according to the time information of the selected trajectory points, the trajectory points are joined together and a new trajectory line is fitted.

Algorithm 4: GMFTP

Input: $\mathcal{P}, \mathcal{T}, l, \theta, \Delta\theta, m$

Output: P'

1. $\theta \leftarrow 0, \phi \leftarrow 0, \xi_{min} \leftarrow +\infty$

2. While $\theta < 2\pi$ **do**

3. **if** $\xi(y = tan\theta \cdot x) < \xi_{min}$ **do**

(continued)

(continued)

Algorithm 4: GMFTP

4. $\xi_{min} \leftarrow \xi(y = tan\theta \cdot x)$

5. $\phi = \theta$

6. $\theta = \theta + \Delta\theta$

7. $k = tan\phi, \mathcal{P}(a) \leftarrow \emptyset, \mathcal{P}(b) \leftarrow \emptyset$

8. for each $p_i \in \mathcal{P}$ do

9. **if** $y_i \geq kx_i$ **then**

10. $\mathcal{P}(a) \leftarrow p_i$

11. **if** $y_i \leq kx_i$ **then**

12. $\mathcal{P}(b) \leftarrow p_i$

13. $P \leftarrow \emptyset, \mathcal{I} \leftarrow \emptyset$

14. if $\sum_{i:p_i \in \mathcal{P}} \frac{|kx_i - y_i|}{\sqrt{k^2+1}} \leq \sum_{i:p_i \in \mathcal{P}} \frac{|x_i + ky_i|}{\sqrt{k^2+1}}$ **then**

15. the intersection set of linear function $y = kx$ **and** $y = \pm l/2, x = \pm l/2$ **is** $\mathcal{I} = \{(x_1, y_1)\ldots\}$

16. for each $I_\varsigma \in \mathcal{I}$ do

17. **//** $(x_\varsigma, y_\varsigma)$ **is the coordinate of** I_ς **with the center of the grid element as its origin.**

18. **if** $x_\varsigma, y_\varsigma \in [-l/2, l/2]$ **then**

19. $p' \leftarrow \arg_{p_\gamma : p_\gamma \in \mathcal{P}} mindis(I_\varsigma, p_\gamma)$

20. $P \leftarrow p'$

21. else

22. $P \leftarrow \mathcal{P}(a) clusteringcenter, P \leftarrow \mathcal{P}(b) clusteringcenter$

23. return P

5 Performance Evaluation

5.1 Simulation Setup

In order to better see the performance of the trajectory analysis model, we use simulation experiments to test. We set up two sets of data. In Set one, the *X*, *Y* coordinates of the trajectory points are randomly within [0, 200], and the incremental step size is 10 at each time. In order to make the effect more real, we make the step size fluctuate within [−8, 2]. For stochastic model and clustering model, we define that the number of substituted trajectory points is 15, the scanning radius of greedy algorithm is 15, and the pixel size of dichotomy grid is 15. In Set two, we expand the data quantity of trajectory points. The *X*, *Y* coordinates of trajectory points are random in [0, 400]. Each random incremental step is 10, and the fluctuation range of step size is [−10,2].

Table 2 Parameter setting

Setting	X	Y	Step	Ware	k	r	l
I	[0, 200]	[0, 200]	10	[−8, 2]	15	15	15
II	[0, 400]	[0, 400]	10	[−10, 2]	30	30	30

The number of representative points selected by random model and clustering model is 30, the scanning radius of greedy algorithm is 30, and the pixel size of dichotomy grid is 30. Detailed parameters are given in Table 2.

5.2 *Simulation Results*

Figure 6 shows the effect of trajectory fitting of random algorithm model, in which original represents the original trajectory. From Fig. 6a, b, we can find that although RAM can simplify trajectory to a great extent, it largely ignores the original characteristics of trajectory. For example, the random model does not capture the broken line feature at the passing point (150, 100) in Fig. 6a and the bending feature at the passing point (50, 100) in Fig. 6b.

Figure 7a, b shows the trajectory fitting effect of the greedy model. From Fig. 7a, we find that if there are no outliers or clutter in the trajectory, the greedy strategy can well show the characteristics of the original trajectory. On the contrary, in Fig. 7b, there will be the same shortcomings as the random model, ignoring some characteristics of the original trajectory. Figure 8 shows the trajectory fitting effect of the clustering model. It is easy to find that the model algorithm can capture the region with dense trajectory points and achieve better trajectory simplification effect. However, the model cannot express the characteristics of discrete points very well. Figure 9 shows the trajectory fitting effect of the dichotomy model. We find that the

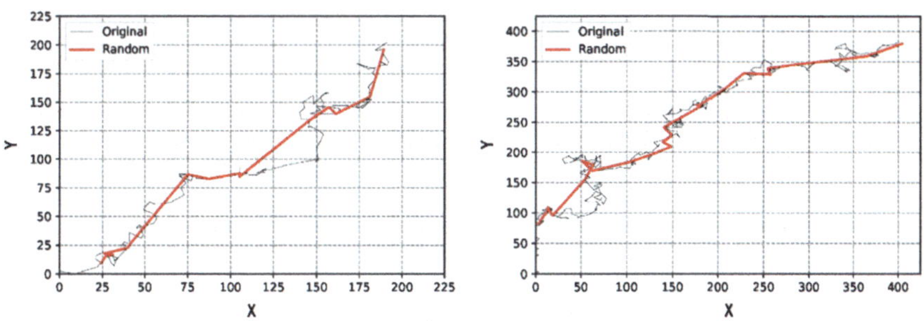

(a) Random Model under setting I (b) Random Model under setting II

Fig. 6 Random model for trajectory fitting

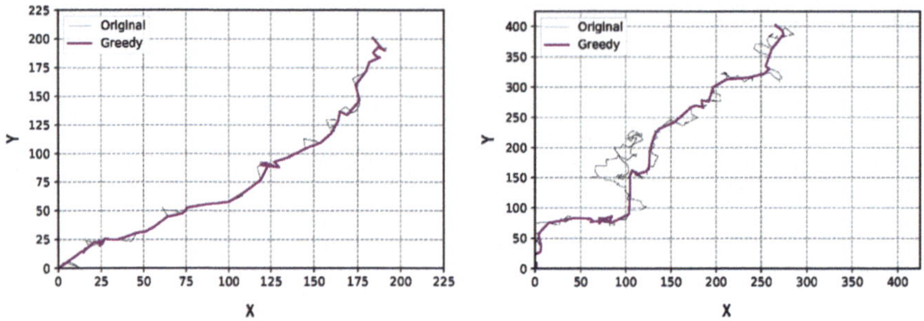

(a) Greedy Model under setting I (b) Greedy Model under setting II

Fig. 7 Greedy model for trajectory fitting

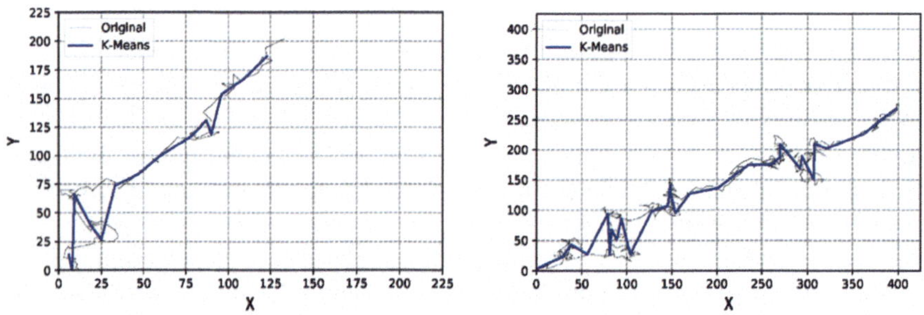

(a) Clustering Model under setting I (b) Clustering Model under setting II

Fig. 8 Clustering model for trajectory fitting

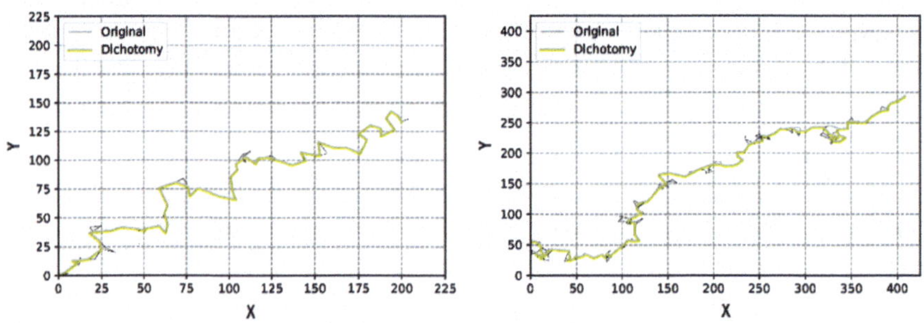

(a) Dichotomy Model under setting I (b) Dichotomy Model under setting II

Fig. 9 Dichotomy model for trajectory fitting

fitting effect is close to the original trajectory and can well reflect the trajectory direction and inflection point characteristics. There is no obvious abnormal phenomenon. Generally speaking, the experimental results show that compared with the other three models, the algorithm model has obvious advantages.

6 Conclusion

In this paper, we design random algorithm model (RAM), greedy algorithm model (GAM), clustering algorithm model (CAM), and dichotomy algorithm model (DAM) for trajectory analysis. Through these four algorithms, representative trajectory point sets are selected. According to the corresponding time information, a new trajectory is fitted. Experiments show that RAM has high efficiency and low-time complexity, but RAM cannot describe the trajectory well. If the trajectory is not complicated, then GAM will be a good choice as it is capable of updating with the current state information, because GAM can make the best strategy that conforms to the current state. CAM has a good performance in dense cases. The DAM is the key research algorithm in this paper. It solves the problem from the point of view of the distribution of trajectory points. This method can combine the ideas from VQ and clustering methods with a dynamic strategy to deal with the change of point density, which allows it to be powerful for studying the trajectory pattern.

References

1. Feng Z, Zhu Y (2017) A survey on trajectory data mining: techniques and applications. IEEE Access 4:2056–2067
2. Kong F, Lin X (2018) The method and application of big data mining for mobile trajectory of taxi based on MapReduce. Cluster Comput 6:1–8
3. Shan J, Ferreira J, Gonzalez MC (2017) Activity-based human mobility patterns inferred from mobile phone data: a case study of Singapore. IEEE Trans Big Data 3(2):208–219
4. Yuan G, Sun P, Zhao J et al (2017) A review of moving object trajectory clustering algorithms. Artif Intell Rev 47(1):123–144
5. Krishnan S, Garg A, Patil S et al (2017) Transition state clustering: unsupervised surgical trajectory segmentation for robot learning. Int J Robot Res 36(13–14):027836491774331
6. Xia D, Lu X, Li H et al (2018) A MapReduce-based parallel frequent pattern growth algorithm for spatiotemporal association analysis of mobile trajectory big data. Complexity 2018:1–16
7. Wen-Bo HU, Wei H, Guo-Chao HU (2017) Trajectory adjoint pattern analysis based on OPTICS clustering and association analysis. Comput Modernization
8. Kong X, Song X, Xia F et al (2017) LoTAD: long-term traffic anomaly detection based on crowdsourced bus trajectory data. World Wide Web-Internet Web Inf Syst (3):1–23
9. Mao JL, Jin CQ, Zhang ZG et al (2017) Anomaly detection for trajectory big data: advancements and framework. J Softw
10. Xu N, Yi W et al (2019) Vector quantization: timeline-based location data extraction and route fitting for crowdsourcing. In: Proceedings of the 5th China High Resolution Earth Observation Conference, CHREOC 2018. Lecture notes in electrical engineering, vol 552, pp 28–36
11. Song Y, Liu LM, Han ZZ (2017) A track data compression method for use on radar. Electron Opt Control 24:89–92 + 98
12. Karimki V (1991) Effective circle fitting for particle trajectories. Nucl Instrum Methods Phys Res 305:187–191
13. Liu FZ, Li HQ, Xiao B (2017) An adaptive track fitting algorithm. J Air Force Early Warning Acad 31:424–426 + 435
14. Liang F, Li H (2019) Research on an improved adaptive fitting algorithm of trajectory information. J Phys Conf Ser 1169:1–6

15. Peng Q, Guo B, Zhu J (2018) Trajectory fitting of aerial bomb based on combination of genetic programming and ant colony optimization. In: Proceedings of the 37th Chinese control conference, 25–27 July 2018, Wuhan, China, pp 4843–4848
16. Lee C, Xu Y (2000) Trajectory fitting with smoothing splines using velocity information. In: Proceedings-IEEE international conference on robotics and automation, vol 3, pp 2796–2801
17. Zhang S, Liu Y (2003) Prediction of moving target trajectory with sliding window polynomial fitting. Opto-Electron Eng 30(4):24–27
18. Ge L, Chen J, Li R (2017) Feedforward control based on Fourier series trajectory fitting method forfor industrialindustrial robot. In: Chinese control and decision conference, 28–30 May 2017

System Design for an Improved SPIDER Imager

Guomian Lv, Yueting Chen, Huajun Feng, Zhihai Xu, and Qi Li

Abstract The segmented planar imaging detector for electro-optical reconnaissance (SPIDER) based on interferometric imaging can greatly reduce the mass and volume of the system while maintaining the same resolution as traditional optical imaging system. However, only in single wavelength operative mode can the system ensure the sampled frequency is an integral multiple of the fundamental frequency. Under this circumstance, the system's imaging quality is poor due to low radial sampling rate of the system's spectrum and a lack of zero frequency. In order to solve this problem, an improved scheme of SPIDER system with interference arms with lens number odd–even alternately distributed is proposed, meanwhile giving its specific lens arrangement and matching mode. The proposed scheme can sample zero frequency and effectively improve radial sampling rate of the system's spectrum. At the same time, it has the advantage of strong versatility, almost no increase in hardware cost and no sacrifice of the field of view. Through computer simulation, it is shown that the improved SPIDER system can achieve better imaging results than the traditional SPIDER system when the number of interference arms is the same.

Keywords Interference imaging · Interference arm optimization scheme · Baseline pairing method · Photonic integrated circuit

1 Introduction

In order to realize the miniaturization and lightweight of imaging system and promote the development of space-based system, Lockheed Martin first proposed the segmented planar imaging detector for electro-optical reconnaissance (SPIDER) [1–3] based on the principle of interferometric imaging in 2012. The system consists of lens array, photonic integrated circuit (PIC) chip and signal processing unit. The front-end small lens array is used to replace the traditional lens group. The optical signals collected by lens pairs are coupled into the photonic integrated circuit chip

G. Lv · Y. Chen (✉) · H. Feng · Z. Xu · Q. Li
State Key Laboratory of Optical Instrumentation, Zhejiang University, Hangzhou, China
e-mail: chenyt@zju.edu.cn

L. Wang et al. (eds.), *Proceedings of the 6th China High Resolution Earth Observation Conference (CHREOC 2019)*, Lecture Notes in Electrical Engineering 657,
https://doi.org/10.1007/978-981-15-3947-3_18

using waveguides and then divided into several narrow bands by the arrayed waveguide grating (AWG). The coherence is guaranteed by phase regulator. The coherent optical signal eventually enters the signal processing unit and the quadrature detector outputs the photocurrent containing the spectrum information of the imaging target. The intensity distribution of the target can be obtained by further processing of the optical current. The working principle of PIC chip is detailed in Ref. [4].

The lens arrangement of SPIDER imaging system [2] is shown in Fig. 1. Odd number of interference arms is distributed in a radial shape with the same angle of interval, and each interference arm has even lenses on it. A pair of lenses symmetrical to the center of an interference arm is matched to compose a baseline. The advantage of this structure is that the baseline design, PIC chip handling and interference arm assembling are easy, and the system can rotate different angles and image many times to collect spectrum information of the target in more directions, and then fuse the information to improve the imaging quality. The simulation imaging results of the above system are given in Ref. [5]. After that, Su et al. [6] built an optical test platform according to the structure of SPIDER system and obtained the actual imaging results of SPIDER system by rotating a single PIC chip for many times, which verified the correctness of the interferometric imaging theory.

Reference [7] studied the distribution of sampled spatial frequency (u, v) in the interferometric imaging system and pointed out that the number, arrangement and matching scheme of lens used in the system determine the spatial frequency that the system can obtain. A pair of lenses with baseline length B can sample spatial frequency of $\mu = B/\lambda Z$, where λ is the working wavelength of the system, and Z is the distance between the target and the system. Obviously, the spoke interference arm distribution of SPIDER system cannot get all frequency information of the target, and

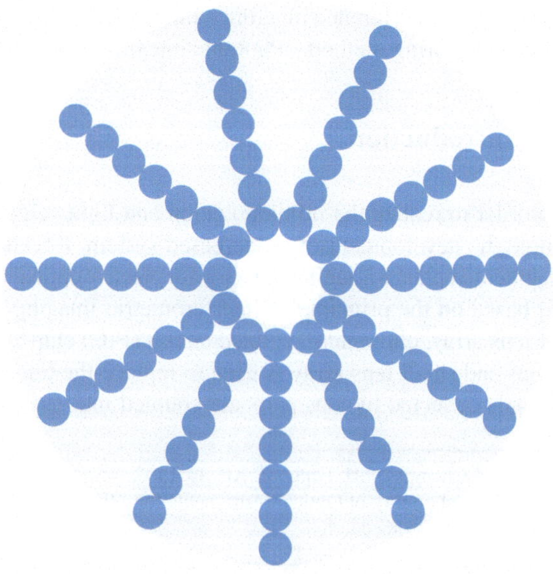

Fig. 1 Lens arrangement of SPIDER imaging system

the uniform baseline distribution will actually cause uneven sampling in frequency domain. In view of this disadvantage, the structure of SPIDER system is improved to have adjustable baseline pairing methods. According to the information distribution characteristics of the target in frequency domain, the baseline pairing method with the best imaging quality is automatically selected among uniform sampling, low-frequency sampling and high-frequency sampling, which expands shootable scenes of SPIDER system [8]. However, this improved scheme of SPIDER system needs to design different pairing methods for different number of lens arrays, which leads to poor versatility. Meanwhile, because of the need to constantly analyze the shooting target and adjust the baseline pairing method, the imaging time is very long and the manufacturing is very difficult. At the same time, zero frequency and frequency information in some directions are still not sampled, and in order to make each spatial frequency sampled integer multiple of the fundamental frequency, it is necessary to make the system operate in single wavelength mode. At this time, the traditional SPIDER system can only sample the information of odd multiple of the fundamental frequency. The improvement scheme of SPIDER system proposed in Ref. [9] can solve the above problems perfectly. Under the single wavelength mode, the system can achieve uniform sampling of the frequency information of the target, while sampling the zero-frequency information to improve the imaging quality. The imaging speed is also very fast. But this array arrangement greatly increases the number of lenses used and the size of PIC chips behind, and because the matching of lenses is no longer only in one dimension, the traditional PIC chips are no longer applicable, and the corresponding three-dimensional PIC chips need to be used [9]. This increases the difficulty and cost of system manufacturing and makes it impossible to further improve the imaging quality by rotating the system with different angles and image multiple times.

In order to solve traditional SPIDER system's problems of zero-frequency missing and even multiple of fundamental frequency information missing under single wavelength mode, while maintaining the advantages of traditional SPIDER system, such as simple structure, easy fabrication, low cost and availability of multi-imaging after rotation, this paper presents an improved scheme of SPIDER system using interference arms with odd–even alternating number of lenses. The improved scheme of SPIDER system is basically consistent with the structure of traditional SPIDER system, and only half of the interference arms have one more lens than before, but it can sample all integer multiples of the fundamental frequency in the direction of the interference arm and obtain zero-frequency information. At the same time, the system can rotate different angles and image multiple times to obtain frequency information in more directions, which can effectively improve the system's imaging quality.

The content of this paper will be as follows: Chap. 2 introduces the specific lens arrangement and baseline pairing method of the proposed SPIDER system improvement scheme and proves that this scheme can realize the sampling of all integer multiples of the fundamental frequency in the direction of the interferometer arm and obtain zero-frequency information; Chap. 3 uses computer program to simulate and compare the imaging results of the traditional SPIDER system and the proposed improved SPIDER system, and it is verified that the proposed improved scheme can achieve better imaging quality.

2 The Improvement Scheme of SPIDER System Using Interference Arms with Odd–Even Alternating Number of Lenses

According to the previous analysis, there are some defects in both the traditional SPIDER imaging system and the improved SPIDER imaging system with a checkerboard lens array. The lens arrangement of SPIDER imaging system using interference arms with odd–even alternating number of lenses presented in this paper is shown in Fig. 2.

Different from the traditional SPIDER system in which all interference arms use the same lens arrangement and pairing method, only half of the interference arms in the improved scheme proposed in this paper still use the same interference arms with even lenses as before, as shown in Fig. 2, the black interference arms, while the other half have one more lens than traditional even lens interference arms, as shown in Fig. 2, the red interference arm, and the two interference arms are alternately distributed. Figure 3a, b shows the baseline pairing methods of classical even lens interference arm *A* and the new odd lens interference arm *B* in Fig. 2, respectively.

It can be seen that the pairing method of the two types of interference arms is the same, both are central symmetrical matching, but there is an additional lens in

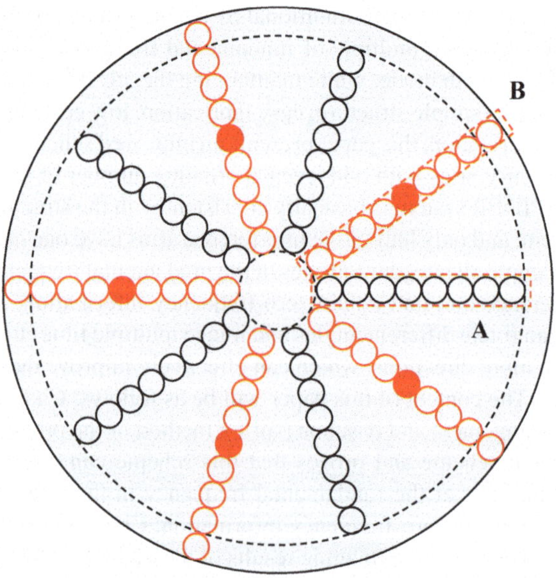

Fig. 2 Lens arrangement of SIDER system using interference arms with odd–even alternating number of lenses

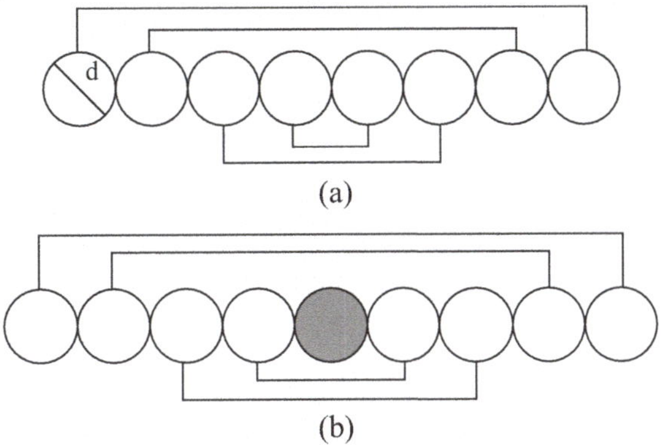

Fig. 3 **a** Baseline pairing method of traditional interference arm; **b** baseline pairing method of new interference arm

the center of the new interference arm. From the deduction in Chap. 1, it can be concluded that the spatial frequency of the spectrum obtained from the processed signal of the paired lenses is as follows:

$$\mu = \frac{L_0}{\lambda Z}. \quad (1)$$

where L_0 is the baseline length, that is, the distance between the two lens centers, λ is the working wavelength, and Z is the distance between the target and the system. Baselines of different lengths in the system can obtain a series of spectrums with discrete spatial frequency values. In order to restore the target image, these spectral values need to be reconstructed into a two-dimensional discrete spectrum matrix, as shown in Fig. 4, where zero frequency is placed at the center of the matrix, as shown

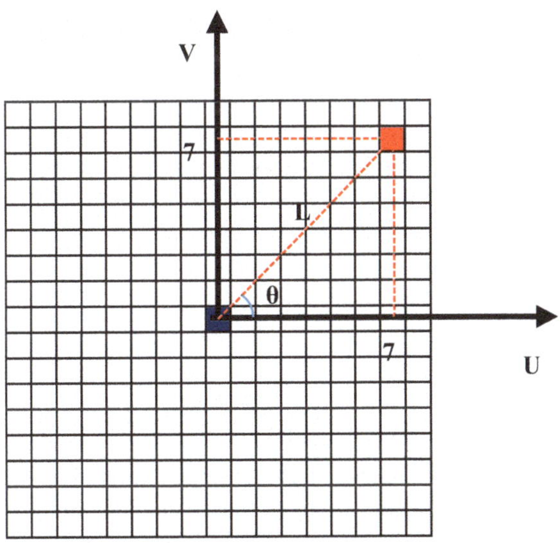

Fig. 4 Schematic diagram of spectrum reconstruction

in the blue square, and L represents the baseline length corresponding to a spectrum point and θ the angle between the baseline and the horizontal axis.

In order to locate the rest of the spectrum points in the matrix, it is necessary to first calculate the minimum spatial frequency that the system can sample. Assuming that the lenses are arranged tightly without intervals (when there are intervals, the same conclusion can be drawn. The assume is made for the sake of describing convenience), the diameter of the lens is d, and the system operates in single wavelength mode and the operating distance of the system remains unchanged, the minimum spatial frequency $\mu_{\min}$ that the system can obtain is determined by the shortest baseline $B_{\min}$:

$$\mu_{\min} = \frac{B_{\min}}{\lambda Z} = \frac{d}{\lambda Z} \tag{2}$$

This very spatial frequency is the fundamental frequency when reconstructing the spectrum, and its position in the matrix is next to zero frequency closely. For baseline with length L, the corresponding spatial frequency is:

$$\mu = \frac{L}{\lambda Z} \tag{3}$$

Divide μ by fundamental frequency $\mu_{\min}$, the distance between the frequency point sampled by this baseline and the origin (zero frequency) in the matrix $\Delta\mu$ can be obtained:

$$\Delta\mu = \frac{\mu}{\mu_{\min}} = \frac{L}{d} \tag{4}$$

Then, we can determine the coordinates Δu, Δv of the spectral point in the matrix:

$$\begin{cases} \Delta u = \Delta\mu \cdot \cos\theta \\ \Delta v = \Delta v \cdot \sin\theta \end{cases} \tag{5}$$

From Fig. 3a, the baseline length of the traditional interference arm is: $L_1 = [d, 3d, 5d, 7d]$. Equation 4 shows that the spatial frequencies it can sample will be fundamental frequency, triple frequency, quintuple frequency and seventh frequency, and the field of view for best imaging is given by the reciprocal of the minimum frequency sampled by the system:

$$\Delta\mu_1 = \frac{\mu_1}{\mu_{\min}} = \frac{L_1}{d} \tag{6}$$

$$\mathrm{Fov}_1 = \frac{1}{\mu_{\min}} = \frac{\lambda Z}{d} \tag{7}$$

This means that if all the interference arms adopt this baseline pairing method, the eventually sampled spectrum in each direction of the interference arm will lose even multiples of fundamental frequency, i.e., spatial frequency information of $\Delta\mu = 2, 4, 6, 8$, resulting in poor imaging quality.

The newly proposed interference arm first has a single lens in the middle, which can collect the zero frequency which contains the average brightness information of the target. Secondly, although the pairing method of the lens remains unchanged, the baseline length of the new interferometer arm is changed to $L_2 = [2d, 4d, 6d, 8d]$, which can acquire even multiples of fundamental frequency. Specific deductions are as follows:

$$\mu'_{\min} = \frac{B'_{\min}}{\lambda Z} = \frac{2d}{\lambda Z} \tag{8}$$

$$\Delta\mu_2 = \frac{\mu_2}{\mu'_{\min}} = \frac{L_2}{2d} = [1, 2, 3, 4] \tag{9}$$

$$\mathrm{Fov}' = \frac{1}{\mu'_{\min}} = \frac{\lambda Z}{2d} = \frac{1}{2} Fov_1 \tag{10}$$

It is clear that if the system only uses the new interference arms, although the system can sample the spatial frequency information of the continuous integer multiples of fundamental frequency, the frequency points sampled by the system do not increase in the radial direction. At the same time, because the fundamental frequency becomes twice larger than before, the field of view for best imaging becomes half as before. Thus, adopting all the new interference arms in the SPIDER system is still not an ideal option.

However, if the new and old interference arms are used alternately in the system, the system will have all the baselines in L_1 and L_2, with $L_3 = [d, 2d, 3d, 4d, 5d, 6d, 7d, 8d]$. At this time, the basic frequency, sampling frequency point and field of view of the system are as follows:

$$\mu''_{\min} = \mu_{\min} = \frac{d}{\lambda Z} \tag{11}$$

$$\Delta\mu_3 = \frac{\mu_3}{\mu''_{\min}} = \frac{L_3}{d} = [1, 2, 3, 4, 5, 6, 7, 8] \tag{12}$$

$$\mathrm{Fov}'' = \mathrm{Fov}_1 \tag{13}$$

It can be seen that the system can sample the spatial frequency information of continuous integer multiples of fundamental frequency, at the same time collect zero frequency, and the imaging field remains unchanged. Meanwhile, the number of baselines in each direction of the interference arm is doubled, the radial sampling rate of the spectrum increases and the final imaging quality becomes higher. Figure 5 compares the spectrum matrix reconstructed from the spectrum points obtained by the SPIDER system using only the traditional interference arms, only the new interference arms and both interferometer arms with alternating distribution under the same interference arm number.

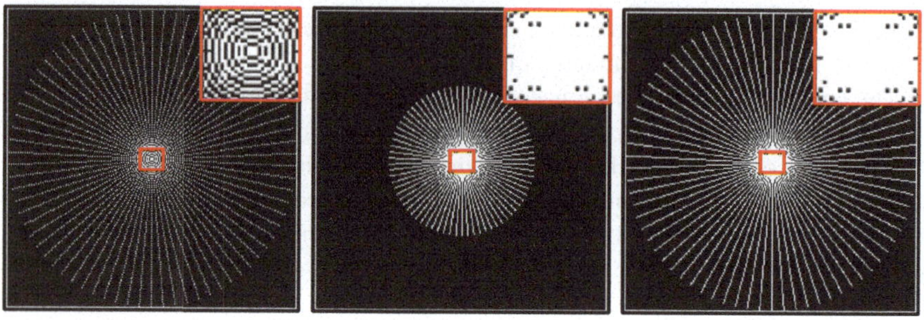

Fig. 5 Spectrum matrix obtained by SPIDER system with different arrangements of interference arms. **a** The traditional interference arm is used; **b** the new interference arm is used; and **c** both types of interference arms are used

It will be proved that in order to achieve the optimal imaging quality of SPIDER system using interference arms with odd–even alternating number of lens, the same number of traditional interference arms and new interference arms are needed, and the number of both interference arms needs to be odd. Firstly, a theorem is quoted that the brightness distribution of an object is generally a real function, and the Fourier transform of the real function satisfies the following properties:

$$F(x, y) = F^*(-x, -y) \tag{14}$$

That is to say, as long as the Fourier transform of one point is obtained, the Fourier transform of its conjugate point can also be obtained. Therefore, in traditional SPIDER system, in order to obtain the spectrum information in as many directions as possible, the total number of interference arms is set to odd number, so that the number of directions of frequency information sampled is twice the number of interference arms. If the improved SPIDER system also uses odd number of interference arms, the number of new and tradition interference arms will be different, and the information sampled by the two interference arms cannot complement each other perfectly. In fact, the number of interference arms in the improved SPIDER system has three situations. The following is a detailed analysis and comparison.

2.1 *The Total Number of Interference Arms Is Odd*

In this situation, the arrangement of interference arms is shown in Fig. 6. The odd and even in the figure represent that the corresponding interference arm has odd or even lenses. The dotted red lines indicate that although there is no lens in this direction, the spectrum information can still be calculated using Eq. 14.

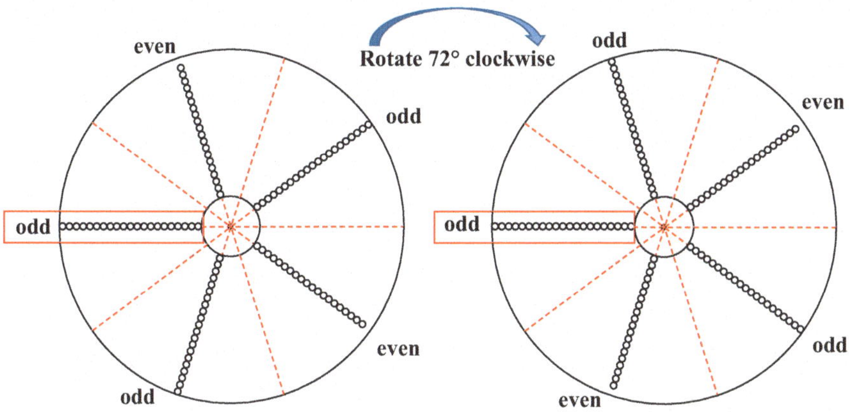

Fig. 6 Lens arrangement when the system's number of interference arms is odd. The right structure is the result of 72° clockwise rotation of the left structure

Here, three interference arms with odd number of lenses (hereinafter referred to as odd interference arms) and another two interference arms with even number of lenses (hereinafter referred to as even interference arms) are used in the system. Although from Eq. 14, the system can sample the frequency information in the direction of 10 interference arms (the combination of lenses and dotted lines), the direction of odd interference arm lacks the information sampled by even interference arm, and so does the latter. In order to complement each other perfectly, the system needs to rotate 72° clockwise, that is, the interval between two interference arms, to make the position interchange of even and odd interference arms. Then, a secondary imaging of the target is implemented, and the information acquired from two imaging process is fused. In this case, the information of all interference arms can be perfectly fused except the interference arm framed in the red box in Fig. 6, where the odd interference arm remains unchanged after the rotation, so the direction of this interference arm still lacks information sampled by even interference arm. It can be seen that when the number of interference arms is odd, the information sampled by even and odd interference arms cannot be perfectly complemented by fusing the imaging information before and after rotation.

2.2 *The Number of Even and Odd Interference Arms Is Both Even*

In this case, it is obvious that the number of even and odd interference arms is the same, and the total number of interference arms can be quadrupled. The interference arm arrangement is shown in Fig. 7.

Here, there are four odd interference arms and four even interference arms, and each kind of interference arm is centrally symmetrical. Therefore, it can be seen that

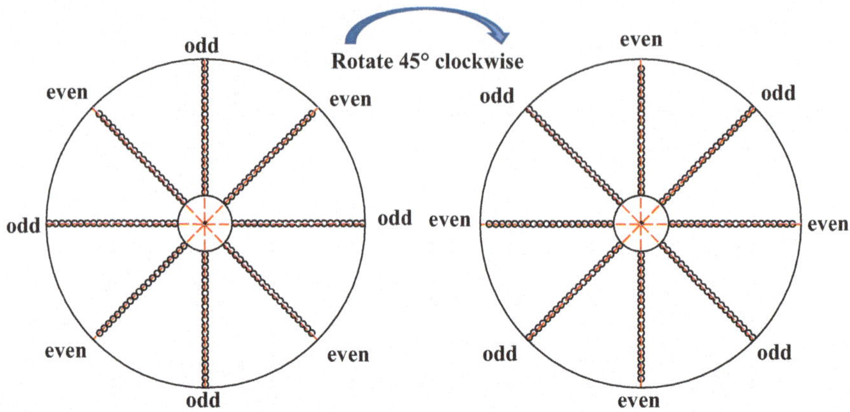

Fig. 7 System's lens arrangement when the number of odd and even interference arms is even. The right structure is the result of 45° clockwise rotation of the left structure

the extended line direction of odd interference arms is still odd interference arms, and so are even interference arms. This means that using this structure, we can neither obtain spectrum information in more directions, nor make the information of even and odd interference arms complement each other using Eq. 14. Compared with the five-interference-arm system in Sect. 2.1, which can sample the spectrum information in ten directions, the structure presented in this section has eight interference arms but can only sample spectrum information in eight directions. At the same time, in order to complement the information of odd and even interferometer arms, it is still necessary to rotate the interval between two interference arms and image a secondary time. Although this structure can integrate the information before and after rotation perfectly, the utilization rate of the interference arms is low, and it is impossible to obtain spectrum information in more directions using Eq. 14. To sample the same number of spectrum points, more interference arms or imaging times are needed than other structures, which increase the time and cost consumption.

2.3 *The Number of Even and Odd Interference Arms Is Odd*

In this case, the number of odd and even interference arms is the same, the total number of interference arms is even and each odd interference arm is centrally symmetrical with an even interference arm, as shown in Fig. 8.

At this time, although the structure cannot obtain spectrum information in more directions using Eq. 14, each odd interference arm is centrally symmetrical with an even interference arm, so that the spectrum information sampled by the two arms can complement each other perfectly in every interference arm's direction, thus realizing the information sampling of continuous integer multiple frequency of

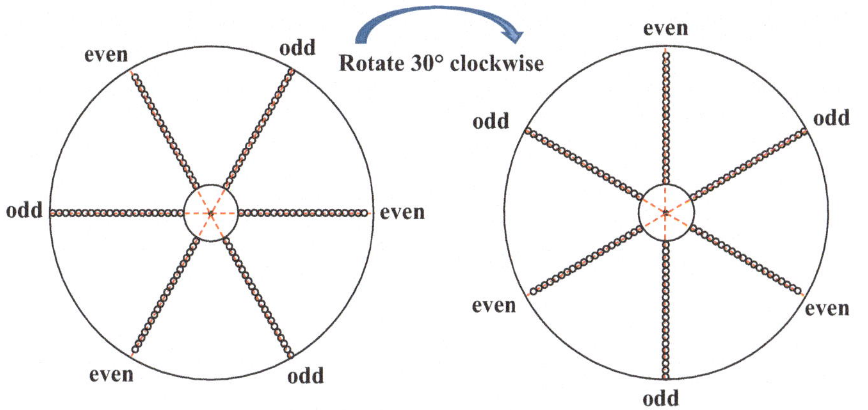

Fig. 8 System's lens arrangement when the number of odd and even interference arms is odd. The right structure is the result of 30° clockwise rotation of the left structure

fundamental frequency. If the frequency information in more directions is needed, we can simply let the system rotate half of the angle between two adjacent interference arms and image again. This structure can not only optimize the frequency sampling in the interferometer arm's direction, but also collect frequency information in more directions by fusing the imaging information before and after rotation. It has the best performance among the above three structures.

2.4 *Summary*

This section describes in detail the improvement scheme of SPIDER imaging system using interference arms with odd–even alternating number of lenses. It proves that the improved SPIDER imaging system can sample the spectrum information of continuous integer multiple fundamental frequencies and at the same time acquire zero frequency without decreasing the field of view or increasing the cost. Sections 2.1–2.3 qualitatively analyzes the three kinds of interference arm arrangements that may be used in the improved SPIDER system and draws the conclusion that when the number of even and odd interference arms is odd and the same, the system can achieve the optimal balance among imaging quality, manufacturing cost and imaging time. Assuming that each even interference arm has N baselines and can sample N frequency points, each odd number interference arm can sample $N + 1$ frequency points. Table 1 compares in detail all the non-repetitive spectrum points the system can obtain by imaging once and twice under the different arrangements of interference arms described in Sects.As shown in bold, scheme 2 owns the maximum baseline number, while scheme 3 has the best spectrum acquisition capability and baseline utilization ratio. 2.1–2.3 and calculates the average spectrum points

Table 1 Spectrum acquisition capability and baseline utilization ratio of the system under different interference arm schemes

Scheme number	Baseline number	Spectrum acquisition capability		Baseline utilization ratio	
		Image once	Image twice	Image once	Image twice
Scheme 1	5N + 3	10N + 6	18N + 10	2	<4
Scheme 2	**8N + 4**	8N + 4	16N + 8	1	2
Scheme 3	6N + 3	**12N + 6**	**24N + 12**	**2**	**4**

sampled by each baseline in each case to measure the system's spectrum acquisition capability and baseline utilization rate. For the convenience of description, the arrangement schemes in Sects. 2.1–2.3 are called scheme 1, scheme 2 and scheme 3, respectively. It can be seen that scheme 3 has both the best spectrum acquisition capability and the highest baseline utilization ratio.

3 Comparison of the Systems' Simulation Imaging Effects

In this chapter, the traditional SPIDER system and the improved SPIDER system are simulated by computer program, and the imaging results are compared and analyzed. The specific structure parameters of the traditional SPIDER system are shown in Table 2. There are 47 interference arms (odd number is chosen here to obtain frequency information in more directions using Eq. 14), and each one has 128 lenses. There is no interval between lenses, so the lenses are closely arranged. The diameter of the lens is set to 1 mm. In order to achieve the highest coupling efficiency between lens and waveguide, the relative aperture of the lens needs to be equal to 0.2, and therefore, the focal length of the lens can be calculated to be 5 mm. Due to the optical waveguide has less attenuate in near-infrared band, the system's operating band is chosen in the near-infrared band, and because the system works in single wavelength

Table 2 Specific structural parameters of the traditional SPIDER system

Parameter	Values
Working wavelength (nm)	1400
Imaging distance (km)	900
Spectral number of AWG	1
Lens diameter (mm)	1
Lens interval	0
Focal length of the lens (mm)	5
Number of interference arms	47
Lens number on each interference arm	128

mode, the spectral number of the arrayed waveguide grating is set to one. The working distance of the system is set to 900 km to simulate remote sensing imaging of the earth.

The number of the improved SPIDER system's interference arms is set to 46 to meet the even requirement. At the same time, half of the interference arms have 129 lenses and the other half have 128 lenses. Other parameters remain unchanged. The imaging target used in the simulation program is the USAF resolution target, as shown in Fig. 9. The image is gray scale with a pixel resolution of 256 * 256.

Figure 10 shows the results of single imaging of traditional SPIDER system and improved SPIDER system using parameters in Table 2 and fusion results of two images before and after rotation of improved SPIDER system. The residual of each reconstructed image with the original target image is shown in Fig. 11.

It can be seen that even if the number of interference arms is one less, the improved SPIDER system can achieve better imaging effect than the traditional SPIDER system, and the image quality of the improved system can be further increased by rotating and imaging multiple times and fusing the results. Next, we compare the final image quality between the traditional SPIDER system and the improved SPIDER system at same rotating imaging times, and draw a curve with the rotating imaging times as abscissa and the peak signal-to-noise ratio (PSNR) of the system's imaging results as ordinate, as shown in Fig. 12.

The imaging quality curves of the two systems show distinct trends. The PSNR of traditional SPIDER system is slightly improved and gradually saturated even after multiple rotational imaging and fusion and finally stabilized at about 13. This is because the interference arms of the traditional SPIDER system can only sample

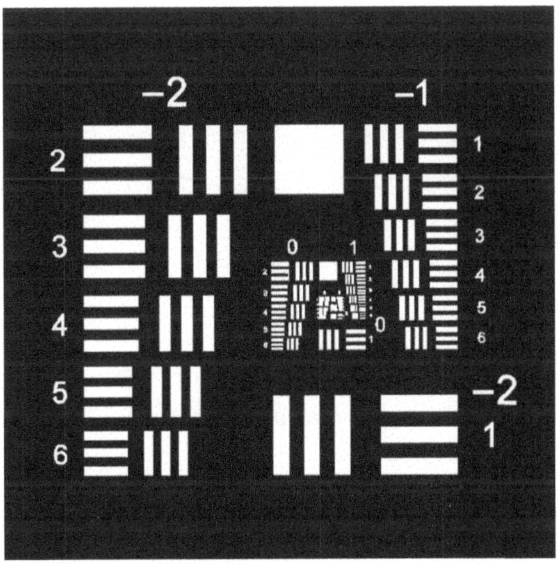

Fig. 9 USAF resolution target image used as imaging target in simulation

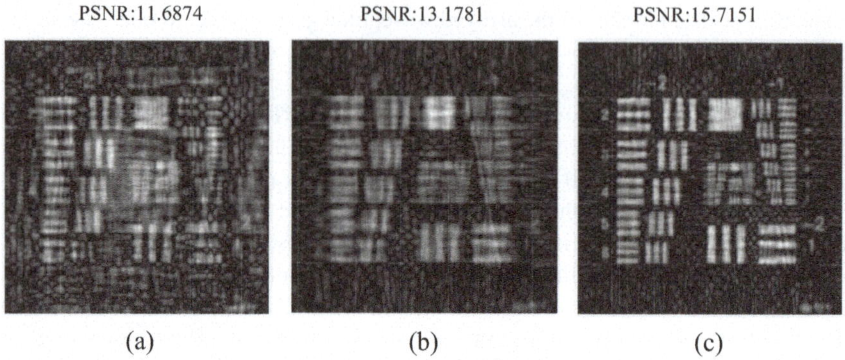

Fig. 10 Single imaging of the traditional SPIDER system and the improved SPIDER system

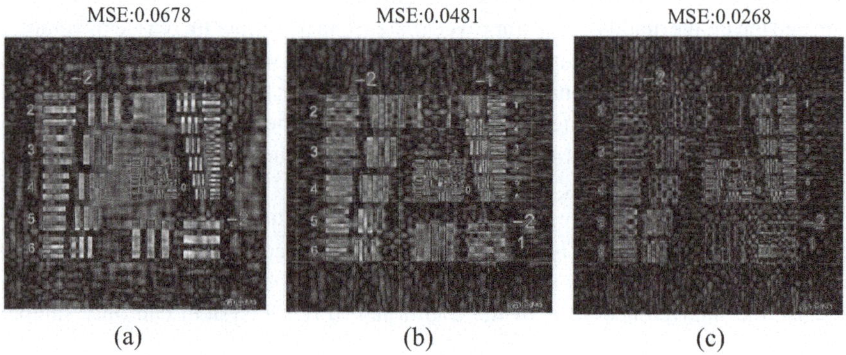

Fig. 11 Residual of each corresponding reconstructed image in Fig. 10 with the original target image in Fig. 9

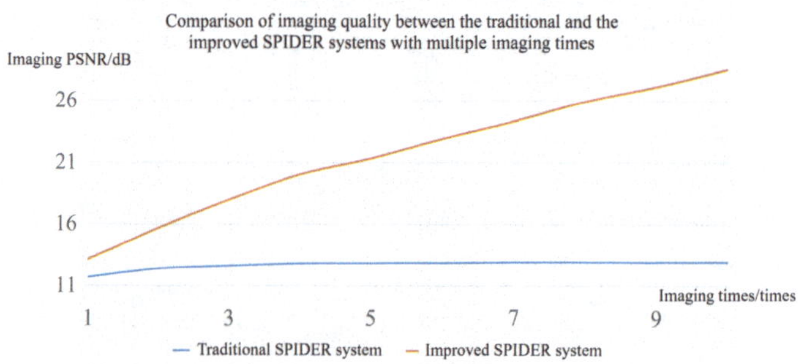

Fig. 12 Comparison of imaging quality between the traditional and the improved SPIDER systems with multiple imaging times

the spectrum information of odd multiple fundamental frequency, and multiple rotational imaging and fusion (equivalent to the number of interferometer arms) cannot solve the problem of under-sampling in the radial direction of the system. Finally, the spectrum information of even multiple fundamental frequency cannot be sampled in all directions, resulting in the imaging quality cannot continue to improve. On the contrary, the imaging quality of the improved SPIDER system increases approximate linearly with the number of rotational imaging. This is because the spectrum information of the target in each direction of the interference arm has been fully sampled, and only the number of interference arms restricts the imaging quality, which can be increased by multi-rotational imaging, so the PSNR of final imaging is gradually improved. The improved SPIDER system not only has better single imaging effect, but also has higher quality upper limit in multiple imaging and fusion. Figure 13 contrasts the imaging results between the traditional SPIDER system and the improved SPIDER system when rotating imaging is performed 10 times.

In addition, several images with different contents are used as imaging targets to further test the applicability of the improved SPIDER system to various imaging targets. The selected target images are shown in Fig. 14. Single imaging of three targets is performed using the traditional SPIDER system and the improved SPIDER system, respectively. The simulation results are shown in Fig. 15.

It can be seen that the improved SPIDER system can achieve better imaging results for various scenes, which proves it applicable to all kinds of imaging targets. Multiple imaging quality of the two systems is further compared. From Fig. 12, we know that the image quality of the traditional SPIDER system has basically reached the highest level when rotational imaging four times. Therefore, the results of the two systems with rotational imaging four times are compared here, as shown in Fig. 16. After the fusion of multiple imaging results, the imaging quality of the two SPIDER systems is both better than that of single imaging results, and the visual effects of the improved SPIDER system's images are better.

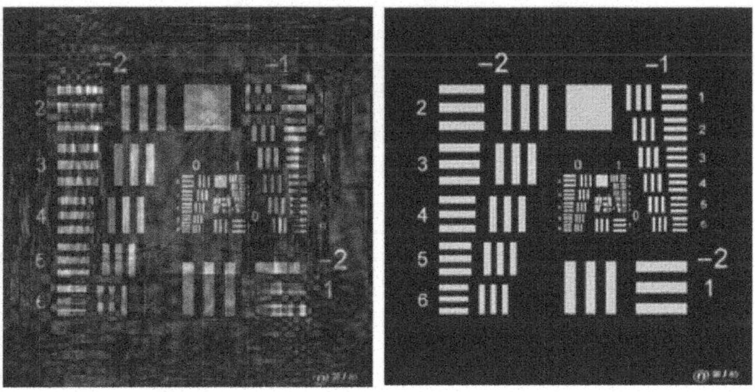

(a) (b)

Fig. 13 Comparison of the traditional SPIDER system and the improved SPIDER system's imaging results when rotational imaging is performed 10 times. **a** Imaging results for traditional SPIDER systems; **b** imaging results for improved SPIDER systems

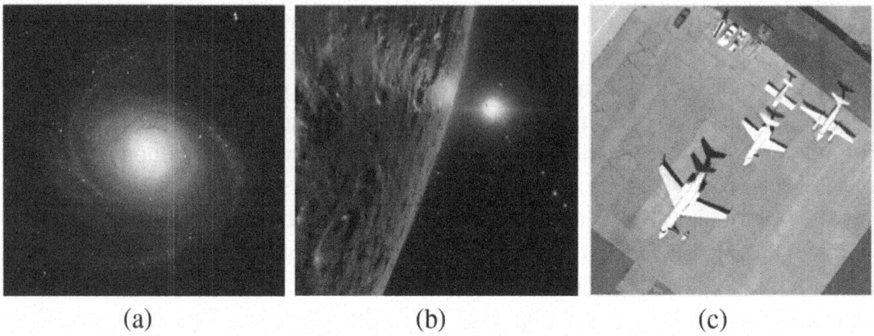

Fig. 14 Target images with different contents. **a** Nebular images; **b** planetary surface images taken by astronomical telescopes; and **c** satellite remote sensing images

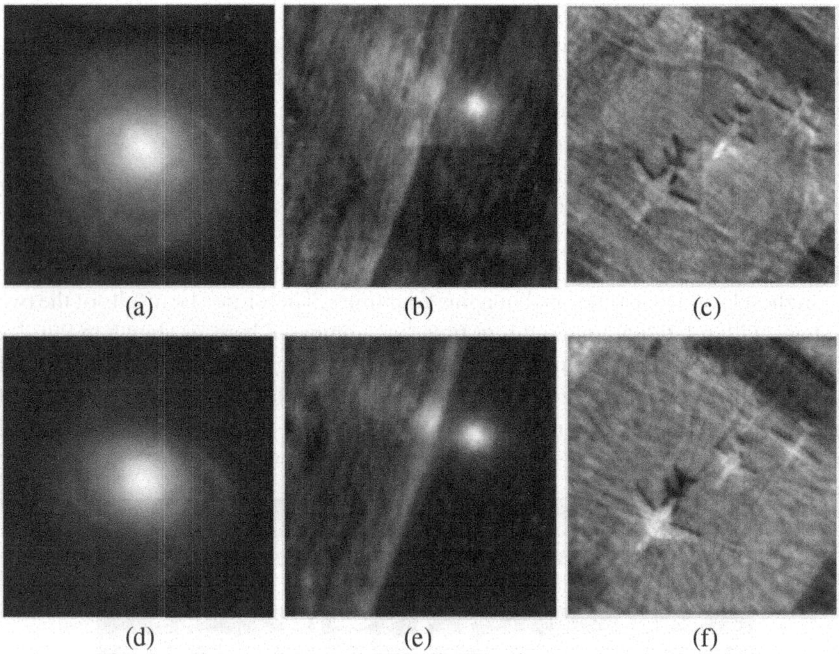

Fig. 15 Simulated imaging results of the traditional and the improved SPIDER system targets in Fig. 14. **a–c** the imaging results of the traditional SPIDER system; **d–f** the imaging results of improved SPIDER system

Further analysis is performed by comparing the two systems' imaging PSNR and the sampled number of spectral points, as shown in Tables 3 and 4. Table 3 gives the PSNR of each result obtained by the traditional SPIDER system and the improved SPIDER system when imaging different targets for once or four times, and Table 4

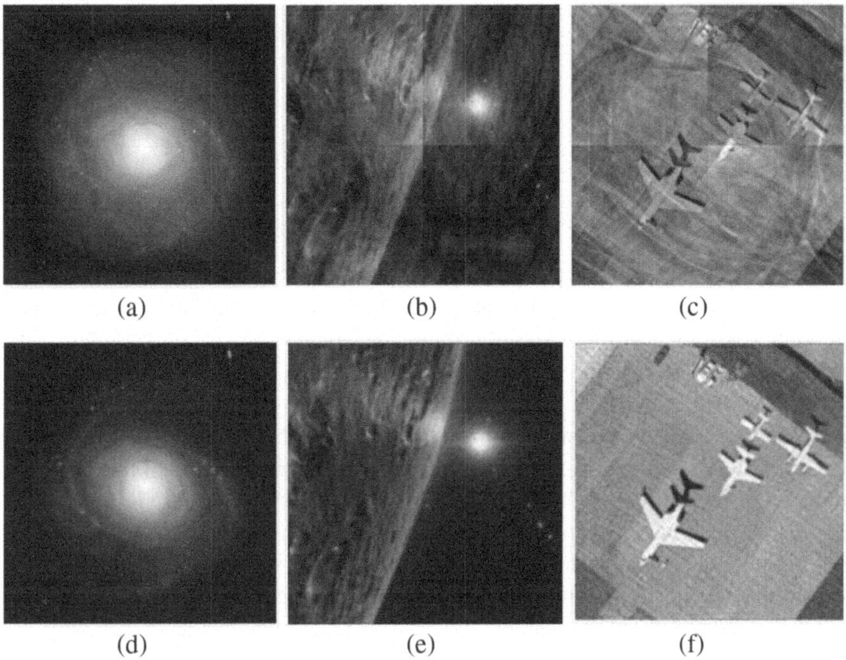

Fig. 16 Traditional and the improved SPIDER system's fusion results of targets in Fig. 14 when rotational imaging four times. **a–c** The imaging results of the traditional SPIDER system; **d–f** the imaging results of the improved SPIDER system

Table 3 Comparison of PSNR between the traditional and the improved SPIDER systems when imaging different targets for one and four times

Imaging target	Imaging times	Traditional SPIDER's PSNR (dB)	Improved SPIDER's PSNR (dB)
Nebular image	Image once	24.6332	**32.3360**
	Image four times	24.8850	**35.1396**
Planetary surface image	Image once	21.6765	**26.0823**
	Image four times	22.2779	**32.9577**
Satellite remote sensing image	Image once	19.9273	**22.1640**
	Image four times	20.9521	**28.5081**
USAF resolution target	Image once	11.6874	**13.1781**
	Image four times	12.8236	**20.0559**

gives the number of spectrum points sampled by the system under corresponding conditions.

Not surprisingly, the results of quantitative indicators are consistent with the visual results of imaging. Bold in Table 3 shows that the improved SPIDER system, which

Table 4 Comparison of sampled frequency points between the traditional and the improved SPIDER systems when imaging once or four times

System structure	Image once	Image four times	Increased multiples
Traditional SPIDER	**5679**	18,801	3.31
Improved SPIDER	5504	**19,995**	**3.63**

can achieve better imaging PSNR, beats the traditional SPIDER system in both single and multiple imaging of various targets. And from bold in Table 4, we can see that although Traditional SPIDER samples more frequency points in single imaging, the improved SPIDER system can acquire more spectrum points than the traditional SPIDER system in multiple imaging, which shows that the improved system has a more reasonable baseline distribution and stronger scalability, together with a higher baseline utilization rate in multiple imaging and a higher upper limit of final imaging quality. This also proves why the improved SPIDER system's PSNR of multiple imaging is much higher than that of single imaging, which the traditional SPIDER system cannot do, as shown in Table 3. In conclusion, the simulation results show that the improved SPIDER system can achieve better imaging quality than the traditional SPIDER system without increasing hardware cost and manufacturing difficulty.

4 Summary

In this paper, an improved scheme of the traditional SPIDER system in single wavelength mode is proposed. A lens is inserted into the center of the traditional interference arm with even number of lenses, which makes traditional one transform into a new type of interference arm. The new and old interference arms are then arranged alternately to form the improved SPIDER imaging system. Firstly, theoretical analysis shows that the improved SPIDER system can not only sample zero frequency which contains the average brightness information of the target, but also realize the information acquisition of continuous integer multiple fundamental frequency in the direction of every interference arm. The proposed scheme improves the system's spectrum radial sampling rate and hardly increases the hardware cost or assembly difficulty, meanwhile maintaining the field of view. What's more, the baseline pairing method can be used for lens arrays of any size, which has strong versatility. Secondly, three possible situations of different number and distribution of interference arms in the improved SPIDER system are compared in detail, and the optimal scheme is given. Finally, a computer program is used to simulate the imaging results of both the improved SPIDER system and the traditional SPIDER system for the same target using the same number of interference arms, which proves that the imaging quality of the improved SPIDER system is significantly higher than that of the traditional SPIDER system.

Innovation Points: Chap. 4 explains the innovation points.

References

1. Kendrick R, Duncan A, Wilm J et al (2013) Flat panel space based space surveillance sensor. In: Advanced maui optical and space surveillance technologies conference
2. Duncan AL, Kendrick RL (2014) U.S. Patent 8,913,859B1
3. Duncan A, Ogden C, Wuchenich D et al (2015) System design for a SPIDER imager. In: Frontiers in optics
4. Scott RP, Su T, Ogden C et al (2014) Demonstration of a photonic integrated circuit for multi-baseline interferometric imaging. In: IEEE photonics conference (IPC)
5. Duncan A, Kendrick R, Thurman S et al (2015) SPIDER: next generation chip scale imaging sensor. In: Advanced maui optical and space surveillance technologies conference
6. Su T, Liu G, Badham KE et al (2018) Interferometric imaging using Si 3 N 4 photonic integrated circuits for a SPIDER imager. J Opt Express 26(10):12801–12812
7. Rhodes WT, Goodman JW (1973) Interferometric technique for recording and restoring images degraded by unknown aberrations. J Opt Soc Am 63(6):647–657
8. Chu Q, Shen Y, Yuan M et al (2017) Numerical simulation and optimal design of segmented planar imaging detector for electro optical reconnaissance. J Opt Commun 405:288–296
9. Yu Q, Ge B, Li Y et al (2018) System design for a “checkerboard” imager. J Appl Opt 57(35):10218–10223

Performance of Support Vector Machines, Artificial Neural Network, and Random Forest for Identifying Banana Fusarium Wilt Using UAV-Based Multi-spectral Imagery

Huichun Ye, Bei Cui, Shanyu Huang, Yingying Dong, Wenjiang Huang, Anting Guo, Yu Ren, and Yu Jin

Abstract The Fusarium wilt of bananas currently threatens to the banana production areas worldwide. Timely monitoring of Fusarium wilt disease is very important for the disease treatment and crop planting adjustment. The aim of this paper was to evaluate the performance of support vector machines (SVM), random forest (RF), and artificial neural network (ANN) with unmanned aerial vehicle (UAV)-based multi-spectral imagery to identify the locations that were infested or not infested with banana Fusarium wilt. A total of 139 ground samples were surveyed to assess the occurrence of banana Fusarium wilt. The results showed that the overall accuracies of SVM, RF, and ANN were higher than 90% for the pixel based. Among the classifiers, SVM had the best performance, followed by ANN and RF. The maps generated by SVM, RF, and ANN appeared a similar distribution trend with regard to the occurrence of Fusarium wilt. The areas of the occurrence of Fusarium wilt were between 5.21 and 5.75 ha, accounting for 36.3–40.1% of the total planting area of bananas in the study area. The results also showed that the inclusion of the red-edge band had 2.9–3.0% increases in overall accuracy. The results of this study indicate

H. Ye (✉) · B. Cui · Y. Dong · W. Huang (✉) · A. Guo · Y. Ren
Key Laboratory of Digital Earth Science, Aerospace Information Research Institute, Chinese Academy of Sciences, Beijing 100094, China
e-mail: yehc@aircas.ac.cn

W. Huang
e-mail: huangwj@aircas.ac.cn

H. Ye · B. Cui · W. Huang
Key Laboratory of Earth Observation, Sanya 572029, Hainan, China

S. Huang
Chinese Academy of Agricultural Engineering Planning and Design, Beijing 100125, China

A. Guo · Y. Ren
University of Chinese Academy of Sciences, Beijing 100049, China

Y. Jin
School of Electronics and Information Engineering, Anhui University, Hefei 230601, China

L. Wang et al. (eds.), *Proceedings of the 6th China High Resolution Earth Observation Conference (CHREOC 2019)*, Lecture Notes in Electrical Engineering 657,
https://doi.org/10.1007/978-981-15-3947-3_19

that the SVM, RF, and ANN with UAV-based remote sensing imagery have the potential to identify and map the banana Fusarium wilt.

Keywords Fusarium wilt · Multi-spectral remote sensing · Support vector machines · Artificial neural network · Random forest

1 Introduction

Banana (*Musa* spp.) is the most popular fruit crop that is widely cultivated in tropical and subtropical climatic regions. Fusarium wilt of banana, also called Panama disease, is a serious soil-borne fungal disease caused by the fungus *Fusarium oxysporum* f. sp. *cubense* race 4 (*Foc*4) [1]. Currently, this disease threatens the banana production areas worldwide [2]. It disseminates either through infected plant material, contaminated soil, tools, or footwear or due to flooding and inappropriate sanitation measures [2]. The first visible signs of the disease are a yellowing or splitting on the oldest leaves, followed by leaf wilt and buckling, forming a 'skirt' around the pseudostem before falling off [3]. Chemical treatment of affected plants is often ineffective. Once a diseased plant is found, 'timely removal' is the best way to avoid the formation of a disease center [4]. Therefore, timely monitoring the occurrence of banana Fusarium wilt disease is very important for the disease treatment and adjustment of crop planting structure.

Traditionally, crop disease data collecting methods from ground surveys are expensive and time-consuming [5]. Remote sensing technology has become a feasible means for crop disease detection and assessment in the past few decades, including for detecting Fusarium head blight and rust infection in wheat [6–10], bacterial leaf blight in rice [11, 12], and gray leaf spot in maize [13]. However, studies using remote sensing technology to monitor Fusarium wilt of banana are scarce.

When plants are infected with diseases, the leaf water, pigment content, and internal structure undergo changes and would be reflected in the spectral signature of the plants [18]. Many spectral features of vegetation were found within the red-edge band that is related to changes in chlorophyll content and leaf area index [14–16], which would be changed significantly when bananas were infected with Fusarium wilt. As to the classification methods, a variety of classification algorithms have been developed and tested for crop monitoring or land cover classification using remote sensing data [17–21]. These methods range from unsupervised algorithms (i.e., K-means) to parametric supervised algorithms (i.e., maximum likelihood) and machine learning algorithms including decision trees (DT), support vector machines (SVM), artificial neural networks (ANN), and random forest (RF). In these methods, the SVM, ANN, and RF were the most increasing classification algorithms in remote sensing of the Earth's surface [22].

The aims of this paper were to (i) evaluate the performance of three machine learning Algorithms with SVM, RF, and ANN in an attempt to classify the locations that were infested or not infested with banana Fusarium wilt and (ii) assess the effect of the imagery with or without red-edge band on the identification accuracy. The results will provide guidance for detecting the disease and adjustment of crop planting structure.

2 Materials and Methods

2.1 Study Area

The study area is located in Long'an County, Guangxi Province, China (23° 7′ 58.8″ N, 107° 43′ 55.2″ E) (Fig. 1). The field crops were the bananas with the variety of 'Williams B6.' For this variety, the leaf number was 34–36, the plant

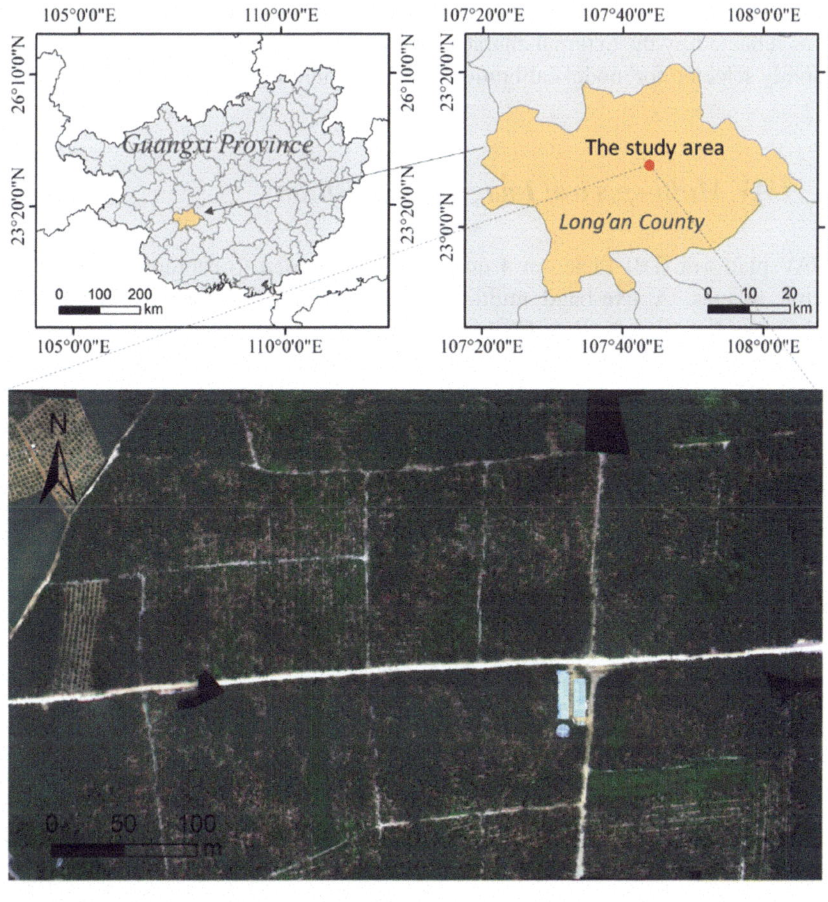

Fig. 1 Location of the study area

height was about 2.4–3 m, the growth period was 10–12 months, and the annual yield was 45,000–60,000 kg ha^{-1}. The farm was developed in September 2015 with the planting distance of 2.0 m by 2.6 m (planting density of 130 plants ha^{-1}). It was harvested for the first time in November 2016. By August 2018 (the time of field investigation in this study), the third generation of bananas was present in the plots. In the study area, nearly 40% of banana plants were infected with Fusarium wilt disease of different severity.

2.2 Field Data Collection

In this study, a total of 139 sample plots were surveyed on August 7–9, 2018, to assess the occurrence of banana Fusarium wilt disease as ground truth data. These samples were classified into two categories: healthy samples (total of 66) and diseased samples (total of 73), representing the occurrence or non-occurrence of Fusarium wilt as reflected by the external characteristics. Finally, a total of 100 samples were randomly selected for model calibration and the remaining were used for validation.

2.3 UAV Multi-spectral Imagery Acquisition

A UAV platform (DJI Phantom 4 quadcopter) was used to capture the multiple-spectral images. A five-band multi-spectral camera MicaSense RedEdge M™ (MicaSense, Inc., Seattle, WA, USA) was mounted in landscape format. The sensor has a spectral range of 400–900 nm with a 47.2° field of view. It has five spectral bands, blue, green, red, red-edge (RE), and near-infrared (NIR). Spectral bands have a ground sampling distance (GSD) of 8 cm at 120 m above ground level (Table 1). A global shutter is aligned with all bands with a 12-bit radiometric resolution at an image capture rate of 1 Hz. In this study, the flight plan ensured cross-track and along-track overlap of 80% and a calibrated reflectance panel was imaged directly

Table 1 Characteristics of MicaSense RedEdge M™

Spectral band	Center wavelength (nm)	Bandwidth (nm)	GSD (cm)
Blue	475	20	8
Green	560	20	8
Red	668	10	8
Red-edge	717	10	8
Near-infrared	840	40	8

before and after each flight and used for reflectance calibration using the empirical line method. The multi-spectral imagery was collected from a flying altitude of approximately 120 m above the ground with a GSD of 8 cm.

2.4 Classification Algorithms

In this study, we compared the performances of RF, ANN, and SVM classifiers for banana Fusarium wilt identification using UAV-based multi-spectral imagery. RF is one of the most popular decision tree-based ensemble models proposed by Breiman [23]. It can be described as an ensemble of classification trees, where each tree votes on the class assigned to a given sample, with the most frequent answer winning the vote [24]. The definition of RF can be found in Refs. [23, 25]. SVM, which is a non-parametric supervised statistical learning classifier, has become increasingly popular in remote sensing classification [26–28]. The SVM algorithm was developed by Vapnik [29]. The SVM classifier tries to find the optimal hyperplane in n-dimensional classification space with the highest margin between classes [22]. Polynomial and radial basis function (RBF) kernels are the most commonly used functions for remote sensing classification [30–32]. ANN classifier can be described as a parallel computing system consisting of an extremely large number of simple processors with interconnections [22]. The ANN was originally designed as pattern recognition and data analysis tools that mimic the neural storage and analytical operations of the brain. It has a distinct advantage in that it is nonparametric and requires little or no a priori knowledge of the distribution model of input data [32].

2.5 Accuracy Assessment

Non-banana areas were masked from final images using a normalized difference vegetation index (NDVI)-based mask. The validation samples were used for accuracy assessment with indicators such as overall accuracy and the Kappa coefficient calculated by confusion matrix method [33, 34]. The overall accuracy is the sum of the correctly classified plots divided by the total number of plots. A Kappa value of 1 represents perfect agreement, whereas a value of 0 represents no agreement.

Table 2 Identification accuracy of banana Fusarium wilt for the pixel based in the conditions with or without red-edge band

Classifier	Red-edge band included		Red-edge band excluded		Contribution of red-edge band	
	Overall accuracy (%)	Kappa coefficient	Overall accuracy (%)	Kappa coefficient	Overall accuracy (%)	Kappa coefficient
SVM	91.4	0.80	88.5	0.75	2.9	0.05
RF	90.0	0.77	87.1	0.72	2.9	0.03
ANN	91.1	0.79	88.1	0.73	3.0	0.06

3 Results

3.1 Results of Accuracy Assessment

In this study, the verification samples were used to verify the classification accuracy of banana Fusarium wilt with SVM, RF, and ANN algorithms (Table 2). The results showed that all three algorithms with original UAV-based multi-spectral imagery can be well used to identify and map the banana Fusarium wilt disease. The overall accuracies of SVM, RF, and ANN were higher than 90% for the pixel based. The SVM had the best performance (with the overall accuracy of 91.4% and Kappa coefficient of 0.80), followed by ANN (with the overall accuracy of 91.1% and Kappa coefficient of 0.79) and RF (with the overall accuracy of 90.0% and Kappa coefficient of 0.77). Furthermore, the performances of the imagery included the red-edge band and without the red-edge band were compared. The results showed that inclusion of the red-edge band has contributed to the increase on identification accuracy of banana Fusarium wilt in the study. It had 2.9%, 2.9%, and 3.0% increases in overall accuracy of SVM, RF, and ANN algorithms, respectively.

3.2 Mapping Disease Distribution

Figure 2 shows the spatial distribution of banana Fusarium wilt infected regions in the study area using SVM, RF, and ANN algorithms in the conditions with or without red-edge band. All the maps appeared a similar distribution trend with regard to the occurrence of Fusarium wilt disease. The results in Table 3 show the areas of the healthy regions and Fusarium wilt diseased regions. When the imagery with red-edge band was used, the areas of the occurrence of Fusarium wilt were between 5.21 and 5.75 ha, accounting for 36.3–40.1% of the total planting area of bananas in the study area. When the imagery without red-edge band was used, the areas of the occurrence of Fusarium wilt were between 4.77 and 5.53 ha, accounting for 33.3–38.6% of the total planting area of bananas in the study area. Among the three classifiers, the results

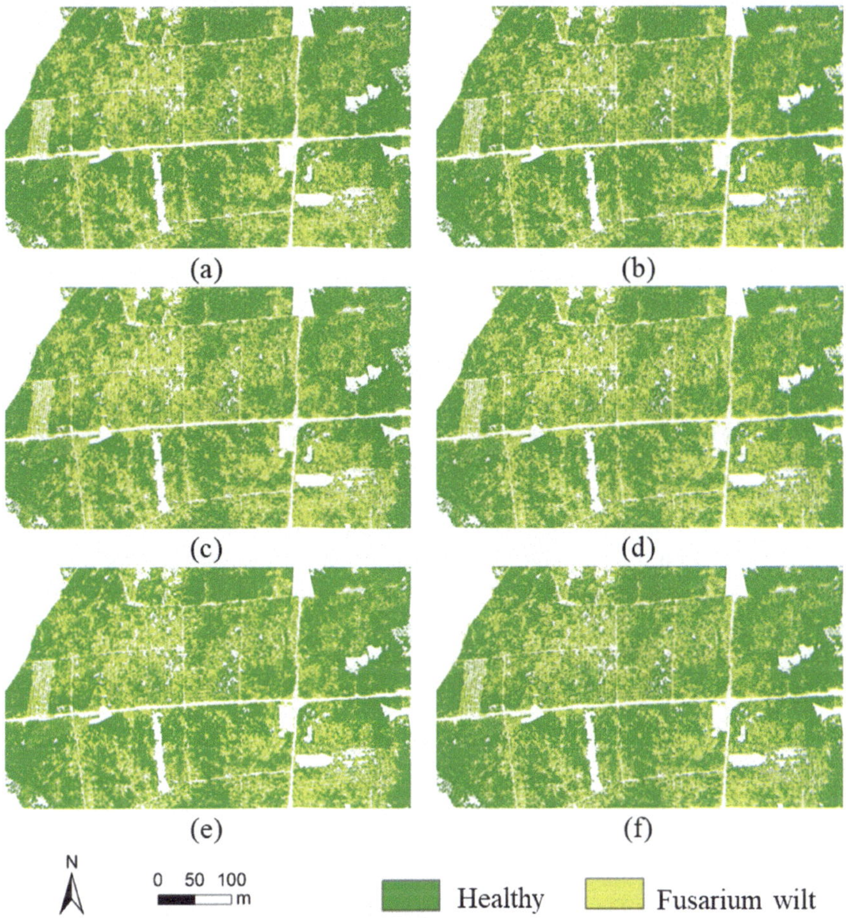

Fig. 2 Maps of the spatial distribution of banana Fusarium wilt infected regions in the study area using SVM, RF, and ANN in the conditions with or without red-edge band. **a** SVM with imagery including red-edge band. **b** SVM with imagery excluding red-edge band. **c** RF with imagery including red-edge band. **d** RF with imagery excluding red-edge band. **e** ANN with imagery including red-edge band. **f** ANN with imagery excluding red-edge band

Table 3 Areas of the healthy regions and Fusarium wilt diseased regions obtained using SVM, RF, and ANN in the conditions with or without red-edge band

Classifier	Red-edge band included			Red-edge band excluded		
	Healthy area (ha)	Diseased area (ha)	Percentage of diseased area (%)	Healthy area (ha)	Diseased area (ha)	Percentage of diseased area (%)
SVM	9.12	5.22	36.4	9.50	4.84	33.8
RF	8.58	5.75	40.1	8.81	5.53	38.6
ANN	9.13	5.21	36.3	9.57	4.77	33.3

generated by SVM are highly consistent with the by ANN, whereas RF obtained a larger diseased area than SVM by 0.53 ha in the condition with red-edge band and 0.69 ha in the condition without red-edge band.

4 Discussion

The results of this study indicate that the three machine learning algorithms (SVM, RF, and ANN) with UAV-based multi-spectral imagery have the potential to identify and map the banana Fusarium wilt disease. All the SVM, RF, and ANN algorithms achieved comparable overall accuracies larger than 90%. SVM produced higher classification accuracy than RF by 1.4% and ANN by 0.3%. However, the operation speed of SVM is much lower than those of RF and ANN. Furthermore, a very high-resolution UAV-based imagery with the resolution of 0.08 m was used in this study. In the further study, it is necessary to simulate the resolutions of satellite-based imagery (i.e., Worldview Series with a resolution of 0.5 m, GF-2 with a resolution of 1 m, GF-1 with a resolution of 2 m, RapidEye with a resolution of 5 m, and Sentinel-2 with a resolution of 10 m) to assess the effects of imagery with different spatial resolution on the identification of disease, which can provide a reference for large-scale applications of satellite-based data.

5 Conclusions

This study evaluated the performance of SVM, RF, and ANN algorithms with UAV-based multi-spectral imagery to identify the locations that were infested or not infested with banana Fusarium wilt. The results showed that the SVM, RF, and ANN algorithms with UAV-based multi-spectral imagery can be well to identify and map the banana Fusarium wilt disease. The overall accuracies of SVM, RF, and ANN were higher than 90% for the pixel based. Among the classifiers, SVM had the best performance, followed by ANN and RF. The maps generated by SVM, RF, and ANN all appeared a similar distribution trend with regard to the occurrence of Fusarium wilt disease. The areas of the occurrence of Fusarium wilt disease were between 5.21 and 5.75 ha, accounting for 36.3–40.1% of the total planting area of bananas in the study area. The results also showed that the inclusion of the red-edge band had 2.9–3.0% increases in overall accuracy. The results of this study indicate that the SVM, RF, and ANN with UAV-based remote sensing imagery have the potential to identify and map the banana Fusarium wilt disease; this provides guidance for disease treatment and crop planting adjustment.

Acknowledgements The author expresses the appreciation of funds received from the Hainan Provincial Key R&D Program of China (ZDYF2018073), National Natural Science Foundation of China (41801352), Agricultural Science and Technology Innovation of Sanya, China (2016NK16),

National Special Support Program for High-level Personnel Recruitment (Ten-thousand Talents Program) (Wenjiang Huang), and Youth Innovation Promotion Association CAS (2017085).

References

1. Shen Z, Xue C, Penton CR, Thomashow LS, Zhang N, Wang B, Ruan Y, Li R, Shen Q (2019) Suppression of banana Panama disease induced by soil microbiome reconstruction through an integrated agricultural strategy. Soil Biol Biochem 128:164–174
2. Ordonez N, Seidl MF, Waalwijk C, Drenth A, Kilian A, Thomma BPHJ, Ploetz RC, Kema GHJ (2015) Worse comes to worst: bananas and Panama disease-when plant and pathogen clones meet. PloS Pathog 11:e1005197
3. Van den Berg N, Berger DK, Hein I, Birch PR, Wingfield MJ, Viljoen A (2007) Tolerance in banana to Fusarium wilt is associated with early up-regulation of cell wall-strengthening genes in the roots. Mol Plant Pathol 8:333–341
4. Lin B, Shen H (2017) *Fusarium oxysporum* f. sp. *Cubense*. In: Wan F, Jiang M, Zhan A, (eds) Biological invasions and its management in China, vol 2. Springer Singapore, Singapore, pp 225–236
5. Shi Y, Huang W, Ye H, Ruan C, Xing N, Geng Y, Dong Y, Peng D (2018) Partial least square discriminant analysis based on normalized two-stage vegetation indices for mapping damage from rice diseases using PlanetScope datasets. Sensors 18:1901
6. Jin X, Jie L, Wang S, Qi HJ, Li SW (2018) Classifying wheat hyperspectral pixels of healthy heads and Fusarium head blight disease using a deep neural network in the wild field. Remote Sens 10:395
7. Mahlein AK, Alisaac E, Al Masri A, Behmann J, Dehne HW, Oerke EC (2019) Comparison and combination of thermal, fluorescence, and hyperspectral imaging for monitoring Fusarium head blight of wheat on spikelet scale. Sensors 19:2281
8. Huang W, Lamb DW, Niu Z, Zhang Y, Liu L, Wang J (2007) Identification of yellow rust in wheat using in-situ spectral reflectance measurements and airborne hyperspectral imaging. Precis Agric 8:187–197
9. Huang W, Guan Q, Luo J, Zhang J, Zhao J, Liang D, Huang L, Zhang D (2014) New optimized spectral indices for identifying and monitoring winter wheat diseases. IEEE J Sel Topics Appl Earth Obs Remote Sens 7:2516–2524
10. Shi Y, Huang W, Gonzalez-Moreno P, Luke B, Dong Y, Zheng Q, Ma H, Liu L (2018) Wavelet-based rust spectral feature set (WRSFs): a novel spectral feature set based on continuous wavelet transformation for tracking progressive host-pathogen interaction of yellow rust on wheat. Remote Sens 10:525
11. Huang J, Liao H, Zhu Y, Sun J, Sun Q, Liu X (2012) Hyperspectral detection of rice damaged by rice leaf folder (*Cnaphalocrocis medinalis*). Comput Electron Agr 82:100–107
12. Yang CM (2010) Assessment of the severity of bacterial leaf blight in rice using canopy hyperspectral reflectance. Precis Agric 11:61–81
13. Dhau I, Adam E, Mutanga O, Ayisi K, Abdel-Rahman EM, Odindi J, Masocha M (2018) Testing the capability of spectral resolution of the new multispectral sensors on detecting the severity of grey leaf spot disease in maize crop. Geocarto Int 33:1223–1236
14. Xie Q, Dash J, Huang W, Peng D, Qin QM, Mortimer H, Casa R, Pignatti S, Laneve G, Pascucci S, Dong Y, Ye H (2018) Vegetation indices combining the red and red-edge spectral information for leaf area index retrieval. IEEE J Sel Topics Appl Earth Obs Remote Sens 11:1482–1493
15. Filella I, Penuelas J (1994) The red edge position and shape as indicators of plant chlorophyll content, biomass and hydric status. Int J Remote Sens 15:1459–1470
16. Dash J, Curran PJ (2004) The MERIS terrestrial chlorophyll index. Int J Remote Sens 25:5403–5413

17. Zheng Q, Huang W, Cui X, Shi Y, Liu L (2018) New spectral index for detecting wheat yellow rust using Sentinel-2 multispectral imagery. Sensors 18:868
18. Ma H, Jing Y, Huang W, Shi Y, Dong Y, Zhang J, Liu L (2018) Integrating early growth information to monitor winter wheat powdery mildew using multi-temporal Landsat-8 imagery. Sensors 18:3290
19. Thanh Noi P, Kappas M (2018) Comparison of random forest, k-nearest neighbor, and support vector machine classifiers for land cover classification using Sentinel-2 imagery. Sensors 18:18
20. Rumpf T, Mahlein AK, Steiner U, Oerke EC, Dehne HW, Plümer L (2010) Early detection and classification of plant diseases with support vector machines based on hyperspectral reflectance. Comput Electron Agr 74:91–99
21. Chemura A, Mutanga O, Dube T (2017) Separability of coffee leaf rust infection levels with machine learning methods at Sentinel-2 MSI spectral resolutions. Precis Agric 18:859–881
22. Raczko E, Zagajewski B (2017) Comparison of support vector machine, random forest and neural network classifiers for tree species classification on airborne hyperspectral APEX images. Eur J Remote Sens 50:144–154
23. Breiman L (2001) Random forests. Mach Learn 45:5–32
24. Sun LY, Schulz K (2015) The improvement of land cover classification by thermal remote sensing. Remote Sens 7:8368–8390
25. Feng Q, Liu J, Gong J (2015) UAV remote sensing for urban vegetation mapping using random forest and texture analysis. Remote Sens 7:1074–1094
26. Cheng G, Han JW, Zhou PC, Guo L (2014) Multi-class geospatial object detection and geographic image classification based on collection of part detectors. ISPRS J Photogramm Remote Sens 98:119–132
27. George R, Padalia H, Kushwaha SPS (2014) Forest tree species discrimination in western Himalaya using EO-1 Hyperion. J Appl Earth Obs Geoinf 28:140–149
28. Ma L, Fu T, Blaschke T, Li M, Tiede D, Zhou Z, Ma X, Chen D (2017) Evaluation of feature selection methods for object-based land cover mapping of unmanned aerial vehicle imagery using random forest and support vector machine classifiers. ISPRS Int J Geo-Inf 6
29. Vapnik VN (1995) The nature of statistical learning theory. Springer, New York
30. Dixon B, Candade N (2008) Multispectral land use classification using neural networks and support vector machines: one or the other, or both? Int J Remote Sens 29:1185–1206
31. Huang C, Davis LS, Townshend JRG (2002) An assessment of support vector machines for land cover classification. Int J Remote Sens 23:725–749
32. Omer G, Mutanga O, Abdel-Rahman EM, Adam E (2015) Performance of support vector machines and artificial neural network for mapping endangered tree species using worldview-2 data in Dukuduku Forest, South Africa. IEEE J Sel Topics Appl Earth Obs Remote Sens 8:4825–4840
33. Congalton RG (1991) A review of assessing the accuracy of classifications of remotely sensed data. Remote Sens Environ 37:35–46
34. Foody GM (2009) Classification accuracy comparison: hypothesis tests and the use of confidence intervals in evaluations of difference, equivalence and non-inferiority. Remote Sens Environ 113:1658–1663

Remote Sensing Image On-Board Restoration Based on Adaptive Wiener Filter

Yunsen Wang, Wenxiu Mu, Xiaohui Du, Chengzhi Ma, and Xuejin Shen

Abstract During the satellite's imaging process, remote sensing images are always degraded due to a variety of factors. To overcome the power spectrum ratio's evaluation problem of traditional Wiener filter methods of image restoration, a novel remote sensing image restoration algorithm based on adaptive Wiener filter is proposed. This algorithm still adopts Wiener filter to restore the degraded remote sensing image. The degenerate function is evaluated by point spread function of uniform linear motion, and the power spectrum ratio of the image is estimated by adaptive iteration. To verify the proposed algorithm's on-board process performance, the method is implemented on embedded graphics processing unit (GPU). Experimentations demonstrate that the proposed algorithm could acquire satisfactory remote sensing image recovering results, and the processing time could be controlled in a relatively short time.

Keywords Remote sensing image restoration · Adaptive wiener filter · Power spectrum ratio · Degenerate function

1 Introduction

During remote sensing image processes, restoration algorithm plays an important role. In the space environment, a remote sensing image is always degraded by different kinds factors, such as atmospheric turbulence effects, atmosphere scatter, and low performance of sensors [1–3]. So the on-board sensing image, reconstruction process aims to restore degraded images on the space platform.

All kinds of filtering algorithms are widely used in remote sensing image restoration field, such as inverse filtering [4], Richardson–Lucy filtering [5, 6], Wiener filter [7], and blind convolution filtering [8]. Most of these methods are implemented based on point spread function (PSF), or modulation transfer function (MTF), which is the essential theoretical support of semi-blind sensing image restoration [9, 10]. Because many useful information of the image's target hidden in the edge portion, those filtering methods need to remove the blur and noise from the image on the premise of

Y. Wang (✉) · W. Mu · X. Du · C. Ma · X. Shen
Shandong Institute of Space Electronic Technology, Yantai 264670, Shandong, China
e-mail: yswang513@163.com

L. Wang et al. (eds.), *Proceedings of the 6th China High Resolution Earth Observation Conference (CHREOC 2019)*, Lecture Notes in Electrical Engineering 657,
https://doi.org/10.1007/978-981-15-3947-3_20

keeping image's details [11, 12]. Many methods were proposed to deal with this competing problem, regularization, total variation [13], and expectation–maximization [14], for example.

Recently, using neural network algorithms to restore remote sensing images has become a new research hotspot along with the rapid development of computer [15]. Zhou et al. firstly adopted Hopfield neural network to realize image restoration by using optimization method [16]. But the network model is huge, and the parameter state would be updated slowly. In order to overcome these disadvantages, Paik et al. put forward an improved Hopfield neural network model for grayscale image restoration [17]. Later, Paik's method was improved by many scholars from the network model and the convergence rate [18–21]. But the authors in the analysis of those algorithms found that the storage of the connection weights is very huge. It is not conducive to the computer processing. Then based on these algorithms, many improved methods are proposed, to simplify the calculation of the connection weights.

As mentioned above, Wiener filter method is still widely used in image restoration. The Wiener filter method is built on the statistics minimization principle. So the Wiener filter method is the best restoration method to a certain extent [22, 23]. But Wiener filter has a problem when intending for on-board application, which is the calculation or evaluation of the power spectrum ratio.

In this paper, a novel remote sensing image restoration algorithm based on adaptive Wiener filter is proposed. The algorithm adopts Wiener filter to restore the degraded remote sensing image. The degenerate function is evaluated by point spread function of uniform linear motion, and the power spectrum ratio of the image is estimated by adaptive iteration. The iterative criterion is formulated by a comprehensive evaluation factor, which integrates two quality factors, GMG and PSNR. Experimental results show that the proposed algorithm could acquire satisfactory remote sensing image recovering results, and the processing time could be controlled in an acceptable period.

2 Remote Sensing Image Degradation and Restoration Models

Before restoration process, several parameters must be known to build exact image degradation model that could reflect the degradation reasons of remote sensing image [24, 25].

Figure 1 shows the image degradation model. In this model, $f(x, y)$ is the original image, $g(x, y)$ is the degrading image, $h(x, y)$ is the degradation function (point spread function, PSF), and $n(x, y)$ is the Gaussian noise function.

Assume $h(x, y)$ is a linear and space-changeless system. The sensing image degrading process could be described as

$$g(x, y) = f(x, y) * h(x, y) + n(x, y) \tag{1}$$

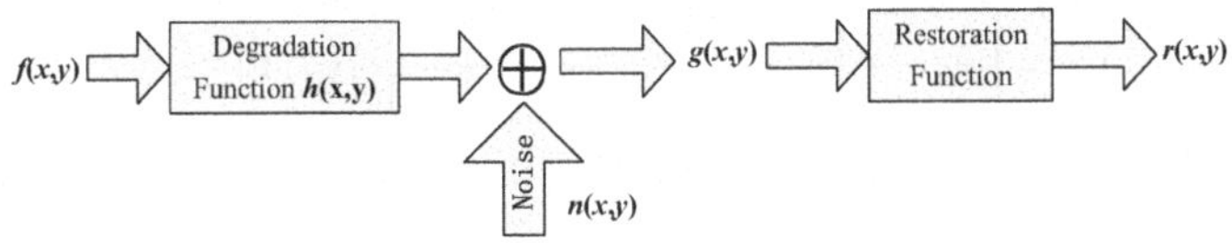

Fig. 1 Image degradation and restoration models

The Fourier transform of Formula (1) is given as

$$G(u, v) = F(u, v)H(u, v) + N(u, v) \tag{2}$$

Then the original image $F(u, v)$ could be calculated by

$$F(u, v) = \frac{G(u, v) - N(u, v)}{H(u, v)} \tag{3}$$

Suppose $r(x, y)$ is the recovering image, and the purpose of image restoration is to make the following equation be established.

$$R(u, v) = F(u, v) \tag{4}$$

3 On-Board Image Restoration Based on Adaptive Wiener Filter

3.1 *Wiener Filter Method*

The restoration principle of Wiener filter method is built on the system degradation functions and noise statistics property. Namely, the degrading image and noise are all stochastic functions. So the goal of the Wiener filter method is to find an apt filter to minimize the difference between the restorative image and the original image.

According to the restoration model, the image restoration problem can be transformed to minimize the following error function.

$$e_{\min}^2 = E\{[r(x, y) - f(x, y)]^2\} \tag{5}$$

where $r(x, y)$ is the restorative image and e^2 is mean square error between the restorative image and the original image. $E\{\cdot\}$ is the mathematical expectation.

While calculating the mean square error, the restorative image's function of frequency domain could be given by

$$R(u, v) = \left[\frac{H*(u, v)S_f(u, v)}{S_f(u, v)|H(u, v)|^2 + S_n(u, v)}\right]G(u, v)$$
$$= \left[\frac{H*(u, v)}{|H(u, v)|^2 + S_n(u, v)/S_f(u, v)}\right]G(u, v)$$
$$= \left[\frac{1}{H(u, v)}\frac{|H(u, v)|^2}{|H(u, v)|^2 + S_n(u, v)/S_f(u, v)}\right]G(u, v) \quad (6)$$

where $H(u, v)$ is the degradation function, $H^*(u, v)$ is the plural conjugate function of $H(u, v)$, $S_n(u, v) = |N(u, v)|^2$ is the power spectrum of the noise, and $S_f(u, v) = |F(u, v)|^2$ is the power spectrum of non-degrading image.

So the transfer function of Wiener filter would be described as

$$H_{weiener}(u, v) = \frac{1}{H(u, v)}\frac{|H(u, v)|^2}{|H(u, v)|^2 + S_n(u, v)/S_f(u, v)} \quad (7)$$

The degradation function would not equal zero only if $u = v$, and $H(u, v) = S_n(u, v) = 0$. When the degradation image's signal-to-noise ratio is relatively high, which means $S_f(u, v) \gg S_n(u, v)$, the specific value $S_n(u, v)/S_f(u, v)$ would become very little. The transfer function of Wiener filter then becomes

$$H_{wiener}(u, v) = \frac{1}{H(u, v)} \quad (8)$$

Here the Wiener filter is retrograded to inverse filter. On the other hand, if $S_f(u, v) \ll S_n(u, v)$, the Wiener filter $H_{wiener}(u, v) \to 0$. In this aspect, Wiener filter could avoid the noises over enlargement problem of inverse filter.

The implement of Wiener filter method needs the power spectrum of the non-degrading image $S_f(u, v)$ and the power spectrum of the noise $S_n(u, v)$. In fact, the two variables are unknown and could not be calculated exactly for remote sensing images.

Suppose $S_n(u, v)/S_f(u, v) = C$, and C is a constant. The transfer function of Wiener filter then becomes

$$H_{wiener}(u, v) = \frac{1}{H(u, v)}\frac{|H(u, v)|^2}{|H(u, v)|^2 + C} \quad (9)$$

The C is the power spectrum ratio, which is significant to image restoration result. When the parameter C is small, which means the signal-to-noise ratio of the relatively small, the restored image would have distinct contours and be more brightsome than degraded image. But the noise could not be restrained effectively. On the other hand, if C is large, the restored results always restrain the enlargement of noise to a certain extent, but the clarity would be comparatively poor.

Therefore, apply Wiener filter method to restore remote sensing image needs two crucial steps: estimating the degradation function $h(x, y)$ and searching the optimal power spectrum ratio C.

3.2 Degradation Estimating

According to the remote sensing image degradation model, it is in evidence that if the point spread function of blurred image or the transfer function of degrading system could be estimated, Wiener filter or inverse filter method could be used to recover the blurred remote sensing images. Suppose the blurred direction is known. Then the blurred images would be rotated to make the blurred direction in horizontal direction. The two-dimensional PSF problem could be changed to one-dimensional problem [26]. In this paper, linear motion-blurred images are premeditated, and the PSF of blurred system satisfies the following equation.

$$h(x) = \frac{1}{d}, \quad 0 \leq x \leq d - 1 \tag{10}$$

where d is blurred scale, and autocorrelation function is quoted to calculate d. The concrete steps are as follows.

(1) Execute difference operation in vertical direction on degrading image $g(x, y)$.

$$g_y(i, j) = g(i, j) - g(i, j - 1) \tag{11}$$

(2) Execute difference operation in horizontal direction on Eq. (11).

$$g_{yx}(i, j) = g_y(i, j) - g_y(i - 1, j) \tag{12}$$

(3) Based on Eq. (12), calculate the autocorrelation function of every row of the degrading image. If the jth row has l pixels, the autocorrelation function could be described as

$$r(k) = \sum_{i=l}^{M-k-l} \sum_{m=-l}^{l} g(i + k + m, j)g(i + m, j) \quad k \in [-1, 1] \tag{13}$$

where M is the image's weight.

(4) Calculate the mean value of autocorrelation function and draw the autocorrelation function curve. Through searching the minimum value of the curve, the blurred scale d could be determined.

3.3 Estimating Optimal Power Spectrum Ratio

As for the restoration of remote sensing image, there is no corresponding distinct image to be consulted. But most of no-reference image quality assessment methods are always focus on the contrast or grayscale change, and the noise suppressing results cannot be reflected. So in this paper, both no-reference method and reference method are adopted to complete image's restoration. Two commonly used assessment methods are used in this paper.

(1) Grayscale Mean Gradient (GMG) [27]

GMG reflects the evaluated image's contrast and texture features to a certain extent. The larger the GMG value is, the more distinct the image's contours are, and the better the image's quality is.

$$\mathrm{GMG} = \frac{1}{(M-1)(N-1)} \sum_{i=1}^{M} \sum_{j=1}^{N} \sqrt{\frac{[g(i, j+1) - g(i, j)]^2 + [g(i+1, j) - g(i, j)]^2}{2}} \tag{14}$$

(2) Peak Signal-to-Noise Ratio (PSNR)

Mean Square Error (MSE) is a commonly used parameter in reference methods. PSNR is calculated on the basis of MSE. PSNR is used to measure the pixels' difference between processed image and reference image. The larger the PSNR value is, the better the noise suppressing results are, and the better the processed image's quality is.

$$\mathrm{PSNR} = 10 \log\left[\frac{(2^t - 1)^2}{\mathrm{MSE}}\right] \tag{15}$$

$$\mathrm{MSE} = \frac{1}{MN} \sum_{i=1}^{M} \sum_{j=1}^{N} [g(i, j) - r_{\mathrm{e}}(i, j)]^2 \tag{16}$$

where t is the image's grayscale order, $r_{\mathrm{e}}(i, j)$ is the reference image's pixel.

To get a reference image, Gaussian filter is adopted to suppress the image random noise.

$$r_{\mathrm{e}}(x, y) = g(x, y) * G_{\mathrm{f}}(x, y) \tag{17}$$

where $G_{\mathrm{f}}(x, y)$ is the Gaussian filter function, which could be described as

$$G_{\mathrm{f}}(x, y) = \frac{1}{2\pi\sigma^2} \mathrm{e}^{-\frac{x^2+y^2}{2\sigma^2}} \tag{18}$$

where σ is the standard deviation.

To make a composite evaluation of restoration quality, a comprehensive evaluation factor Q_c is proposed in this paper. Q_c is described through GMG and PSNR.

$$Q_c = w_1 \times \frac{GMG_r - GMG_g}{GMG_g} + w_2 \times \frac{PSNR_r - PSNR_g}{PSNR_g} \tag{19}$$

where w_1 and w_2 are the weight parameters of two evaluation indexes above mentioned. GMG_r and $PSNR_r$ are restored image's evaluation indexes, and GMG_g and $PSNR_g$ are blurred image's evaluation indexes.

Obtain the optimal power spectrum ratio C means the comprehensive evaluation factor Q_c must be as large as possible.

In this paper, an automatic iterative Wiener filter algorithm is proposed to calculate optimal power spectrum ratio.

Step 1: Initialize the values of power spectrum ratio C_0, alternative step ΔC, and alternative times T_c. If the image is the first one that needs to be processed, then $C_0 = 0.01$. Otherwise $C_0 = C_{pre}$, where C_{pre} is the previous image's power spectrum ratio. $C = C_0$;
Step 2: Implement Gaussian filter on the original image.
Step 3: Update the power spectrum ratio C_1 and C_2, $C_1 = C_0 + \Delta C$, and $C_2 = C_0 - \Delta C$. Calculate corresponding comprehensive evaluation coefficients Q_{c1} and Q_{c2} by Eq. (19). Compare Q_{c1} and Q_{c2}. If $Q_{c1} > Q_{c2}$, go to Step 5. Otherwise go to Step 6.
Step 4: $C = C_0 + \Delta C$ and $t = t + 1$. Circulate implementing the Wiener filter process and calculating a series of comprehensive evaluation factor values until $t = T_c$.
Step 5: $C = C_0 - \Delta C$ and $t = t + 1$. Circulate implementing the Wiener filter process and calculating a series of comprehensive evaluation factor values until $t = T_c$ or $C = 0$.
Step 6: Find the corresponding power spectrum ratio C_{opt} with the maximum value of Q_c. If C_{opt} is not the initial value or the last value, then C_{opt} would be the optimal factor of Wiener filter, and $C_{pre} = C_{opt}$. The algorithm ends.
Step 7: If the largest value of Q_c corresponds to the initial value of C, then $\Delta C = \Delta C/10$ and repeat Step 2–Step 5. On the other hand, if the corresponding value is the last one, $\Delta C = \Delta C \times 10$ and repeat Step 3–Step 7.

The main reason to implement Step 7 is to avoid the partly maximum value caused by unfit initial value of C, and Step 3 could find the comprehensive evaluation factor change's trend, which could reduce iterative times and calculation time.

(a) Original image (b) Degraded image

Fig. 2 Original "Lena" image and degraded image

4 Experimental Results and Analysis for Image Restoration

To validate the performance of our proposed restoration model, we developed software implementations in an embedded GPU.

4.1 *Simulations*

In this part, simulation experiments are carried out to verify the restoration results of improved Wiener filter and comprehensive evaluation factor. Classical image of "Lena" is chosen as the testing image, as shown in Fig. 2a. The size of this image is 512 × 512. This image is degraded by the uniform motion PSF and Gaussian noise function, as shown in Fig. 2b, and the blurred scale is 5, and $C_0 = 0.01$, $\Delta C = 0.01$, $C_t = 20$. The weight parameters of two evaluation indexes are set as 0.6 and 0.4.

To illustrate the superiority of the proposed algorithm, another two different methods are implemented to compare with this paper's Wiener filter method. The two methods are conventional Wiener filter method and inverse filtering method, respectively. The restored results are shown in Fig. 3. It is in evidence that the adaptive Wiener filter's restored result is better than the other two methods through observing with naked eyes. At the condition of noise pollution, inverse filter's restored results are not satisfied, so does conventional Wiener filter. The optimum power spectrum ratio C is calculated as 0.04.

4.2 *Experimental Restoration Effects*

In this part, some degraded remote sensing images with size 2000 × 2000 are chosen to validate the proposed method, and three images' restored results are shown in Fig. 4a1–c1. These images are degraded by the uniform motion PSF and Gaussian

(a) Inverse filter (b) Conventional Wiener filter (c) Adaptive Wiener filter

Fig. 3 Restored results of three methods

noise, but the blurred scale is unknown. According to the previous experience, the initial parameters are set as $C_0 = 0.01$, $\Delta C = 0.01$, $C_t = 10$. The weight parameters of three evaluation indexes are set as 0.65 and 0.35.

Figure 4a2–c2 is restored results of conventional Wiener filter. It is obvious that conventional Wiener filter's restored results are not satisfied, just like the previous section's experiments. New noises are drawn into the images, and the details are fuzzier than the original images.

Figure 4a3–c3 is restored results of the proposed adaptive Wiener filter. From the comparison between original images and restored images, it is in evident that the restored image's contours and details are more distinct, and the surface features seem more prominent and seem more brightsome than degraded image. The time statistics result shows that the average spending time of one image's restoration is just 0.252 s, which is acceptable of on-board process. Table 1 shows three evaluation indexes' comparison of 3 images. It is obvious that through the adaptive Wiener filter, all of the remote sensing images' quality is improved to a certain extent.

5 Summary

This paper presents a novel remote sensing image restoration method based on adaptive Wiener filter. Compared to traditional Wiener filter or inverse filter methods, the proposed method mainly has two advantages: (1) Iteration method based on comprehensive evaluation factor could acquire more accurate value of power spectrum ratio, which is significant to image restoration results. (2) The proposed restoration method integrates two quality factors, which represents the original and non-degraded image's property to a certain extent. Experimental results demonstrate that the proposed algorithm could acquire satisfactory remote sensing image recovering results, and the proposed adaptive Wiener filter algorithm could be used to restore remote sensing images on-board.

(a1) Original image 1 (a2) Conventional Wiener filter (a3) Adaptive Wiener filter

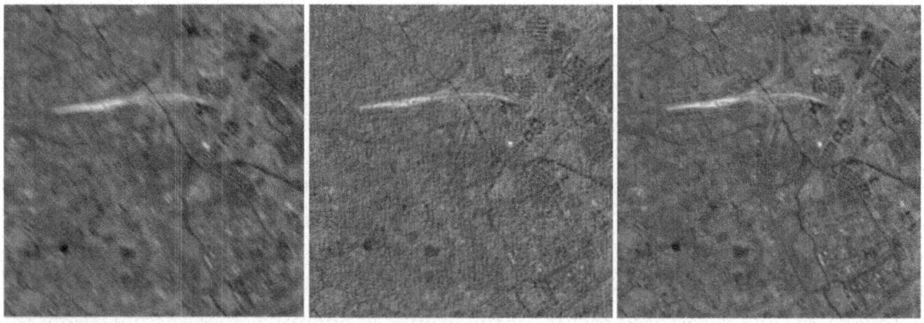

(b1) Original image 2 (b2) Conventional Wiener filter (b3) Adaptive Wiener filter

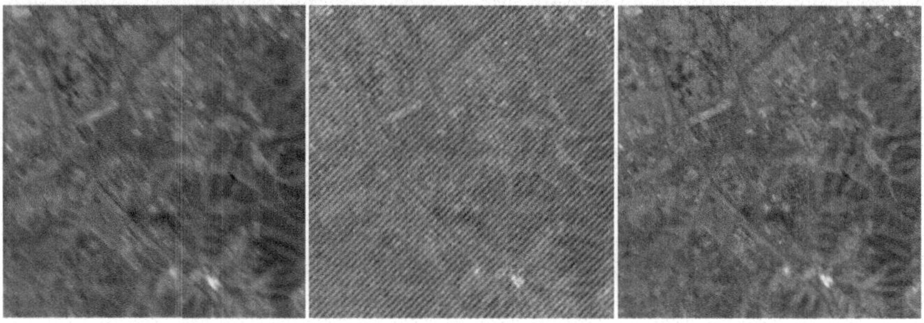

(c1) Original image 3 (c2) Conventional Wiener filter (c3) Adaptive Wiener filter

Fig. 4 Restored results of the proposed adaptive Wiener filter method

Table 1 Three methods' process results

Image No.	GMG		PSNR		Q_c
	Original	Restored	Original	Restored	
1	0.0037	0.0054	16.68	20.87	0.710659
2	0.0040	0.0048	18.67	21.65	0.359614
3	0.0037	0.0049	19.46	22.29	0.469751

References

1. Jia T, Shi Y, Zhu Y, Wang L (2016) An image restoration model combining mixed L1/L2 fidelity terms. J Vis Commun Image R 38:461–473
2. Cai X, Chan R, Zeng T (2013) A two-stage images segmentation method using a convex variant of the Mumford-Shah model and thresholding. SIAM J Imag Sci 6:368–390
3. He L, Cui G, Feng H, Xu Z, Li Q, Chen Y (2015) The optimal code searching method with an improved criterion of coded exposure for remote sensing image restoration. Opt Commun 338:540–550
4. Wu X, Wu J, Zhang H (2011) Research on image restoration techniques based on inverse filtering algorithm. Inf Technol 35(10):183–185
5. Richardson WH (1972) Bayesian-based iterative method of image restoration. J Opt Soc Am 62(1):55–59
6. Lucy L (1974) An iterative technique for the rectification of observed distributions. Astron J 79(6):745–753
7. Khan M, Nizami IF, Majid M (2019) No-reference image quality assessment using gradient magnitude and wiener filtered wavelet features. Multimedia Tools Appl 78(11):14485–14509
8. Wang H, Anthony TSH, Li S (2014) A novel image restoration scheme based on structured side information and its application to image watermarking. Sig Process Image Commun 29:773–787
9. Chen S (2009) Development of space remote sensing science and technology. Spacecraft Eng 18(2):1–7
10. Peng Q (2010) Study about motion-blurred image restoration. University of Electronic Science and Technology of China, Chengdu
11. Chen BH, Huang SC, Ye JH (2015) Hazy image restoration by bi-histogram modification. ACM Trans Intell Syst Technol 50(7):17
12. Gong Z, Shen Z, Toh KC (2014) Image restoration with mixed or unknown noises. Soc Ind Appl Math 12(2):458–487
13. Bioucas-Dias José M (2006) Bayesian wavelet-based image deconvolution: a GEM algorithm exploiting a class of heavy-tailed priors. IEEE Trans Image Process 15(4):937–951
14. Figueiredo MAT, Nowak RD (2003) An EM algorithm for wavelet-based image restoration. IEEE Trans Image Process 12(8):906–916
15. Yang B, Zhang Z, Dai S, Xiao Z (2012) Modified image restoration algorithm using neural network based on harmonic model. AASRI Procedia 1:196–206
16. Zhou YT, Chellappa R (1988) Image restoration using a network. IEEE Trans Acoust Speech Signal Process 36(7):1141–1151
17. Paik JK, Katsaggelos AK (1992) Image restoration using a modified Hopfield network. IEEE Trans Image Process 1:49–63
18. Han Y, Wu L (2002) Image restoration using a modified Hopfield neural network of continuous state change. Sig Process 20(5):431–435
19. Perry SW, Guan L (2000) Weight assignment for adaptive image restoration by neural networks. IEEE Trans Neural Netw 11(1):156–170
20. Erler K, Jernigan E (1994) Adaptive image restoration using recursive image filters. IEEE Trans Signal Process 42(7):1877–1881
21. Wu W, Kundu A (1992) Image restoration using fast modified reduced update Kalman filter. IEEE Trans Signal Process 40(4):915–926
22. Helstrom CW (1967) Image restoration by the method of least square. Josa 57(3):297–303
23. Slepian D (1967) Linear least-squares filtering of distorted images. J Opt Soc Am 57(7):918–919
24. Li J, Gong W, Li W (2015) Dual-sparsity regularized sparse representation for single image super-resolution. Inform Sci 298:257–273
25. Liu J, Huang T, Selesnick I, Lv X, Chen P (2015) Image restoration using total variation with overlapping group sparsity. Inform Sci 295:232–246

26. Di H, Yu Q (2006) Autocorrelation based identification the blur extent of uniform motion blurred images. J Nat Univ Def Technol 28(5):123–125
27. Jin H, Wang Y (2014) A fusion method for visible and infrared images based on contrast pyramid with teaching learning based optimization. Infrared Phys Technol 64(3):134–142

Yunsen Wang male, Ph.D. in computer science and technology, senior engineer, research areas includes artificial intelligence and remote sensing image process.

Design of Space-Based Information Service Architecture Based on Mobile Cloud

Ying-chun Hou, Chao Yang, and Yan Hou

Abstract In view of the practical problems of space-based information application, this paper analyzes the design idea of space-based information application service on the basis of the functional requirements of space-based information service and puts forward the design of space-based information application service based on mobile cloud. In this study, the detailed design of the overall structure of space-based information service is carried out from presentation layer, service layer, mobile cloud resource layer and core infrastructure layer. Efficient space-based information service system provides some theoretical reference.

Keywords Mobile cloud · Space-based information service · Architecture design

Through decades of construction and development, China has established a complete space-based information acquisition, processing and application system, and accumulated a large amount of information resources. However, at present, the application of space-based information in China is mainly aimed at the application of various professional departments, with different distribution areas, heterogeneous formats, different data development platforms, lack of unified specifications with external user interfaces, low efficiency of resource use, difficulties in data sharing, low timeliness and relevance of information, which restrict the performance of the whole space-based information system in China [1]. To do this, we need new methods and techniques to establish a unified service architecture, unified organization, management, access, sharing, integration and analysis of distributed space-based information and function, establish a unified, efficient and on-demand service synergy space-based information application environment, realize the multi-level, the space-based information analysis processing and application of high performance services. Since cloud computing system abstracts software and hardware into the form of resources and provides services to the customer service in the form of service, the customer service does not need to pay attention to its internal structure and operation mode. Cloud computing makes the customer service acquire better computing and storage capacity by adopting the way of service, effectively improving the flexibility and effectiveness

Y. Hou (✉) · C. Yang · Y. Hou
Space Engineering University, Beijing, China
e-mail: winter_hyc@163.com

L. Wang et al. (eds.), *Proceedings of the 6th China High Resolution Earth Observation Conference (CHREOC 2019)*, Lecture Notes in Electrical Engineering 657,
https://doi.org/10.1007/978-981-15-3947-3_21

of information service, and providing a technical architecture construction idea for the design of space-based information service system.

1 Functional Requirements of Space-Based Information Service System

Faced with wide distribution, complex heterogeneous, functional diversity of space-based information resources, the space-based information service system should not only realize the seamless connection of all kinds of resources in the space-based information system through the integrated network of space and ground, also need to take full advantage of the space-based information system use new technologies such as the cloud, big data, virtualization and network resources, to provide users with value-added services [2]. Space-based information application service system needs to have different functions for different customer service terminals. For resource providers, it needs to provide functional support of resource interconnection, management and integration. For system managers, it needs to provide users with security protection and unified function support in time and space. For information service-oriented users, it needs to provide information on demand services, fusion and sharing of functional support. Functional requirements of space-based information service system mainly include the following aspects (as shown in Fig. 1).

Among them, the interconnection function of heterogeneous resources realizes physical interworking of resource providers in space-based information systems in the form of network nodes, such as space-based information processing center, data storage and management center, and operation and control center. Users of information services, such as those of combat forces, can access the space-based information database system within their scope of authority through the mobile cloud network,

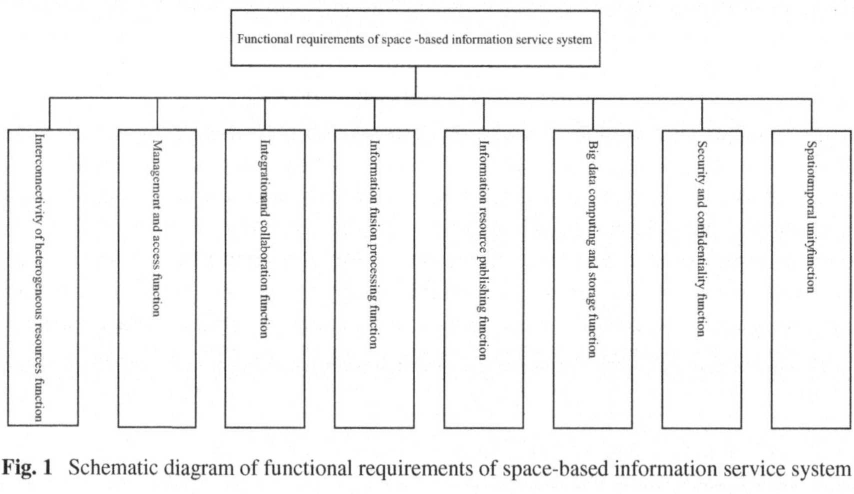

Fig. 1 Schematic diagram of functional requirements of space-based information service system

or put forward information support requirements and communication requirements to the space-based information system through the access to the mobile cloud network. The space-based information system can complete command and control, information transmission and distribution of information acquisition and processing resources through the mobile cloud, etc.

The management and access function realize the unified management of space-based information resource providers and users of information application services, such as the monitoring and management function of land observation, ocean observation, meteorological observation and other resources. Space-based information service is capable of providing safe and reliable user authority and resource management for information service users, classifying users according to priority, and providing users with space-based information service computing, storage, resource sharing and other functions.

The integration and collaboration function organically combine space-based information acquisition systems, space-based information transmission systems and space-based information fusion processing systems distributed in different spaces to form a unified and efficient service guarantee system. It not only has the functions of each subsystem, but also has the overall functions that each subsystem does not have. It can realize the organic aggregation and efficient coordination of multi-platform, multi-sensor and multi-resolution space-based information acquisition, processing and transmission capacity.

The information fusion processing function supports the information fusion of electronic signals and communication signals, various imaging information fusion, electronic signals and imaging information fusion and other types of information fusion, as well as the information association mining of target association, image product association, knowledge data association, geo-location association and so on.

The information resource publishing function provides the publishing function of space-based information services and directory information at different levels. Users of information services can query, browse and download relevant space-based information resources according to their authorization levels, and query and store space-based information services within their authority. The unauthorized space-based information service can be inquired according to the catalog information. If necessary, the users of the information service can apply to the space-based information resource manager, who will transmit the information according to the level of the users of the information service according to the approval rules.

The big data computing and storage function can support large data massive structured, semi-structured, and unstructured data, environment, electromagnetic, humanistic geography data modeling, collection, storage, analysis, mining, offline and real-time distributed storage and calculation, to achieve the organization of the armed forces day data management and analytic foundation deeply.

The security and confidentiality function checks and limits the qualification and authority of users by formulating unified security policies to prevent illegal users from accessing data or legitimate users from accessing data beyond their rights, so that space-based information can be accessed by legitimate users and used according to the required characteristics.

The spatiotemporal unity function provides time and space reference standards, provides military users with time reference and navigation and positioning services, and achieves the purpose of ensuring the global situation consistency of application scenarios, implementing joint operations and improving the efficiency of joint operations.

2 Design Idea of Space-Based Information Service Architecture

Space-based information service system is an open and dynamic system. In the design of architecture, besides considering the static structure of the system, it also needs to consider the dynamic factors such as the change of battlefield environment, service resources and users, and optimizes or reconstructs the service process according to the different operational requirements.

When designing space-based information service architecture, the first is to solve the integration of heterogeneous space-based information resources and give full play to the overall advantages of the system. The space-based information service system is a space-ground integration system composed of space-based information acquisition system, information processing system and distribution system. In the structural design, the various components involved in the system and the relationship between them should be considered systematically. The application service system should realize seamless connection of various resources through the space-ground integration information network, eliminate information island and make full use of cloud computing, big data technology, integration of heterogeneous systems, unified management and control [3] of access resources and available states, to give full play to the overall advantages of space-based information system.

The second is to solve the on-demand sharing and collaboration of service resources. Based on the concept of distributed mobile cloud data center, by means of virtual integration and service encapsulation of resources, mapping resources such as surveying and mapping data, navigation data, meteorological data, hydrological data, optical remote sensing, microwave remote sensing, infrared remote sensing, etc. scattered in different geographical locations into a service collection with standard interface, shielding the heterogeneity and distribution of underlying resources, and realizing resources source virtualization and middleware technology, through unified resource management, service release, information exchange and other ways, realize the sharing of various service resources on demand.

Third is to solve the interworking openness of the system. The space-based information service system can meet dynamic and variable characteristics such as scalable scale, reconfigurable process, portable platform, extensible function, customizable service and updated technology. Through unified interface standards, service interface standards and information interaction standards, all kinds of service resources can be easily accessed, moved out and applied. When the service resources and objects are damaged, they can be supplemented at any time to enhance the dynamic and openness of the system, realize plug and play of service resources, and adapt to the dynamic changes of battlefield resources and users.

3 Overall Design of Space-Based Information Service Architecture

The space-based information service system integrates computing resource, storage resource, communication resource, information resource, knowledge resource, software resource and command and control systems flexibly through the space-integrated network, realizing end-to-end connection and providing comprehensive resource sharing and dynamic information processing capacity. The architecture of space-based information service system includes presentation layer, service layer, mobile cloud resource layer and core infrastructure layer, as shown in Fig. 2.

3.1 The Presentation Layer

The presentation layer provides human–computer interaction services at the customer service end, mainly to build a bridge between space-based resource providers and users of information services, and to realize information interaction services in the form of Web browser, image browser, text browser and other functions. Users of information services include system management users and end users. System management users are mainly responsible for the maintenance of space-based information service system and the unified management of customer service users, including the establishment, modification, single login, authorization of available resources, use of resources, etc. The interactive services provided by the presentation layer achieve platform-independent, location-independent, secure, easy-to-use and other goals. Users of information services can use the authorized resources in the form of virtual organizations or directly.

3.2 The Service Layer

The service layer is the main part of space-based system architecture. The application of services runs through the whole process of space-based information acquisition, sharing, analysis and processing, encapsulates various business functions into services, and provides an application program interface for other services on the network to call and integrate, so as to realize dynamic space-based information services. It can be divided into core service layer and basic application service layer.

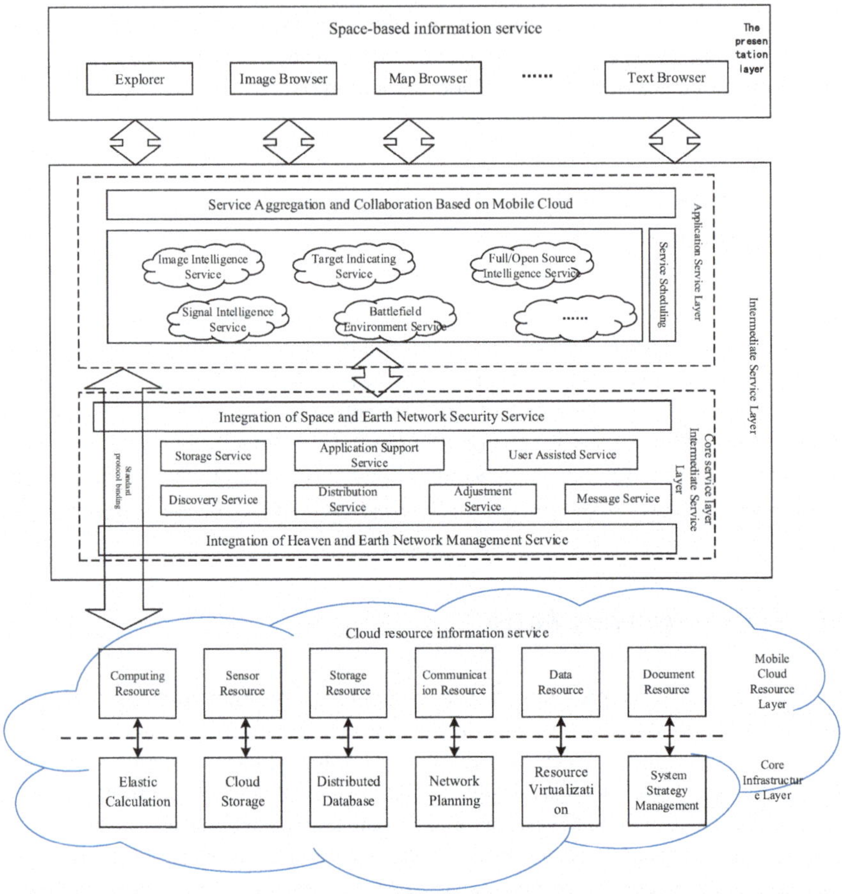

Fig. 2 Space-based information service architecture based on mobile cloud

3.2.1 Core Service Layer

The core service layer is located at the bottom of the space-based information service system. It is responsible for the management of space-based information service system and provides the bottom and universal support for space-based information service services, such as network management, security, message, storage, user assistance, information discovery, distribution and mediation. Core services include a set of standards, specifications, guidelines, architectural definitions, software infrastructure, reusable parts, application program interfaces, runtime environment definitions, reference tools and methods for building system environments. Core services allow greater flexibility in the selection and configuration of parts. Core services achieve security and management services through the integration of space and earth network foundation, and are relatively independent from the application service layer.

The function upgrading and expansion of the application service layer will not affect the technical infrastructure of the core service layer, and can solve the problems of heterogeneity and interconnection of space-based information system platform. The core services layer includes service functions as shown in Table 1.

3.2.2 Application Services Layer

Through the service aggregation and collaboration engine of mobile cloud, the application service layer provides an organic and integrated service set for the application of portal layer and provides functional support for the application, including image intelligence service, signal intelligence service, battlefield environment service, full source/open source intelligence service and other service sets. Each type of service can be used as a mobile cloud to provide corresponding service services, including the reception, processing, storage and other service functions of various business information [4].

3.3 Mobile Resource Layer

The mobile resource layer covers various resources that can be deployed by the space-based information system, including computing resource, sensor resource, storage resource, communication resource, data resource and file resource. These information resources are provided to combat users as a kind of combat resources in the battlefield environment based on network information system, thus realizing the combat effectiveness of space-based information.

Cloud resource information service realizes resource configuration and heterogeneous mobile cloud resource scheduling according to user requirements. All kinds of mobile cloud resources are composed of a large number of heterogeneous servers. Cloud resource information service provides services to users with virtual machine as the basic unit. The flexible scheduling of resources can be realized through the migration and deployment of virtual machines, and the system's flexibility can be enhanced. Cloud resource information servers are usually configured in different geographical locations according to different business requirements. The decentralized configuration mode reduces the difficulty of resource management and can better adapt to the characteristics of future information battlefield environment. The cloud task scheduling of space-based information service system can be divided into batch scheduling and real-time scheduling [5]. Batch scheduling refers to that the required tasks are first put into the task set after they are submitted, and they will not be executed immediately until the scheduling cycle or a unified scheduling occurs. Online scheduling refers to the scheduling and execution of tasks immediately after they are submitted. The real-time performance of task execution is an important indicator to measure the scheduling effect. Batch scheduling gets more information due to waiting, so the scheduling is more targeted, while the online scheduling is more real-time.

Table 1 Core services of space-based information service system

Ordinal	Service name	Content
1.	Integration of space and earth network management service	Ensure the normal operation of the space-based information application service system of the background support and management system, mainly through the system of centralized management and the real-time monitoring for space and earth resources, accurately grasp the operation of the equipment and application service system and the using condition, maximize reduce failure and improve service reliability, availability and continuity
2.	The discovery service	Provides the ability to express and perform lookup, enabling users to locate data resources and services in the shared space-based information repository, and helping users quickly find the required specialized information
3.	Message service	Provide synchronous and asynchronous information exchange capabilities for entities and users on the space-based application services infrastructure
4.	Distribution service	Responsible for summarizing various requirements information from different combat units, and after comprehensive analysis, space-based resource coordination service department will coordinate the response and provide corresponding requirements summary, information fusion, integrated data and other functions
5.	Distribution service	To achieve the goal of transferring appropriate information to appropriate users, and to be responsible for transferring information from space-based information service system to various information users at all levels
6.	Storage service	Provides users with various services to support data information, and realizes the addition, deletion, modification, data access, extraction and other services of space-based data information

(continued)

Table 1 (continued)

Ordinal	Service name	Content
7.	Application support service	Provides the capabilities and resources needed to develop, operate, and maintain space-based information services applications, and provides reliable computing capabilities to users
8.	User support service	Provides automated assistive functions to reduce the workload required for human-intensive tasks
9.	Integration of space and earth network security service	Be responsible for detecting the vulnerability of network, service or system, and provide communication protocol service, security authentication service, anti-virus intrusion service, etc. in accordance with the prescribed policies, to ensure the security of service system resources and user information

The two scheduling modes are configured with each other, and different scheduling modes are adopted according to different tasks. Through resource scheduling, real-time processing algorithm is adopted to assign tasks to the server resources with the fastest completion of tasks, and virtual machines are used to achieve optimal matching of target system, constraint conditions, space-based information resources, task requirements and other factors, as shown in Fig. 3.

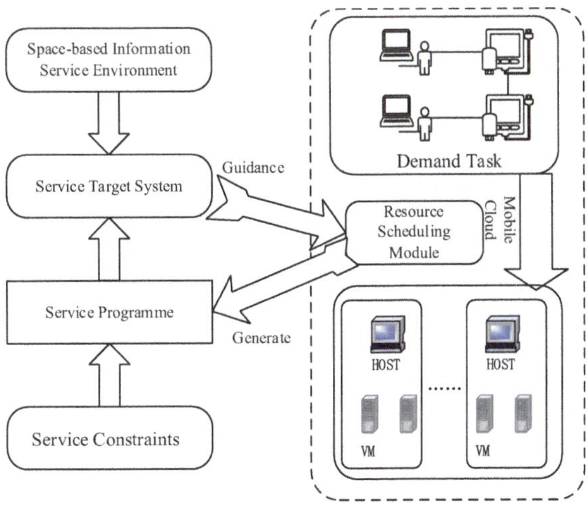

Fig. 3 Scheduling process of space-based information service based on mobile cloud

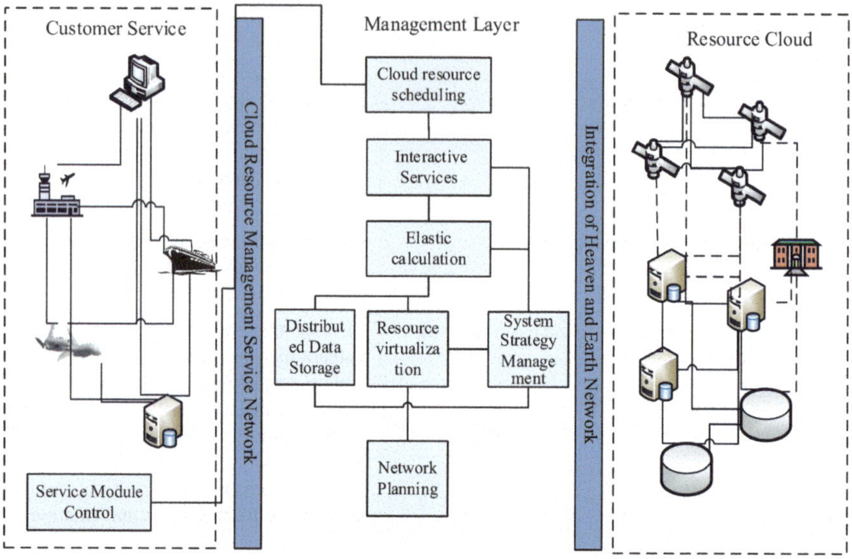

Fig. 4 Physical view of space-based information system based on mobile cloud pattern

3.4 Core Infrastructure Layer

The core infrastructure layer includes various hardware and software basic equipment needed to ensure the normal operation of the cloud resource service system, including cloud resource computing module, resource scheduling module, information resource storage equipment and security equipment, etc., as shown in Fig. 4. Through the cloud computing platform, a unified space-based information resource pool is integrated to achieve the unified management of networked computing and storage environment of information. All kinds of mobile space-based information resources in the cloud are carriers of cloud computing services. Only through centralized control of space-based information resources, cloud computing's efficiency advantages can be reflected, and the organization of space-based information resources also determines the elastic support ability of space-based information resources for operations. Current cloud computing can meet the needs of military applications. Specifically, space-based information service system focuses more on heterogeneous information resources distributed physically and in different locations, and adopts mobile cloud mode to effectively support combat tasks. The resource scheduling module is capable of task processing and resource scheduling, mainly responsible for managing various space-based information mobile resource clouds, and realizing the configuration of information resources based on the integration of heaven and earth network environment. Space-based information data can be stored in a distributed manner according to the requirements of customer service users, and classified data management and storage of meteorological, reconnaissance, communication, navigation, early warning and nuclear explosion detection can be realized. Through network planning and resource virtualization, business services in information domain are mapped and associated with multi-type information devices in physical domain, so as to realize cooperative use of resources.

4 Conclusion

Based on the heterogeneous characteristics of space-based information system, this paper analyzes the functional requirements and design ideas of space-based information application service, proposes the architecture of space-based information service system based on mobile cloud and designs the overall architecture of space-based information service from the presentation layer, service layer, mobile cloud resource layer and core infrastructure layer. The space-based information service system built on the mobile cloud service architecture realizes the collection and comprehensive processing of heterogeneous space-based information resources and features such as service flexibility and scalability, enabling convenient and fast sharing of space-based information.

References

1. Guan Q-b, Feng S-x, Ma Y-h (2012) Research on space-based information service model. J Equip Coll 6:66–70
2. Liu L-x (2015) Integration of space and earth network. Science Press, Beijing
3. Wang J, Yang J-p, Liang W-t, Mao X-b (2016) Construction of integrated space-earth network information system. Command Inf Syst Technol (4):61–65
4. Liu G, Lu Z et al (2018) Design of space-based information application service system based on cloud architecture. J China Acad Electron Sci (5):527–544
5. Ji H-r (2015) Research on command information system architecture based on mobile cloud model. National University of Defense Technology

Monitoring of a Subgrade Subsidence by Means of Ground-Based SAR Interferometry

Pingping Huang, Fang Liu, Weixian Tan, Wei Xu, Qi Lin, and Huifang Ren

Abstract Ground-based synthetic aperture radar (GB-SAR) has already been recognized as a powerful tool, complementary or alternative to spaceborne SAR interferometry, for terrain deformation monitoring, and building deformation monitoring. In this paper, the application of GB-SAR interferometry for subgrade subsidence monitoring is analyzed. The experiment carried out in China in 2018 was aimed at finding potential subgrade subsidence risks of a railway. Interferometric data were collected through a GB-SAR system operating at Ku-band. A slight subgrade subsidence is found, and the relationship between periodicity of displacement and temperature is discussed. Results show that the GB-SAR can be attractive and effective in the monitoring of subgrade subsidence.

Keywords Ground-based synthetic aperture radar · Differential interferometry · Subgrade subsidence · Periodicity of displacement

1 Introduction

Radar differential interferometry (D-InSAR) from space in a leader position within the remote sensing community is a powerful tool in many application fields [1]. Terrain mapping [2], alpine glacier motion [3, 4], the monitoring of open-pit quarries [5] and landslide are some of the investigated topics. Satellite, airborne and GB-SAR interferometry are derived from the same principles, but they are often characterized by specific problems, mainly due to the difference of the geometry of the observation. GB-SAR installations are usually at their best when monitoring small-scale phenomena, while satellite and airborne radar are able to monitor a very large area. Different kinds of signals are usually employed as well: step frequency continuous

P. Huang · F. Liu (✉) · W. Tan · W. Xu · Q. Lin · H. Ren
College of Information Engineering, Inner Mongolia University of Technology, Hohhot 010051, Inner Mongolia, China
e-mail: liufang0471@163.com

Inner Mongolia Key Laboratory of Radar Technology and Application, Inner Mongolia University of Technology, Hohhot 010051, Inner Mongolia, China

L. Wang et al. (eds.), *Proceedings of the 6th China High Resolution Earth Observation Conference (CHREOC 2019)*, Lecture Notes in Electrical Engineering 657,
https://doi.org/10.1007/978-981-15-3947-3_22

wave (SFCW) instead of frequency-modulated continuous wave (FMCW) is usually preferred in GB-SAR.

The first paper is about GB-SAR interferometry from 1999, and the potentials of GB-SAR in the field of civil engineering were demonstrated studying the displacements of a dam [6]. Within the last twenty years, the development of GB-SAR interferometry has demonstrated their usefulness for the precise monitoring of ground displacement episodes. It can be considered a useful technique, particularly when the use of optical sensors is not possible or compromised by weather conditions.

This paper discusses the results of a railway subgrade subsidence experiment carried out in the year 2018. The paper is organized as follows: Sect. 2 is devoted to the description of the observation geometry and basic principle of GB-SAR. In Sect. 3, after a brief description of the experimental site, the data collected by means of a GB-SAR system operating at Ku-band is analyzed, and the experimental results are discussed. Finally, some conclusions are drawn.

2 Principle

2.1 The SFCW and SAR Technique

The employed GB-SAR instrumentation consists of a SFCW transceiver unit working at Ku-band, a linear horizontal track where the antennas move for creating a synthetic aperture, and a personal computer controlling the motion of the antennas and data recording. The SFCW and SAR technique improve the resolution of range and azimuth, respectively. The sketch of GB-SAR is shown in Fig. 1.

The SFCW technique is based on the transmission of a burst of N monochromatic pulses equally spaced in frequency (with frequency step of Δf Hz) within a bandwidth, and the transmitted signal can be expressed as [7, 8]:

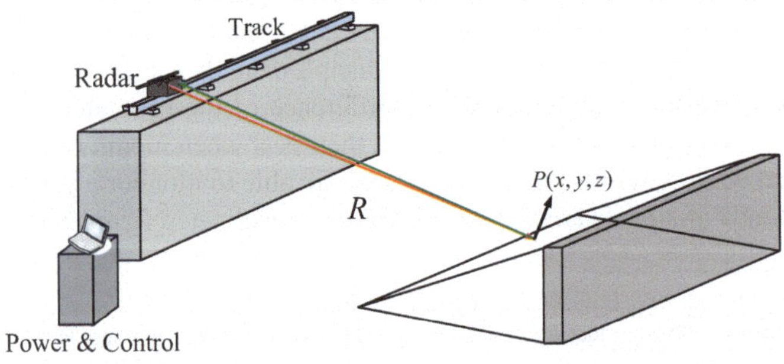

Fig. 1 Sketch of GB-SAR

$$S_{\rm rt}(t) = \sum_{n=0}^{N-1} A_n \exp\{j2\pi(f + n\Delta f)t\} \tag{1}$$

where A_n is the amplitude of each narrow-band pulse, Δf is the frequency stepping quantity, f is the initial frequency and the duration of t is one pulse cycle. Its effective bandwidth B is $N\Delta f$, and its range resolution is [8]:

$$\Delta R = \frac{C}{2N\Delta f} \tag{2}$$

where C is the speed of light.

When the antenna moves along the straight track, radar echoes from the illuminated scene are collected. Images containing amplitude and phase information are formed by means of standard SAR focusing techniques. The resolution of azimuth is related to the length of synthetic aperture, which can be expressed as [8]:

$$\Delta\theta = \frac{C}{2Lf_{\rm c}} = \frac{\lambda_c}{2L} \tag{3}$$

where $f_{\rm c}$ is the central frequency of the SFCW, $\lambda_{\rm c}$ is the wavelength corresponding to $f_{\rm c}$ and L is the length of synthetic aperture. In order to avoid azimuth ambiguity, the sampling interval must satisfy equation [8]

$$\delta \le \frac{C}{2f_{\rm c}\sin(\theta_{\rm a})} \tag{4}$$

where $\theta_{\rm a}$ is the antenna's half-beam width.

2.2 The GB-SAR Interferometric Technique

GB-SAR interferometry is based on exploiting the phase-differences. If decorrelation sources are negligible, the phase φ of each pixel of a SAR image contains information about its distance from the sensor. Phase will vary accordingly when displacement of the observed surface occurs, a map can be obtained, and it is referred to as an interferogram. GB-SAR with a null spatial baseline, assuming that the atmospheric phase and other error sources are negligible, the displacement can be obtained by using the following simple equation [9]:

$$\Delta R = \frac{\Delta\varphi C}{4\pi f_{\rm c}} \tag{5}$$

where $\Delta\varphi$ is the phase difference.

Fig. 2 Actual monitoring scene of the railway subgrade

3 Experimental Campaign and Data Analysis

3.1 Experimental Site and Instrumentation

The subgrade settlement monitoring experiment of a railway was carried out from September 15 to September 25, 2018. It is located at the junction between the subgrade and the railway bridge, during the experiment the railway was under a normal traffic. Figure 2 shows the experimental site, most of the area is covered by weeds. Above the weed-covered area, there are a lot of scattered gravel blocks and electric wire towers. The subsidence of this section was found during the routine inspection by railway staff, and then, GB-SAR was deployed for subsidence monitoring experiment.

In the monitoring process, the MPDMR-05-LSA1701 GB-SAR system developed by Inner Mongolia Mypattern Technology Co. Ltd. is used, and it consists of a SFCW radar, a linear horizontal track, which is 1.6 m long, and two diagonal horn antennas, one for transmitting the radar signal and the other for receiving the echo signal, moving at discrete incrementing positions for creating a synthetic aperture. The system can be set up in about 30 min, provided that a stable plinth is available. It is frequency agile radar equipped with a capability of Ku-band and a spatial resolution of 0.2 m × 5.4 m rad.

3.2 Data Collection and Analysis

The data collection was performed in mid-autumn. The experiment took place in the area corresponding to the west of the railway, and a total of 1265 groups of interferometry data were collected. Synthesized image is shown in Fig. 3a, points with high brightness in the image are gravel blocks or electric wire towers, those points have

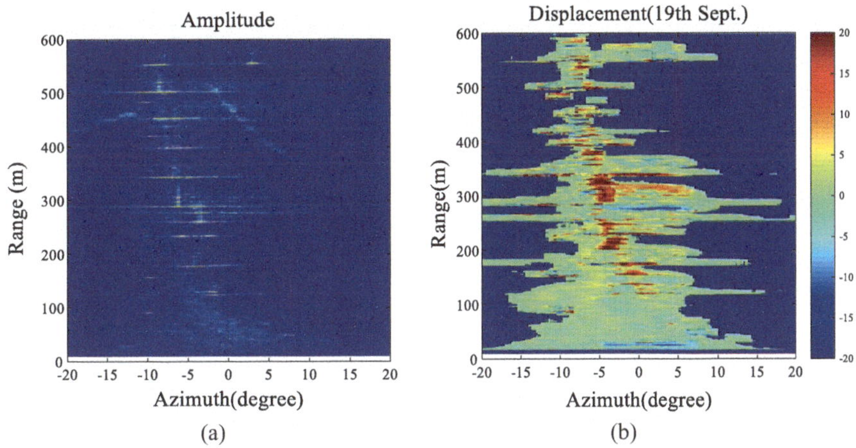

Fig. 3 **a** Synthesized image; **b** displacement image

a strong reflected energy, and some of them can be regarded as permanent scatterers (PS). As shown in Fig. 3b, blue indicates that the displacement of the monitoring area is away from the radar line of sight, and red indicates the displacement is close to the radar line of sight.

In order to make the settlement trend clear, seven monitoring points were selected for analysis and discussion. Those points were consistent with the characteristics of PS points. Figure 4a shows the position of those PS points in synthesized image. The accumulated displacement values of those PS points can be obtained in Fig. 4b. All the seven selected PS points show an opposite settlement trend, in which the displacement

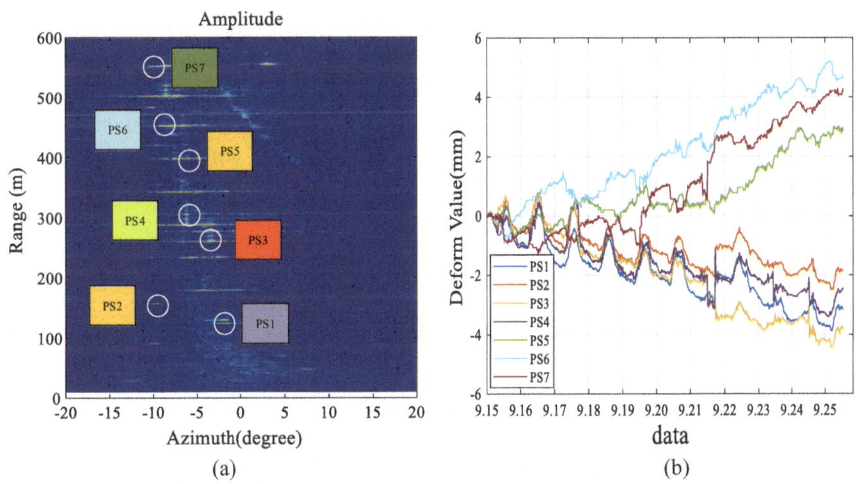

Fig. 4 **a** Position of PS points in synthesized image; **b** displacement curve

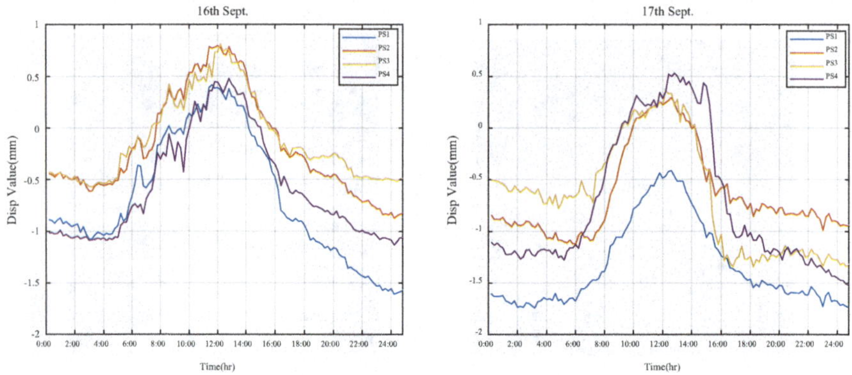

Fig. 5 Periodicity displacement of PS points 1–4

at PS points 1, 2, 3 and 4 shows a downtrend, while the displacement detected at PS points 5, 6 and 7 shows an uptrend. The maximum negative displacement at PS point 3 is about −4 mm, and the maximum positive displacement at PS point 6 is about 5 mm. Different geological conditions lead to this phenomenon. The displacement is within safe limits according to *Code for design on subgrade of railway* [10].

The displacement shows a periodicity in each day that was particularly obvious in the point ps1–ps4, as shown in Fig. 5. The daily displacement of each point shows the same variation trend. At 0:00–6:00, displacement shows a slight downtrend. After that, it shows an uptrend, and it rose to maximum at noon, and then began to fall, and it gets back to a slight downward after 18:00.

The monitoring accuracy of GB-SAR is mainly affected by two factors: frequency deviation of the system and atmospheric phase, but the errors caused by either the frequency deviation of the system or the atmospheric phase are random and will not have periodic influence on the displacement. Considering the corresponding relationship between temperature and time, it is found that the variation tendency of displacement is consistent with the variation tendency of temperature. Large temperature difference between day and night in autumn, resulting in concrete structures thermal expansion and cold contraction, the displacement will have a periodicity variation with it.

4 Summary

The MPDMR-LSA GB-SAR system was used to carry out a monitoring experiment, and its potential use in detecting the subgrade settlement has demonstrated. The GB-SAR operates from a distance, without requiring any contact with the subgrade, and the measurement procedure is simple. Dedicated signal processing software allows the displacement to be obtained practically in real time. Displacement tendency can be obtained from the GB-SAR displacement image. Under the effect of thermal expansion and cold contraction, the displacement shows a same periodicity variation in each day. It is the first time that the periodicity displacement with temperature variation detected by GB-SAR. However, here, it is a preliminary measurement campaign and encouraging further experimentation on this field.

References

1. Luzi G, Pieraccini M et al (2004) Ground-based radar interferometry for landslides monitoring: atmospheric and instrumental decorrelation sources on experimental data. IEEE Trans Geosci Remote Sens 2454–2466
2. Noferini L, Pieraccini M, Luzi G et al (2006) Ground-based radar interferometry for terrain mapping. In: IEEE international conference on geoscience and remote sensing symposium. IEEE, pp 2569–2572
3. Rudolf H, Leva D, Tarchi D, Sieber A (1999) A mobile and versatile SAR system. In: Proceedings IEEE IEEE International Geoscience and Remote Sensing Symposium (IGARSS'99), pp 592–594
4. Mecatti D, Noferini L, Macaluso G et al (2007) Remote sensing of glacier by ground-based radar interferometry. In: IEEE international geoscience and remote sensing symposium. IEEE
5. Iglesias R, Aguasca A, Fabregas X (2015) Ground-based polarimetric SAR interferometry for the monitoring of terrain displacement phenomena–part II: applications. IEEE J Sel Top Appl Earth Observations Remote Sens
6. Tarchi D, Rudolf H, Luzi G et al (1999) SAR interferometry for structural changes detection: a demonstration test on a dam
7. Fratini M, Pieraccini M, Dei D et al (2007) An experimental comparison of interferometric radar vs. accelerometers for monitoring of large structures. In: European radar conference. IEEE
8. Zhang X (2011) Study on subgrade SFCW SAR differential interferometry. National University of Defense Technology, pp 10–13. (in Chinese)
9. Luzi G, Pieraccini M, Mecatti D et al (2007) Monitoring of an alpine glacier by means of ground-based SAR interferometry. IEEE Geosci Remote Sens Lett 4(3):495–499
10. TB 10001-1999, Code for design on subgrade of railway, pp 13

A Polarization Calibration Solution Method of Airborne SAR Based on Point Target

Zhida Xu, Shucheng Yang, Chunquan Cheng, and Jianwei Tan

Abstract Polarization calibration is a necessary prerequisite for quantitative analysis and application of polarized synthetic aperture radar images. The polarization calibration method based on point target has the characteristics that the model is relatively simple and easy to understand. However, the existing point target polarization calibration method has the problem of relatively poor stability and accuracy. In this paper, a polarization calibration optimization method combining genetic algorithm and Levenberg–Marquard method is proposed. The genetic algorithm solution result is taken as the initial value into the Levenberg–Marquard method to obtain more accurate polarization distortion parameters. Experiments with airborne X-band polarization synthetic aperture radar images show that the proposed method has higher calibration accuracy than the classic Whitt algorithm and unconstrained algorithm, and can obtain more accurate and stable calibration results.

Keywords Polarization calibration · Point target · Genetic algorithm · Levenberg–Marquard

1 Introduction

Polarized synthetic aperture radar (PolSAR) can obtain rich feature information by acquiring echo information of ground objects under different polarization modes, and has been widely studied and applied in agriculture, forestry, and ocean [1, 2]. Polarization calibration is an essential pretreatment step in various quantitative applications

Foundation support: The National Key Research and Development Program of China, No. 2018YFF0215303; Chinese Academy of Surveying and Mapping Fundamental Research Project under Grant, No. 7771808.

Z. Xu · J. Tan
College of Geomatics, Shandong University of Science and Technology, Qingdao, China
e-mail: 536450196@qq.com

Z. Xu · S. Yang (✉) · C. Cheng · J. Tan
Chinese Academy of Surveying and Mapping, Beijing, China
e-mail: yangsc@casm.ac.cn

L. Wang et al. (eds.), *Proceedings of the 6th China High Resolution Earth Observation Conference (CHREOC 2019)*, Lecture Notes in Electrical Engineering 657,
https://doi.org/10.1007/978-981-15-3947-3_23

of PolSAR. Due to the PolSAR transceiver system, system distortion factors such as crosstalk and channel imbalance are mixed in the polarization scattering matrix obtained by the system. Therefore, it is necessary to obtain polarization distortion parameters including crosstalk and channel imbalance by polarization calibration to correct the system distortion effect existing in the image [3]. The current polarization calibration methods are mainly divided into three categories: calibration methods based on point targets, calibration methods based on distributed targets, and methods combining point targets with distributed targets [4]. The polarization calibration algorithm based on point target has the characteristics that the model calculation is relatively simple and easy to understand, and so, it has been widely studied and applied.

The polarization calibration method based on point target constructs a polarization calibration model based on the point target observation polarization scattering matrix and the point target theoretical polarization scattering matrix and solves the crosstalk and channel imbalance distortion existing in the system. Classic point target polarization calibration methods include Whitt algorithm, unconstrained algorithm, etc. [5, 6]. Among them, the Whitt algorithm has high versatility, but the algorithm has certain restrictions, and the three polarization scattering matrices used for calibration are required to have at least one reversible [7]. In addition, the current polarization calibration algorithm based on point target still has the problem of relatively poor accuracy and stability [8]. Therefore, this paper proposes a genetic algorithm (GA) and a damped least squares method (also known as Levenberg–Marquard method, LM) combined polarization calibration optimization method. Genetic algorithm has the characteristics of being able to solve the global optimal solution or approximate optimal solution of nonlinear equations. The LM method has the characteristics of fast convergence speed and high accuracy of calculation, but requires high initial values. Therefore, combining the characteristics of the two algorithms, the nonlinear equations constructed by the polarization calibration model are jointly solved to improve the accuracy of polarization calibration. At the same time, the method of this paper can solve the problem with only two point targets. Compared with the commonly used polarization calibration algorithm with three point targets, the dependence on the number of point targets is reduced.

2 Polarization Calibration Model

A polarization calibration method based on a point target is performed by arranging a corresponding corner reflector as a calibration point target in the calibration field. The basic expression of the polarization calibration model is [9, 10]:

$$\begin{bmatrix} O_{\mathrm{hh}} & O_{\mathrm{hv}} \\ O_{\mathrm{vh}} & O_{\mathrm{vv}} \end{bmatrix} = Ae^{\varphi j}\begin{bmatrix} 1 & \delta_1 \\ \delta_2 & f_1 \end{bmatrix}\begin{bmatrix} S_{\mathrm{hh}} & S_{\mathrm{hv}} \\ S_{\mathrm{vh}} & S_{\mathrm{vv}} \end{bmatrix}\begin{bmatrix} 1 & \delta_3 \\ \delta_4 & f_2 \end{bmatrix} + \begin{bmatrix} N_{\mathrm{hh}} & N_{hv} \\ N_{\mathrm{vh}} & N_{\mathrm{vv}} \end{bmatrix} = Ae^{\varphi j}RST + N \tag{1}$$

In the formula, observed scattering matrix O and real scattering matrix S can be linked by (1), in which T and R are distortion matrices in transmit channel and receive channel; δ_1, δ_2, δ_3, δ_4 are crosstalk parameters; f_1, f_2 are channel imbalance parameters; and N indicates the system noise. The theoretical value of A after completion of absolute radiation calibration is 1, ignoring the system noise N, and the polarization calibration model is simplified to:

$$O = \begin{bmatrix} O_{\text{hh}} & O_{\text{hv}} \\ O_{\text{vh}} & O_{\text{vv}} \end{bmatrix} = \begin{bmatrix} 1 & \delta_1 \\ \delta_2 & f_1 \end{bmatrix} \begin{bmatrix} S_{\text{hh}} & S_{\text{hv}} \\ S_{\text{vh}} & S_{\text{vv}} \end{bmatrix} \begin{bmatrix} 1 & \delta_3 \\ \delta_4 & f_2 \end{bmatrix} = RST \tag{2}$$

The key to polarization calibration is to solve the polarization distortion parameters of δ_1, δ_2, δ_3, δ_4, f_1, f_2 in (2). Equation (2) is transformed and expanded to construct a system of equations. For any point target, a system of equations containing four nonlinear equations can be constructed [11]:

$$f(x) = \begin{cases} (\delta_1\delta_4 O_{\text{vv}} - \delta_4 f_1 O_{\text{hv}} - \delta_1 f_2 O_{\text{vh}} + f_1 f_2 O_{\text{hh}})/((f_1 - \delta_1\delta_2) \times (f_2 - \delta_3\delta_4)) - S_{\text{hh}} = 0 \\ -(\delta_1 O_{\text{vv}} - f_1 O_{\text{hv}} - \delta_1\delta_3 O_{\text{vh}} + \delta_3 f_1 O_{\text{hh}})/((f_1 - \delta_1\delta_2) \times (f_2 - \delta_3\delta_4)) - S_{\text{hv}} = 0 \\ -(\delta_4 O_{\text{vv}} - \delta_2\delta_4 O_{\text{hv}} - f_2 O_{\text{vh}} + \delta_2 f_2 O_{\text{hh}})/((f_1 - \delta_1\delta_2) \times (f_2 - \delta_3\delta_4)) - S_{\text{vh}} = 0 \\ (O_{\text{vv}} - \delta_2 O_{\text{hv}} - \delta_3 O_{\text{vh}} + \delta_2\delta_3 O_{\text{hh}})/((f_1 - \delta_1\delta_2) \times (f_2 - \delta_3\delta_4)) - S_{\text{vv}} = 0 \end{cases} \tag{3}$$

To accurately solve the six polarization distortion parameters, at least two point targets are needed to construct a nonlinear equation system containing eight equations. Therefore, the problem of solving the polarization calibration parameters is transformed into an optimal solution problem for complex nonlinear equations [11].

3 Optimization Solution Method

For the optimal solution of complex nonlinear equations, a multivariate function is transformed into an extremum problem on a certain set of sets, that is, transformed into an optimization problem. At present, when solving complex nonlinear equations, the commonly used methods are least squares method, genetic algorithm, and simulated annealing algorithm [12].

According to the polarization calibration model, the polarization distortion parameters are solved to solve the numerical solution of the nonlinear equations constructed by the polarization calibration model. The genetic algorithm has the characteristics of jumping out of the trap of local optimal solution and obtaining the global optimal solution or approximate global optimal solution of complex nonlinear equations. The Levenberg–Marquard method has a fast convergence speed and high accuracy for solving complex nonlinear equations, but it has higher requirements for the selection of initial values. Therefore, combined with the characteristics of genetic algorithm and Levenberg–Marquard algorithm, the genetic algorithm solving result is

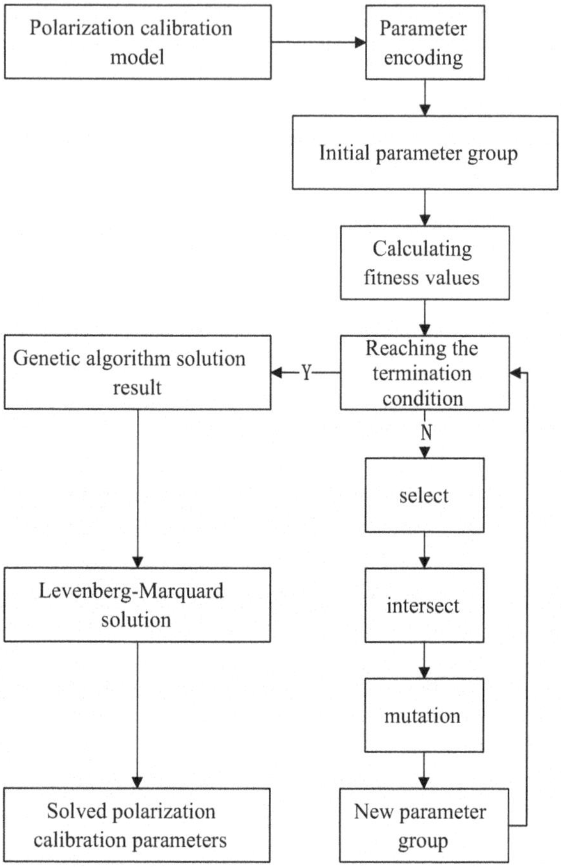

Fig. 1 Point target polarization calibration optimization algorithm flowchart

used as the initial value iterative solution of the damped least squares algorithm to obtain the exact solution of the polarization distortion parameter. The flowchart of the optimization solution algorithm is shown in Fig. 1.

4 Introduction Experimental Data

In order to verify the calibration method described in this paper, the X-band airborne SAR full polarization data was used for polarization calibration experiments. The experimental data parameters are shown in Table 1.

In order to ensure that the corner reflector as the target of the calibration point can be effectively applied to the polarization calibration work, the corner reflector should be disposed in a uniform background area. According to the classical point target polarization calibration algorithm, three different angle reflectors are used as the calibration target. Therefore, the calibration experiment is carried out by setting

Table 1 Parameters of experimental data

Parameter type	Value
Radar wavelength	0.031228 m
Polarization mode	Full
Distance resolution	0.187370 m
Azimuth resolution	0.228278 m
Central incident angle	43.97739°
Flight altitude	4016.928 m

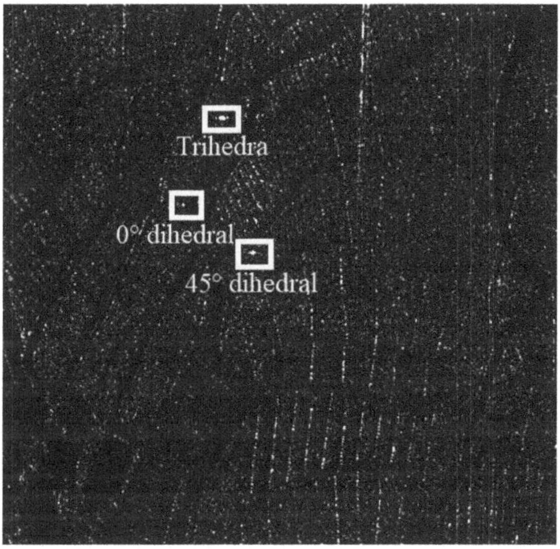

Fig. 2 Experimental area and corner reflector position map

the trihedral angle, the 0° dihedral angle, and the 45° dihedral angle in the calibration field. The experimental area is located in the Danling area of Sichuan Province, and the experimental area and corner reflector layout are shown in Fig. 2.

5 Polarization Calibration Experiment and Result Analysis

According to the calibration model, two corner reflectors can be used to construct a system of nonlinear equations containing eight equations for solving six polarization distortion parameters. The theoretical scattering matrices of the trihedral angle, the 0° dihedral angle, and the 45° dihedral corner reflector are $S_3 = \begin{bmatrix} 1 & 0 \\ 0 & 1 \end{bmatrix}$, $S_0 = \begin{bmatrix} -1 & 0 \\ 0 & 1 \end{bmatrix}$ and $S_{45} = \begin{bmatrix} 0 & 1 \\ 1 & 0 \end{bmatrix}$. According to the theoretical scattering matrix of the three corner reflectors, the trihedral angle is similar to the 0° dihedral theoretical scattering matrix. If a system of nonlinear equations is constructed from trihedral angles and 0° dihedral angles, the 0° dihedral angle provides only one a priori condition of co-polarization, and it is still impossible to construct a positive definite or overdetermined equation to solve six polarization distortions parameter. Therefore, this paper selects the trihedral angle and the 45° dihedral angle to construct nonlinear equations containing eight equations for solving.

The current airborne SAR characteristics are analyzed. The amplitude range of crosstalk is in [0, 0.5], the crosstalk phase value is in $[-\pi, \pi]$, the channel unbalance amplitude range is in [0.5, 2], and the channel imbalance phase value range is in $[-\pi, \pi]$. If the crosstalk and channel imbalance parameters exceed this range, it indicates that the system has serious problems and cannot be used. The complex number in the observed scattering matrix is decomposed into the real part and the imaginary part, respectively, into the genetic algorithm for solving, wherein the crosstalk parameter has a value range of [−0.5, 0.5], and the channel imbalance parameter has a value range of [0.5, 2]. The group size is 40, the crossover probability is 0.8, and the mutation probability is 0.2. After the genetic operation, the initial solution values of the six polarization distortion parameters are obtained. The polarization distortion parameter solved by the genetic algorithm is solved as the initial value of the Levenberg–Marquard method, and the final solution of the polarization distortion parameter is obtained. The obtained polarization distortion parameters are brought into the polarization calibration model to complete the polarization calibration of the SAR image.

After the polarization calibration is completed, the accuracy is evaluated from both qualitative and quantitative aspects. In the qualitative evaluation, the polarization response map of the corner reflector is used for analysis and evaluation. In the quantitative evaluation, the ratio of the commonly used HH and VV terms and the residual polarization distortion parameter in the post-calibration image were used for evaluation. The polarization response of the trihedral reflector before and after calibration is shown in Fig. 3.

The polarization response diagram of the trihedral reflector after calibration is compared with the theoretical trihedral reflector polarization response diagram, which is very similar, indicating the validity of the calibration results.

The calibration results are analyzed from a quantitative perspective, and the calibration results are compared with the calibration accuracy of the classical Whitt algorithm and the unconstrained algorithm. The calibration results were evaluated using the commonly used HH and VV terms, and the results are shown in Table 2.

According to the theoretical polarization scattering matrix of the trihedral corner reflector, the amplitude ratio and the phase difference theoretical value of the HH and VV terms should both be 0. As can be seen from Table 2, the three calibration methods can achieve better calibration result. Comparing the three calibration results, the calibration algorithm combining genetic algorithm and Levenberg–Marquard method has an accuracy of 0.18 dB in amplitude term and 12.12° in phase term accuracy, which are better than Whitt algorithm and unconstrained algorithm. In addition, the polarization distortion parameters remaining in the image after calibration by three

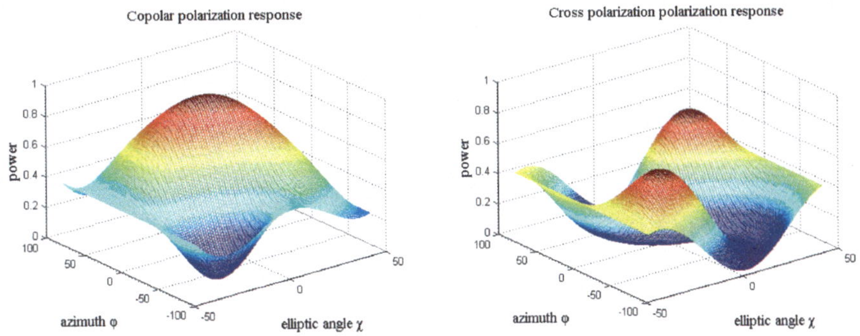

(a) Polarization response map before polarization calibration

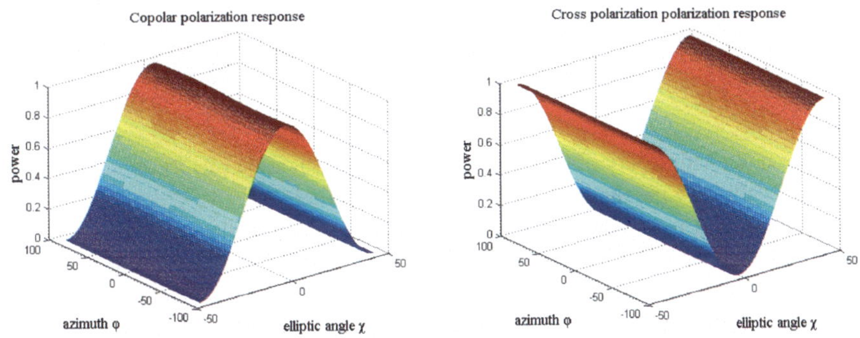

(b) Polarization response map after polarization calibration

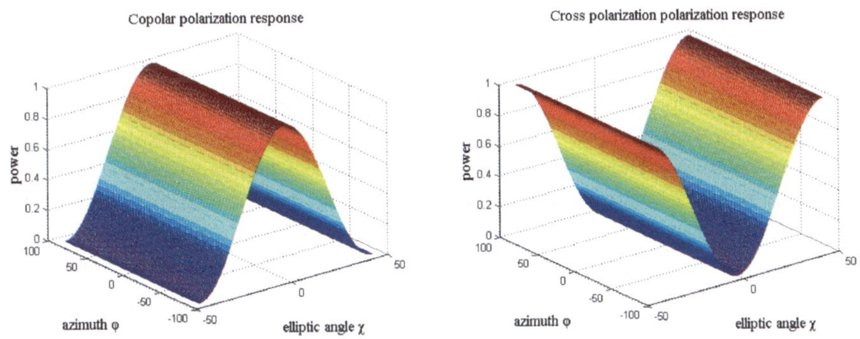

(c) Theoretical polarization response map

Fig. 3 Before and after calibrations and theoretical polarization response of the trihedral corner reflector

Table 2 Trihedral corner reflector HH and VV terms verification result table

Index	Before calibration	GA + LM	Whitt	Unconstrained
VV/HH (dB)	1.93	0.18	0.52	1.11
VV − HH (°)	71.56	−12.12	12.33	−12.21

Table 3 Statistical table of residual polarization distortion parameters after calibration by different methods

Parameters	GA + LM (dB)	Whitt (dB)	Unconstrained (dB)
δ_1	−38.45	−28.67	−30.95
δ_2	−37.82	−29.95	−34.76
δ_3	−37.87	−29.97	−30.77
δ_4	−38.51	−27.49	−30.91
f_1	−0.029	−0.152	0.378
f_2	−0.027	−0.081	0.393

calibration algorithms are obtained. The residual polarization distortion parameters after calibration by the three methods are shown in Table 3.

It can be seen from Table 3 that the residual crosstalk of the calibration results of the proposed method is less than −37 dB, the absolute value of the residual channel imbalance is less than 0.03 dB, and the crosstalk and channel imbalance residuals are smaller than the Whitt algorithm and the unconstrained algorithm. The calibration results show that the method has better calibration accuracy. According to the polarization calibration model, a nonlinear equation system composed of a trihedral corner reflector and a 45° dihedral corner reflector are constructed, which fully utilizes the polarization information of each channel of the two corner reflectors. At the same time, according to the genetic algorithm, the global optimal solution or the approximate optimal solution of the nonlinear equations can be obtained, and combined with the Levenberg–Marquard method to obtain the exact solution to the nonlinear equations. Therefore, although, the method of this paper reduces the use of a corner reflector relative to the Whitt algorithm and the unconstrained algorithm, it can still obtain more accurate and stable calibration result than other polarization calibration algorithms by fully utilizing the characteristics of genetic algorithm and Levenberg–Marquard method.

6 Conclusion

Aiming at the problem that the accuracy and stability of the existing point target polarization calibration algorithm are relatively poor, this paper proposes a polarization calibration optimization solution based on genetic algorithm and damped least squares. The method uses two corner reflectors to construct nonlinear equations, and uses the genetic algorithm and the damped least squares algorithm to solve the polarization distortion parameters. Experiments with airborne X-band full polarization data show that the proposed method has higher polarization calibration accuracy than the classic Whitt algorithm and unconstrained algorithm. At the same time, the method of this paper also reduces the dependence on the number of point targets.

References

1. Jung YT, Park S-E (2018) Comparative analysis of polarimetric SAR calibration methods. Remote Sens 10(12):2060
2. Zhang H, Lu W, Zhang B et al (2013) Improvement of polarimetric SAR calibration based on the Ainsworth algorithm for chinese airborne PolSAR data. IEEE Geosci Remote Sens Lett 10(4):898–902
3. Chen X, Wu T, Zhong X (2012) Airborne polarimetric SAR experiments with different crosstalk calibration techniques. In: IEEE international conference on radar. IEEE
4. Sun G (2017) Polarimetric calibration algorithm of airborne SAR using corner reflectors. Shandong University of Science and Technology, Qingdao
5. Whitt MW, Ulaby FT, Polatin P et al (1991) A general polarimetric radar calibration technique. IEEE Trans Antennas Propag 39(1):62–67
6. Sun G, Huang L, Chen K et al (2017) An efficient polarimetric SAR calibration algorithm using corner reflectors. Can J Remote Sens
7. Xi Y, Yang R (2002) Experiment on Whitt 's polarimetric SAR calibration algorithm using point targets. Remote Sens Technol Appl 17(04):220–223
8. Liao L (2017) Optimization research on airborne SAR polarimetric calibration methods. Acta Geodaetica Cartogr Sin 46(6):804
9. Freeman A, Shen Y, Werner CL (1990) Polarimetric SAR calibration experiment using active radar calibrators. IEEE Trans Geosci Remote Sens 28(2):224–240
10. Freeman A, Van Zyl JJ, Klein JD et al (1992) Calibration of Stokes and scattering matrix format polarimetric SAR data. IEEE Trans Geosci Remote Sens 30(3):531–539
11. Liao L (2015) Optimization research on airborne SAR polarimetric calibration methods. Wuhan University, WuHan
12. Chen H, Yan L, Dai X et al (2016) The joint inversion of simulated annealing method for IP spectrum parameters of rock. Chin J Eng Geophys 13(2):170–174

Zhida Xu Male, Master student, the main research direction is research on radar remote sensing.

Recognition Hydropower Stations from Remote Sensing Images by Multi-stage CNN Detection and Segmentation

Xiaowei Tan, Zhifeng Xiao, and Weiping Shao

Abstract Global energy Internet is China's national strategy. However, due to the lack of public information of power energy infrastructure, it is difficult to investigate foreign power facilities through public information. Therefore, it is a feasible way to recognize power facilities through remote sensing images. In this paper, we aim to recognize hydropower stations efficiently and accurately from high-resolution remote sensing images. Traditional target detection methods need to extract target features from images manually, so the designed features are not very robust. In this paper, we propose a method to recognize hydropower stations by multi-stage DCNN. Concisely, our task includes two parts: (1) Preliminary detection of hydropower station. DCNN models are used as feature extractor to get a rough detection result of hydropower stations. (2) Extract water surface by segmentation. We utilize the spatial relationship between hydropower station and water surface to obtain the final recognition results. Compared with other detection methods of deep learning, our method uses multi-stage CNN to detect hydropower stations and water surface, respectively, and uses the spatial relationship between hydropower stations and water surface, so as to further improve the recognition accuracy. The results show that our proposed method has high accuracy and confidence.

Keywords Object detection · Water surface segmentation · Multi-stage CNN · Spatial relations

---

This article is supported by "Science and Technology Project of State Grid Corporation of China," No. SGTYHT/18-JS-206.

X. Tan · Z. Xiao (✉)
State Key Laboratory of Information Engineering in Surveying, Mapping and Remote Sensing, 129 Luoyu Road, Wuhan, China
e-mail: xzf@whu.edu.cn

W. Shao
State Grid Zhejiang Electric Power Co. LTD., 8 Huanglong Road, Hangzhou, China

L. Wang et al. (eds.), *Proceedings of the 6th China High Resolution Earth Observation Conference (CHREOC 2019)*, Lecture Notes in Electrical Engineering 657,
https://doi.org/10.1007/978-981-15-3947-3_24

1 Introduction

Building a global network of electricity infrastructures plays a significant role in the construction of One Belt One Road. However, the traditional manual ways to collect hydropower stations and transmission towers information cost money and time. What's more, there is a great difficulty to get the worldwide geographic information of transmission towers by manual measurement. In order to collect the geographic information of electricity infrastructures all over the world efficiently, our work makes use of the advantage of high-resolution remote sensing images and deep convolutional neural networks (DCNN). Additionally, we take into account the domain knowledge of electricity infrastructures by calculating the spatial relations for the purpose of similar target distinction and extracting the water surface around hydropower stations.

On the one hand, with the development of sensor technology and aerospace remote sensing technology, the quality and quantity of remote sensing images have undergone great improvement. Therefore, it is possible to capture worldwide high-resolution remote sensing images which include abundant spatial information of electricity infrastructures. The problem followed is how to exploit the rich information and the details of objects contained in remote sensing image. In addition, compared with natural images, remote sensing images have unique features with a more complicated scene and smaller objects. Consequently, detecting objects from remote sensing images will be different from detecting objects from natural images. On the other hand, a detection result with higher accuracy and efficiency has been achieved by numerous algorithms which based on DCNN. These methods of object detection are mainly divided into two types: two-stage methods and one-stage method. Possibly, recent advances in object detection are driven by the success of region proposal methods [1, 2]. Due to the existence of the RPN structure, the two-stage methods represented by the R-CNN algorithm have reached a higher and higher detection accuracy, but the detection speed is difficult to meet the real-time requirements of some scenes. Consequently, the detection algorithms based on regression were proposed to deal with the challenge of calculation speed. The one-stage method [3, 4] can share features in a complete training and improve the detection speed greatly on the premise of guaranteeing a certain accuracy.

In the general process of remote sensing image object detection, it can be summarized as obtaining the interest object information and processing these information to achieve object positioning. Furthermore, we can acquire the qualitative and quantitative description of interest objects. Remote sensing images can quickly acquire the true surface information on the Earth. These objective remote sensing data have a wide range of application prospects in many areas such as military reconnaissance, environmental detection, natural resource investigation and evaluation, and change detection. Nowadays, object detection from remote sensing images has a wide range of applications in many fields such as defense security, traffic monitoring, and urban management. Hydropower station is one of the most important power and energy infrastructures. Undoubtedly, detecting hydropower station from remote sensing images plays a vital role in work of electrical information surveying.

The rest of this paper is organized as follows: Sect. 2 reviews related works on object detection, scene segmentation, and applications of CNN in remote sensing images. Section 3 presents the details of our method for hydropower station recognition, including the method of object detection, the method of water surface extraction, and the method for refining the recognition results. Section 4 discusses the entire procedure of the proposed method, including datasets introduction, a more detailed process of our experiments, and results of each experimental procedure. Finally, Sect. 5 concludes the paper.

2 Related Work

In the field of object detection and scene segmentation, lots of significant achievements have been improved day by day.

Generally, a traditional object detection method is divided into three stages: region proposal selective, feature extraction, and classifier selection. Selecting region proposals aims to realize object positioning. Due to the diversity of the object size, it is a tough problem to determine the position of objects. A common method for region proposal selective is sliding window, an exhaustive strategy that traverses the entire image with different scales and aspect ratios. Undoubtedly, sliding windows will cause a very high time complexity and computational complexity, affecting the detection speed and algorithm performance. In fact, the aspect ratio of the sliding window is artificially set to control the time complexity of the target recognition in region proposals selection. Therefore, the sliding window method is not ideal when the size of the identified objects fluctuates. As to feature extraction, many researchers have used local features to extract characteristics such as SIFT [5], HOG [6], and Saliency [7, 8]. However, manually designed features are not so robust that the changes in diversity of illumination changes and background can be easily distinguished. This situation will be more obvious especially in dealing with the remote sensing images. The key issues mentioned before seriously affect the performance of detection results. The third step is to determine the classifier for classification. And the classifiers mainly include SVM, Adaboost, Softmax, and so on. There are two main problems with the traditional target recognition method. The first problem is the high computational complexity caused by the sliding window method. It also produces a large number of invalid windows. The second problem is that the features extracted by manual which based on prior knowledge are not robust and university for objects detection. Especially when the objects changes, it is more difficult to achieve better detection results.

With regard to the problems existed in selecting region proposals, a series of region proposal method appeared, such as selective search and edge boxes. These methods take advantage of the color and the edge information of objects to obtain the objects'possible position in the image in advance. In this case, it not only ensures a higher recall rate of the detection result, but also produces fewer windows with higher quality. More importantly, it can reduce the computational burden and save time for the whole process of object detection.

Additionally, based on the features used for scene segmentation, the existing methods for remote sensing image scene segmentation can be roughly categorized into three main categories: handcrafted feature-based methods, unsupervised feature learning-based methods, and deep feature learning-based methods. In practical applications, the handcrafted descriptors largely limited the performance of the rich semantic information description, especially in remote sensing images. When compared with handcrafted feature-based methods, unsupervised feature learning methods have obtained good performance. However, most of the unsupervised feature learning methods do not make use of the scene semantic information and so cannot guarantee the best discrimination ability between different scene classes. Therefore, in the field of object detection and scene segmentation, there are so many problems that the traditional algorithms cannot cope with.

Recently, D-CNNs are widely used in the works of object detection [9, 10] from remote sensing images in recent years. For example, Ge et al. [11] detected inshore ships in optical satellite images by using shape and context information that are extracted in the segmented image. Yang et al. [12] obviously tackled the problem of automatic accurate localization of detected objects in high-resolution remote sensing images. Cheng et al. [13] provided a review of the recent progress in object detection in remote sensing images and proposed two promising research directions, namely deep learning-based feature representation and weakly supervised learning-based geospatial object detection. Bazi et al. [14] proposed a novel CSVM network for the classification of UVA imagery, providing better results compared to recent solutions based on knowledge transfer from pretrained CNNs. Zhu et al. [15] used DCNN features from combined layers to perform orientation robust aerial object detection. Salberg [16] extracted features from a pretrained deep CNN and used it for automatic detection of seals in aerial remote sensing images. Zhang et al. [17] reconstructed an iterative weakly supervised learning framework to automatically mine and augment the training dataset from the original image and combined the candidate region proposal network and a localization network to extract the proposals and locate aircraft in large-scale very high-resolution (VHR) images.

Deep learning models also have achieved great success in remote sensing image scene classification task [18, 19]. There is few evidence of the good performance of deep architectures in remote sensing image classification. For the problem of dense semantic labeling, Nicolas et al. [20] investigated various methods to deal with semantic labeling of very high-resolution multimodal remote sensing data and especially studied how deep fully convolutional networks can be adapted to deal with multimodal and multi-scale remote sensing data for semantic labeling. Tarabalka et al. [21] designed new specific architectures to provide fine-grained classification maps with convolutional neural networks (CNNs) method. In the task of scene classification, Michele et al. [22] proposed a system to cope with the high spatial and geometrical information contained in ultrahigh-resolution images (<10 cm), usually coupled to little spectral information. Cheng et al. [23] proposed a simple but effective method to learn discriminative CNNs (D-CNNs) to boost the performance of remote sensing image scene classification, addressing the problems of within-class diversity and between-class similarity which are still two big challenges. Maggiori et al. [24] proposed an end-to-end framework for the dense, pixelwise classification of satellite imagery with convolutional neural networks (CNNs). Romero et al. [25] introduced single-layer and DCNN for remote sensing data analysis and proposed greedy layerwise unsupervised pretraining coupled with a highly efficient algorithm for unsupervised learning of sparse features. Vaduva et al. [26] introduced a deep learning algorithm for classification of (low-dimensional) VHR images. Chen et al. [27] introduces hybrid deep neural networks to enable the extraction of variable-scale features for detecting vehicles in satellite images. Wang et al. [28] proposed a hybrid framework based on stacked autoencoders for classification of HS data.

In the search of object detection and scene segmentation, although great success has been obtained so far, specific strategies are needed to solve different problems in various applications. Detecting hydropower stations from remote sensing images faces challenges when dealing with the problems of within-class diversity. As to the domain issue, a detection strategy which combines with prior knowledge needs to be designed to solve the existing problem.

In this paper, a suitable feature extractor based on the CNN model is designed to extract the essential features of objects from remote sensing images. These features are applied for object detection and scene classification. What's more, we adopt domain knowledge of hydropower station to improve the detection results, trying to get a higher detection accuracy while guaranteeing the detection recall.

3 Method

In this section, we mainly describe the proposed method for hydropower station recognition with superior precision and credibility. Concisely, our task includes three parts, as illustrated in Fig. 1: (1) preliminary detection of hydropower station; (2)

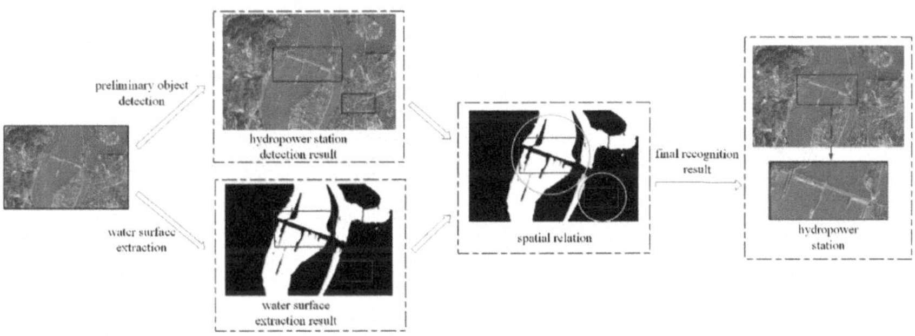

Fig. 1 Overall architecture of our method

extract water surface by segmentation; (3) combine the spatial relations among the two types of features.

3.1 Object Detection

The traditional object detection methods generally consist of three steps, region proposal selective, feature extraction, and classification and location. Obviously, when compared to selective search, one of the most popular methods, the algorithm for region proposal selective based on an exhaustive search method is computationally expensive. Additionally, extracting object features from images by manual is not very robust to manage the changes in diversity. By considering the problem of the time spent and accuracy, an economical inference scheme is selected to accomplish hydropower station efficient detection with a higher accuracy.

Comparing various algorithms for region proposal generation, we select a popular, deep learning-based object detection system, Faster R-CNN, as shown in Fig. 2, which use a region proposal network strategy enabling nearly cost-free region proposals. The learned RPN [2] also improves region proposal quality and thus the overall object detection accuracy. The detection methods based on region proposals developed from R-CNN to SPP-Net, Fast R-CNN, and Faster R-CNN. These algorithms have achieved faster detection speed and better detection results.

The main process of Faster R-CNN is as follows. The first step is to input any size pictures to DCNN network. After the convolutional and pooling operation, the feature map is obtained for RPN network and a high-dimensional feature map. The function of RPN network is to generate region proposals and regional scores. And then, non-maximum suppression (NMS) algorithm is employed to reduce the amount of region proposals. The second step is that input the high-dimensional feature map

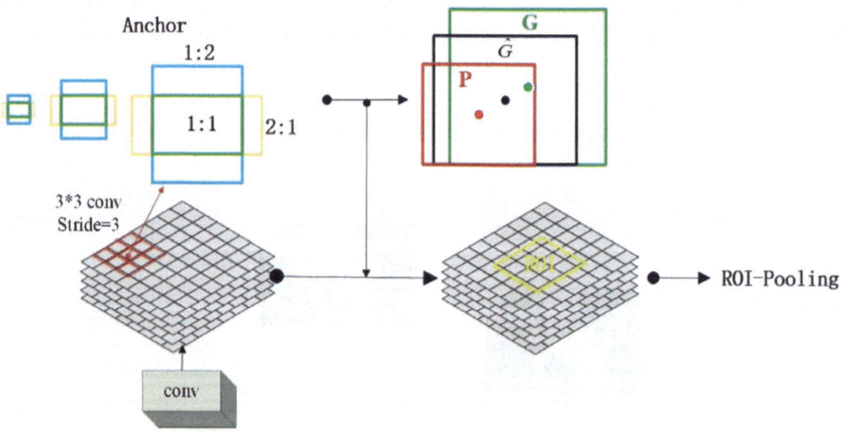

Fig. 2 Region proposal network (RPN)

and region proposals to ROI pooling layer and extract the feature of region proposals. Lastly, the feature of the region proposals is input to the fully connected layer. We can get the classification score of the region proposals and the position of object boxes after regression.

Faster R-CNN completes the end-to-end training for object detection with the DCNN methods. One of the key tricks in Faster R-CNN is RPN, as shown in Fig. 2. RPN [2] shares full-image convolutional features with the detection network, thus enabling nearly cost-free region proposals. Another trick is anchor strategy. An anchor is centered at the sliding window in question and is associated with a scale and aspect ratio. Because of this multi-scale design based on anchors, the convolutional features which computed on a single-scale image can be simply used. Concisely, Faster R-CNN can be seen as the combination of RPN and Fast R-CNN. By further merging RPN and Fast R-CNN into a single network, sharing their convolutional features, using the recently popular terminology of neural networks with "attention" mechanisms [2], the RPN component tells the unified network where to look.

Traditional CNN models increase the accuracy of object detection by increasing the number of network layers. A series of problems follows while increasing the number of network layers in the process of model training, such as degradation. The degradation becomes more and more obvious as the number of network layers deepens. In this case, it will not allow the shallow layers to learn the parameters, resulting in model training failure. In addition, when the number of networks exceeds a certain number of layers, the BP algorithm cannot effectively update the network before the gradient. Studies have shown that in training the DCNN, when the DCNN reaches a certain depth, the deep network training error is greater than the shallow network.

In other words, deeper neural networks are more difficult to train. The presence of ResNet eases the training of networks that are substantially deeper than those used previously. Residual nets with a depth of up to 152 layers-8 $\times$ deeper with lower complexity than VGG nets evaluate the ImageNet dataset [29]. An ensemble of these residual nets achieves 3.57\% error on the ImageNet test set. This result won the first place on the ILSVRC 2015 classification task.

In ResNet [29], the degradation problem can be addressed by introducing a deep residual learning framework. Instead of hoping each few stacked layers directly fit a desired underlying mapping, ResNet explicitly let these layers fit a residual mapping. Formally, denoting the desired underlying mapping as $H(x)$, the stacked nonlinear layers fit another mapping of $F(x) = H(x) - x$. The original mapping is recast into $F(x) + x$. It is easier to optimize the residual mapping than to optimize the original, unreferenced mapping. To the extreme, if an identity mapping were optimal, it would be easier to push the residual to zero than to fit an identity mapping by a stack of nonlinear layers. The formulation of $F(x) + x$ can be realized by feedforward neural networks with "shortcut connections" (Fig. 3).

In our method, we combine Faster R-CNN [2] (Fig. 4) with ResNet [29] for preliminary detection. Res-Net, a residual learning framework, is easier to optimize and can gain accuracy from considerably increased depth [29]. Nevertheless, the preliminary results cannot be applied directly because of the phenomenon of misjudgments

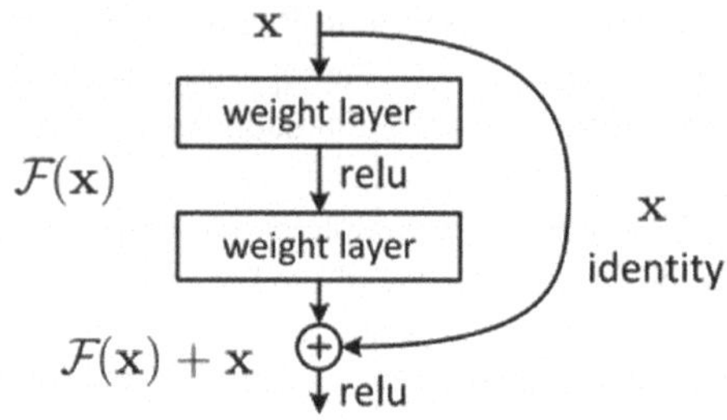

Fig. 3 Residual learning: a building block

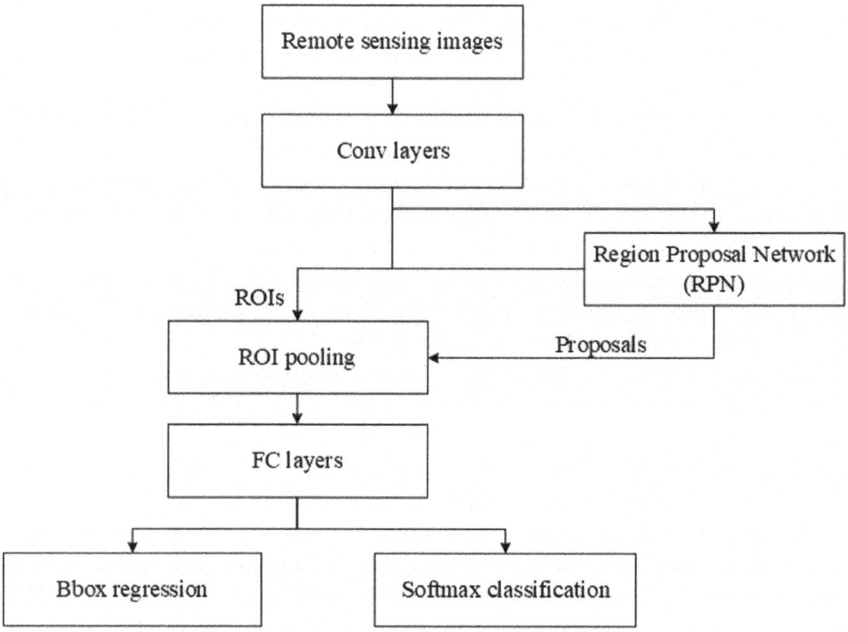

Fig. 4 Object detection by faster R-CNN

and missed judgments. Consequently, we give a smaller confidence value to evaluate the target category to acquire a larger number of potential objects. In addition, the results of water surface extraction will be used to verify the confidence values of hydropower stations.

3.2 Water Surface Extraction

Hydroelectric power station is an integrated engineering facility capable of converting water energy into electricity. There is no doubt that where there is a hydropower

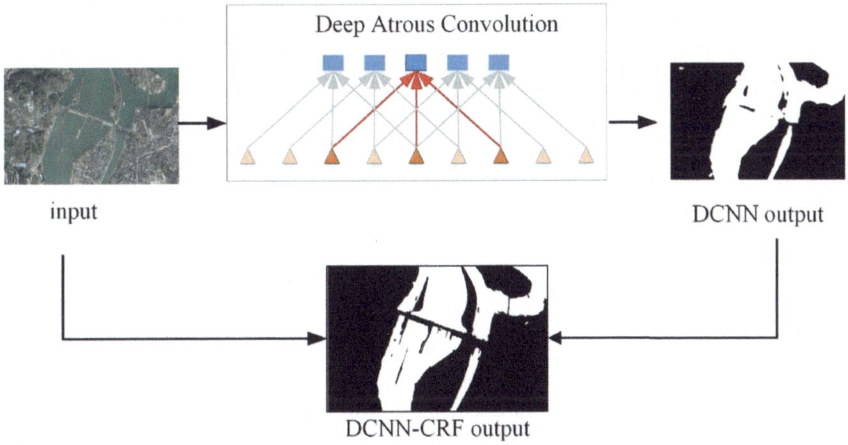

Fig. 5 Architecture of water surface extraction

station there is water. Thus, we accomplish the task of water surface extraction with atrous convolutional model demonstrated in Fig. 5, a deep learning method [30].

So that we use deep atrous convolutional neural network to extract water surface. The model can be considered as a pair of encoder and decoder. The encoder uses a sequence of convolutional and max-pooling layers to compress the raw input into lightweighted but informative features. The model adopts the ResNet-101 [29] as the encoder. The decoder applies a sequence of atrous convolution with up-sampled filters for the encoded feature to restore its spatial details and size.

In order to make the size of the feature map consistent with the input image, and the convolution kernel has a larger receptive field. The network presents an atrous convolution in order to maintain the range of the receptive field. The atrous convolution transforms the structure of the traditional convolutional kernel and fills the cavity in the convolutional kernel, thus enlarging the feeling field of the convolutional layer while keeping the number of convolutional kernel parameters unchanged. To produce semantically accurate predictions and detailed segmentation maps along object boundaries, we also combine ideas from fully connected conditional random fields (CRF).

3.3 *Recognition*

In this part, the results of preliminary detection and water surface extraction are used to assist the hydropower station judgment. The whole process to refine the detection results is indicated in Fig. 1. With spatial relationship of electrical power infrastructure, the difficulty in similar features differentiation can be alleviated to a certain extent.

In order to obtain the candidates of hydropower station as many as possible, we lower the threshold which determines the candidate category. In preliminary detection results, almost all the hydropower stations can be located in a test image. Considering the spatial relationship of electric power infrastructure, the confidence score of hydropower station is recalculated according to the results of water surface extraction, as shown in Eq. 1

$$S_t = c * S_0 \tag{1}$$

$$c = \begin{cases} -1, x = 0 \\ 0, x = 1 \end{cases} \tag{2}$$

S_0 is the initial score of a candidate hydropower station, getting from the primary detection results, while S_t is the final score. S_t is the final score hydropower station detection judgment. S_t is calculated by the above formulas. c is the impact factor of water surface on the hydropower station. x represents the result of water surface segmentation. The value of x is 1 if there exists water surface. On the contrary, the value of x is -1, while the segmentation result shows no water surface. We assume that the higher score the candidate gets, the more likely the hydropower station will be.

4 Experiments and Results

4.1 Datasets

In CNNs applications, extensive experimentation shows that the more training data you have, the better experiment results you will get. The supervised learning models need large amounts of labeled data and many iterations to train their large number of parameters. In the case of lack of training data, Yang et al. [31] presented a conceptually simple, flexible, and general framework for few-shot learning, where a classifier must learn to recognize new classes given only few examples from each.

In our work, however, the electrical datasets are not very easy to collect. Since the lack of public datasets designed for object detection in remote sensing, we collected 256 images of hydropower station downloaded mainly from Google Earth by manual. These positive images are labeled as hydropower station. Both the size and quality of training datasets are critical to detection performance in CNNs methods. Dataset enhancement can be seen as a method of preprocessing the training set. To address the object diversity in remote sensing images and the difficult in collecting data for some object classes, we augmented all the positive samples by rotation transforms, re-sampling, color transforms, and mirror transforms. Furthermore, 1572 (500 * 500) remote sensing images collected for scene segmentation are from GF2 satellite. Some training samples are illustrated in Fig. 6. More details of our datasets are listed in Tables 1 and 2.

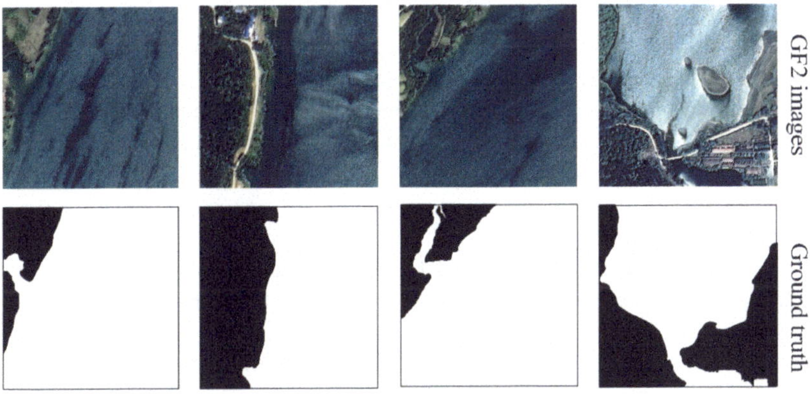

(a) training examples for scene segmentation

(b) training examples for object detection

Fig. 6 Remote sensing images for training

Table 1 Image resolution of each class

Class	Hydropower station	Water surface images
Resolution (m)	3–5	0.8

Table 2 Statistics the training and validation datasets

Class	Image number	Original object number	Augment object number
Hydropower station	256	261	8352
Water surface image	1572	/	/

4.2 Experiment Procedures

The experiment procedures mainly include two parts as illustrated in Fig. 7: the training process and the testing process. Specifically, the test process consists of two procedures: the detection process and segmentation process. The well-known ResNet [11] and DeepLab [12] architecture are employed for robust DCNN feature extraction, object detection, and scene segmentation, respectively. The training process of feature extraction used the GPU and the Compute Unified Device Architecture (CUDA) to speed up computation. We combine Faster R-CNN with ResNet to complete preliminary object detection. Simultaneously, a well-known framework is applied to extract water surface.

In the work of object detection, we used small patches to train our DCNN models through the backpropagation algorithm, employing the GPU and CUDA. We set the

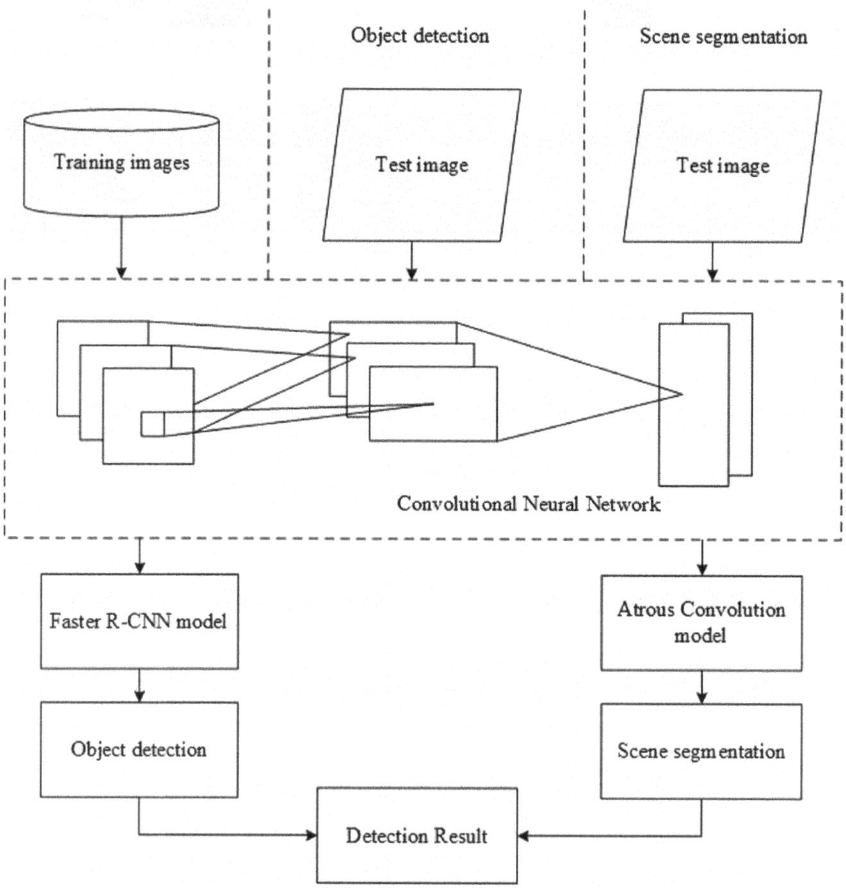

Fig. 7 Experiment procedures

Table 3 Arguments for training

Argument	Object detection	Scene segmentation
Learning rate	0.001	0.001
Batch size	256	5
Momentum	0.9	0.9
Weight decay	0.0005	0.0005
Max iter	200,000	50,000

learning rate to 0.001 and set the batch size to 256 for ResNet. Due to the resolution of remote sensing images and the object size, the anchor scales for RPN are set as [4, 8, 16, 32,]. Besides, regarding to the water surface extraction, we set the learning rate to 0.001 and set the batch size to 5 for atrous convolutional model. Moreover, some tricks such as local response normalization, momentum, overlapping pooling, and dropout have been used in these networks to improve their properties. The arguments for DCNN training are listed in Table 3. The training process was run on a Linux server with an Nvidia GTX Titan X GPU with 12 GB ram.

4.3 Experiment Results

The baseline values for object detection and scene segmentation based on our method are listed in Table 4.

In Fig. 8, images are the water extraction results, respectively. In Figs. 9 and 10, the detection results of using two different methods are displayed. Obviously, Faster R-CNN method misjudges two images out of six images. In Figs. 9a and 10b, a dam and a bridge are misjudged as hydropower stations. It seems using Faster R-CNN directly for hydropower station detection is not so robust that the similar target can be distinguished, although it is efficient.

Compared to Faster R-CNN, our method takes advantage of the professional knowledge of electrical infrastructure by calculating the spatial relations and extracting the water surface. We apply the spatial relations between transmission towers

Table 4 Results

Class	Recall (%)	Precision (%)
Water surface	95.04	96.89
Hydropower station with DCNN	92.34	85.03
Hydropower station with our method	92.34	**89.48**

Bold representes using our method to extract the hydropower station, using the extraction results of the water surface can reduce the false detection to a certain extent and improve the precision of the hydropower station

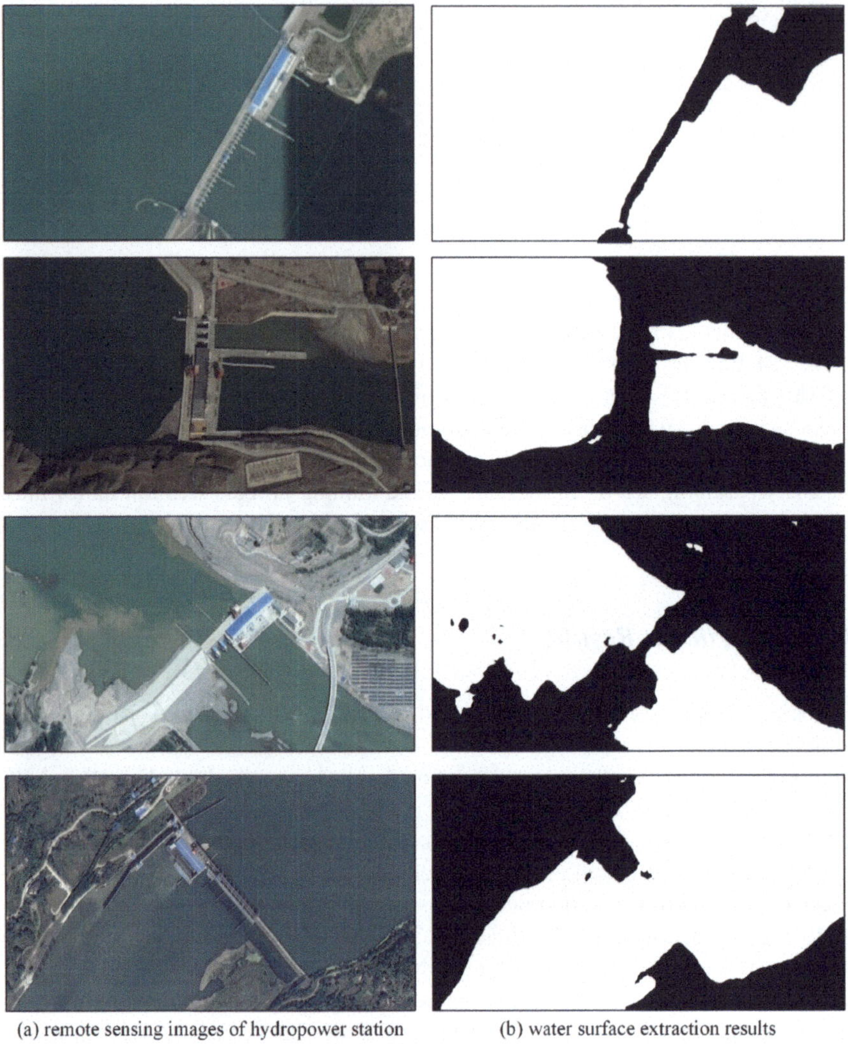

Fig. 8 Water surface extraction results

with hydropower station and the result of water surface extraction to improve the detection accuracy. Clearly, in the final detection results, some images that include similar object are filtered by our integrated method. In other words, our method performed better.

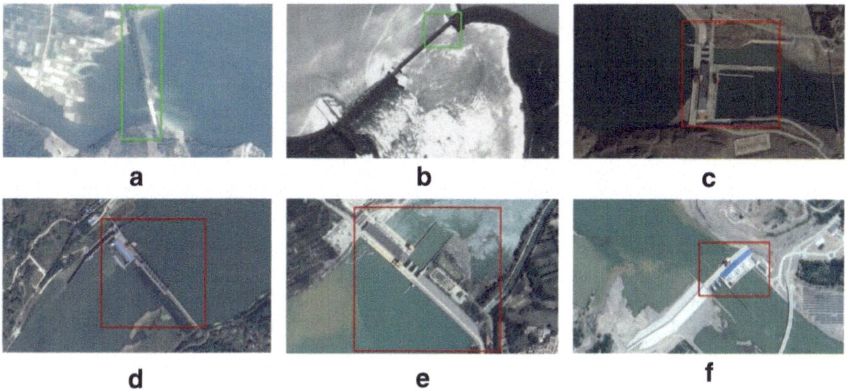

Fig. 9 Detection results of DCNN

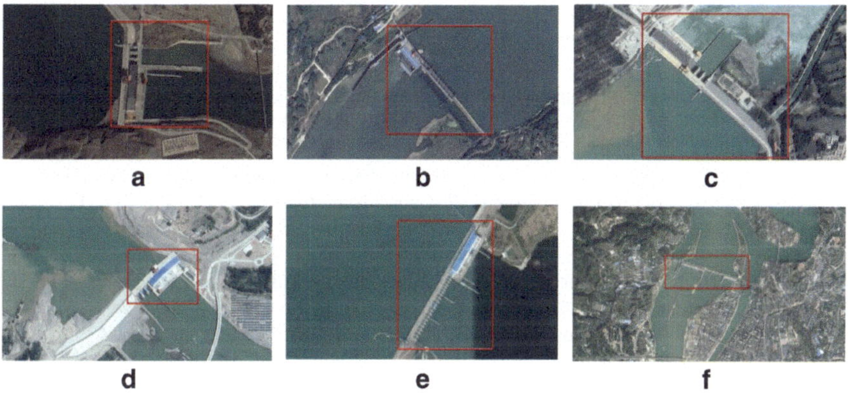

Fig. 10 Detection results of our method

5 Conclusion

In this work, we take advantage of the spatial relationship of power facilities to assist the hydropower station recognition in deep learning method. The results suggest that integrating with spatial relationship of power facilities and DCNN features for hydropower station detection is not only an effective way with a reliable detection accuracy, but also a cost-efficient solution. In further work, by taking into account the localization precision and computational efficiency, we will explore a more enhanced detection scheme for other electrical infrastructures.

References

1. He K, Zhang X, Ren S et al (2015) Spatial pyramid pooling in deep convolutional networks for visual recognition. IEEE Trans Pattern Anal Mach Intell 37(9):1
2. Ren S, He K, Girshick R et al (2017) Faster R-CNN: towards real-time object detection with region proposal networks. IEEE Trans Pattern Anal Mach Intell 39(6):1137–1149
3. Redmon J, Divvala S, Girshick R et al (2015) You only look once: unified, real-time object detection
4. Liu W, Anguelov D, Erhan D et al (2016) SSD: single shot multibox detector. In: European conference on computer vision
5. Han J, Zhang D, Cheng G et al (2015) Object detection in optical remote sensing images based on weakly supervised learning and high-level feature learning. IEEE Trans Geosci Remote Sens 53(6):3325–3337
6. Shao W, Yang W, Liu G et al (2012) Car detection from high-resolution aerial imagery using multiple features. In: Geoscience and remote sensing symposium. IEEE
7. Zhang F, Du B, Zhang L (2015) Saliency-guided unsupervised feature learning for scene classification. IEEE Trans Geosci Remote Sens 53(4):2175–2184
8. Han J, Zhou P, Zhang D et al (2014) Efficient, simultaneous detection of multi-class geospatial targets based on visual saliency modeling and discriminative learning of sparse coding. ISPRS J Photogram Remote Sens 89:37–48
9. Ouyang W, Zeng X, Wang K et al (2017) DeepID-Net: object detection with deformable part based convolutional neural networks. IEEE Trans Pattern Anal Mach Intell 39(7):1320–1334
10. Ren S, He K, Girshick R et al (2015) Object detection networks on convolutional feature maps. IEEE Trans Pattern Anal Mach Intell 39(7):1476–1481
11. Liu G et al (2013) A new method on inshore ship detection in high-resolution satellite images using shape and context information. IEEE Geosci Remote Sens Lett 11(3):617–621
12. Long Y, Gong Y, Xiao Z et al (2017) Accurate object localization in remote sensing images based on convolutional neural networks. IEEE Trans Geosci Remote Sens 55(5):2486–2498
13. Cheng G, Han J (2016) A survey on object detection in optical remote sensing images. ISPRS J Photogram Remote Sens 117:11–28
14. Bazi Y, Melgani F (2018) Convolutional SVM networks for object detection in UAV imagery. IEEE Trans Geosci Remote Sens 1–12
15. Zhu H, Chen X, Dai W et al (2015) Orientation robust object detection in aerial images using deep convolutional neural network. In: 2015 IEEE international conference on image processing (ICIP). IEEE
16. Salberg AB (2015) Detection of seals in remote sensing images using features extracted from deep convolutional neural networks. In: 2015 IEEE international geoscience and remote sensing symposium (IGARSS). IEEE
17. Zhang F, Du B, Zhang L et al (2016) Weakly supervised learning based on coupled convolutional neural networks for aircraft detection. IEEE Trans Geosci Remote Sens 1–11
18. Farabet C, Couprie C, Najman L et al (2013) Learning hierarchical features for scene labeling. IEEE Trans Pattern Anal Mach Intell 35(8):1915–1929
19. Long J, Shelhamer E, Darrell T (2014) Fully convolutional networks for semantic segmentation. IEEE Trans Pattern Anal Mach Intell 39(4):640–651
20. Audebert N, Le Saux B, Lefèvre S (2017) Beyond RGB: very high resolution urban remote sensing with multimodal deep networks. ISPRS J Photogram Remote Sens S0924271617301818
21. Maggiori E, Tarabalka Y, Charpiat G et al (2017) High-resolution aerial image labeling with convolutional neural networks. IEEE Trans Geosci Remote Sens PP(99):1–12
22. Volpi M, Tuia D (2016) Dense semantic labeling of subdecimeter resolution images with convolutional neural networks. IEEE Trans Geosci Remote Sens 1–13
23. Cheng G, Yang C, Yao X et al (2018) When deep learning meets metric learning: remote sensing image scene classification via learning discriminative CNNs. IEEE Trans Geosci Remote Sens 1–11

24. Maggiori E, Tarabalka Y, Charpiat G et al (2016) Convolutional neural networks for large-scale remote-sensing image classification. IEEE Trans Geosci Remote Sens 55(2):645–657
25. Romero A, Gatta C, Camps-Valls G (2015) Unsupervised deep feature extraction for remote sensing image classification. IEEE Trans Geosci Remote Sens 54(3):1349–1362
26. Vaduva C, Gavat I, Datcu M (2012) Deep learning in very high resolution remote sensing image information mining communication concept. In: Signal processing conference. IEEE
27. Chen X, Xiang S, Liu CL et al (2014) Vehicle detection in satellite images by parallel deep convolutional neural networks. In: Pattern recognition, pp 1797–1801
28. Wang Z, Nasrabadi NM et al (2014) Spatial-spectral classification of hyperspectral images using discriminative dictionary designed by learning vector quantization. IEEE Trans Geosci Remote Sens 52(8):4808–4822
29. He K, Zhang X, Ren S et al (2015) Deep residual learning for image recognition
30. Chen LC, Papandreou G, Kokkinos I et al (2014) Semantic image segmentation with deep convolutional nets and fully connected CRFs. Comput Sci 4:357–361
31. Sung F, Yang Y, Zhang L et al (2017) Learning to compare: relation network for few-shot learning

The Intelligent Planning for Spacecraft Autonomy in On-Orbit Servicing Mission

Jing Yu, Dong Hao, Hongyang Liu, and Xiaoqian Chen

Abstract In On-orbit servicing (OOS) mission planning problem, mission sequences and orbital rendezvous trajectories for servicing spacecraft (SSc) servicing multiple targets would be optimized, aiming at reducing the mission cost, the time duration, or some other merits. The onboard intelligent mission planning for OOS mission is studied in this paper. Each SSc is supposed as an autonomy or agent. They could perform planning onboard, communicate, and negotiate with other SScs. A multi-agent system (MAS) model is firstly introduced. And then, based on the MAS model, an intelligent planning method based on contract net protocol (IPCNP) is developed. Simulations demonstrate the effectiveness of IPCNP.

Keywords Optimization · On-orbit servicing mission · Mission planning · Multi-agent system

1 Introduction

On-orbit servicing (OOS) is the process of improving a space-based capability through a combination of in-orbit activities which may include inspection, rendezvous and docking, and value-added modifications to a satellite's position, orientation, and operational status [1]. Mission planning and scheduling is one of its key technologies, in which mission sequences and orbital rendezvous trajectories for

J. Yu · D. Hao (✉) · H. Liu
China Aerodynamics Research and Development Center, Mianyang 621000, Sichuan, China
e-mail: haodongcardc@outlook.com

J. Yu
e-mail: yujinghd@hotmail.com

H. Liu
e-mail: lhy_flow_nudt@163.com

X. Chen
National Innovation Institute of Defense Technology, Chinese Academic of Military Science, Beijing, China
e-mail: chenxiaoqian@nudt.edu.cn

L. Wang et al. (eds.), *Proceedings of the 6th China High Resolution Earth Observation Conference (CHREOC 2019)*, Lecture Notes in Electrical Engineering 657,
https://doi.org/10.1007/978-981-15-3947-3_25

servicing spacecraft (SSc) servicing multiple targets would be optimized, aiming at reducing the mission cost, the time duration, or some other merits.

So far, there have been a number of reported works devoted in OOS mission planning. Alfriend [2] studied the SSc rendezvousing with satellites in GEOs with small inclinations. It is shown that (Law-1) "the minimum fuel solution is proportional to the minimum distance path through the set of points that are the projections of the angular momentum vectors on the equatorial plane." It is also pointed out that (Law-2) "if sufficient time is allowed, the plane change Δv dominates the total Δv, thus the minimum Δv is found by minimizing the Δv for the plane change." Using two-impulse maneuver for orbital transfer, Shen et al. [3] studied the problem of servicing multiple satellites in a circular constellation. The heuristic study showed that (Law-3) when the SSc is required to go back, the best (minimum-Δv) mission sequence is either the orbit-wise or counter-orbit-wise sequential order; when not required, the best is sequential or partially sequential. Based on Shen's study, Yu et al. [4] studied the OOS problem for GEO active debris (ADR) removal and found that (Law-4) the optimal sequence is the combination of the counter-orbit-wise segments for most cases. Afterward, Yu et al. [5] studied LEO ADR mission with long-duration and figured out that (Law-5) some of the suboptimal sequences are equivalent since the Δv differences between them are sufficient small. To further study the influences of the SSc initial location on the mission sequence, Yu et al. carried out their research in [6] and found that (Law-6) "the optimal initial position of the SSc prefers the area where the targets are concentrated." More work can be found in [7–11].

However, all of the mission planning modes mentioned above should be carried out on ground and then send the planning results to the satellites. With the advancement of the space technology, some researchers paid attention to the spacecraft autonomy, which has the potential for effecting significant cost savings in mission operations by reducing the need for dedicated ground staff. To help the SSc function as an autonomous agent, onboard intelligent planning and scheduling methods are presented, assisting the SSc to perform its own planning and scheduling by itself, not the ground staff [12]. In [12–14], the concept of distributed satellite constellation is proposed by NASA, as well as the intelligent planning system Distributed Spacecraft Coordination Planning and Scheduling (D-SpaCPlanS).

As a matter of fact, the computation and communication capabilities of the onboard system are much worse than those on ground. So, complex algorithms are not suitable for onboard intelligent planning. To simplify the planning mechanism, specialization knowledge is required. So far, there are only few studies focused on the OOS onboard intelligent planning [15]. In contract, plenty of heuristic conclusions have been concluded by on-ground OOS mission planning studies (see Law-1–Law-6). Based on these conclusions, the onboard intelligent planning for OOS mission will be studied in this paper.

Multi-agent system (MAS) would be used for this study. So, Sect. 2 details the concepts of agent and MAS, and then, the MAS model for OOS mission planning would be established. Based on the MAS model, the mission scenario and the corresponding optimization model are developed in Sect. 3. Section 4 presents the onboard intelligent planning method for the MAS model based on contract net protocol, while Sect. 5 gives the numerical simulation, and Sect. 6 concludes the whole paper.

2 MAS Model for SSc Autonomies

In this paper, each SSc is supposed as an autonomy. They could perform planning onboard, communicate, and negotiate with other SScs. Depending on this assumption, a MAS model is employed for intelligent planning. In this section, the MAS concept is introduced, and then, the MAS model for planning is established.

2.1 Agent and Multi-agent System

The concept of agent is firstly developed by McCathy [16] in the 1950s. Agent is supposed to be an entity, which exhibits, in some significant measure, autonomy, intelligence, and environmental awareness and which interacts with its environment to achieve internal goal. With the development of artificial intelligence (AI), more researchers paid attention to the MAS, initially for distributed artificial intelligence (DAI) study [17, 18]. In general, MAS consists of several mutually independent agents. These agents gather together for certain specified goal. A typical MAS would be social and collaborative.

2.2 Agent Model

In this paper, a single SSc is viewed as a simple agent with low intelligence. Each agent mainly consists of analyzer, mission planner, trajectory generator, memorizer, actuator, and transceiver. Transceiver is used for sending and receiving messages; analyzer is to evaluate the feasibility for servicing mission; mission planner is used for onboard mission planning; trajectory generator is used for path planning for orbital rendezvous; memorizer is employed for data storage, while actuator is in charge of performing OOS activities. Mission planner is the key of the whole agent. In this paper, it can only process the tasks for itself, not for any other agent (see Fig. 1).

2.3 MAS Model

In a fully distributed MAS model (Fig. 2a), each agent in the system has "full group intelligence," where any agent has intelligence equal to any other agent. There is no hierarchy, but in order to achieve a goal, there must be an extensive communication

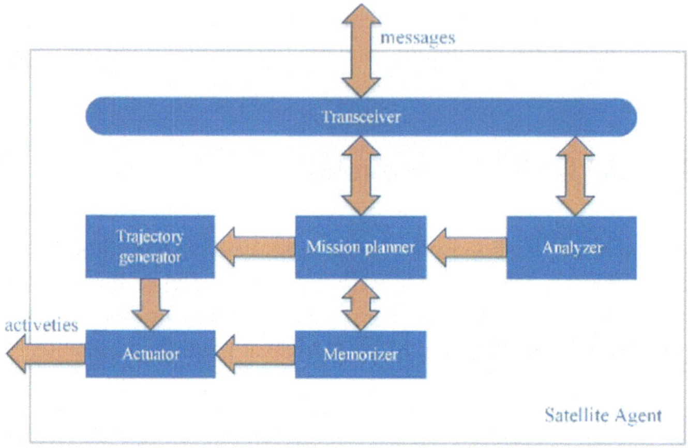

Fig. 1 Agent model

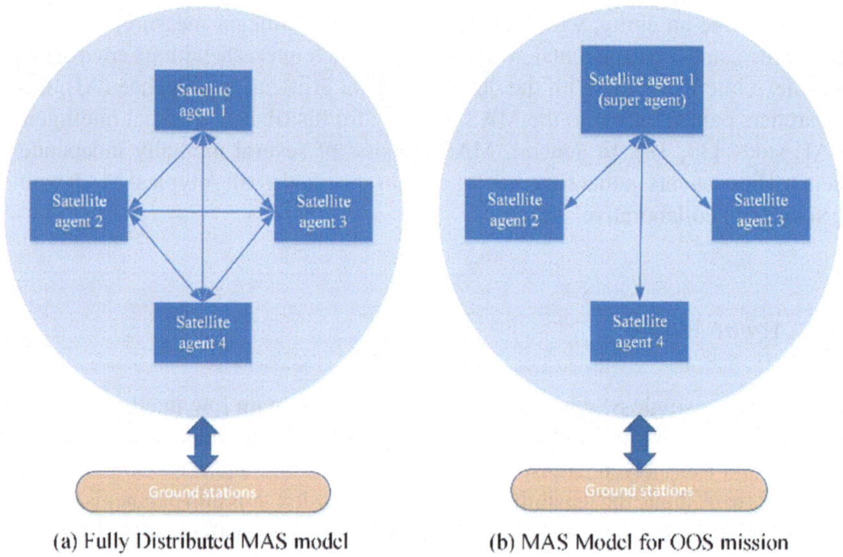

Fig. 2 MAS models

between all agents. To reduce the communication cost, another distributed model is considered in this paper for the SSc MAS, as shown in Fig. 2b. The intelligence between any two agents is still equal, but one of them is authorized as a super-agent which should take charge of other agents. As the mission is issued, all requests are sent to the super-agent. The super-agent publishes tasks to all agents, collects their responses, and makes decisions about task allocation. Afterward, each SSc agent tries to improve the plan for itself so that it can achieve its goals more efficiently. In the whole model, the super-agent can communicate with any other agent, while the normal agent could only communicate with the super-agent.

3 Mission Description

3.1 Mission Scenario

The mission studied in this paper could be stated as: n tasks are issued by the malfunction satellites, and they send the requirements to the SScs. After receiving the requirements, the super-agent SSc negotiates with others, discussing the mission scheme by intelligent planning mechanism. Finally, the mission scheme for each other is carried out onboard, and the SScs perform OOS operations in order. The optimization goal is to minimize the Δv cost for rendezvous, while the planning goal is to decide which SSc should serve which task, as well as its mission sequence (see Fig. 3).

In this paper, the rendezvous trajectories will not be designed and optimized. We assume that there is a trajectory generator embedded in the SSc, and it works only after the mission sequences are determined. During the whole planning process, the trajectory generator would not be called.

Assumptions:

(1) The time for OOS mission is sufficient long;
(2) Only targets with the same orbital altitude and in circular orbits are considered.
(3) One target can and only can be served by only one SSc.

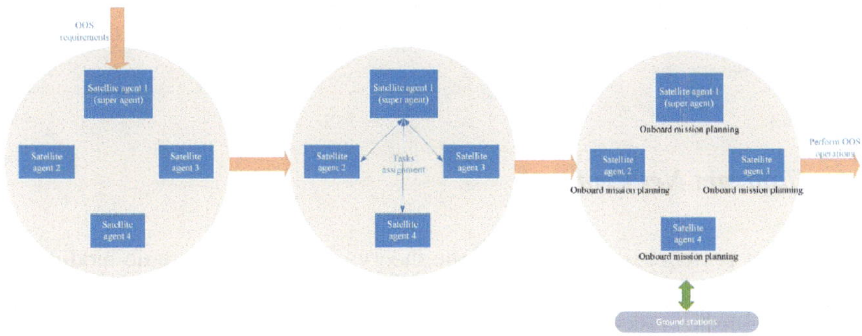

Fig. 3 Mission scenario

3.2 Optimization Model for Intelligent Planning

Denote $T = \{\text{tsk}_1, \text{tsk}_2, \ldots, \text{tsk}_i \ldots \text{tsk}_n\}$ as the task set. For each task, $\text{tsk}_i = \{\text{ID}_i, \text{loc}_i, \text{inf}\}_i$, where ID_i is the ID code, $\text{loc}_i = \{r_i, a_i, e_i, \Omega_i, w_i, f_i\}$ describes the six orbital elements, and inf_i gives some other information. For example, if it is an on-orbit refueling task, inf_i would give the required fuel mass.

Denote $S = \{\text{SSc}_1, \text{SSc}_2, \ldots \text{SSc}_j, \ldots \text{SSc}_m\}$ as the SSc set. For each SSc, $\text{SSc}_j = \{\text{ID}_j, \text{loc}_j, Q_j, \text{inf}_j\}$, where loc_j describes its six orbital elements, Q_j presents its mission sequence, and inf_j gives some other information, e.g., the remaining fuel mass.

In this paper, the design variable is $X = \{Q_1, Q_2, \ldots, Q_m\}$, and it is assumed that one task can and only can be served by one SSc.

The total cost is evaluated by Δv, the velocity increment for orbital rendezvous. Summing up, the whole problem can be stated as Eq. (1):

$$\begin{aligned} \text{find} \quad & X = \{Q_1, Q_2, \ldots Q_m\} \\ \min \quad & J = \sum_i \Delta v(Q_i) \end{aligned} \tag{1}$$

4 Intelligent Planning Based on Contract Net Protocol

To deal with practical problems, the coordinating mechanism of MAS is especially significant. And the foundation of the mechanism is the coordination protocol. The contract net protocol is an excellent example for utilizing the MAS model. In this section, the coordination mechanism is developed based on the contract net protocol, and an intelligent planning based on contract net protocol (IPCNP) method would be developed.

4.1 Contract Net Protocol

Contract net protocol is motivated by the theory of real-life economic bidding. A contract net consists of manager, bidder, and contractor (see Fig. 4). Firstly, the manager publishes tasks to the bidders (called task announcement); secondly, the bidders receive the information, evaluate the capabilities and preferences for itself, and then send bidding information for auction (called bidding); lastly, the manager collects information and then decides who will be the contractor (call awarding).

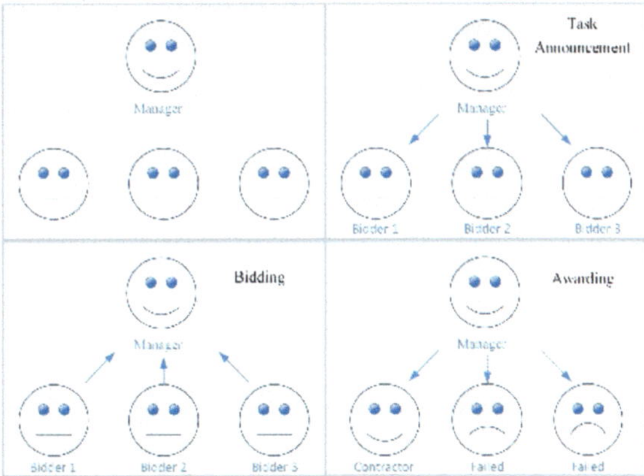

Fig. 4 Contract net protocol

4.2 An Auction Algorithm for MAS OOS Mission

(1) Task Announcement.

Malfunction satellites issue their requirements to the manager. Then, the manager publishes the task tender to bidders (other SSc agents) as $\text{tsk}_i = \{\text{ID}_i, \text{loc}_i, \text{inf}_i\}$, where ID_i is the ID code, $\text{loc}_i = \{r_i, a_i, e_i, \Omega_i, i_i, f_i\}$ describes the six orbital elements, and inf_i gives some other information. For different tasks, the format for inf_i is different. Note that, in this paper, the super-agent is not only the manager, but also the bidder, since it also can perform planning and OOS activities.

(2) Bidding.

After receiving the tenders, analyzer is called by each agent. It analyzes each task for the SSc, checking if it is feasible to complete the task and evaluating the possible cost. After analyzing, each agent would send back their tender as $B_i = \{\text{ID}_i, \text{Flag}_i, \text{d}v_i\}$, where ID_i is the ID code of the task, $\text{Flag}_i = 1$ means the SSc is able to complete the mission, and $\text{d}v_i$ gives the evaluated Δv cost. If $\text{Flag}_i = 0$, then the SSc is unable to complete the mission, and $\text{d}v_i = -1$.

For $\text{d}v$ evaluation, based on Law-2, only velocity increment for plane change is considered. Assume that the inclination and right ascension of SSc and target are (i, Ω) and (i', Ω'), respectively. Let the angle between the two orbit planes be γ, then

$$cos\gamma = \cos i \cos i' + \sin i \sin i' \cos(\Omega' - \Omega) \tag{2}$$

For circular orbits, where the final velocity is equal to the initial velocity, the $\text{d}v$ for a plane change of γ can be calculated by

$$\text{d}v = 2v \sin(\gamma/2) \tag{3}$$

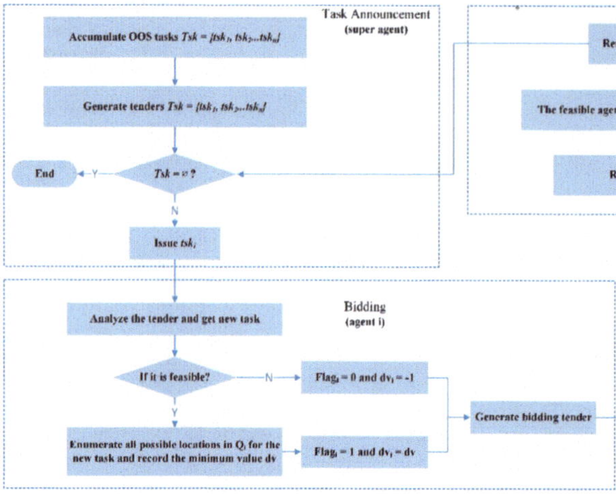

Fig. 5 Auction process

After SSc_j receiving the tender, if the missions sequence Q_j is not empty, then enumerate all possible locations for inserting the new task and record the minimum dv as dv_j.

(3) Awarding

For all bidders, the feasible agent with the lowest cost would be the contractor (see Fig. 5).

4.3 *Intelligent Planning for Single Agent*

Assume that the onboard computation ability for each agent is not very good. So, the onboard mission planning algorithm is very simple here.

For each task received:

(1) Enumerate all possible locations for inserting the new task and record the minimum dv and the corresponding location;
(2) Insert the new task to that location;
(3) If the new task and its neighbors are in the same coplanar circular orbit, order them in counter-orbital-wise direction according to Law-3.

Table 1 Selected satellites with properties

Target	Name	Inclination (°)	Right ascension (°)	Mean anomaly (°)	Mass (kg)
1	AGILA_2_24901	0.0597	116.354	196.343	2200
2	INTELSAT_602_20315	7.6349	62.3031	77.5328	1700
3	INTELSAT_1R_26608	0.118	77.9958	187.552	3300
4	IS-603_20523	7.1157	63.8733	197.089	2830
5	JCSAT-R_24732	2.2649	75.3772	84.8655	1383
6	SHENTONG-1_28082	0.0757	62.8655	214.089	900
7	SKYNET_4C_20776	10.4757	44.2337	182.74	2566
8	YAMAL_102_25897	6.1595	66.3288	207.771	2600

Table 2 Selected SScs with properties

SSc	Inclination (°)	Right ascension (°)	Mean anomaly (°)
1	0	0	0
2	7	60	80

5 Simulations

5.1 Configurations

To demonstrate the method, eight objects of GEO belt were randomly selected as targets (see Table 1). The inclination, right ascension, and the mean anomaly are those from satellite database. It is assumed that all the objects and SScs are in circular orbits with the orbit altitude of 35,786 km. Two SScs are taken into account with the orbital parameters listed in Table 2.

5.2 Simulations for Single Agent

In this section, only SSc 1 is considered. The optimization results of IPCNP and exhaustive search (ES, i.e., enumerate all feasible sequences, and the optimal choice is the one with the best cost) would be compared and discussed. The optimization results are listed in Tables 3 and 4.

In Table 4, the optimization results of IPCNP are equivalent with those of ES, since the starting and ending positions are the same and the mission sequence of IPCNP is the inverted sequence of ES. In Table 3, the mission sequence of IPCNP is slightly different from that of ES, and the total cost of IPCNP is a little higher than that of ES. It shows that the proposed method IPCNP is efficient in dealing with

Table 3 Simulations for single agent with the first 5 targets

Method	Mission sequence	Rendezvous cost (m/s)	Total cost (m/s)
IPCNP	(1, 3, 5, 4, 2)	[1.62, 2.18, 58.30, 133.48, 15.12, 207.18]	417.89
ES	(3, 2, 4, 5, 1)	[3.20, 204.10, 15.12, 133.48, 60.29, 1.62]	417.82

Table 4 Simulations for single agent with eight targets

Method	Mission sequence	Rendezvous cost (m/s)	Total cost (m/s)
IPCNP	(1, 3, 5, 8, 4, 2, 7, 6)	[1.62, 2.18, 58.30, 106.94, 27.08, 15.12, 108.25, 282.14, 2.06]	603.70
ES	(6, 7, 2, 4, 8, 5, 3, 1)	[2.06, 282.14, 108.25, 15.12, 27.08, 106.94, 58.30, 2.18, 1.62]	603.70

single SSc OOS mission planning, though in some cases, it just presents suboptimal results.

5.3 Simulations for MAS OOS Mission

In this section, two SScs are considered. The optimization results of IPCNP and particle swarm optimization (PSO) would be compared and discussed. The optimization results are listed in Tables 5 and 6.

In Tables 5 and 6, the optimization results of IPCNP and PSO are all the same. It proves that the proposed method IPCNP could effectively address MAS OOS mission planning problem.

Table 5 Simulations for multiple agents with the first 5 targets

Method	Mission sequence	Rendezvous cost (m/s)	Total cost (m/s)
IPCNP	(1, 5, 3) (2, 4)	[1.62, 60.29, 58.30, 3.20] [18.99, 15.12, 13.30]	170.82
PSO	(1, 5, 3) (2, 4)	[1.62, 60.29, 58.30, 3.20] [18.99, 15.12, 13.30]	170.82

Table 6 Simulations for multiple agents with eight targets

Method	Mission sequence	Rendezvous cost (m/s)	Total cost (m/s)
IPCNP	(1, 5, 3, 6) (8, 4, 2, 7)	[1.62, 60.29, 58.30, 1.33, 2.06] [30.11, 27.08, 15.12, 108.25, 113.76]	417.93
PSO	(1, 5, 3, 6) (8, 4, 2, 7)	[1.62, 60.29, 58.30, 1.33, 2.06] [30.11, 27.08, 15.12, 108.25, 113.76]	417.93

6 Conclusions

The onboard intelligent mission planning for on-orbit servicing (OOS) mission is studied in this paper. Each SSc is supposed as an autonomy. They could perform planning onboard, communicate, and negotiate with other SScs. A multi-agent system (MAS) model is firstly introduced. And then, based on the MAS model, an intelligent planning method based on contract net protocol (IPCNP) is developed. In simulations, the optimization results of IPCNP are almost the same with those of ES or PSO. It demonstrates the effectiveness of IPCNP in some degree.

Note that the mission scenario studied in this paper is ideal to some extent. This research is just an exploration for onboard cooperated mission planning. In our near future, more practical factors would be considered, e.g., the communication environment, issuing multiple tasks simultaneously, and so on.

Acknowledgements This work was supported by National Natural Science Foundation of China under Grant No. 71601179.

References

1. Richards MG (2006) On-orbit serviceability of space system architectures. Massachusetts Institute of Technology
2. Alfriend KT, Lee D, Creamer NG (2006) Optimal servicing of geosynchronous satellites. J Guid Control Dyn 29(1):203–206
3. Shen H, Tsiotras P (2002) Optimal scheduling for servicing multiple satellites in a circular constellation. In: AIAA/AAS astrodynamics specialists conference and exhibit, (Monterey, CA), Aug 2002
4. Yu J, Chen XQ, Chen LH, Hao D (2014) Optimal scheduling of GEO debris removing based on hybrid optimal control theory. Acta Astronaut 93:400–409
5. Jing Y, Chen X-q, Chen L-h (2015) Optimal planning of LEO active debris removal based on hybrid optimal control theory. Adv Space Res 55(11):2628–2640
6. Yu J, Yu Y-g, Hao D, Chen X-q, Liu H-y Bio objective mission planning for geosynchronous satellites on-orbit refueling. Proc IMechE Part G: J Aerosp Eng. https://doi.org/10.1177/0954410017740383
7. Zhang J, Parks GT, Luo YZ et al (2014) Multispacecraft refueling optimization considering the J2 perturbation and window constraints. J Guid Control Dyn 37(1):111–122
8. Yu J, Hao D, Liu H, Chen X (2018) Mission rescheduling for GEO active debris removal. In: The 4th IAA conference on dynamics and control of space systems (DyCoSS2018), Changsha, China, 21–23 May 2018
9. Liu Y et al (2017) Multi-objective optimal preliminary planning of multi-debris active removal mission in LEO. Sci China Inf Sci 60(7):072202
10. Chen X-q, Jing Y (2017) Optimal mission planning of GEO on-orbit refueling in mixed strategy. Acta Astronaut 133:63–72
11. Zhou Y et al (2015) Optimal scheduling of multiple geosynchronous satellites refueling based on a hybrid particle swarm optimizer. Aerosp Sci Technol 47:125–134
12. Das S, Knights D, Wu C et al (2001) Distributed intelligent planning and scheduling for enhanced spacecraft autonomy. In: Proceedings of the AAAI 2001 spring symposium series, California, USA

13. Schetter T, Campbell M, Surka D (2003) Multiple agent-based autonomy for satellite constellations. Artif Intell 145:147–180
14. Richards R, Houlette R, Mohammed J et al (2001) Distributed satellite constellation planning and scheduling. In: The proceedings of the 14th international Florida artificial intelligence research society conference, Key West, USA
15. Salazar A, Tsiotras P (2006) An auction algorithm for optimal satellite refueling. In: Proceedings of the 2006 american control conference. 2006. Minneapolis, Minnesota, USA, 14–16 June 2006
16. McCarthy J, Hayes P (1969) Some philosophical problems from the standpoint of artificial intelligence. Edinburgh University Press, Edinburgh
17. Wooldridge M, Jennings N (1995) Intelligent agents: theory and practice. Knowl Eng Rev 10(2):115–152
18. Debugging M (1995) Multi-agent system. Inf Softw Technol 37(2):102–112

Numerical Aerodynamic and Design Analysis of Combined Saucer-Shaped Buoyancy-Lifting Airships

Yanxiang Cui, Yanchu Yang, Jinggang Miao, and Xiangqiang Zhang

Abstract Saucer-shaped buoyancy-lifting airships have attracted worldwide attention due to their huge potential for stratospheric platform and cost-effective operation, especially in large load transportation. Combined saucer-shaped airship is a high-altitude, maneuverable, high buoyancy efficiency lighter-than-air (LTA) vehicle for providing continuous communications and surveillance capabilities over a wide geographical area. In this work, combined saucer-shaped airships are designed different in thickness-to-radius ratio but the same in semi-major axis. Based on the Reynolds-averaged N-S equations and k-ω SST turbulent model, numerical aerodynamic investigations on combined saucer-shaped stratospheric airships are conducted to confirm the influence of thickness-to-radius ratio on the aerodynamic forces, stability, and payloads. It is found that with the increased of thickness-to-radius ratio, the volume-to-area ratio increases, and the positive lift at the negative angles and the static instability in pitching channel is improved, while the positive lift at the positive attack angles and the lift-to-drag ratio decrease, and drag is significantly enhanced. That means the larger the thickness-to-radius ratio, the more the payload, but the worse the aerodynamic characteristic. The airship B, with the thickness-to-radius ratios of 1/2, provides average performance in buoyancy efficiency and drag characteristics. This investigation serves to provide reasonable reference for new concept design of load airships.

Keywords Combined saucer-shaped airship · Thickness-to-radius ratio · Aerodynamics · Superpressure system stratospheric airship

1 Introduction

Near space is the area of atmosphere between 20 and 100 km above the earth. With the tailored payloads that are part of the design, air vehicles operating at near space provide persistent communications, cost-effective operational capabilities, and higher sensor performance that cannot be duplicated by platforms at other altitudes or

---

Y. Cui (✉) · Y. Yang · J. Miao · X. Zhang
The Academy of Opto-electronics, Chinese Academy of Sciences, Beijing 100094, China
e-mail: yxcui@aoe.ac.cn

L. Wang et al. (eds.), *Proceedings of the 6th China High Resolution Earth Observation Conference (CHREOC 2019)*, Lecture Notes in Electrical Engineering 657,
https://doi.org/10.1007/978-981-15-3947-3_26

operating regimes [1]. Among various air vehicles, saucer-shaped buoyancy-lifting airship is a new concept platform with flat envelope, simple structure, uniform stress distribution, significant lift characteristics, and high buoyancy efficiency obtained through different combinations of volume-to-area ratio. It is insensitive to the wind direction, enabling better control by changing flexible power configurations instead of the rudder. Therefore, it has attracted worldwide attention due to these advantages and huge potential for platform and transport utility in recent years [1].

Some advances have been made in both theory and manufacture of the saucer-shaped buoyancy-lifting airships. NASA developed a new generation of pumpkin-shaped superpressure balloon project named Ultra-Long Duration platforms. NASA developed a new generation of pumpkin-shaped superpressure balloon project named Ultra-Long Duration Balloon (ULDB) [2] and conducted a number of test flights for carrying a 2721 kg payload to 33.5 km for up to one hundred days. Cees Bil [3] presented a lens-shaped superpressure balloon carrying a substructure that houses the observation instruments and researched drag forces acting on LTA in various element sizes. Ilieva et al. [4] introduced British MAAT HALE saucer-shaped airship program and compared aerodynamic characteristics of traditional shape with cylindrical ones. Gnanaprakash et al. [5] focus their study on the key issues in hull design, manufacture, and fly process of Indian flat sphere OSIRCA airship. Graham [6] estimated drag of the spheroid-cone shaped airship through formula, wind tunnel, and the flight test and then summarized drag of early airships with different shapes. Suefuku et al. [7] proposed a torus-type airship to improve the airworthiness at mooring condition and maneuverability for movement of any direction and carried out towing experiments about aerodynamic characteristics of the model in upper water space of huge towing tank instead of in a wind tunnel. Chu et al. [1] conducted a comprehensive study on a saucer-shaped buoyant-lifting airship named Star Light (formerly known as MaXflyer [8]), including system design, material selection, aerodynamics and finite element analysis, thermal analysis, launch and deployment and compared the cross-sectional areas of saucer, blimp, and sphere. However, there are few works to research the effect of the thickness-to-radius ratio and shape of airships on aerodynamics and practicality especially to perform quantitative treatment in the same flight conditions.

This study aims to explore a saucer-shaped airship with the optimal comprehensive performance. To meet the needs of operating in the stratosphere, the saucer-shaped airships generally adopt combined shape, which consists of the up, sphere or ellipsoid part and the below, saucer part sharing the same cross section. First, the new concept hull design of three combined saucer-shaped airships is presented; secondly, numerical aerodynamic characteristic analysis is proposed based on Reynolds-averaged N-S equations, and then, the impact of thickness-to-radius ratio on the aerodynamics and practicality is analyzed.

2 Envelope Design and Analysis

Conventional stratospheric airship is generally streamlined, and it has relatively smaller volume-to-area ratio and large slenderness, and it has difficulties in residing at higher altitudes. For achieving higher altitude in near space, buoyancy efficiency should be improved. Based on the streamlined configuration, shortening the length and increasing the thickness of the hull can improve the volume-to-area ratio, thereby enhancing the buoyancy of the airship in the same volume. Meanwhile, extending the lateral dimension of the hull to the major horizontal axis can gradually shift to improve the natural stability of the gravity. Then, a lifting body, saucer-shaped configuration is obtained through the method of increasing the volume-to-area ratio and decreasing of slenderness ratio. Though the saucer-shaped configuration mainly relies on helium buoyancy in the hull as conventional airships, it can produce appreciable lift, which distinguishes itself from conventional airships. Meanwhile, this new configuration will cause larger drag and lead to increase in propulsion power demand, but it generates more advantages over the conventional hull features, such as simpler and more stable structure, higher buoyancy efficiency. More importantly, it can be expected to explore the higher altitude.

Based on the above considerations, the three combined saucer-shaped airships of different thickness-to-radius ratios are designed as the upper ellipsoid and lower saucer parts, which share the same equatorial plane and the semi-major axis (radius) of *a*. The semi-minor axis of the upper part in vertical plane is *b*, and the semi-minor axis of the lower part is bl. All the thickness-to-radius ratios (bl/*a*) of three lower parts are 1/4, while the thickness-to-radius ratios (*b*/*a*) of three upper parts are set as 1/4, 1/2, and 1, respectively. The sharing radius (*a*) of equatorial plane is set as 50 m; correspondingly, the huge volume of three combined saucer-shaped airships is 130, 900 m^3, 196, 350 m^3, and 327, 248 m^3. For expressing easier, three combined saucer-shaped airships are replaced with airship A, B, and C, and their cross-sectional plots are shown in Fig. 1. The combined saucer-shaped airship design

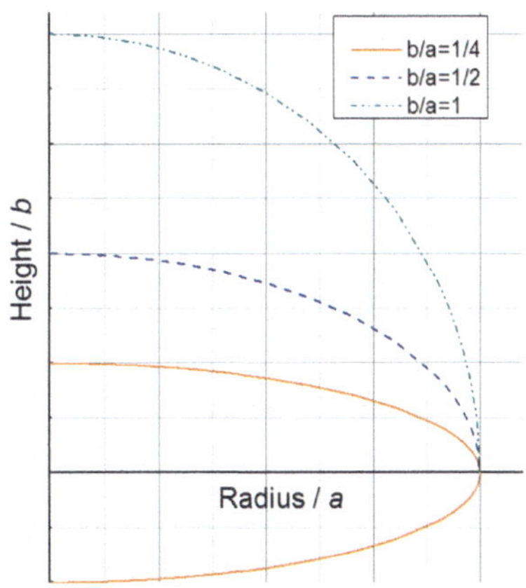

Fig. 1 Cross-sectional plot of three saucer-shaped airships

mainly drives from large buoyancy efficiency and low drag characteristics demands; therefore, the volume-to-area ratios of typical slender body with the fineness ratio of 4 and spherical designs are compared with three combined saucer-shaped airships of different thickness-to-radius ratios (Fig. 2). The saucer shape improves the volume-to-area ratio and maximizes maneuverability against changing winds than slender body, while reducing frontal cross-sectional area (Fig. 3) and rear wake size than

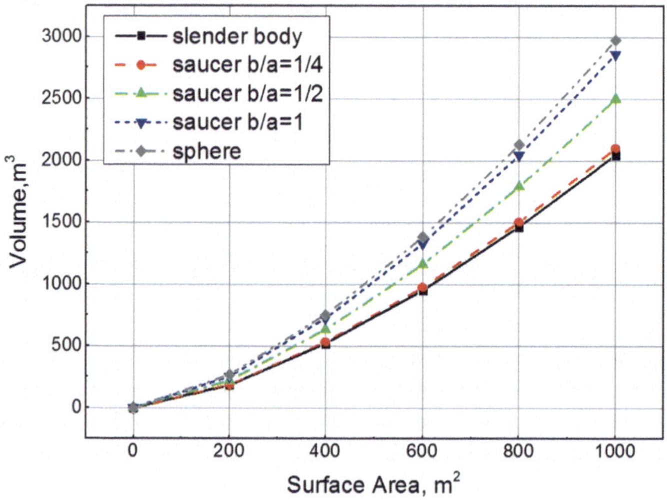

Fig. 2 Volume-to-area ratio of different shape airship

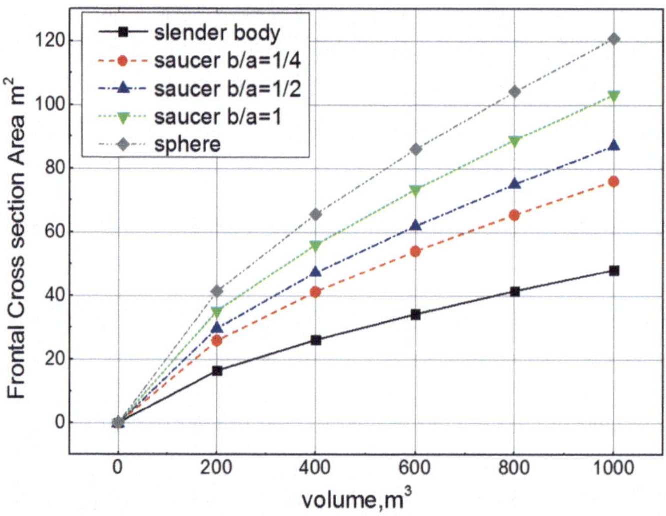

Fig. 3 Frontal cross-sectional area-to-volume ratio of different shape airships

sphere, which means the saucer shape reflects comprehensive advantages. Moreover, airship B, with the thickness-to-radius ratios of 1/2, provides average performance in buoyancy efficiency and drag characteristics, which will also be validated by the following CFD analysis.

The three airships are designed as superpressure system and utilize a small proportion ballonet to save weight and improve the effective lift. The altitude control and maintenance function of the airship mainly depend on superpressure technology instead of a huge ballonet. For keeping the ideal aerodynamic shape, the superpressure balloon envelope is strengthened with attachment lines and associated load patches, which are shown in Figs. 4 and 5. The airship consists of a fuselage, solar arrays, fuel cells, propulsion motor braces, flight control functions, payload boxes, and landing structures, as illustrated in Fig. 6. The fuselage is suspended at a prescribed distance below the hull for the requirement of the environmental conditions of high altitude and the pitch stability of the hull. The fuselage houses and supports all the functional operating systems except solar arrays on the top surface and four yaw propulsion around the hull. In addition, the other two propulsions mounted on the fuselage ensure pitch and forward movement control operability.

Fig. 4 Superpressure envelopes in side view

Fig. 5 Superpressure envelopes in bottom view

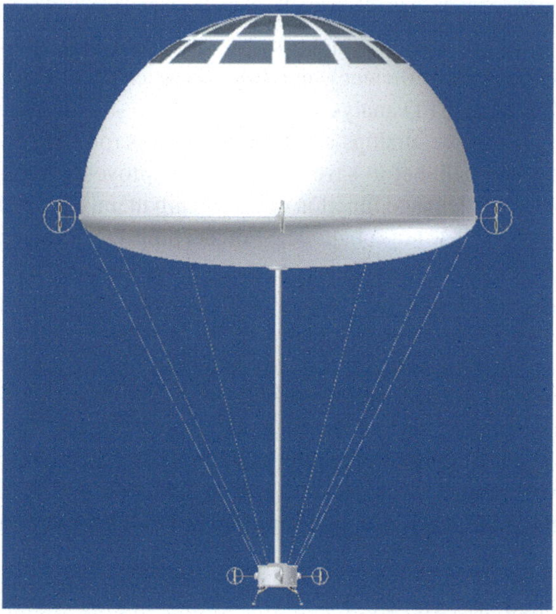

Fig. 6 Components of the saucer-shaped airship

3 Numerical Aerodynamic Methods

3.1 Models and Methods

High fidelity analysis of the platform aerodynamics during research is critical to validating the design and improving the probability of successful achievement of performance requirements [1]. Numerical simulations based on analysis on three combined saucer-shaped airships of different thickness-to-radius ratios is employed to validate design characteristic.

The computation is based on the body axis system *Oxyz* which is fixed to the airship, with the origin *O* at the equatorial plane center, *Ox* axis as the longitudinal axis of the airship pointing downstream, *Oy* axis on the longitudinal symmetry plane and pointing up, and *Oxyz* axis forming the right-hand Cartesian coordinate system. Both the body and wind axis system are shown in Fig. 7. For facilitating calculation, the half models of three airships are utilized due to their axially symmetry. The structure grid is applied for better simulation in the boundary layer and computational field, the field is 10 times the airship length away from the surface, and the downstream surface is 15 times away. Three configurations and their grids are shown in Figs. 8, 9, 10, 11, 12 and 13, respectively.

As the flow around airships belongs to the low-speed flow, the three-dimensional flow solver, based on the incompressible Reynolds-averaged Navier–Stokes equations and SST k-w turbulence model, is employed for simulating the turbulence.

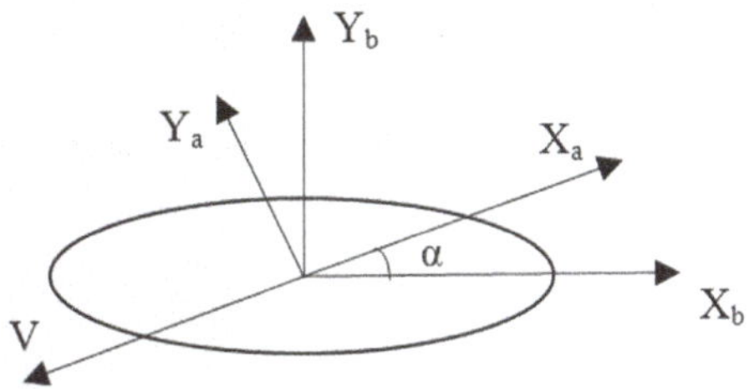

Fig. 7 Body- and wind-axis system

Fig. 8 Airship A

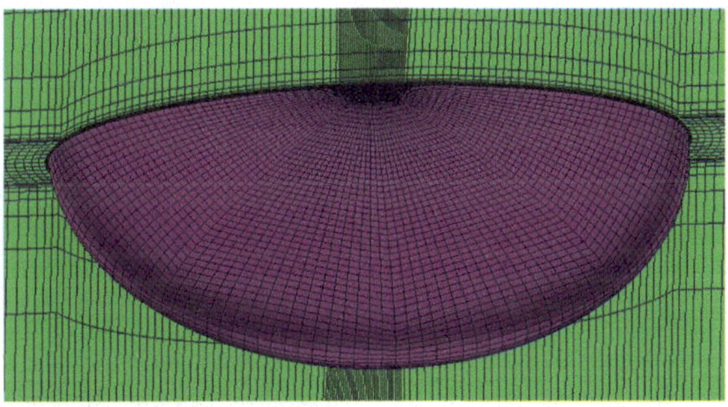

Fig. 9 Grid of airship A

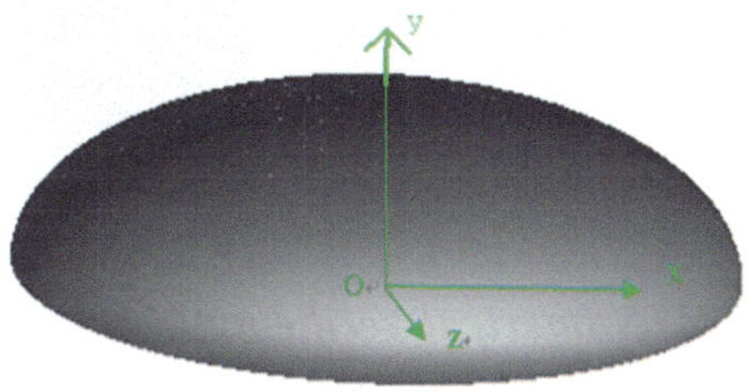

Fig. 10 Airship B

All far-field boundaries are set as velocity inlet except the downstream surface as pressure outlet. The non-slip condition is applied for the airship surface.

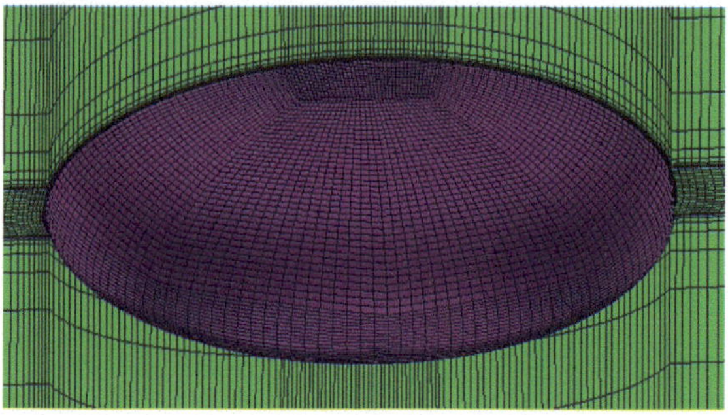

Fig. 11 Grid of airship B

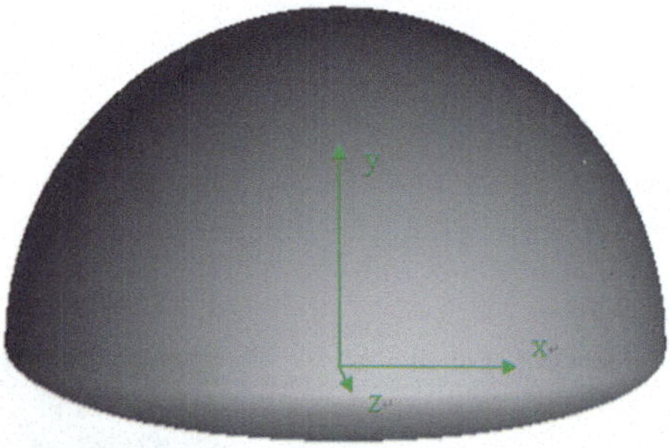

Fig. 12 Airship C

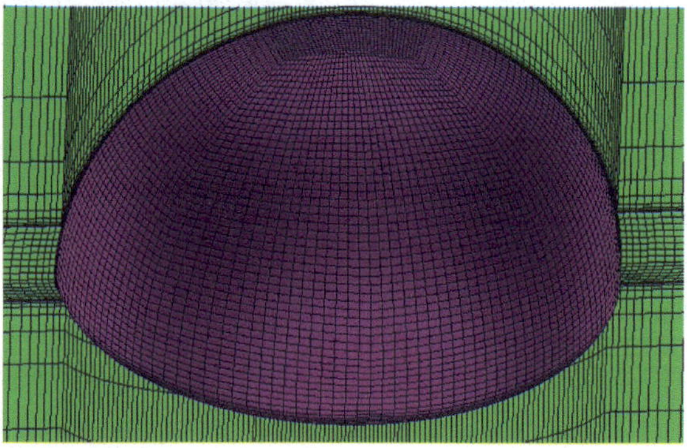

Fig. 13 Grid of airship C

3.2 Numerical Verification

The aerodynamic empirical formulas of conventional airship are mainly based on slender bodies; therefore, they cannot be applied to the combined saucer-shaped configurations which consist of the ellipsoid and saucer shape and are quite different from the slender bodies.

To validate the numerical method, calculation results of LOTTE are compared to experimental data in MWT quantified by Funk et al. [9, 10]. Although combined saucer-shaped airships are still different from the airship LOTTE, the same grid precision based on saucer-shaped airships can be ensured and even improved conveniently. The main dimensions of LOTTE are a volume of $V = 109\ \text{m}^3$, an overall length of $L = 16$ m, and a fineness ratio of $L/D = 4$. The test model is 1:20 scale model, in the condition of the wind speed of $V = 24$ m/s, and Reynolds number of $\text{Re} = 3.9 \times 10^5$ based on reference length of $V^{1/3}$. The hexahedral meshes are shown in Fig. 14. The comparison of experiment data and calculation results using different mesh dimensions is shown in Figs. 15, 16, 17 and 18, in which *2M, 3M* stands for 2 million and 3 million meshes results, respectively, and *exp* stands for test results.

The comparison shows that numerical aerodynamic coefficients are in good agreement with experiment data, and the increase of mesh dimensions from 2 to 3 million does not improve the simulation accuracy significantly. The maximum error of lift coefficient is about 13.81%. The error of drag coefficient after 10° is about 13.93% but larger disagreement at small angles. This is mainly due to the sharp tail of airship LOTTE and the consequent relative sparse grids, which are difficult to accurately simulate the flow separation emerging at small angle of attack. The curve of numerical lift-to-drag ratio is similar to the experiment data. The maximum error of pitching moment coefficient choosing the volume center as reference point is about 4.39%. This numerical method and grid precision are proved applicable for the aerodynamic simulation of airships, and the combined saucer-shaped configurations are not excluded.

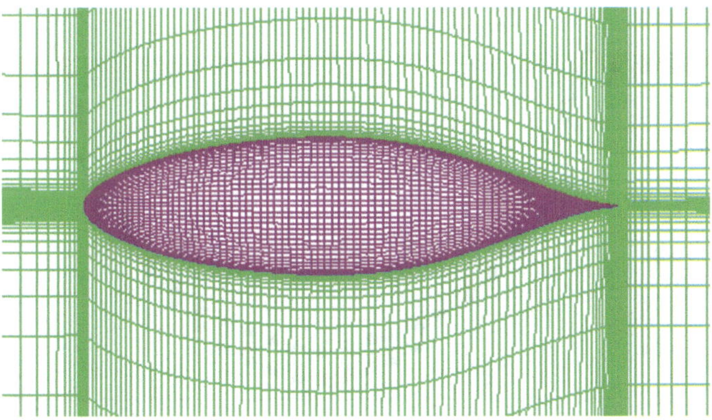

Fig. 14 Grid of airship LOTTE

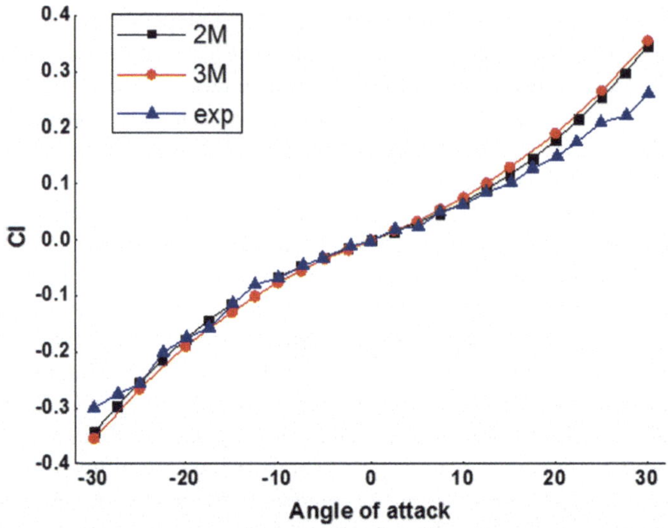

Fig. 15 Lift coefficient of LOTTE

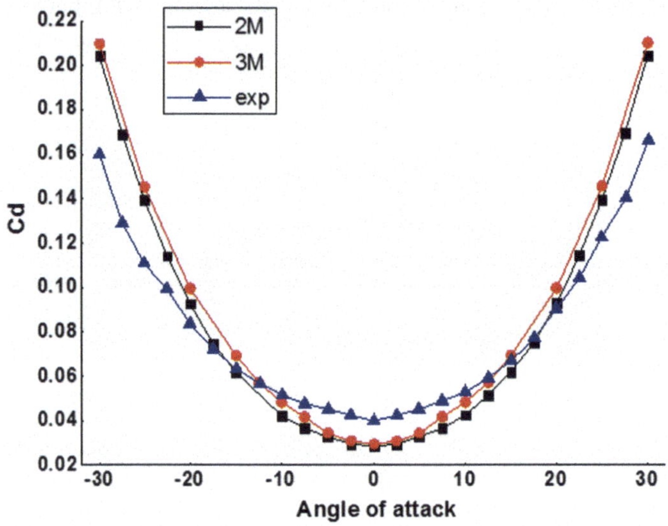

Fig. 16 Drag coefficient of LOTTE

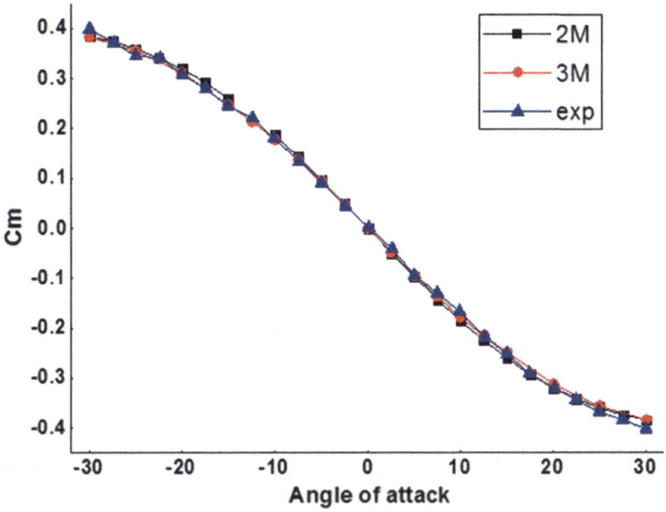

Fig. 17 Lift-to-drag ratio of LOTTE

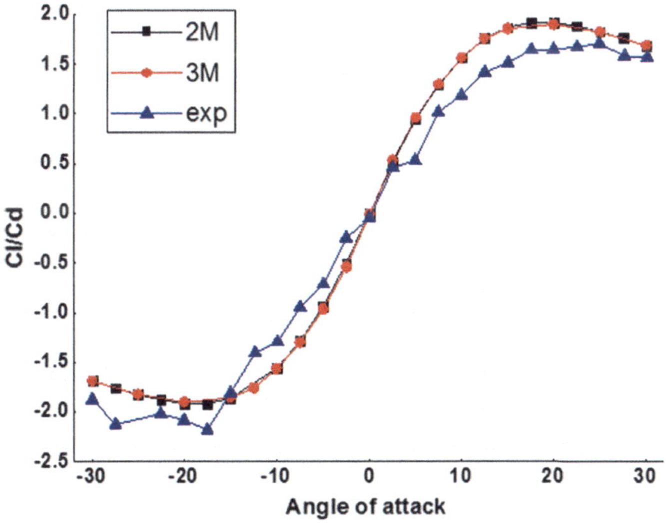

Fig. 18 Pitching moment coefficient of LOTTE

4 Aerodynamic Analysis

4.1 Aerodynamic Forces Characteristics

The CFD analysis focuses on forces and moment in pitching channel of the aerodynamic lifting body. Three combined saucer-shaped airships are set in the condition of flight altitude of 26 km, the free stream velocity of $v = 10$ m/s, and Reynolds number of $Re = 2.419 \times 10^6$, based on reference length of overall length. The volume center is chosen as reference point for the moment. Aerodynamic forces and pitching moment of three thickness-to-radius ratio airships are shown in Figs. 19, 20, 21 and 22, respectively.

As Fig. 19 shows, for configuration A, it produces positive lift only at positive attack angles. However, for B and C configurations, their upper parts are thicker and lack vertically symmetry; hence, they produce positive lift at most attack angles including negative degrees. This means that airship A can climb above neutral buoyancy using aerodynamic lift, while airship B and C may also climb even pitching down [1].

Figure 20 shows the drag component for attack angles. Notice that the drag remains the lowest between $-10°$ and $5°$. With the decrease of the thickness-to-radius ratio, the angle providing the least amount of drag to maximize thrust increases from $-10°$ to $0°$, and the drag of three configurations at the same angle decreases significantly, which is mainly because both the pressure and friction drag decrease due to the reduction of the upwind and surface area.

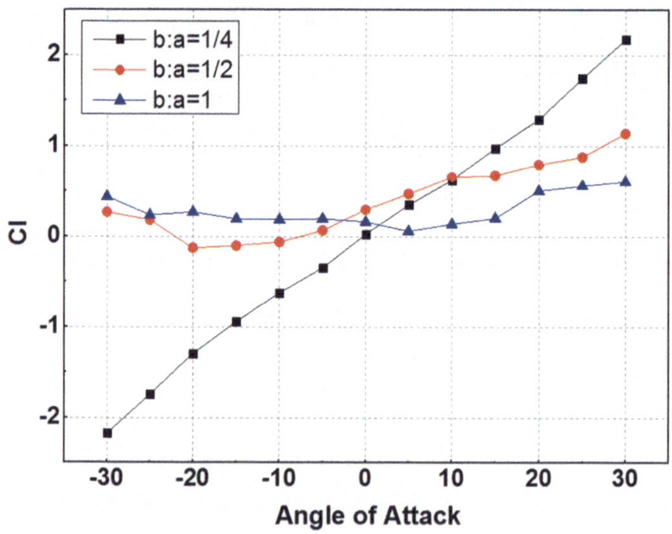

Fig. 19 Lift coefficient of three airships

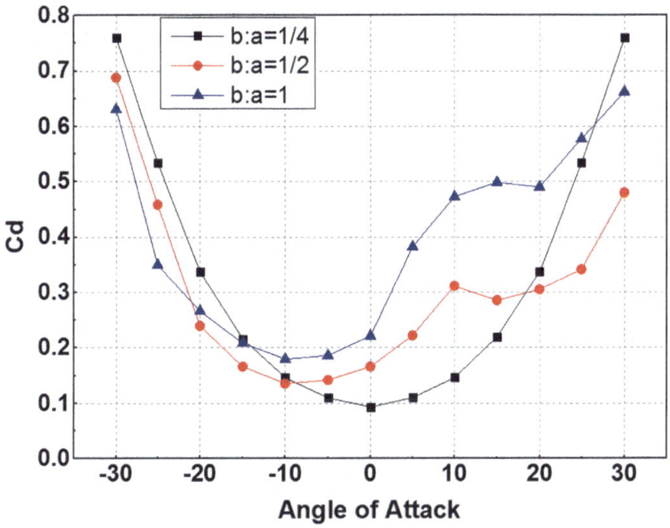

Fig. 20 Drag coefficient of three airships

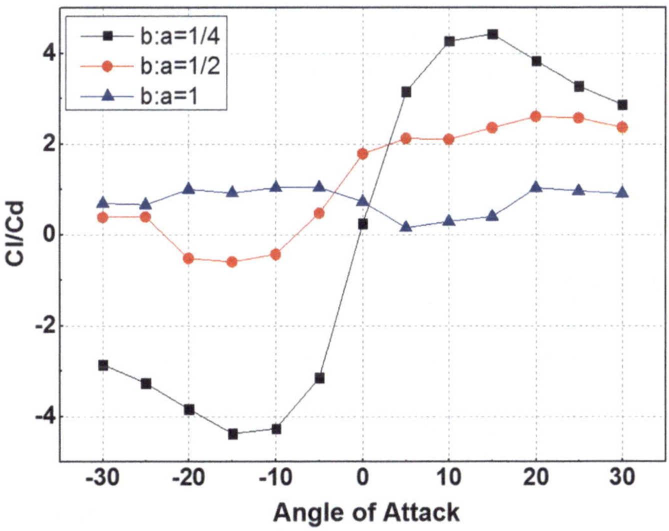

Fig. 21 Lift-to-drag ratio of three airships

Three lift-to-drag ratio curves are shown in Fig. 21. A positive lift-to-drag ratio occurs when the angle is at 0° or higher. The lower the thickness-to-radius ratio is, the greater the lift-to-drag ratio at the same angle. The maximize lift-to-drag ratios of three configurations approximate 4.5, 2.5, and 1, and corresponding angles are

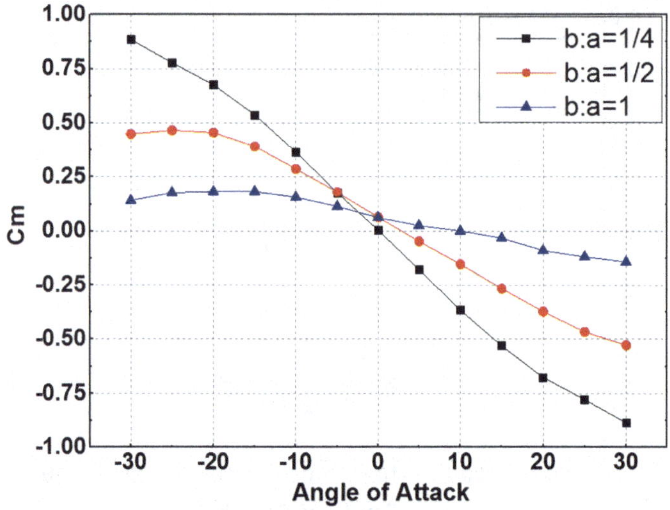

Fig. 22 Pitching moment coefficient of three airships

15°, 20°, and 20°, respectively. With the greater the lift-to-drag ratio, combined saucer-shaped airships can climb above neutral buoyancy higher.

The pitching moment curves of three combined saucer-shaped airships are compared in Fig. 22. They are instable in pitching channel in the range of computational angles. The thinner airship (configuration A) generates the greater instable moment at the same angle.

4.2 *Flow Behavior Analysis*

The flow field on and around three combined saucer-shaped airship's hull is compared at the same angle of attack $\alpha = 15°$ in side view to support their aerodynamics performance. In the vicinity of attack angle of 15°, three airships can reach the maximum lift-to-drag ratio.

The pressure distribution and surface limiting streamlines on three airship's hulls are shown in Figs. 23a, 24a and 25a, respectively. The flow on the surface in different configurations is very complex. Obviously, the airships' hulls are strongly affected by side edge flow, which led to momentum decrease and subsequent boundary layer separation at the rear walls. The flow from the bottom surface crosses with the top surface flow in the vicinity of the middle side edge. When the confluence comes to the rear region, the turbulent boundary layer on the hull is not able to overcome the high adverse pressure gradient and brings on strong three-dimensional flow. With the increase of thickness-to-radius ratios, the separation on the rear surface tends to

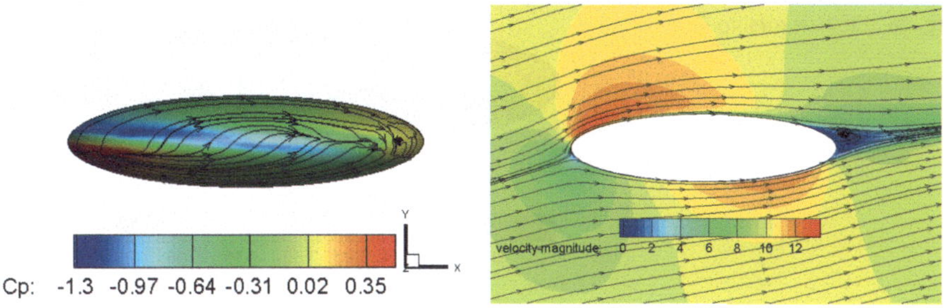

(a) (b)

Fig. 23 Flow on and around airship A ($\alpha = 15°$)

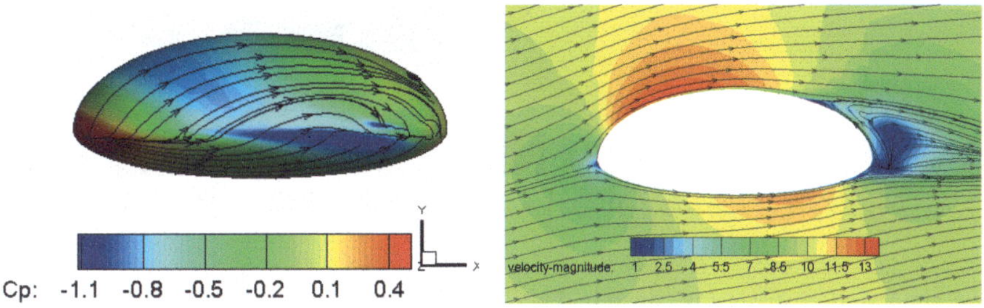

(a) (b)

Fig. 24 Flow on and around airship B ($\alpha = 15°$)

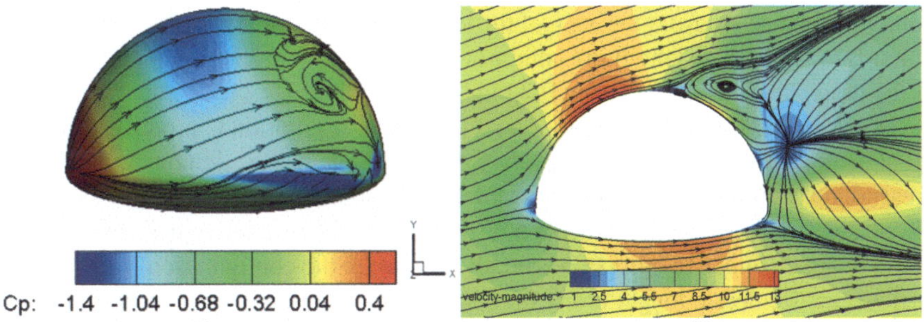

(a) (b)

Fig. 25 Flow on and around airship C ($\alpha = 15°$)

expand and involves drag rise; meanwhile, negative pressure gradually strengthens on the top surface, which helps to produce greater positive lift at the angle.

The flow field and velocity magnitude distribution around three airships' hulls are shown in Figs. 23b, 24 and 25b, respectively. The overall hull, especially the frontal part, acting as an obstacle on the airflow, causes great pressure difference in head and rear region, which induces the airflow separation and vortexes generation.

Fig. 26 Flow on airship A ($\alpha = -15°$)

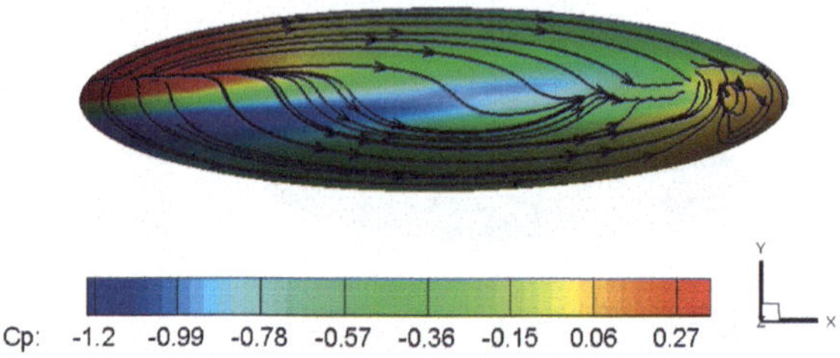

Fig. 27 Flow on airship B ($\alpha = -15°$)

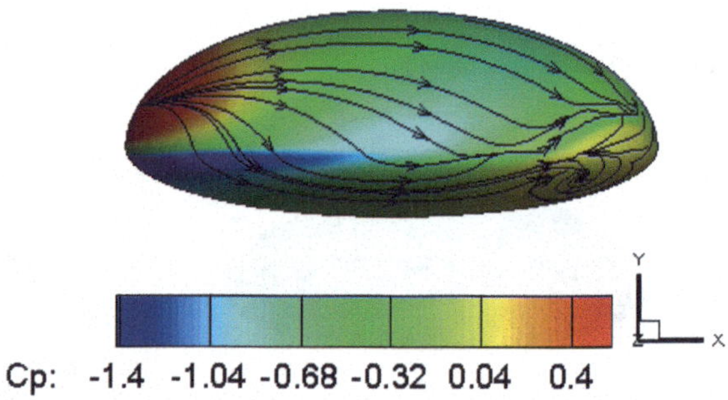

The generated vortices can cause bad aerodynamic performance, especially for drag increase. Moreover, this can lead to increased thrust requirements and a strong system for control and stability. With the increase of the thickness-to-radius ratios, the wake significantly expands, the vortex structure at the backside of the hulls widens, consequently the flow velocity drops (pressure rises) distinctly, and the separation point moves forward from the 85% of the length in configuration A, 80% in configuration B to 50% of the length in configuration C. For this case, a significant increase of the drag coefficient value from 0.22 to 0.50 is observed.

As the airship lift curves shows, (see Fig. 19), configuration C produces positive lift at most attack angles even at negative degrees. This will be demonstrated by the pressure distribution of three airships at the angle of attack $\alpha = -15°$, (see Figs. 26, 27 and 28). In the case of configuration A, its pressure distribution is just similar to the graph of attack angle 15° flipped vertically. While for the configuration B and C, the airflow can accelerate longer along the top surface than that in the configuration A, which causes distinct negative pressure at the backside of top surface in configuration C. Obviously, the top negative region of this hull has absolute predominance in pressure distribution and hence determinates the positive lift at these negative angles.

5 Experimental Verification

To verify manufacturing and flight ability, a small-scale model of configuration C was fabricated and tested with the fuselage, propulsion motor braces, flight control

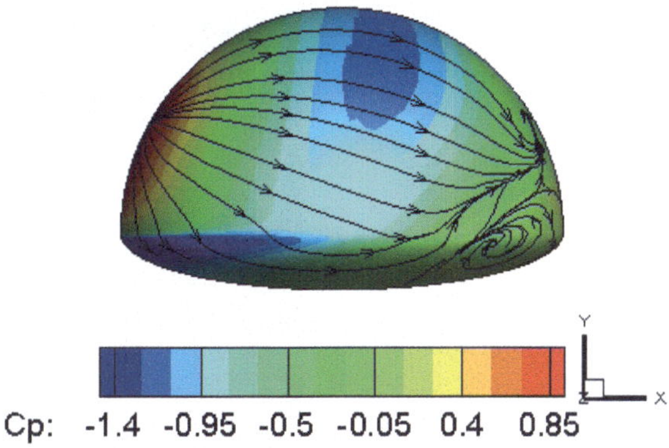

Fig. 28 Flow on airship C ($\alpha = -15°$)

functions, and landing structures at the altitude of 100 m, as illustrated in Figs. 29 and 30. The model flight realizes straight fly and the pitch control under hover and low-speed conditions. However, there are still many issues in yaw control, low leakage material, and launch for high altitude.

Fig. 29 Scale model of airship C

Fig. 30 Scale model of airship C in flight ($h = 100$ m)

6 Conclusion

The combined saucer-shaped airships can provide continuous communications and cost-effective operational capabilities at 20–30 km than comparable space-based platforms. As lifting bodies, they produce larger buoyancy efficiency but greater drag over the conventional LTA vehicles. The payloads and aerodynamic analysis of different thickness-to-diameter ratio configurations have been conducted to advance the development of the combined saucer-shaped airships. Conclusions from this study are as follows:

With the increase of thickness-to-radius ratios, the positive lift and drag are significantly enhanced, and the static instability in pitching channel is slightly improved. The vortex structures formed behind the airship provoke bad aerodynamic performance and lead to increased thrust requirements and a strong system for control and stability.

Considering material, structural, and aerodynamic characteristics, airship C, which provides larger buoyancy efficiency and payloads, probably advances the practical use of the new concept LTA vehicles.

Acknowledgements This study was supported by the Strategic Priority Research Program of China Academy of Sciences under Grant No.Y80A06A16Y.

References

1. Chu A, Blackmore M, Oholendt RG (2007) A novel concept for stratospheric communications and surveillance: star light. Am Inst Aeronaut Astronaut 1–14
2. Smith MS, Rainwater EL (2004) Optimum designs for superpressure balloons, PSB1–0032-02
3. Bil C (204) Lighter-than-air stationary platform concept. In: 52nd aerospace sciences meeting, 13–17 Jan 2014, National Harbor, Maryland. AIAA 2014-0029
4. Ilieva G, Páscoa J, Dumas A (2014) MAAT–promising innovative design and green propulsive concept for future airship's transport. Aerosp Sci Technol (35):1–14
5. Gnanaprakash K, Kamalraj KP (2011) Design-build-fly of OSIRCA: oblate spheroid indoor remotely controlled airship. AIAA 2011–6914
6. Dorrington GE (2006) Drag of spheroid-cone shaped airship. J Aircraft 43(2):363–371
7. Suefuku H, Hirayama T, Hirakawa Y, Takayama T (2010) Torus-type airship aiming at high airworthiness quality. In: 27th international congress of the aeronautical sciences, pp 1–10
8. Scott WB (2006) Novel hybrid near-space platform awaits a visionary financial angel. Aviat Wkly Space Technol 60
9. Lutz T, Funk P, Jakobi A, Wagner S (2002) Summary of aerodynamic studies on the Lotte airship. In: Proceeding of the 4th international airship convention and exhibition, Cambridge, UK
10. Funk P, Lutz T, Wagner S (2003) Experimental investigations on hull-fin interferences of the Lotte airship. Aero Space Sci Technol 7:603–610

Monitoring the Surface Subsidence of Liupanshui City Using ALOS-2 Images and NSBAS-InSAR Technology

Huan Chen, Guoman Huang, Guoqi Cheng, and Yuting Sheng

Abstract In recent years, the land subsidence of Liupanshui has begun to attract people's attention. In order to grasp the current situation of land subsidence in Liupanshui city, ALOS-2 data and NSBAS-InSAR technology were selected to monitor the surface deformation of Liupanshui city. The L-band ALOS-2 SAR data is very suitable for land subsidence monitoring based on InSAR technology due to its high ground resolution, long wavelength, and short revisit period. The results show that the surface cumulative deformation ranged from −72 to 12 mm in most regions of study area. The maximum settlement is −553 mm, which is near the Shangjia Town. The areas where the settlement is obvious include Shuangdong Village, South of Muqiao Village, Dahe Town, and Junlong Mountain.

Keywords ALOS-2 · NSBAS-InSAR · Land subsidence · Liupanshui

1 Introduction

Surface deformation is a complex engineering and environmental geological phenomenon, and it is also a geological disaster. Its formation factors can be divided into two categories: natural factors and human factors. Natural factors include structural factors and natural consolidation of soil. Human factors include a large amount of construction engineering and mining work carried out by humans, including the construction of high-rise buildings in towns, the extraction of oil and gas, and the extraction of groundwater. These human activities will cause different degrees of surface deformation. Surface deformation, due to its wide range of influence, long duration, slow development, hard to detect, and ineffective governance, poses a great threat to people's survival and social and economic development [1].

In recent years, the ground subsidence in Liupanshui City is very serious, and the large-scale differential settlement on the surface has seriously damaged the surface

H. Chen (✉) · G. Huang · G. Cheng · Y. Sheng
Shandong University of Science and Technology, Qingdao 266590, China
e-mail: chcx0329@163.com

Key Laboratory of Geo-Informatics of State Bureau of Surveying and Mapping, Chinese Academy of Surveying and Mapping, Beijing 100830, China

L. Wang et al. (eds.), *Proceedings of the 6th China High Resolution Earth Observation Conference (CHREOC 2019)*, Lecture Notes in Electrical Engineering 657,
https://doi.org/10.1007/978-981-15-3947-3_27

structure and the buildings above it, threatening the lives and property of urban residents. Interferometric synthetic aperture radar (InSAR) is a new type of space-to-ground observation technology. It displays outstanding advantages such as high precision, large area, and high efficiency in surface deformation monitoring [2].

The conventional differential interferometric synthetic aperture radar (D-InSAR) technology has certain limitations, which are mainly affected by the spatiotemporal decoherence and atmospheric delay. The existence of these limitations causes the surface micro-deformation to be masked by errors, which affects the application of conventional D-InSAR technology [3]. In order to overcome the limitations of D-InSAR technology, time series InSAR technology came into being. This technology can effectively overcome the shortcomings of traditional D-InSAR technology and obtain millimeter-level surface deformation monitoring accuracy. Among them, small baseline subset (SBAS) InSAR technology is widely used because of its high computing efficiency and the relatively small number of SAR images required, which is suitable for large-scale ground deformation monitoring [4].

However, in the case where the SBAS network is incomplete and a disconnected subset occurs, the results obtained by the SBAS will have a large error. In 2009, Lopez-Quiroz [5] and others added constraints in SBAS timing inversion instead of using singular value decomposition (SVD) method and proposed a small baseline interferogram network (NSBAS) method. In 2011 and 2012, Doin [6], Jolivet [7], and others conducted further research on the NSBAS method, integrating a processing chain from raw data to NSBAS calculations. A large number of scholars have verified that NSBAS-InSAR technology has high monitoring accuracy in urban settlement monitoring, up to millimeter level.

2 Study Area and Experimental Data

2.1 Introduction of Study Area

The study area of this experiment is Liupanshui City and surrounding counties, which is located in the west of Guizhou Province. The analog image of the study area is shown in Fig. 1. The Liupanshui city terrain is high in the west and low in the east and high in the north and low in the south. The landscape is mainly composed of mountains and hills, as well as landforms such as basins, mountains, plateaus, and terraces. The climate in the study area belongs to the north subtropical monsoon humid climate, and the water resources are mainly derived from rainwater, and plants are flourishing. The study area has a variety of mineral resources, including coal, iron, lead, zinc, copper, etc., with coal reserves ranking first in the province.

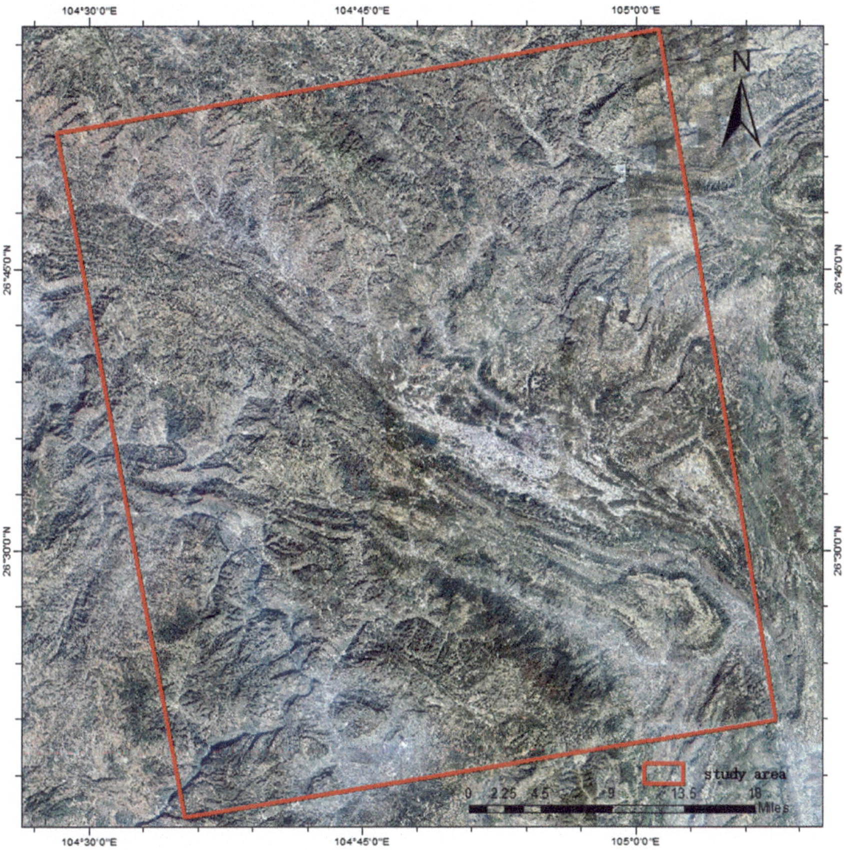

Fig. 1 Location of study area

2.2 *Experimental Data*

L-band Synthetic Aperture Satellite-2 (ALOS-2) images can obtain high signal-to-noise interference fringes in areas with poor coherence, easy phase unwrapping, and subsequent interpretation. ALOS-2 images have been used in multiple SBAS-InSAR processing cases [8–10]. This paper obtains 14 ALOS-2 data for 2017–2019. The data information is shown in Table 1. The selected external reference DEM is SRTM data at a resolution of 30 m.

Table 1 ALOS-2 data summary

Serial number	Image date	Sensor	Incidence angle
1	20170416	ALOS-2/PALSAR-2	36.1725
2	20170528	ALOS-2/PALSAR-2	36.1643
3	20170820	ALOS-2/PALSAR-2	36.1643
4	20171029	ALOS-2/PALSAR-2	36.1740
5	20171126	ALOS-2/PALSAR-2	36.1723
6	20180218	ALOS-2/PALSAR-2	36.1644
7	20180513	ALOS-2/PALSAR-2	36.1722
8	20180624	ALOS-2/PALSAR-2	36.1651
9	20180805	ALOS-2/PALSAR-2	36.1707
10	20180916	ALOS-2/PALSAR-2	36.1702
11	20190120	ALOS-2/PALSAR-2	36.1641
12	20190303	ALOS-2/PALSAR-2	36.1643
13	20190414	ALOS-2/PALSAR-2	36.1706

3 Basic Theory and Workflow of NSBAS-InSAR Technology

3.1 Basic Theory of NSBAS-InSAR Technology

The SBAS-InSAR technology was originally proposed by Berardino et al. [11]. This technology uses small baseline interference pairs composed of multiple main images to inverse temporal deformation [11]. The principle of SBAS-InSAR technology is to combine SAR images into several sets of interference pairs, and then, the least-square method is used to solve the deformation sequence of each set. Finally, the singular value decomposition method is used to solve multiple small baseline sets to obtain the surface deformation rate of the study area.

However, in the case where the SBAS network is incomplete and a disconnected subset occurs, the results obtained by the SBAS will have a large error. During the inversion process, some inversion pixels, at least a critical link in the interferogram network, are missing.

In the case where there is no temporal (B_t) or geometrical ($B_\perp$) overlaps in the groups of independent images, SVD sets the incremental phase delay between successive image groups to zero. It causes a deviation in the subsidence time, which causes a deviation in the subsidence rate. To overcome this problem, a constraint is added in the timing inversion instead of using the SVD method. First, the cumulated phase delay ϕ_k^l of pixel l at time t_k is defined by the following formula:

$$\phi_k^l = \sum_{i=1}^{k-1} m_i^l, \quad 2 \le k \le m_l l, \quad \phi_1^l = 0 \tag{1}$$

where m_i^l is the phase delay increment of the two SAR images at pixel l. Inspection of the previous delay time series, for pixels for which $G_l^T G_l$ [5] is invertible, shows that modeling ϕ_k^l by a quadratic behavior in time is reasonable. Therefore, the following constraint can be added to the inversion:

$$\phi_k^l = \alpha_t^l(t_k - t_1) + b_t^l(t_k - t_1)^2 + e^l B_\perp^k + c_t^l \tag{2}$$

where $e^l B_\perp^k$ denotes the phase due to DEM error correlated with the perpendicular baseline of each acquisition, $B_\perp^K$. We then solve by least square inversion the system $d^c = G^c m^c$.

$$\begin{bmatrix} d_1^l \\ \vdots \\ d_k^l \\ \vdots \\ d_{N_l}^l \\ 0 \\ \vdots \\ 0 \end{bmatrix} = \begin{bmatrix} & \begin{bmatrix} G_1 \end{bmatrix} \begin{bmatrix} 0 & 0 & 0 & 0 \\ \vdots & \vdots & \vdots & \vdots \\ 0 & 0 & 0 & 0 \end{bmatrix} \\ \gamma \cdot & \begin{bmatrix} 0 \cdots & 0 & \cdots 0 & 0 & 0 & -B_\perp^1 & -1 \\ 1 & \ddots & \vdots & -(t_2 - t_1) & -(t_2 - t_1)^2 & -B_\perp^2 & -1 \\ \vdots & \ddots \; 0 & 0 & \vdots & \vdots & \vdots & \vdots \\ \vdots & \ddots & \ddots \vdots & \vdots & \vdots & \vdots & \vdots \\ 1 \cdots\cdots & 1 & 0 & -(t_{Ml} - t_1) & -(t_{Ml} - t_1)^2 & -B_\perp^{Ml} & -1 \end{bmatrix} \end{bmatrix} \begin{bmatrix} m_1^l \\ \vdots \\ m_{Ml}^l \\ a_t^l \\ b_t^l \\ e_t^l \\ c_t^l \end{bmatrix} \tag{3}$$

The weight γ for scaling the additional matrix is small enough to ensure that: (a) if $G_l^T G_l$ is invertible, the increments m_k^l are not affected by the additional constraint, (b) if not, the additional constraint only sets the relative delays between independent image groups. In all cases, a_t, b_t, and e are the best-fit coefficients to all parts of cumulated phase delays that are constrained by the interferometric data set. The use of Eq. (2) allows to strongly reduce the artifacts associated with the singularity of $G_l^T G_l$ provided that Eq. (2) models most of the phase delay signal.

The average subsidence rate or velocity is then obtained by a linear fit of $(\phi_k^l - e^l B_\perp^k)$. Under the premise that the SBAS network is complete, this method does not affect the inversion and is applicable to the data. If the SBAS network is incomplete and a disconnected subset occurs, the method joins these subsets.

3.2 Data Processing

This experiment used NSBAS-InSAR technology to process ALOS-2 data and monitor the surface deformation of Liupanshui City. D-InSAR processing is implemented in the Gamma software, and the NSBAS time series inversion is implemented in the Giant software. The process of NSBAS-InSAR technology is shown in the Fig. 2.

The data processing steps include data preprocessing, image co-registration, baseline estimate, interferogram, filtering, phase unwrapping, and NSBAS time series inversion. The main parameter settings and method choices in these steps are as follows:

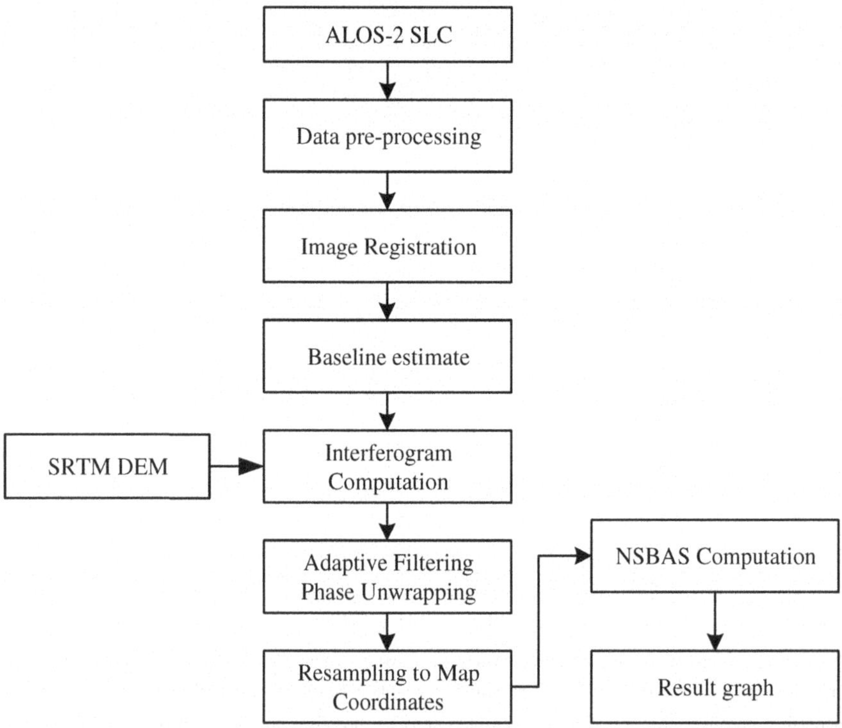

Fig. 2 NSBAS technology processing flow

1. In order to ensure the quality of the interference pairs, the time interval of the interference pair in the baseline estimation does not exceed 340 days, and the length of the spatial baseline does not exceed 500 m.
2. The phase unwrapping method is a minimum cost flow (MCF) method that takes both speed and accuracy into account. The method can limit the whole error to a small range, prevent the retransmission of the error, and the unwrapping result is more accurate.
3. The radar wave used in the ALOS-2 satellite is located in the L-band, which belongs to the low-frequency band. The study area belongs to the low-latitude region. Considering the temporal and spatial distribution of the ionosphere, the ionospheric state in the low-latitude region is more active than the middle and high latitudes. The interferograms are susceptible to the ionosphere, so we performed ionospheric correction on all interferograms.
4. In the NSBAS time series inversion, the multi-scale method is used to remove the atmospheric phase, and the network deramping method is used to perform orbit error correction.

4 Results and Analysis

The connection of the interference pairs is shown in Fig. 3. The cumulative deformation of the study area is shown in Fig. 4.

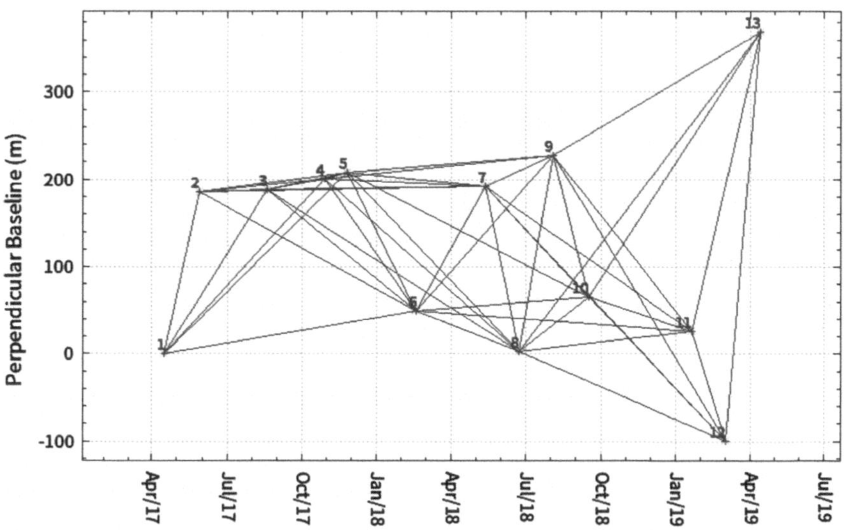

Fig. 3 Spatial-temporal baselines of interferograms

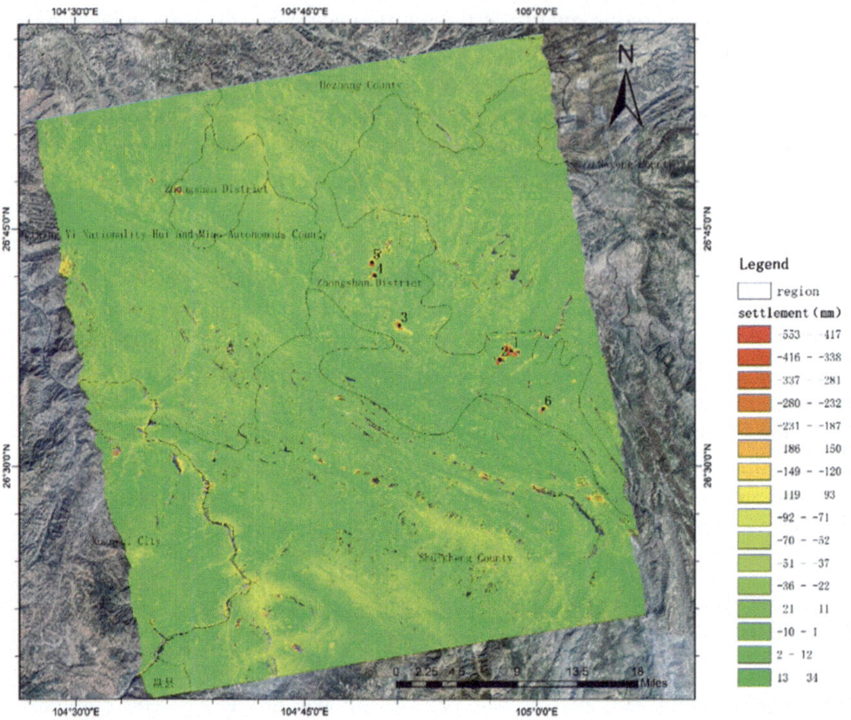

Fig. 4 Cumulative deformation diagram

It can be seen from Fig. 4 that the most serious surface subsidence in Liupanshui City from April 2017 to April 2019 is near Shangjia Town. Significant surface subsidence occurred in Shuangdong Village, South of Muqiao Village, Dahe Town, and Junlong Mountain. These four places belong to Zhongshan District. According to statistics, about 60% of the areas in the study area is converted into surface subsidence, and the area with settlement over 10 mm accounts for about 15% [12]. In this paper, six points with obvious settlement were selected in the research area for detailed analysis. The time series deformation curve of the selected points and the corresponding ground conditions are shown in Figs. 5 and 6, respectively.

Point 1 is located in Niu Hanpo, and the actual ground situation corresponds to Fig. 6(1). Point 2 is located in Cannabis Village, and the actual ground situation corresponds to Fig. 6(2). Point 3 is near Dadi Village, and the actual ground situation corresponds to Fig. 6(3). Point 4 is located in Hanging Rock, and the actual ground situation corresponds to Fig. 6(4). Point 5 is near Shagoutou Village, and the actual ground situation corresponds to Fig. 6(5), and Point 6 is located in Weijia Mawo Village, and the actual ground situation corresponds to Fig. 6(6). It can be seen from Fig. 6 that the settlement of these six places during the monitoring period is relatively serious, and the sedimentation rate is also relatively large.

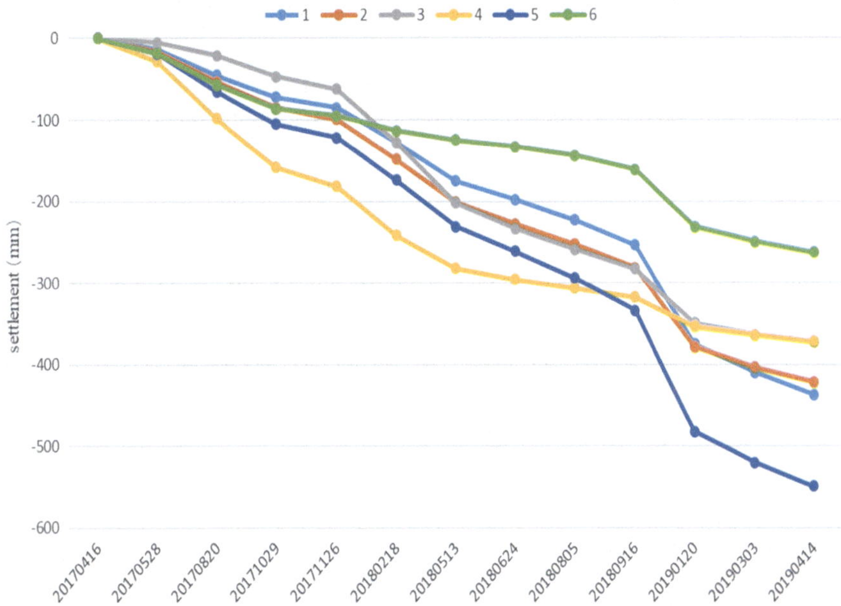

Fig. 5 Deformation curve of selected points

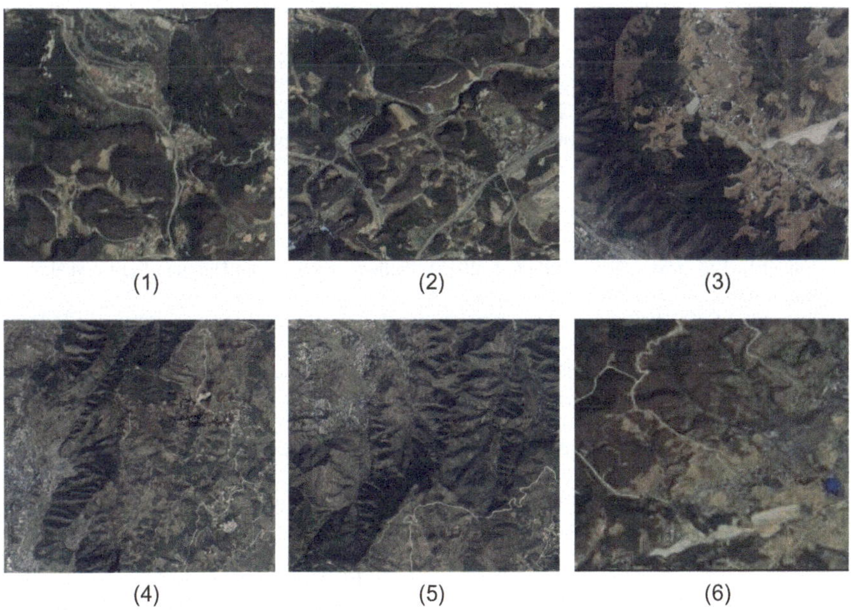

Fig. 6 Actual ground conditions correspond to the selected points

These six places belong to Zhongshan District. Zhongshan District belongs to the important part of Liupanshui Mining Area. By 2011, the proven minerals include more than 20 kinds of resources such as coal, iron, lead, dolomite, marble, limestone, and barite. The reserves of lead and zinc mines are 450,000 tons. The iron ore industry has a reserve of 20 million tons and the coal industry has a reserve of 13.4 billion tons. So, these six places are most likely mines that have been mining. It can be seen from the corresponding satellite imagery that these six places are all around the hills, and the foot of the mountain is a gathering place for residents. Therefore, once it rains, it is easy to cause geological disasters such as landslides and collapses.

Acknowledgements This work was supported in part by The National Key Research and Development Program of China (No. 2018YFF0215303, 2017YFB0503004) and Chinese Academy of Surveying and Mapping Fundamental Research Project under Grant (No. 7771808).

References

1. Zhou H, Wang Y, Yan S et al (2017) Land subsidence monitoring and analyzing of cangzhou area. Bull Surv Mapp 07:89–93
2. Hanssen RF (2001) Radar Interferometry: data interpretation and error analysis. Springer, Netherlands, pp 67–68
3. Zebker HA (1992) Decorrelation in interferometric radar echoes. IEEE Trans Geosci Remote Sens 30(5):950–959
4. Zhu J, Li Z, Hu J (2017) Research progress and methods of InSAR for deformation monitoring. J Geodesy Geoinf Sci 46(10):1717–1733
5. Lopez-Quiroz P, Doin M-P, Tupin F et al (2009) Time series analysis of Mexico City subsidence constrained by radar interferometry. J Appl Geophys 69:1–15. https://doi.org/10.1016/j.jappgeo.2009.02.006
6. Doin MP, Lodge F, Lodge S et al (2011) Presentation of the small baseline NSBAS processing chain on a case example: the ETNA deformation monitoring from 2003 to 2010 using ENVISAT data. European Space Agency Special Publication. ESA SP-697, pp 1–7
7. Jolivet R, Lasserre C, Doin MP et al (2012) Shallow creep on the Haiyuan fault revealed by InSar. JGR (in Press, 2012)
8. Samsonov S (2010) Topographic correction for ALOS PALSAR interferometry. IEEE Trans Geosci Remote Sens 48(7):3020–3027
9. Samsonov S, Tiampo K, Joyce K, et al (2010) Modified stacking and SBAS algorithms for mapping of ground deformation in Taupo volcanic zone, New Zealand With Alos Palsar
10. Jiangtao Q, Liangyu Z, Suisui W (2018) InSAR study on the current structural deformation mode of the middle section of the Altyn Tagh Fault Zone. J Geodesy Geodyn 38(08):17–20 + 51
11. Berardino P, Fornaro G, Lanari R et al (2002) A new algorithm for surface deformation monitoring based on small baseline differential sar interferograms. IEEE Trans Geosci Remote Sens 40(11):2375–2383
12. Huang G, Chen H, Li X, Cheng G, Yu Z, Gu H (2019) Monitoring the surface subsidence of Handan city using SENTINEL-1A images and SBAS-InSar technology. Int Arch Photogramm Remote Sens Spatial Inf Sci XLII-2/W13, 1933–1937. https://doi.org/10.5194/isprs-archives-XLII-2-W13-1933-2019, 2019

L^1-Norm Method for Dual-Wavelength Phase Unwrapping

Guoqi Cheng, Jixian Zhang, Huan Chen, Jiaqi Chen, and Lijun Lu

Abstract According to the relationship between the absolute phase and the baseline in multiple baselines, the relationship between the absolute phase and the band in the multi-band is designed analogously. This method constructs the equation that the wavelength and the absolute phase are positively correlated with the phase relationship of any pixel in different band images. Using the equivalence relation, two inequalities of adjacent pixels are constructed, the gradient information is obtained, the ambiguity is minimized, and the phase unwrapped 2π fuzzy number is obtained. This method avoids the limitation of the Itoh hypothesis. In this paper, two experiments are designed with two wavelengths. The results effectively improve the unwrapping ability of the phase inconsistent region compared with single wave unwrapping, which proves the feasibility of this analogy method.

Keywords L^1-norm · The gradient information · Dual-wavelength · Incoherent

1 Introduction

Synthetic aperture radar (SAR) is a two-dimensional high-resolution radar imaging technology.

In the earlier literature, it was also called "synthetic array radar." In most cases, synthetic aperture radar is often used to image fixed scenes on the ground. The working mechanism is to obtain the image phase information by acquiring two complex image data of the ground scene target and then obtain the ground elevation, which is widely used in the military and civilians. The current SAR images are mainly

G. Cheng (✉) · J. Zhang · H. Chen · J. Chen
Shandong University of Science and Technology, Qingdao 266590, China
e-mail: cgq999@126.com

G. Cheng · H. Chen · J. Chen · L. Lu
Key Laboratory of Geo-Informatics of State Bureau of Surveying and Mapping, Chinese Academy of Surveying and Mapping, Beijing 100830, China

J. Zhang
National Quality Inspection and Testing Center for Surveying and Mapping Products, Beijing 100830, China

L. Wang et al. (eds.), *Proceedings of the 6th China High Resolution Earth Observation Conference (CHREOC 2019)*, Lecture Notes in Electrical Engineering 657,
https://doi.org/10.1007/978-981-15-3947-3_28

acquired through onboard or airborne platforms, with resolutions ranging from tens of meters to a few inches.

Phase unwrapping has always been a core step in synthetic aperture radar interferometry (InSAR), which plays a key role in nuclear magnetic resonance imaging, optical interference, and digital holographic imaging of cancer cells. The phase difference obtained from the interferogram is called the winding phase, which is only the main value of the real phase, but is calculated by the inverse trigonometric function, and the value range is between $[-\pi, \pi)$. To obtain the true phase, it is necessary to increase or decrease the integer multiple of 2π based on the main value of the phase difference. This integer is called the ambiguity degree. The process of phase unwrapping can also be said to solve the whole ambiguity degree. We call this process phase unwrapping and phase unwinding [1]. The traditional single-band InSAR data processing flow is mainly divided into main and auxiliary image registration, interferogram generation, noise filtering, baseline estimation, phase unwrapping, elevation information extraction, and correction (geocoding). Especially for the image data of mountainous terrain, there is a sudden phase in the data, and the image coherence is poor, which makes phase unwrapping difficult, or the unwrapping effect is extremely unsatisfactory. According to the upcoming new synthetic aperture radar satellite NiSAR, the satellite is the world's first dual-band synthetic aperture radar satellite, built by National Aeronautics and Space Administration (NASA) and the Indian Space Research Organization (ISRO) [2]. The L and S bands are 24 and 12 cm in size, respectively. The satellite NiSAR will reveal the motion of the Earth's surface to better understand glacial movements, earthquakes, volcanoes, and vegetation changes in agriculture and forests.

In recent years, dual-frequency synthetic aperture radar (SAR) for monitoring earth has begun research. In the field of optics, phase unwrapping is performed using dual-wavelength interference; compared to numerical algorithms, dual-wavelength interference can effectively process step-type objects and restore the original phase [3]. Using domestic airborne dual-frequency data for reconstruction of interfering terrain elevation, [4]; because the dual-band joint unwrapping does not need to follow the Itoh hypothesis, the terrain is fluctuating, and the mountain high slope can effectively obtain the terrain elevation information. In the medical field [5], the author applied the dual-band linear regression phase expansion algorithm to cancer cells, which can effectively deal with the complex topology in holographic images. The minimum norm unwrapping algorithm is a global algorithm that uses the relationship between the absolute phase of the multi-baseline interferogram to avoid the limitation of the adjacent phase continuity hypothesis [6]. In theory, it can effectively deal with the unwrapping problem of the undulating terrain, the relationship between the wavelength of the two interferograms and the absolute phase. The feasibility of the two-wave phase unwrapping based on the L^1-norm is verified by experiments.

2 Principle of the Paper

The traditional single baseline phase unwrapping is based on the assumption that the winding phase gradients of adjacent pixels of the same baseline interferogram are equal to the unwrapping phase gradient [7]. Due to factors such as noise, in most cases, the winding phase difference and the unwrapping phase difference may not be completely equal. More scholars have used the L^1-norm to represent the magnitude of the distance by minimizing some “distance” between the winding phase difference and the unwrapping phase difference. The single baseline L^1-norm optimization model is [8]:

$$\min \sum_{i,j} \omega^y_{i,j} \left|\nabla^y \phi_{i,j} - \nabla^y \psi_{i,j}\right| + \sum \omega^x_{i,j} \left|\nabla^x \phi_{i,j} - \nabla^x \psi_{i,j}\right| \tag{1}$$

where ∇^y is the ambiguous gradient difference operator for adjacent pixels in the vertical direction. ∇^x is the ambiguity gradient operator for adjacent pixel in the horizontal direction, and $\omega^y_{i,j}$, $\omega^x_{i,j}$ are weighting coefficients in the vertical and horizontal directions.

Compared with the dual-baseline L^1-norm phase unwrapping principle, a pixel s at the same position in the same interferogram has the same elevation, and the two types can be obtained in combination:

$$B_1 \psi^2(s) = B_2 \psi^1(s) \tag{2}$$

where B_1, B_2 represent two different length baselines and $\psi^1(s)$, $\psi^2(s)$ represent different baselines that correspond to the absolute phase of the pixel s in the interferogram; this paper studies the dual-band phase unwrapping based on the L^1-norm. Similarly, the relationship between the absolute phases of the two interferograms in the same target scene at different wavelengths is obtained:

$$\lambda^1 \psi^1(s) = \lambda^2 \psi^2(s) \tag{3}$$

where λ^1 and λ^2 represent different wavelength. According to the previous studies on multiple baselines, the same reason, if it can describe different bands in the same region in the optimization model of an L^1-norm, it is also possible to avoid the single-baseline/single-band following the limitation of the continuous phase hypothesis, and the winding phase can be given by:

$$\varphi(s) = \psi(s) - 2k(s)\pi \tag{4}$$

where (3)–(4) can be combined:

$$\lambda^1(\varphi_1(s) + 2k_1(s)\pi) = \lambda^2(\varphi_2(s) + 2k_2(s)\pi) \tag{5}$$

where the unknowns $k_1(s)$ and $k_2(s)$ represent the integer ambiguity of the interferogram pixel s. There are two unknowns in an equation, and in general, it is difficult to solve. However, assuming a certain ratio between the wavelengths, the integer ambiguities $k_1(s)$ and $k_2(s)$ can be obtained to satisfy the above formula. The above assumption is based on the ideal situation and cannot satisfy the general situation. This paper attempts to find the full-circumference ambiguity of the above formula according to the scholar's research method, that is, complete the dual-band phase unwrapping.

It can be known from the multi-baseline L^1-norm principle that the core is to minimize the absolute phase difference between the pixels at the same position of the two interferograms, that is, the difference between the unwrapped phase gradient and the estimated phase gradient is the smallest, thereby achieving multi-baseline phase unwrapping. Similarly, the equation for the adjacent pixel $(s-1)$ of the pixel s at the same position satisfying the formula (5) in the dual-band is given as follows:

$$\lambda^1(\varphi_1(s-1)+2k_1(s-1)\pi)=\lambda^2(\varphi_2(s-1)+2k_2(s-1)\pi) \tag{6}$$

Union (5)–(6) can be obtained:

$$\begin{aligned}&\lambda^1(\varphi_1(s)-\varphi_1(s-1)+2(k_1(s)-k_1(s-1))\pi)\\&=\lambda^2(\varphi_2(s)-\varphi_2(s-1)+2(k_2(s)-k_2(s-1))\pi)\end{aligned} \tag{7}$$

In (7), define $k_1(s)-k_1(s-1)$ and $k_2(s)-k_2(s-1)$ is the theoretical ambiguity difference of adjacent pixels in the two-wavelength interferogram, $\nabla_{k_1}(s,s-1)$ and $\nabla_{k_2}(s,s-1)$ are the estimated values of the ambiguity difference, that is, the new gradient information . In order to solve the gradient information [9], we define the function:

$$\begin{aligned}&R(\nabla_{k_1}(s,s-1),\nabla_{k_2}(s,s-1))\\&=\lambda^1(\varphi_1(s)-\varphi_1(s-1)+2(k_1(s)-k_1(s-1))\pi)\\&\quad-\lambda^2(\varphi_2(s)-\varphi_2(s-1)+2(k_2(s)-k_2(s-1))\pi)\end{aligned} \tag{8}$$

Estimating the ambiguity gradient information using the Chinese remainder theorem (CRT) [10],get the estimate of $\nabla_{k_1}(s,s-1)$, $\nabla_{k_2}(s,s-1)$. In practical applications, the ambiguity difference estimation value is not equal to the adjacent pixel ambiguity difference of the interferogram. The traditional single-baseline L^1-norm optimization model is used to minimize the difference between the estimated value and the true value. The mathematical expression is as follows:

$$\begin{aligned}&\min\sum(f(m_1(s,s-1))+f(m_2(s,s-1)))\\&\text{s.t}\ k_1(s),k_1(s-1),k_2(s-1),k_2(s)\in\text{integer}\end{aligned} \tag{9}$$

$m_1(s, s-1)$ and $m_2(s, s-1)$ represent the free variables introduced, which are independent of each other. $f(\cdot)$ represent L^1-norm function. To simplify the research process, the free variables of an interferogram are solved.

$$\begin{gathered} \min \sum f(m_1(s, s-1)) \\ \text{s.t}\ m_1(s, s-1) = k_1(s) - k_1(s-1) - \nabla_{k_1}(s, s-1) \\ k_1(s), k_1(s-1) \in \text{integer} \end{gathered} \tag{10}$$

Introduce the L^1-norm optimization model to solve the principle:

$$\min \sum \omega_1(s, s-1) \cdot |m_1(s, s-1)| \tag{11}$$

$\omega_1(s, s-1)$ represent the optimization weighting coefficient of adjacent pixels in the interferogram; according to the idea of inequality clamping, it is converted into linear operation, avoiding the difficulty of the absolute value to the calculation and transforming into the constraints of the optimization model, as follows:

$$\begin{gathered} \min \sum \omega_1(s, s-1) \cdot m_1'(s, s-1) \\ \text{s.t}\ -m_1'(s, s-1) \le k_1(s) - k_1(s-1) - \nabla_{k_1}(s, s-1) \le m_1'(s, s-1) \\ k_1(s), k_1(s-1) \in \text{integer} \end{gathered} \tag{12}$$

Use the linear calculation rule (12) to improve the calculation efficiency of the above optimization model, and the solution of the minimum cost flow in the analog single baseline is obtained by the linear optimization model of the above formula.

3 Algorithm Performance Analysis

In order to verify the feasibility of dual-wave phase unwrapping based on L^1-norm, this paper uses MATLAB (R2014a) to realize the dual-band phase unwrapping principle of L^1-norm. Since the dual-band SAR radar has not been launched, the simulation data is used to test verification. The phase map data in both the flat and mountainous terrains was simulated separately. The winding phase size is 512×512, and the noise with a mean of 0 is added. The main simulation parameters of the dual-wave system are shown in Table 1.

According to the above parameters, the interferograms and coherence maps of the bands X and Ku are, respectively, obtained. This paper mainly starts from two

Table 1 Main simulation parameters of dual-wave system

Baseline (m)	Look angle (rad)	Baseline horizontal angle (rad)	Initial point slant range (m)
100	0.3316	0.1745	83,000

aspects, which are divided into ideal inclined flat phase and more complex mountain phase. The preprocessed phase data such as the inclined flat phase DEM, the *X*-band interferogram, and the Ku-band interferogram are, respectively simulated.

The longer the wavelength, the sparser the interference fringes, and conversely, the interference fringes are tight. The simulation data is ideal, and the coherence coefficient is set to a certain value. It mainly tests the feasibility of the L^1-norm dual-band unwrapping algorithm. The test is as follows.

3.1 Flat Phase Test

The first line shows the illustration before unwrapping, the second line shows the result analysis after unwrapping, and Fig. 1b–d, respectively, represent the long-band *X* interferogram, short-wavelength Ku Interferogram and coherence coefficient map corresponding to the inclined plane DEM. The coherence coefficients of the pixels in the above figure are all 0.8836. It can be seen from Fig. 1a DEM that the elevation is gradually increased from left to right and appears as a slope. As can be seen from the figure, the number of stripes in the *X*-band interferogram is 10; the number of stripes in the Ku-band interferogram is 21, and the fringes are relatively closely spaced.

According to the unwrapping results of (e) and (f) in Fig. 1, the result is smooth, and there is no unwrapped black hole region. In this paper, the *X*-band is taken as an example. (g) and (h), respectively, represent the *X*-band winding phase profile and the *X*-band unwrapped phase profile. Due to the selection of filtering and unflatten methods, some phase jump points are skipped during the unwrapping process, resulting in a few phase jumps in the unwrapped phase profile.

3.2 Simulation Decoherence Zone Test

For this part of the study, the simulated DEM is a four-week and high-rise rising terrain, as shown in Fig. 2a. In order to verify the ability of the dual-band unpacking method to decoherent regions, a poor coherence region is artificially constructed in the interferogram, as shown in Fig. 2b, c. the result after unwrapping is shown in Fig. 2d, e, and the *X*- and Ku-band unwrapping results are shown in order.

The result of the incoherent area unwrapping is shown in Fig. 2f, g. The incoherent area can also be well unwrapped. The winding phase profile of any row and its corresponding unwrapping profile are given, as shown in Fig. 2h, i; the phase continuity is better after unwrapping, and the severe phase abrupt point is corrected by the 2π fuzzy number. In the decoherent region, the test also gives a phase profile of the partial column (187–223) and the row (226–260) before and after unwrapping.

In order to verify that the double-wave unwrapping has better unwrapping effect in the decoherent region, the author performs a single-band contrast test based on the branch-cutting method. The band X/Ku is arbitrarily selected, and the single-band

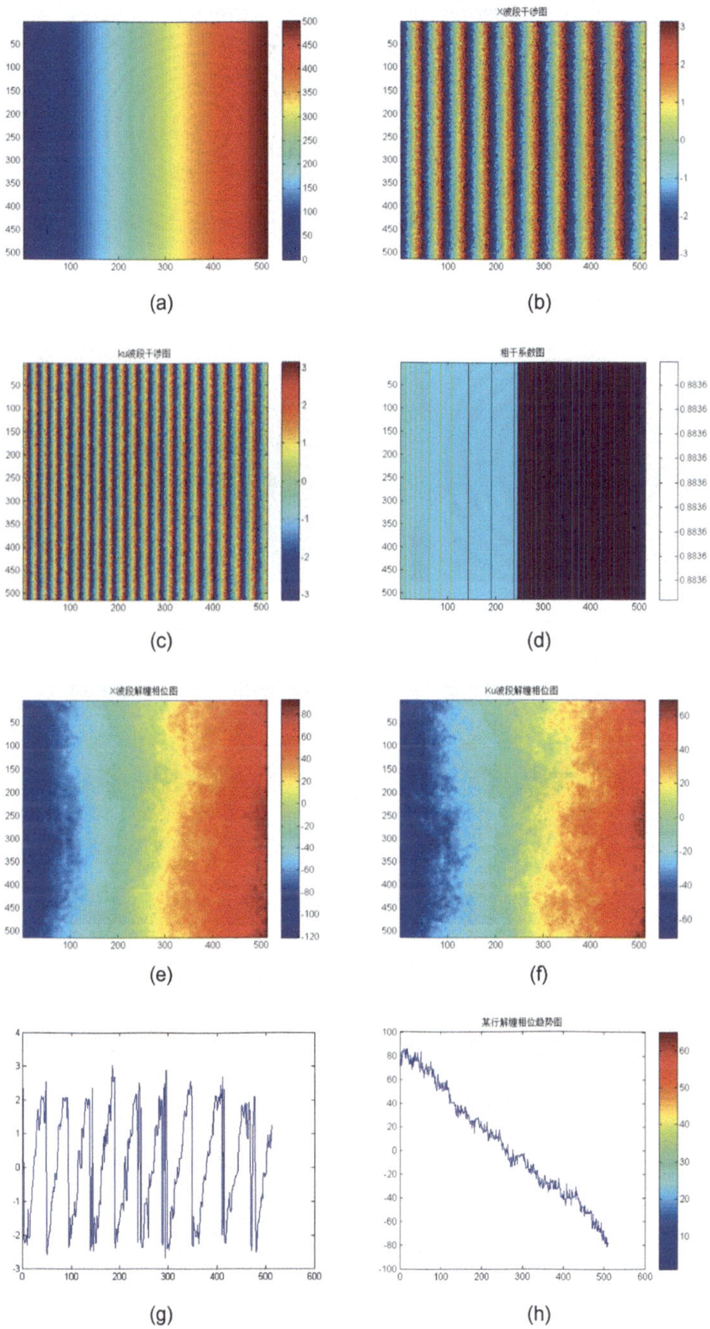

Fig. 1 Raw data and unwrapping results

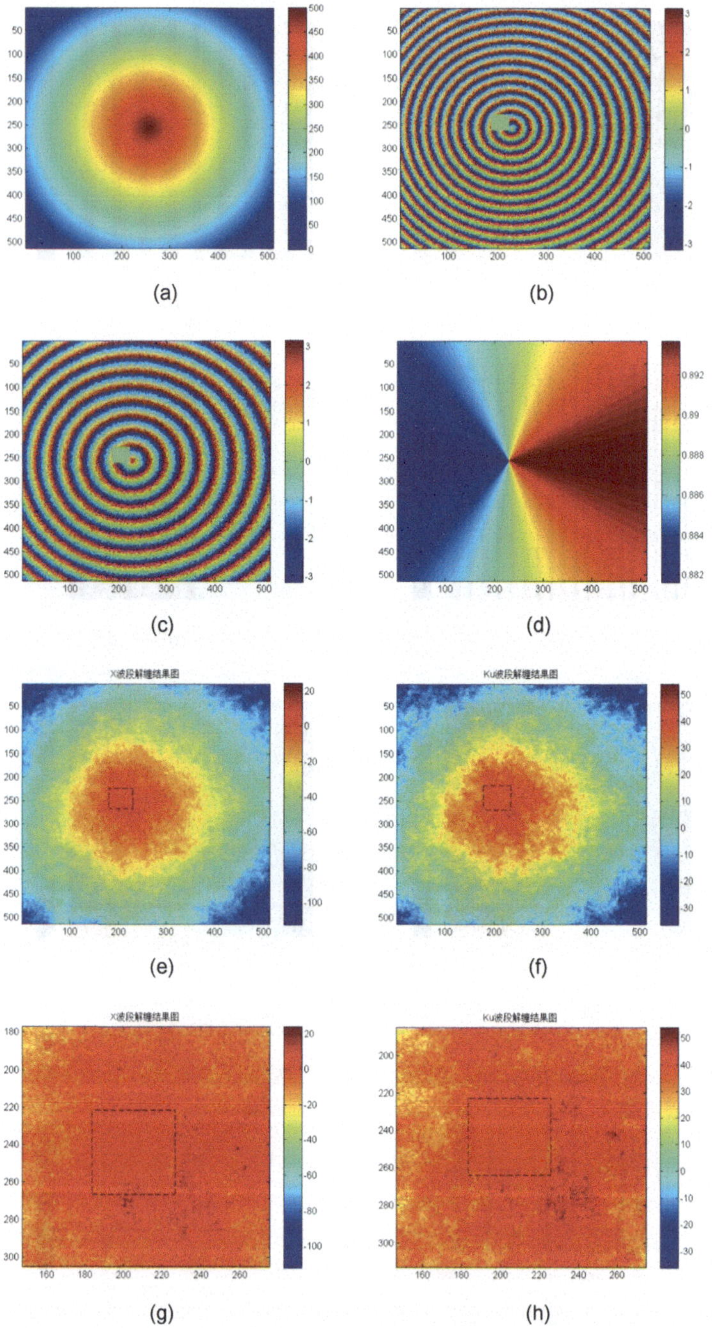

Fig. 2 Raw data, result graphs, and profile analysis

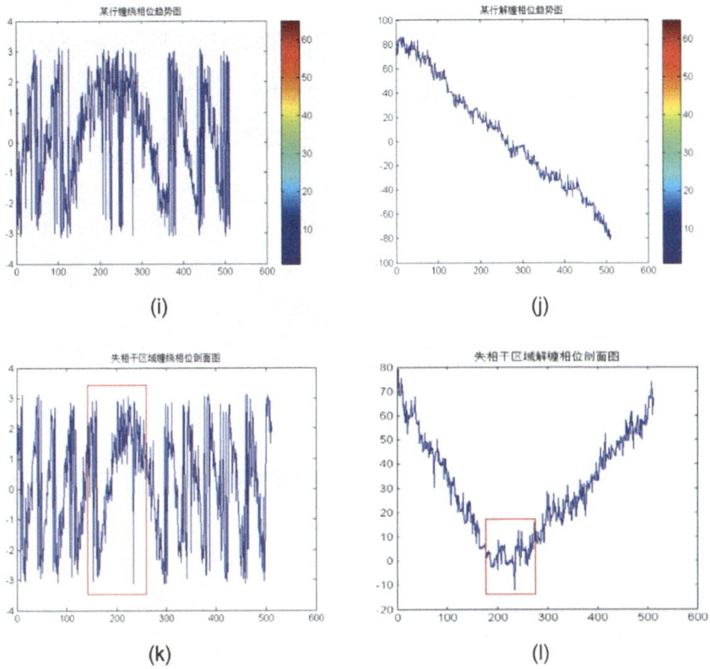

Fig. 2 (continued)

phase unwrapping method selects the two-dimensional phase unwrapping method based on Goldstein. Knowing the coordinates of the real phase reference point $(x, y) = (203.7746, 294.6347)$, the maximum search box radius is defined as $r = 4$. Single-band X unpacking results are as follows:

In Fig. 3a–d represent phase residues (charged), brant cuts, wrapped phase, unwrapped phase respectively; comparing the results of double-wave unwrapping, single-band unwrapping in the incoherent coherence region is not ideal. As shown in Fig. (d), the method of unwrapping in the simulated incoherent region is relatively ideal.

4 Conclusion

The dual-band phase unwrapping obtains the interference fringe pattern of the same terrain by using different bands. Based on the L^1-norm dual-baseline phase as the unwrapping principle, the author improves a study on the wavelength. Based on the L^1-norm, the dual-band unwrapping is how to obtain the global maximum, fitting the winding phase data by looking for the global optimal solution. This paper achieves the derivation and algorithm based on the L^1-norm principle.

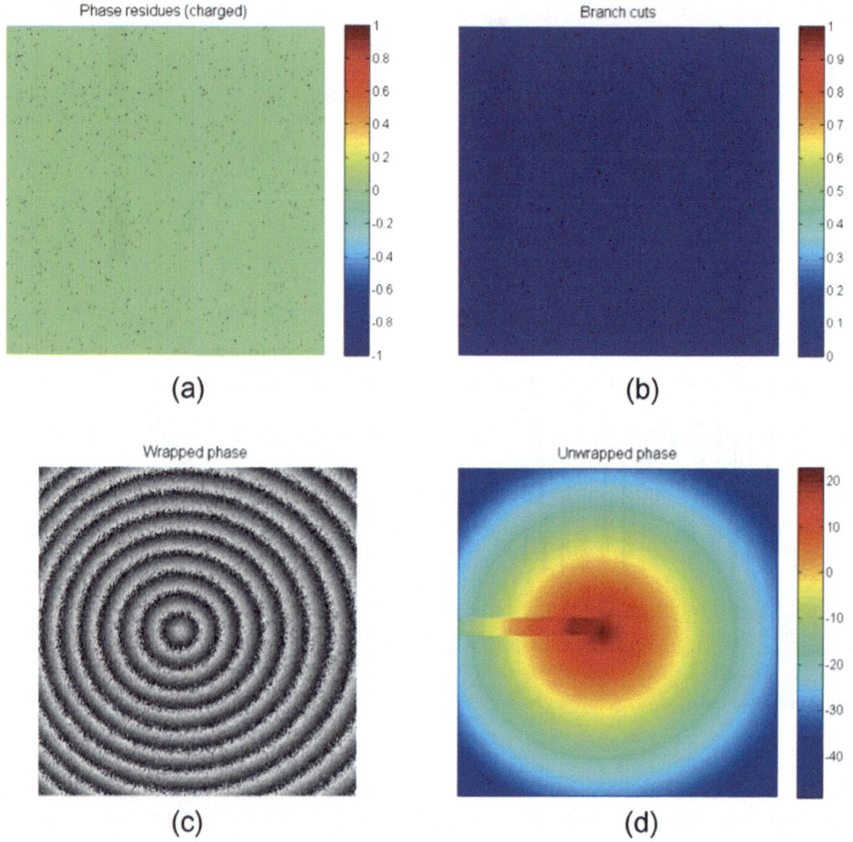

Fig. 3 Single-band unwrapping contrast experiment

The feasibility of the method is proved from two aspects: ideal terrain and more complex terrain. However, the shortcoming of this paper is that the accuracy of the unwrapping result of the method is not evaluated, and only the feasibility of the method is verified. In future research, DEM is extracted based on the gradient information obtained by the method and the simulated radar parameters (baseline B, look angle, slant range, wavelength), compared with the simulated DEM, and the DEM root mean square error is calculated as the main evaluation content of the method. And the noise robustness of the L^1-norm dual-band unwrapping algorithm is discussed by using the filtered conventional unwrapping method and the L^1-norm method.

Acknowledgements Foundation items: The National Key Research and Development Program of China, No. 2018YFF0215303, 2017YFB0503004; Chinese Academy of Surveying and Mapping Fundamental Research Project under Grant, No. 7771808.

References

1. Pingxiang L, Jie Y (2006) Synthetic aperture radar interferometry principle and application, vol 12. The Mapping Publishing Company, Beijing, pp 62–70
2. Information on https://www.jpl.nasa.gov/missions/nasa-isro-synthetic-aperture-radar-nisar/
3. Ni NA (2016) Phase unwrapping based on path-following and two-wavelength interferometry. Anhui University, pp 32–43
4. Huayou L, Mingjie Z, Heng Z et al (2018) An effective method of airborne dual-frequency interferometric terrain elevation reconstruction. J Radars 7(4):475–486
5. Khmaladze A, Matz RL, Zhang C, Wang T, Banaszak Holl MM, Chen Z (2011) Dual-wavelength linear regression phase unwrapping in three-dimensional microscopic images of cancer cells. Opt Lett 36(6):912
6. Mingsheng L, Hui L (2003) Synthetic aperture radar interferometry-principle and signal processing, vol 8. The Mapping Publishing Company, Beijing, pp 106–111
7. Ghiglia DC, Romero LA (1996) Minimum Lp-norm two-dimensional phase unwrapping. J Opt Soc Am A 13(10):1999
8. Hanwen Y, Zheng B (2013) L^1-norm method for multi-baseline InSAR phase unwrapping. J XiDian Univ 40(04):37–41
9. Yu H, Lan Y (2016) Robust two-dimensional phase unwrapping for multi-baseline SAR interferograms: A two-stage programming approach. IEEE Trans Geosci Remote Sens 54(9):5217–5225
10. Jin G, Zhang H, Xu Q et al (2011) Phase unwrapping algorithm with CRT for multi-band InSAR. J Xidian Univ 38(6):97–102

High-Efficiency Flexible GaAs Solar Modules Used for Unmanned Aerial Vehicles and Stratospheric Airships

J. K. Yang, X. Y. Jiao, Y. Yang, X. S. Wang, L. L. Song, J. Xue, and Z. C. Chen

Abstract High-efficiency flexible solar modules are fabricated with thin-film GaAs solar cells to satisfy the demand for power generation systems of unmanned aerial vehicles (UAVs) and stratospheric airships. The polyimide substrate inverted metamorphic triple junction (IMM 3J) GaAs thin-film solar cells are interconnected using the electrically conductive films by low temperature hot press welding technology, and the solidity of welding and efficiency loss ratio of the cells reach the level of rigid GaAs triple junction solar cells used in space now. Two different size solar cells are integrated and encapsulated separately using the thermo-sensitive adhesive supporting films to achieve a good laminating performance. The 0.2 m^2 flexible module consisting of unfilled corner cells has the specific power over 750 W/kg (AM0, 1353 W/m^2), and the 0.5 m^2 flexible module consisting of rectangle cells has higher specific power over 800 W/kg benefit from a higher cell-to-module ratio (CTM). The two modules have an areal density of less than 500 g/m^2. Several near space reliability tests have been done to verify that the flexible solar modules could meet the special environment application need.

Keywords Flexible solar module · High-efficiency solar cell · Thin-film GaAs · IMM 3J

1 Introduction

For low altitude unmanned aerial vehicles (UAVs) that use electric propulsion are limited in flight duration now and due in part to the weight of energy storage, the ability to supplement available battery power with solar energy can allow for longer mission times and extended range [1]. For high altitude long endurance (HALE) vehicles and stratospheric airships, which are designed to maintain its station continuously for months or even years, sufficient solar energy supply is crucial important for its power generation system [2, 3]. Flexible amorphous silicon (a-Si) and copper indium gallium selenide (CIGS) solar cells are light weight, but they also have

J. K. Yang (✉) · X. Y. Jiao · Y. Yang · X. S. Wang · L. L. Song · J. Xue · Z. C. Chen
Shanghai Institute of Space Power Sources, Shanghai, China
e-mail: yjk0926@126.com

L. Wang et al. (eds.), *Proceedings of the 6th China High Resolution Earth Observation Conference (CHREOC 2019)*, Lecture Notes in Electrical Engineering 657,
https://doi.org/10.1007/978-981-15-3947-3_29

relatively low conversion efficiencies, causing low power density (W/m^2). The crystalline silicon (c-Si) and rigid GaAs solar cells have high efficiency, but they are fragile and have low specific power (W/kg). All these cells mentioned above could not meet the energy need of the UAVs and the large stratospheric airships in the future.

The inverted metamorphic triple junction (IMM 3J) thin-film GaAs solar cells, which have high efficiency, light weight, and flexible characteristics, can meet the needs of the high power density and high specific power of the power generation system. High-efficiency modules made by IMM solar cells can provide enough energy to stay aloft indefinitely up to 26 days have been demonstrated with Airbus ZephyrHALE in 2018 [4]. Besides, the IMM cells could be used for future space application; the IMM solar modules have been flight demonstrated on the scientific satellite and been verified in the international space station (ISS) now [5, 6].

In this paper, high-efficiency flexible solar modules are fabricated by using two different size IMM GaAs solar cells, one module consists of unfilled corner cells has the specific power over 750 W/kg (AM0, 1353 W/m^2), the other flexible module consists of rectangle cells has higher specific power over 800 W/kg which benefit from a higher cell-to-module ratio (CTM), these modules have areal density less than 500 g/m^2. Several reliability tests have been done to verify that the flexible solar modules could meet environment application need of the UAVs and stratospheric airships.

2 The IMM 3J Solar Cells and Modules Fabrication Process

2.1 *The IMM 3J Solar Cells*

The high-efficiency GaInP/GaAs/InGaAs IMM 3J solar cells are first fabricated to use metal organic chemical vapor deposition (MOCVD), and then transferred to flexible polyimide substrates using the epitaxial lift-off (ELO) technology.

The solar cells have a best average efficiency (*Eff*) of 32% measured by Spectrolab X-25 Mark II simulator (AM0, 1353 W/m^2), with V_{oc} 3050 mV, J_{sc} 17 mA/cm^2, and FF 0.84. The total thickness of the cells is 70 μm, including 50 μm thickness flexible polyimide (PI) substrates, the specific powers is over 2000 W/kg, the areal density is about 200 g/cm^2, and the section diagram of the cell is shown in Fig. 1a. We design two size cells, both have a same 12 cm^2 area but different appearance as shown in Fig. 1b, one has 40.5×30.5 mm^2 size with a unfilled corner, which can enhance the use ratio of the expensive GaAs material, and another has rectangle appearance, 43 $\times$ 28 mm^2 size, which can achieve a higher cell filling rate.

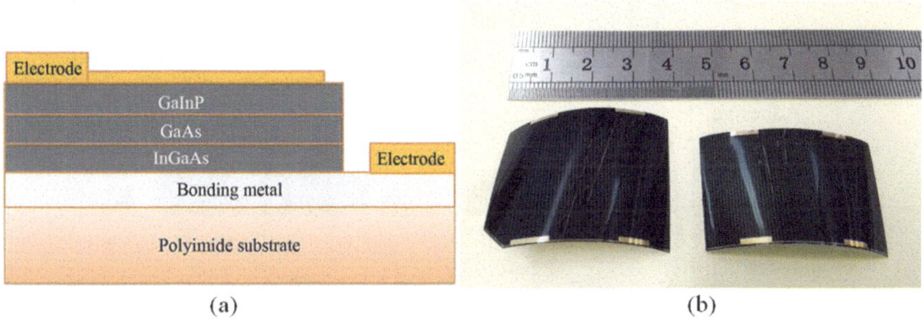

Fig. 1 IMM 3J solar cell: **a** section diagram **b** real products

2.2 Modules Fabrication Process

The solar cells interconnection is the first step to accomplish a whole module. There are many different methods to get the cells connected, including developed parallel gap resistance welding [7], ultrasonic welding, and electrically conductive adhesive connection [8]. We use a different welding method to achieve a good interconnection performance, which has many merits, such as simple, effective, and low cost.

The cells are connected use the electrically conductive films. The conductive films are attached to the electrode and peeling the release film at first, then attach the 20 μm silver foils on the conductive films, use the low temperature hot press welding technology to get the ribbon connected at last. The weld temperature is about 200 °C. The tensile strength of the welding cells is more than 4 N/mm^2, and the efficiency loss ratio is less than 2% as shown in Table 1, which reaches the level of rigid substrate GaAs solar cells used in space now.

A typical PV encapsulation sandwich structure is used to produce the solar modules as shown in Fig. 2. The transparent encapsulate film on the top of the modules have excellent transmission in the blue and near-UV regions of the solar spectrum. Polyimide film materials are chosen for the bottom substrate, which have good flexibility and environmental suitability. A standard vacuum laminator is utilized to cure the encapsulate stack below 150 °C. The total thickness of the flexible module is less than 300 μm.

Since the flexible IMM 3J solar cells are curving due to the stress mismatch between the GaAs material and the PI substrate, the large area module lamination is not going to be easy. We draw inspiration from the ELO process of the IMM 3J

Table 1 Electrical property loss ratio of the cell after welding

	V_{oc} (mV)	J_{sc} (mA)	FF	Eff (%)
No welding	3038.68	17.28	0.809	31.40
Welded	3008.50	17.22	0.805	30.82
Loss ratio (%)	−0.99	−0.35	−0.49	−1.85

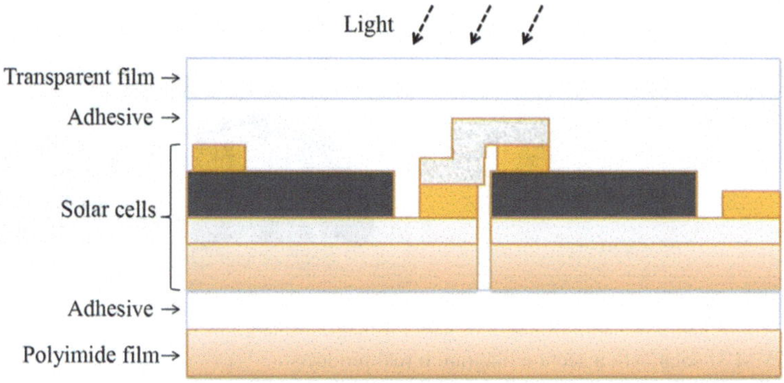

Fig. 2 Section diagram of the encapsulated solar module

solar cells, and optimize the welding and encapsulation process to adapt the cells. All the cells are attached to the thermos-sensitive adhesive supporting film precisely at first, and then the ribbons between each cell and bus bars between each string are interconnected use the hot press weld mentioned above, after that, the transparent film and adhesive film are laminated on the front side of the welded cells use the laminator, peeling off the supporting film, laminate the adhesive film, and polyimide film to the lower surface by the laminator second time, a whole module fabrication progress is accomplished. This large area flexible encapsulation technology ensured the flexible IMM 3J cells can be laminated reliably and uniformly.

3 Modules Performance

3.1 The Module with Unfilled Corner Solar Cells

We fabricate a flexible module at first using the unfilled corner solar cells. This module consists of two sub-modules, one sub-module consists of 88 cells, 22 cells are interconnected in series and then 4 strings in parallel; all cells in series and parallel are spaced 1 mm apart. After that, two sub-modules are coupled into one module in series having a 5 mm space between each other. There are 176 cells in total and the active area is 0.21 m^2, but the inner frame area of the module is 0.23 m^2 due to the gaps and spaces. The weight of the module is 110 g, having no bypass diodes and block diodes. Figure 3 is the picture of this module.

This module was tested use the Spectrolab Large Area Pulsed Solar Simulator (LAPSS), Fig. 4 shows the I–V performance of this module. The P_{max} of the module is 83.6 W, with 760 W/kg specific power and areal density less than 490 g/m^2. The V_{oc} is 133.4 V and I_{sc} is 0.79 A, but the FF is only 0.79, and this may be due to the current mismatch and the lack of diodes.

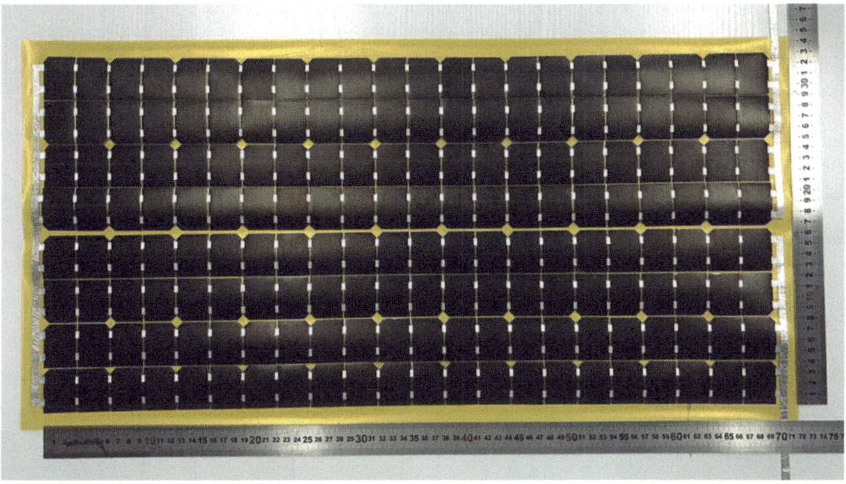

Fig. 3 Flexible module with the unfilled corner IMM 3J solar cells

3.2 The Module with Rectangle Solar Cells

In order to improve the power density and high specific power of the GaAs module further, we design a different layout using rectangle cells, which has a higher cell filling rate. The cells are interconnected into strings of 42 cells like "S" form, each string includes three rows with 14 cells per row, then ten strings are connected in parallel. The electroluminescence (EL) image of one string cells is shown in Fig. 5. Being different from the module with the unfilled corner solar cells, there is no gap between the adjacent cells and adjacent rows, and effective margin insulation protection of each cell is applied to prevent short circuit and high-voltage breakdown. Although there are few cracks in some cells observed by EL image, it will not influence the electrical performance of the cells and modules [9].

Figure 6 shows the module using rectangle IMM 3J solar cells. The module consists of 420 cells in total, has a 0.51 m^2 inner frame area, and tested using the PASAN High Light SAT AM0 simulator. Each cell has a V_{oc} over 3 V, resulting in over 125 V per 42-cell string. The I_{sc} of each string is about 0.2 A, equivalent to that of a single cell. All ten strings are connected in parallel with bus bar including block diodes on each series string, and the final I_{sc} measures over 2 A. The FF is over 0.825, much higher than 0.2 m^2 module, benefited from the interconnect optimization and the block diodes employment.

Figure 7 shows the I–V and P–V curves of the 0.5 m^2 module. The P_{max} of this module is 212 W, has a high power density 415 W/m^2, and specific power over 800 W/kg due to the higher cell filling rate. The module efficiency is over 30% (inner frame area) and the cell-to-module (CTM) efficiency ratio is over 95%.

In order to verify that the solar module made by the hot press weld and encapsulated technology as previously mentioned can successfully work in the application

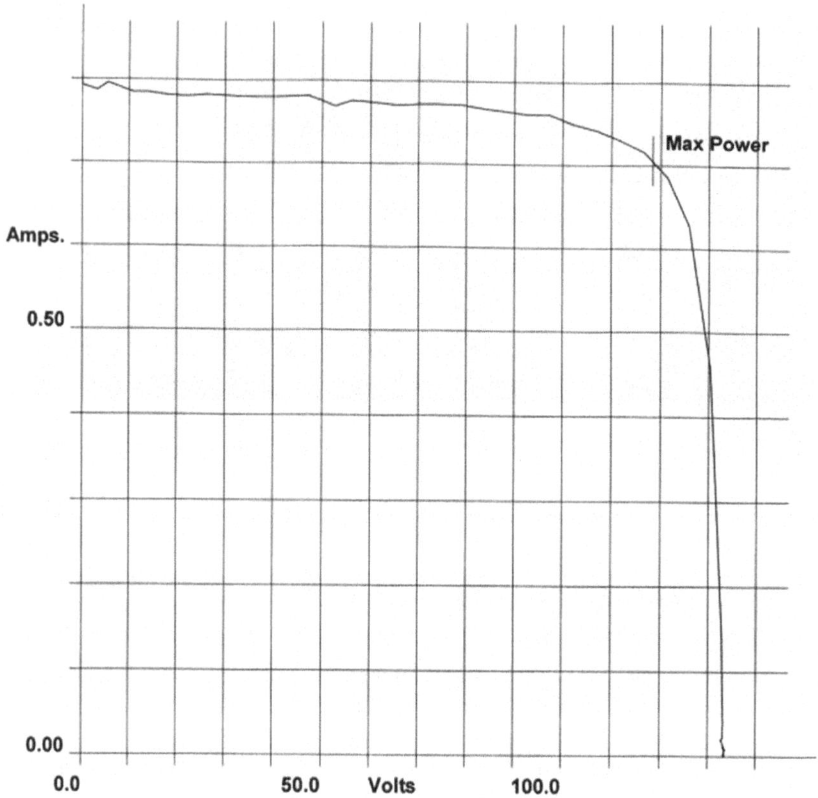

	Volt	Amp.	Watt	Fill Factor
Isc	---	0.7917	---	---
Voc	133.4	---	---	---
MaxP	118.4	0.7064	83.63	0.7918
Test	116.0	0.7180	83.29	---

Temperature: Ambient 18.0 °C Corrected 25.0 °C

Fig. 4 I–V performance of the 0.2 m^2 module

environment of the UAVs and stratospheric airships, several near space reliability tests and qualification tests have been done, including thermal shock, high temperature and vacuum, ultraviolet radiation, humidity, and the combined tests. The result of the reliability tests is shown in Table 2. As a result, the efficiency loss ratio of the cells after weld and encapsulated is only about 1% under the several near space environment tests. And the efficiency and appearance of the test samples have no change after hundreds of bends on 5 cm radius cylinder. So the modules could meet the energy supply demand of the stratospheric airships and UAVs.

Fig. 5 EL image of one string with 42 cells

Fig. 6 Flexible module using rectangle IMM 3J solar cells

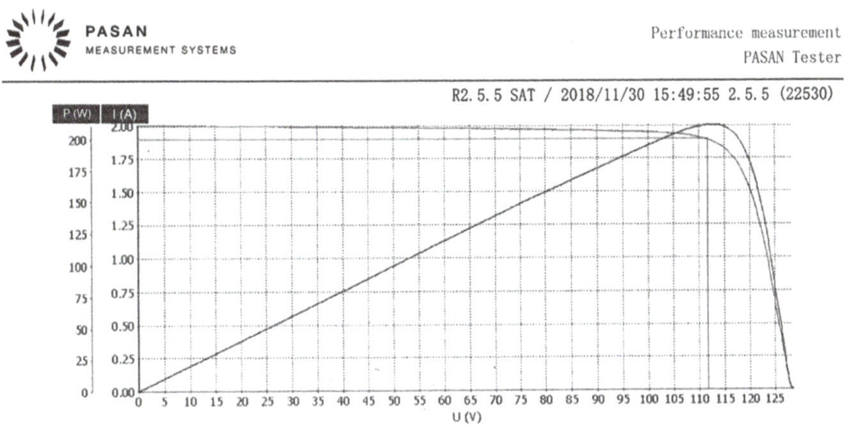

Fig. 7 I–V and P–V curves of the 0.5 m^2 module

Table 2 Reliability test of the solar cells and modules

Test	Test condition	*Eff* loss ratio (welded) (%)	*Eff* loss ratio (encapsulated) (%)
Thermal cycling	−60 to +90 °C, 16 cycles	−0.21	+1.04
High temp. vac.	4 kPa, 100 °C, 100 h	+0.80	+0.78
Ultraviolet radiation	10 W/m^2, 100 h	−0.18	+0.44
Humidity test	85 °C, 85% Rh, 100 h	−0.56	−0.72
Combined test	Ultraviolet 10 W/m^2, 4 kPa, −50 to + 90 °C, 16 cycles, 100 h	−1.08	+1.00

4 Summary

The IMM 3J GaAs solar cells can be integrated into high-efficiency flexible modules using the electrically conductive films interconnection and large area flexible encapsulation technology. The module with the rectangle size cells could achieve a higher specific power over 800 W/kg due to the higher cell filling rate. Several near space environment reliability tests have been done, and verified that the modules using these welding and encapsulate technology could meet the application need of the UAVs and stratospheric airships.

References

1. Carey PG, Aceves RC, Colella NJ (1994) A solar module fabrication process for hale solar electric UAVs. In: Proceeding of 1st world conference on photovoltaic energy conversion, pp 1–7
2. Scheiman D, Hoheisel R, Edwards D (2018) Solar technology comparison for wing integration in unmanned aerial vehicles. In: Proceeding of 45th IEEE photovoltaics specialists conference, pp 3539–3544
3. Stern T, Steele K, Yang L (2018) High efficiency wing integrated solar arrays for UAV applications. In: Proceeding of AIAA joint propulsion conference, pp 1–5
4. Information on http://www.airbus.com/defence/uav/zephyr.html
5. Shimazaki K, Kobayashi Y, Takahashi M (2014) First flight demonstration of glass-type space solar sheet. In: Proceeding of 40th IEEE photovoltaics specialists conference, pp 2149–2154
6. Information on http://www.altadevices.com/space/
7. Scheiman D, Jenkins P, Walters R (2014) High efficiency flexible triple junction solar panels. In: Proceeding of 40th IEEE photovoltaics specialists conference, pp 1376–1380
8. Mattos LS, Scully SR, Syfu M (2011) New module efficiency record: 23.5% under 1-sun illumination using thin-film single-junction GaAs solar cells. In: Proceeding of 37th IEEE photovoltaics specialists conference, pp 3187–3190
9. Takamoto T, Washio H, Juso H (2014) Application of InGaP/GaAs/InGaAs triple junction solar cells to space use and concentrator photovoltaic. In: Proceeding of 40th IEEE photovoltaics specialists conference, pp 0001–0005

Landslide Risk Zoning Method Based on Improved Water Seepage Capacity Model

Zhaohua Wang, Jixian Zhang, Zheng Zhao, and Haiying Gao

Abstract In order to solve the problem that traditional rainfall infiltration model does not take into account human factors and rainfall infiltration mechanism, an improved rainfall infiltration model for loess landslide is proposed. Firstly, the paper improves the traditional model through the stratified hypothesis infiltration theory. Secondly, the slope, slope direction, vegetation index, impervious layer, different lithological water seepage capacity and highway data are the key influencing factors of landslide occurrence. Thirdly, the experimental area is selected. Correlation data is used to calculate the weight function through SPSS. Finally, the logistic regression model is used to calculate the probability of landslide occurrence, and the risk zoning map of loess landslide under different rainfall is obtained. The model results show that the area with higher risk of landslides in the study area increases with the increase of rainfall and has a close relationship with the stability of highway slopes under low rainfall, under high rainfall and lithology. The water seepage capacity has a great relationship, and the early warning rainfall is 30 mm. It is verified by the historical landslide data. The proportion of landslide points with the landslide probability is more than 87.1%. The verification results show that the prediction results of the model are more in line with the distribution of historical landslide points.

Zhaohua Wang (1994–), Male (Han), Master, studies in the analysis of GIS big data and spacial.

Z. Wang (✉)
Faculty of Geomatics, Lanzhou Jiaotong University, Lanzhou 730070, China
e-mail: wzh940526@163.com

Gansu Provincial Engineering Laboratory for National Geographic State Monitoring, Lanzhou, Gansu 730070, China

Z. Wang · Z. Zhao
Chinese Academy of Surveying & Mapping, Beijing 100830, China

J. Zhang
National Quality Inspection and Testing Center for Surveying and Mapping Products, Beijing 100830, China

H. Gao
College of Geomatics, Shandong University of Science and Technology, Qingdao, China

L. Wang et al. (eds.), *Proceedings of the 6th China High Resolution Earth Observation Conference (CHREOC 2019)*, Lecture Notes in Electrical Engineering 657,
https://doi.org/10.1007/978-981-15-3947-3_30

Keywords Water permeability model · Impervious layer · Rainfall · Logistic regression model · Landslide prediction

1 Introduction

Rainfall infiltration is a very complicated geological phenomenon. The infiltration capacity of different geological types is different, and the rainfall infiltration model is also different. Landslides are a common natural disaster, mostly occurring in the backing area. According to the relevant literature, the relationship between landslide and rainfall infiltration is very close [1, 2]. Therefore, the study of quantitative relationship between rainfall infiltration and landslide is of great significance for the study of landslide prediction.

In recent years, many scholars at home and abroad have done a lot of research work on water seepage capability model and landslide risk zoning. In terms of water permeability model, foreign scholar Lee et al. [3] studied slope stability based on rainfall, and Kale and Sahoo [4] conducted in-depth research on Green-Ampt model for different geological conditions. Domestic scholars Zhenchun [5] proposed a rainfall infiltration model based on gravity, resistance, capillary force and inertial force. Tan Xiaohua et al. studied the infiltration mechanism based on rainfall intensity [6]. Because the rainfall infiltration situation is very complicated, the traditional infiltration model needs to be based on actual conditions, and it needs to be corrected in practical applications. However, the existing models have less to consider the influence of human factors on rainfall infiltration.

In terms of landslide risk zoning: Li Huanqi et al. conducted a statistical study on the relationship between rainfall and landslide [7], Zhao Fangli et al. Analysis of the critical rainfall leading to landslide [8]. The loess soil structure is loose and the water seepage ability is strong. The rainfall-induced loess landslide is more sudden and destructive, so it has become the focus and hotspot of the current research. Scholars have done a lot of research work in this area. For example, Zhang Zelin and others studied the deformation modes of earthquakes and loess landslides [9]; Niu Quanfu et al. studied the selection criteria of topographic factors for loess landslides [10]; Zhang Shan et al. studied the spatial distribution characteristics of loess landslides [11].

In summary, although the existing methods make important contributions to the landslide warning research, there are few literature systems to study the risk zoning of loess landslide under different rainfall. To this end, this paper takes Lanzhou and its surrounding areas as an example, selecting slope [12], slope direction [13], impervious layer, vegetation index [14], water seepage capacity of different geology and road influence. As an influencing factor, a logistic regression model was established for prediction, and the change of actual seepage capacity of lithologic strata with rainfall was taken as the entry point, and the layout of landslide hazard area under different rainfall in the study area was obtained. The results show that the model prediction results are in good agreement with the actual landslide hazard points.

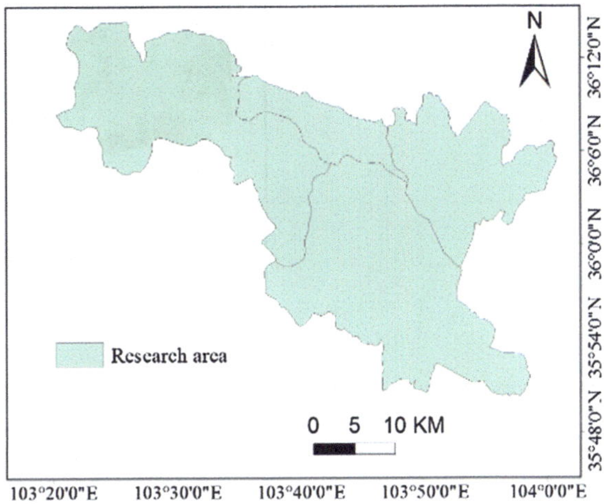

Fig. 1 Research area location

2 Experimental Area and Data

2.1 Location of the Study Area

The study area is located in Lanzhou City, Gansu Province, including Xigu District, Anning District, Chengguan District and Qilihe District. Lanzhou City is located in the Yellow River Valley basin on the northeastern side of the Qinghai–Tibet Plateau, surrounded by mountains. The city center is about 1520 m above sea level, and the relative heights of the two mountains are 600 m. The Yellow River runs through the city from west to east. It forms a river valley basin with a shape of about 35 km from east to west and 2–8 km from north to south. The average annual precipitation is 292.53 mm. Its location is shown in Figs. 1 and 2.

2.2 Research Area Lithologic Stratum

The lithologic strata in the study area are relatively well developed, and they are exposed from the pre-Sinian to the Quaternary. They are mainly composed of aeolian loess, accompanied by shale, quartzite, limestone and other lithology types.

Fig. 2 Research area DEM

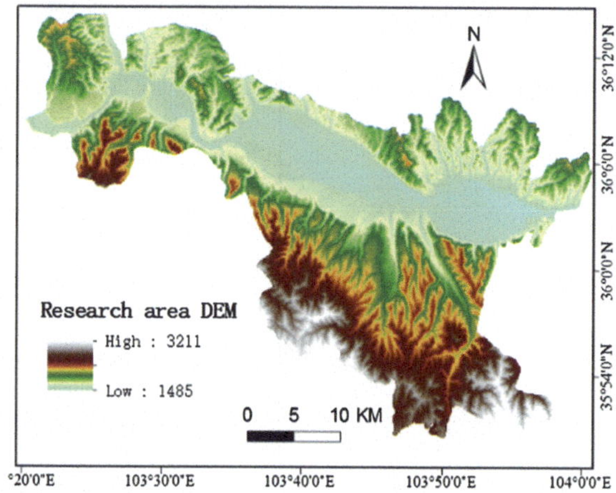

2.3 Data Source and Processing

In the research of the thesis, the data involved mainly includes DEM data, Landsat 8 remote sensing images, lithologic stratigraphic data and some geographical conditions monitoring data. Among them, the DEM data comes from the 30 m DEM data of Lanzhou City downloaded from the geospatial data cloud, and the slope and aspect of the study area are extracted by the slope and slope extraction tools in ArcGIS, respectively. Vegetation index and impermeable layer data in the study area were extracted from the Landsat 8 image. The lithology data is the geological survey data of Lanzhou City, and it is graded according to its water seepage capacity. Highway data is the result of geographical national conditions monitoring. The landslide point data is the rainfall-induced landslide point for historical data and field research.

The logistic regression model is very useful for classification problems, especially based on probability classification. Correspondingly, the logistic regression model also requires classification of the impact factors in the process of modeling. By referring to other literatures and the existing data, the specific processing results are as follows:

Figure 4a is the slope information extracted from the DEM data of the study area, wherein the DEM data is divided into five levels according to the maximum likelihood method.

Figure 4b is the aspect information extracted from the DEM data of the study area, which is divided into nine levels according to the plane and the eight directions (one level every 45°).

Figure 4c shows the vegetation index in the study area extracted from the Landsat image and is divided into two levels, namely vegetation area (0–1) and no vegetation area (−1 to 0).

Figure 4d is the data of the impervious layer extracted from the remote sensing image in the study area and is divided into two levels, namely the water permeable layer (0) and the water impermeable layer (1).

Figure 4e is the water permeability capacity grade according to the degree of influence of lithology water seepage capacity on the landslide and is divided into five grades, namely extremely low, low, medium, high and extremely high.

Figure 4f is the freeway information in the study area, and is divided according to a buffer zone of 600 m around the road, within 600 m (1), and outside 600 m (0).

3 Water Permeability Model

3.1 Traditional Model

The calculation of water seepage capacity in this paper is based on the relevant model proposed in the literature [5]:

$$f = K\left(1 + \frac{Bs}{F + F_{\mathrm{A}}}\right) \tag{1}$$

In formula (1), f represents the water permeability, K represents the maximum water permeability, B represents the water absorption rate, s represents the capillary suction expressed by the height of the water column, F_{A} represents the total amount of the old water body participating in the infiltration in the soil layer and F represents the soil layer. Newly increased infiltration water.

3.2 Improved Model

3.2.1 Model Improvement Theory

Wenyan [7] analyzed the variation characteristics of soil moisture profile through the infiltration experiment of loess water accumulation and proposed that the saturated layer and the conduction layer are treated as a saturated region, and the water content change of the unsaturated wet layer is regarded as an application. The matrix potential in the upper saturated zone is shown in Fig. 3. At the same time, Wenyan [7] proposed a layered hypothesis infiltration model based on the infiltration relationship between rainfall and loess. The specific model is as follows.

In Fig. 4, θ shows the infiltration rate, z represents the infiltration depth, s represents the effective rainfall and A and B are critical rainfall, respectively.

$$\text{Saturated layer: } \theta_{(z)} = \theta_s,\ 0 \le z \le L_1 \tag{2}$$

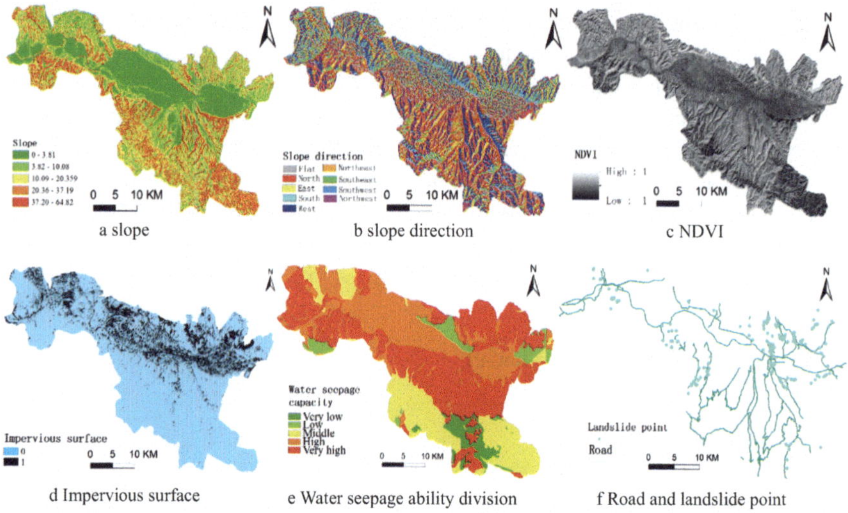

Fig. 3 Distribution of influencing factors of rainfall landslides in the research area

Fig. 4 Penetration model of stratification hypothesis

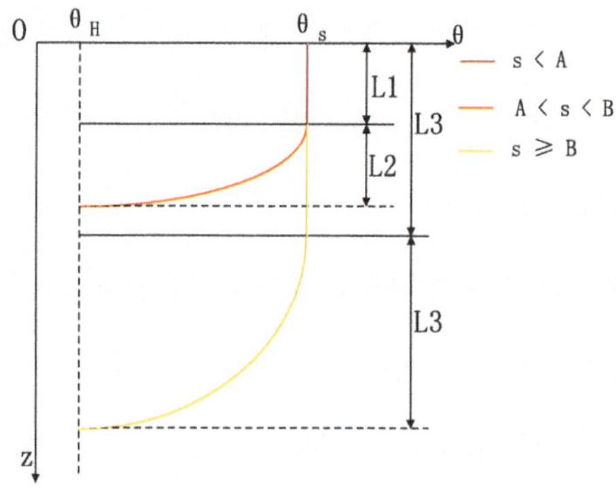

$$\text{Transition layer: } \theta_{(z)} = \theta_H + \frac{2(\theta_s - \theta_H)}{L} \sqrt{z(L-z)},\ L_1 \le z \le L_2 \tag{3}$$

$$\text{Dry soil layer: } \theta_{(z)} = \theta_H,\ z \ge L_2 \tag{4}$$

The characteristics of the water cut curve of the unsaturated wet layer are very similar to the elliptic curve, so the elliptic curve is used for fitting. The fitting result is that L represents the depth of infiltration where the water absorption rate changes, because the factors that affect the suction of the capillary are mainly precipitation and air moisture, but in the northern arid regions, the influencing factors of air moisture can be ignored [17], so in this model s is effective Rainfall.

3.2.2 Improved Model

Based on the multi-layer infiltration model and the impervious surface data, this paper proposes a quantitative model for studying the water seepage capacity of Lanzhou area. Based on the layered hypothesis infiltration model, this formula is combined with the impervious layer data for coupling calculation. The specific formula is as follows:

$$\text{Rainfall} < \text{Threshold } A\text{: } f = K\left(1 + \frac{Bs}{L_1(\theta_s - \theta_H)}\right) + D * U \tag{5}$$

$$\text{Threshold } C > \text{rainfall} \geq \text{threshold } A\text{: } f = K\left(1 + \frac{BA}{L_1(\theta_s - \theta_H)}\right) + K\left(1 + \frac{4B(s - A)}{\pi L_2(\theta_s - \theta_H)}\right) + D * U \tag{6}$$

$$\text{rainfall} \geq \text{threshold } C\text{: } f = K\left(1 + \frac{BC}{(L_1 + L_2)(\theta_s - \theta_H)}\right) + K\left(1 + \frac{4B(s - C)}{\pi L_3(\theta_s - \theta_H)}\right) + D * U \tag{7}$$

where f is the water seepage capacity, K is the maximum water seepage speed, B is the water absorption rate, s is the effective rainfall, L_1 is the saturated layer infiltration depth, L_2 is the unsaturated layer infiltration depth, L_3 is the dry soil depth, A, C is the rainfall threshold, D is the impervious surface coefficient and U is the impervious layer value (generally 0 and 1).

Among them, the effective rainfall of single-day rainfall is expressed by the rainfall of the day. The effective rainfall of continuous rainfall is represented by a linear regression equation. The specific coefficient of linear regression needs to be determined according to the actual rainfall situation in the study area.

4 Experimental Methods

4.1 Theoretical Methods

Since landslide occurrence is a classification problem, traditional numerical analysis methods are not suitable for such analysis. Therefore, the parameters of the logistic regression model can be analyzed when verifying the relationship between variables.

Logistic regression is a generalized linear model with $B_0 + B_1X_1 + \cdots + B_nX_n$ as the weight function, where B_i and B_0 are the parameters to be sought, and X_i is the influence factor. The model uses an implicit function $P(x)$ corresponding to the weight function, namely $P(x) = L(B_0 + B_1X_1 + \cdots + B_nX_n)$, and then according to $P(x)$ and $1 - P(x)$ The size determines the value of the dependent variable. The logistic regression model is currently widely used in risk prediction, and the prediction results are good, so this article uses the logistic regression model for prediction. The expression of the logistic regression model is as follows.

The weight function is:

$$Z = B_0 + B_1X_1 + \cdots + B_nX_n \tag{8}$$

The probability implicit function is:

$$\ln\left(\frac{P(x)}{1 - P(x)}\right) = B_0 + B_1X_1 + \cdots + B_nX_n \tag{9}$$

The probability function obtained by (4), (5) is:

$$P(x) = \frac{\mathrm{Exp}((B_0 + B_1X_1 + \cdots + B_nX_n))}{1 + \mathrm{Exp}((B_0 + B_1X_1 + \cdots + B_nX_n))} \tag{10}$$

where $P(x)$ represents the probability of occurrence of landslide, Z represents the sum of weights after superposition, B_i represents the regression coefficient, X_i represents the influence factor and the output of logistic regression is 0–1, 0 represents the probability of occurrence of landslide is 0% 1 means that the probability of occurrence of landslide is 100%. Therefore, according to the probability of landslide occurrence in each region, the landslide risk level of the study area is finally divided into: very low ($0\% < P \leq 20\%$), low ($20\% < P$ Five risk levels: $\leq 40\%$), medium ($40\% < P \leq 60\%$), high ($60\% < P \leq 80\%$) and extremely high ($80\% < P \leq 100\%$).

Based on the logistic regression model, the regression coefficients of each influencing factor are determined by SPSS, and then, the regression equation is determined. The probability of occurrence of the landslide is calculated by the probability implicit function.

During the experiment, the relevant data of 45 landslide points and 45 non-landslide points in the study area were selected and imported into SPSS for analysis, and the parameters were determined (Tables 1 and 2).

The coefficients of each parameter fitted in spss can determine the logistic weighting weight function used in the calculation:

$$P = \frac{e^Z}{1 + e^Z} \tag{11}$$

In the formula: $Z = 0.008$ * seepage capacity + 0.085 * slope − 0.002 * aspect + 0.257 * NDVI + 1.37 * impervious surface value + 2.002 * road − 5.791.

Table 1 Logistic model coefficients and significance table

	Coefficient	Sig.	Exp (*B*)
30 mm water permeability	0.008	0.028	1.070
Slope direction	−0.002	0.019	0.998
Slope	0.085	0.043	1.089
NDVI	0.257	0.021	1.293
Impervious surface	1.370	0.045	79.011
Road	2.002	0.033	1.263
Constant	−5.791	0.028	0.003

Table 2 Logistic model evaluation R square table

Cox and Snell R^2	Nagelkerke R^2
0.885	0.947

4.2 *Experimental Results and Analysis*

From the landslide risk zoning map, the following conclusions can be drawn:

1. Areas with higher landslide risk increase with increasing rainfall.
2. The probability of landslides with low rainfall is highly correlated with highway excavation; under high rainfall, the probability of occurrence of landslides is highly correlated with geological water uptake capacity.
3. It can be seen from Fig. 5e that when the rainfall is $\geq$30 mm, the probability of occurrence of landslide is high and the rate of increase of the region is significantly increased. Therefore, it can be determined that 30 mm is the early warning rainfall for landslide occurrence.
4. Under high rainfall, the central, northern and northwestern part of the study area is the hardest-hit area of the landslide. The geological types of the above areas are all aeolian loess. The wind-induced loess has a loose structure, strong water seepage capacity and is prone to landslides disaster. And the spatial distribution characteristics of landslide risk zoning are basically consistent with the distribution characteristics of historical landslide points.

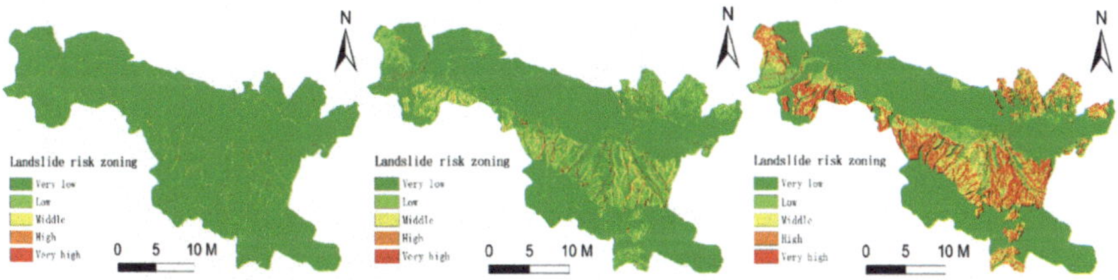

a 10mm Landslide risk zoning b 25mm Landslide risk zoning c 40mm Landslide risk zoning

Fig. 5 Landslide risk zone map

5. The distribution of extremely low and extremely high areas under high rainfall is wide, and the distribution in low, medium and high areas is relatively small, which is caused by the geographical conditions of the two mountains in the study area (Table 3).

The maximum continuous is selected. The rainfall was 40 mm as the verification rainfall, and the results are shown in Figs. 6 and 7.

Figure 6 shows that there are more landslides in the northeast, northwest and central areas of the study area, and the disaster situation is serious. The disasters in the southwest are less, mainly in mountainous areas, and rainfall is the main cause of landslides. Historical landslide point and 40 mm landslide risk zoning map, a total of 179 landslide points in the study area, 97 landslide points occurred in the area with high risk zoning, occurred in the risk zone. There are 37 landslide points in the area; 22 landslide points occur in the area where the risk is zoned, 16 landslide points occur in the area where the risk zone is low and 7 landslide points occur in the area where the risk zone is extremely low. According to the 33 landslide points surveyed in the field (as shown in Fig. 7), there are 15 areas in which the risk zoning is extremely high, of which 7 occur in areas with high risk zoning, including in the

Table 3 Relationship between rainfall and landslide probability

Rainfall (mm)	Very low (%)	Very low (%)	Middle (%)	High (%)	Very high (%)
10	97.41	1.88	0.51	0.19	0.01
15	94.68	4.04	0.77	0.41	0.11
20	88.12	8.86	2.07	0.61	0.34
25	76.82	15.74	5.17	1.61	0.66
30	66.08	18.60	10.02	4.01	1.30
35	59.27	14.46	14.06	9.15	3.06
40	58.74	5.28	13.97	14.71	7.31
45	59.13	0.85	7.38	16.83	15.81
50	59.51	0.86	0.74	13.07	25.82

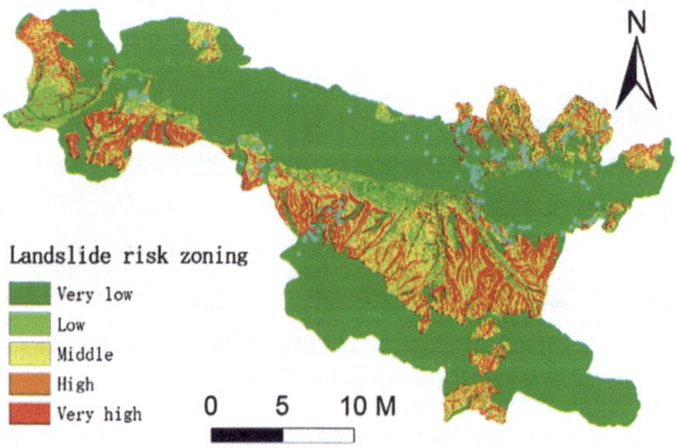

Fig. 6 Historical landslide point and 40 mm landslide risk zoning

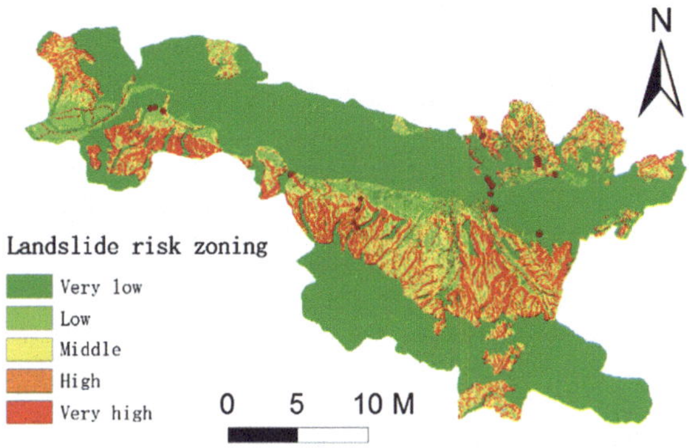

Fig. 7 Survey of landslide points and 40 mm landslide risk zoning

risk zoning. There are 6 regions, of which 3 occur in areas with low risk zones, of which 2 occur in areas where the risk zone is extremely low.

5 Summary

In this paper, the landslide hazard conditions are analyzed, and the mathematical model of landslide occurrence in the study area is established by logistic model. The case study is carried out. The main conclusions are as follows:

1. By analyzing the landslide hazard factor, the landslide hazard assessment model is established by logistic model. The model can predict the area where landslide occurs well under rainfall conditions.
2. Calculate the landslide risk zoning under different rainfall and provide a mathematical model for landslide prediction in the future study area.
3. Taking the landslide point induced by historical rainfall in Lanzhou as an example, the current model is used to evaluate the risk of landslides under rainfall conditions, and the results are tested. The results show that the landslides in the study area are mostly located in high and very high areas. The landslide hazard area (high and very high area) increases with the increase of rainfall, and the evaluation results are in good agreement with the actual landslide occurrence.

Acknowledgements This work was supported in part by The National Key Research and Development Program of China (No. 2018YFF0215303, 2017YFB0503004) and Chinese Academy of Surveying and Mapping Fundamental Research Project under Grant (No. 7771808).

References

1. Wenxing J, Qiang X, Longyun T (2013) Study on rainfall infiltration model of loess slope landslide in three gorges reservoir area. Geotechn Mech 34(12):3527–3534
2. Jin W, Xianlun L, Hang R, Shiwei L, Zhen C (2016) Infiltration and stability of shallow landslides under heavy rainfall. J Southeast Univ (Nat Sci Edn) 46(S1):153–158
3. Lee LM, Gofar N, R H (2009) A simple model for preliminary evaluation of rainfall-induced slope instability. Eng Geol 108:272–285
4. Kale RV, Sahoo B (2011) Green-ampt infiltration models for varied field conditions: a revisit. Water Resour Manage 14:236–243
5. Zhenchun H. Preliminary study on rainfall infiltration model in loess region. Adv Water Sci 5(3):186–192
6. Xiaohua T, Dongsheng L, Qianghui S, Xu W, Runze W, Jianping X (2016) Infiltration model and stability of bedrock layered slope under heavy rainfall conditions. Geotechn Mech 37(11):3156–3164
7. Wenyan WANG, Zhirong WANG, Quanjiu WANG (2003) Improvement and verification of green-ampt infiltration model in loess. J Water Resour 35(1):30–35
8. Huanzhen L, Zhaohui C, Wenliang F, Jiahai W (2018) The statistical risk analysis model of rainfall-induced landslide in large areas. J Nat Disast 27(4):103–111
9. Fangli Z, Huaying C, Lina L et al (2017) A determination method of critical launching precipitation. Eng Constr 12:30–32
10. Zelin Z, Shuren W, Tao W et al (2018) Study on acceleration depth amplification effect and deformation model of loess landslide under earthquakes. Chin J Civil Eng (4)
11. Shan Zhang, Shuwen Yang, Meng Yang et al (2016) Spatial distribution characteristics of precipitation-induced loess landslide hazards in Lanzhou. J Survey Mapping 41(12):142–146
12. Xina W, Huabing H, Ya B et al (2015) GIS-assisted drawing landslide zonation map—A case study of Panzhihua City Sichuan Province. Bull Surv Mapping 2:46–50
13. Gonghao D, Ruiqing N, Ling P et al (2017) A Landslide displacement prediction research based on optimization parameter ARIMA model under the inducing factors. J Wuhan Univ (Inf Sci Edn) 42(4):531–536
14. Park DW, Nikhil NV, Lee SR (2013) Landslide and debris flow susceptibility zonation using TRIGRS for the 2011 Seoul landslide event. Nat Hazards Earth Syst Sci 13(11):2833–2849

Semi-supervised Classification of PolSAR Image Based on Self-training Convolutional Neural Network

Xianxiang Qin, Wangsheng Yu, Peng Wang, Tianping Chen, and Huanxin Zou

Abstract Convolutional neural networks (CNNs) have been successfully used in the field of polarimetric synthetic aperture radar (PolSAR) image classification. However, it is still a big challenge to perform them with limited labeled training samples. To address this problem, a semi-supervised scheme of PolSAR image classification based on self-training CNN is proposed in this letter. Our basic idea is that the PolSAR image to be classified is actually an important source of training samples, from which we iteratively extract reliable samples to enhance the CNN. In our scheme, a CNN is initially trained by the given training samples. Then, the PolSAR image is classified by the trained CNN, of which the pixels with high predicted probability, along with their predicted labels, are selected as new training samples. The operations of retraining CNN, reclassifying PolSAR image and reselecting training samples, are iteratively performed until a stop condition is met. Two actual PolSAR images acquired by AIRSAR and Gaofen-3 systems are employed to verify the effectiveness of the proposed algorithm. Experiment results demonstrate the superiority of our method to the general CNN-based classifiers.

Keywords Polarimetric synthetic aperture radar · Semi-supervised classification · Self-training convolutional neural network

1 Introduction

Synthetic aperture radar (SAR) is an active microwave imaging system that can work day and night, regardless of weather conditions. The polarimetric SAR (PolSAR) is an advance SAR that can operate in different polarization modes, so it has a stronger ability to acquire information. This makes the PolSAR image classification one of the most important parts of PolSAR image interpretation [1–10].

X. Qin (✉) · W. Yu · P. Wang · T. Chen
Information and Navigation College, Air Force Engineering University, Xi'an, China
e-mail: qinxianxiang@126.com

H. Zou
College of Electronic Science, National University of Defense Technology, Changsha, China

L. Wang et al. (eds.), *Proceedings of the 6th China High Resolution Earth Observation Conference (CHREOC 2019)*, Lecture Notes in Electrical Engineering 657,
https://doi.org/10.1007/978-981-15-3947-3_31

In the past decades, various classification algorithms for PolSAR images have been developed. Most early algorithms mainly focus on two contents: feature extraction and classifier design [1–5]. Recently, deep learning technologies, especially the convolution neural networks (CNNs), have been successfully applied to the PolSAR image classification [6–10]. For example, in [7], a CNN of four layers is proposed for classifying PolSAR images. To fit the complex format of PolSAR data, a complex-valued CNN is designed and provides better results in [8]. In [9], a polarimetric-feature-driven CNN that introduces expert knowledge is developed, providing positive results for some PolSAR images. These CNN-based approaches have shown much superior performance to conventional ones.

However, to ensure the performance of these CNN-based classifiers, a large amount of training samples are generally required, which may be invalided in practice. Therefore, it is still a big challenge to classify PolSAR image with CNN when the labeled training samples are limited. Focusing on this problem, we proposed a semi-superpixel classification algorithm by introducing a self-training scheme to the CNN-based classifier. The basic idea is that the PolSAR image to be classified itself can be regarded as an important source of training samples. Therefore, we can select some reliable samples from the PolSAR image for training the CNN. Experimental results obtained from the PolSAR images have validated the superiority of our algorithm.

2 Related Work

2.1 CNN-Based Classifier

Figure 1 presents an illustration of the architecture of CNN. It usually consists of

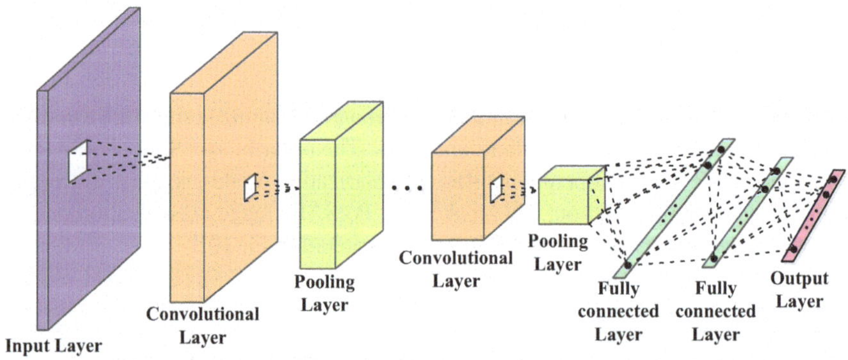

Fig. 1 Illustration of the architecture of CNN

multiple cascaded layers, including an input layer, several alternations of convolutional (Conv) and pooling layers, several fully connected (FC) layers and an output layer [11, 12].

In the most CNN-based classifiers, a softmax function is often employed to transfer the output of CNN to the classification results. Let $\mathbf{s} = \left[s_1, s_2, \ldots, s_{N_C}\right]^{\mathrm{T}}$ be the output of CNN, where N_{C} is the number of classes, then the predicted classification probability of assigning an input image to class k can be calculated by a softmax function as [12]

$$P_k = \frac{\exp(-s_k)}{\sum_t \exp(-s_t)}, \quad k = 1, 2, \ldots, N_{\mathrm{C}}. \tag{1}$$

The predicted label of the input image can be given as the sequence number of the element with the largest probability

$$L = \underset{k=1,2,\ldots N_{\mathrm{C}}}{\arg\max} \{P_k\}. \tag{2}$$

Different to most traditional classification algorithms that perform the feature extraction and classification separately, the CNN is an end-to-end technology and has shown a much outstanding performance.

2.2 CNN-Based PolSAR Image Classification

For the general monostatic PolSAR that employs the same antenna for both transmitting and receiving, the data of each pixel of a PolSAR image is often represented by a polarimetric coherency matrix [1]

$$\mathbf{T} = \left\langle \mathbf{k}_P \mathbf{k}_P^{\mathrm{H}} \right\rangle = \begin{bmatrix} T_{11} & T_{12} & T_{13} \\ T_{21} & T_{22} & T_{23} \\ T_{31} & T_{32} & T_{33} \end{bmatrix}, \tag{3}$$

where $\mathbf{k}_P = [S_{\mathrm{HH}} + S_{\mathrm{VV}}, S_{\mathrm{HH}} - S_{\mathrm{VV}}, 2S_{\mathrm{HV}}]^{\mathrm{T}}/\sqrt{2}$ is the Pauli scattering vector, S_{HH}, S_{HV} and S_{VV} denote the scattering coefficients of HH, HV and VV polarization modes, respectively, $\langle\cdot\rangle$ is the temporal or spatial ensemble averaging [1] and the superscripts “T” and “H” are the transpose and conjugate transpose operators, respectively. Besides, the polarimetric coherency matrix is a complex conjugate matrix.

In many optical image classification tasks, each image is usually assigned a class label as a whole, such as dog, cat and so on [13]. Unlike these tasks, PolSAR image

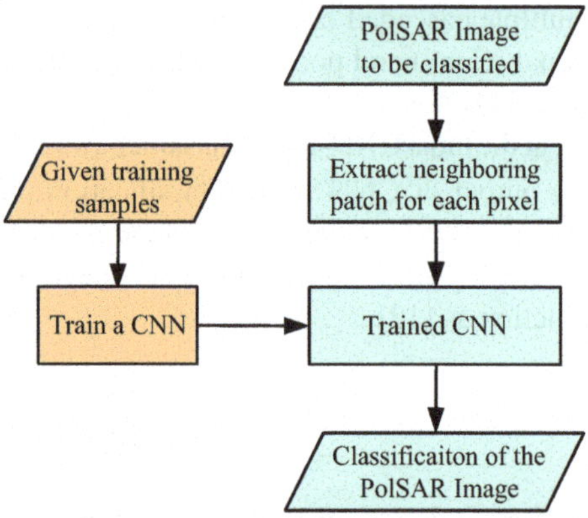

Fig. 2 Flowchart of conventional CNN-based algorithms for PolSAR image classification

classification treats each pixel rather than the entire image as the basic unit of classification. However, to ensure the depth and performance of the network, CNN generally requires its input data to have a certain size, which makes the individual pixels unsuitable as the input of CNN. In [7–9], in order to use CNN for PolSAR image classification, pixels are firstly represented by their neighboring image patches. In this case, the CNN can be trained with such labeled image patches, and the trained CNN can be further used to classify pixels by performing on the corresponding patches. Figure 2 shows the flowchart of the conventional CNN-based classification algorithm for PolSAR images.

3 Proposed Algorithm

In [7–9], the CNN-based classifiers have shown their superiority to most conventional classification algorithms. However, it is still a big challenge to perform them when the given labeled samples are limited. To alleviate this problem, in this section, a data-adaptive training framework of CNN is proposed.

The motivation of our work is simple. In the PolSAR image classification task, since each image to be classified is a collection of numerous basic classification units, it can be also regarded as an important source of training samples. Therefore, it would be benefit to improve the adaptability and reliability of CNN by using new training samples extracted from the PolSAR image itself. For this consideration, a framework of iteratively training CNN as shown in Fig. 3 is designed. The key point is to determine which pixels to select and how to get their labels simultaneously.

Different to the flowchart of general CNN-based classification algorithm, in our framework, new training samples are selected from the testing PolSAR image itself and then employed to train the CNN. More details of the procedure of our framework are summarized as follows:

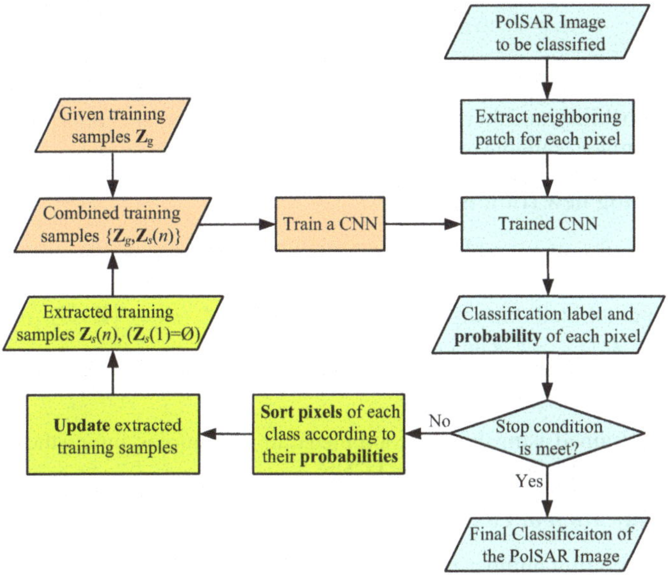

Fig. 3 Flowchart of the proposed algorithm for PolSAR image classification

(1) **Initial classification of PolSAR image**

Set the iteration number n as 1. Then, a CNN can be trained with the initial given training samples $\mathbf{Z}_g$. Then, the PolSAR image can be classified by this trained CNN, yielding both the predicted probability and label of each pixel.

(2) **Selecting new training samples**

In the previous classification step, all pixels of the PolSAR image have their own predicted labels, which can be then used as new samples for training CNN formally. However, many of these labels may be unreliable especially when the number of correct training samples is small. Therefore, to avoid selecting too many incorrectly labeled pixels as training samples, two constraints are adopted.

Firstly, a classification probability threshold P_T is set, and only the pixels whose predicted probability greater than the threshold are treated as candidate training samples. In our algorithm, we set P_T dynamically to the iteration number n as

$$P_T = 1/N_C + a \times n. \tag{4}$$

where a is a given constant coefficient used to increase the threshold as the increase of iteration.

Secondly, a maximum number of selected samples N_{MAX} is set, requiring that the number of samples of each class selected from the image should not be greater than N_{MAX}. Let $N_S(k)$ be the number of samples of the kth class satisfying the first condition, then the number of selected samples of the kth class is

$$N_{\mathrm{E}}(k) = \min(N_{\mathrm{S}}(k), N_{\mathrm{MAX}}). \tag{5}$$

Thus, the pixels of the kth **class** are sorted from large to small according to their predicted probability, and then, the $N_{\mathrm{E}}(k)$ pixels ranked high, meaning their predicted labels are relatively reliable and are selected as new training samples.

(3) **Combining new training samples**

Since the given training samples are generally reliable, they are preserved for training the CNN. Let $\mathbf{Z}_s(n)$ be the set of selected training samples in the nth iteration, then a new set of combined training samples is formed as $\mathbf{Z}_c(n) = \{\mathbf{Z}_g, \mathbf{Z}_s(n)\}$. For the first iteration, $\mathbf{Z}_s(1) = \emptyset$ is an empty set.

(4) **Retraining the CNN**

The CNN is retrained using the new combined training samples, tuning the parameters of CNN based on the previous trained CNN.

(5) **Reclassification of PolSAR image**

Increase the iteration number n as $n = n + 1$. The PolSAR image is reclassified by the newly trained CNN.

(6) **Repeat steps (1)–(5) until a stop condition is meet**

A transferring ratio of pixel label is defined as $R = N_A/N_B$, where N_A is the number of pixels whose labels have changed in the current iteration and N_B is the total number of pixels. If this ratio is less than a given small threshold, meaning the classification result becomes stable, the iteration procedure stops. Besides, the algorithm would also stop when the iteration number reaches a given maximum iteration number N_I.

4 Experimental Results and Analysis

Since the CNN-based classifiers have shown superior performance to most traditional classification algorithms, in the experiments of this section, we focus on the performance of our algorithm by introducing a self-training scheme to the CNN-based classifier.

4.1 Employed PolSAR Images and CNN

In our experiments, two actual PolSAR images [13] are employed. The first one is a PolSAR image acquired by the NASA/JPL AIRSAR system over an agriculture field of Flevoland, Netherlands. The Pauli RGB image of this PolSAR data is shown in Fig. 4a, and the ground truth of this image is shown in Fig. 4b [4, 8]. The size of the image is 750 × 1024 pixels. There are 15 identified classes in this scene, including

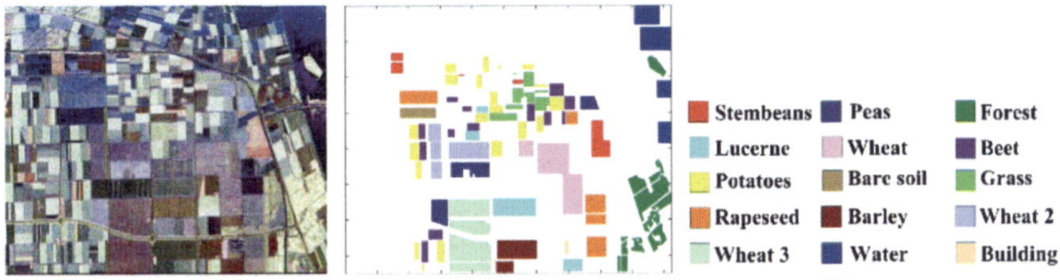

(a) (b)

Fig. 4 **a** Pauli RGB image of AIRSAR PolSAR data and **b** ground truth

(a) (b)

Fig. 5 **a** Pauli RGB image of Gaofen-3 PolSAR data and **b** ground truth

stembeans, peas, forest, lucerne, wheat, beet, potatoes, bare soil, grass, rapeseed, barley, wheat 2, wheat 3, water and buildings, which are called classes 1–15 for simplicity in this paper. The other employed data is a PolSAR image acquired by the Gaofen-3 system over the area of San Francisco. The Pauli RGB image and ground truth are shown in Fig. 5. The size of this image is 600×1000 pixels. There are 5 categories in this scene, including the water, vegetation, three kinds of building areas of low-density, high-density and slanted buildings.

In our work here, due to the relatively simple classification tasks, a CNN slightly modified from the LeNet [15] is used. The architecture of the employed CNN is shown in Fig. 6. The size of input data is $12 \times 12 \times 9$, where 12×12 is the patch size, and 9 is the channel number corresponding to the 9 real components of the coherent matrix $\{T_{11}, T_{22}, T_{33}, \mathrm{Re}(T_{12}), \mathrm{Im}(T_{12}), \mathrm{Re}(T_{13}), \mathrm{Im}(T_{13}), \mathrm{Re}(T_{23}), \mathrm{Im}(T_{23})\}$ ($\mathrm{Re}(\cdot)$ and $\mathrm{Im}(\cdot)$ mean the real and imaginary parts, respectively). The size of convolution kernel is 3×3, with the stride as 1. Besides, the size of pooling kernel is 2×2, with the stride as 2.

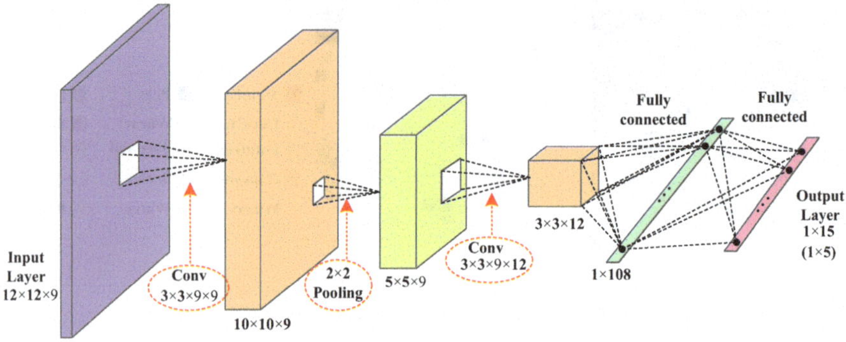

Fig. 6 Architecture of the employed CNN

4.2 Classification Results and Discussion

To reduce the influence of intrinsic speckle, the PolSAR images are firstly filtered by the refined Lee filter [8] with the window of 7 × 7. Besides, pixels of each class are randomly selected from the image according to the ground truth, of which the 12 × 12 neighboring patches are used as the initial given training samples. Moreover, these given samples will be excluded in subsequent procedure of selecting new training samples. For training the CNN, the backpropagation algorithm-based on stochastic gradient descent is employed [8, 14, 15], where the learning rate is 0.5 and the batch size is 100 with 100 training epochs. Moreover, the maximum number of selected samples N_{MAX} for each class is set as 200, the maximum iteration number N_I is set as 15 and the transferring ratio is set as 0.001.

For the first image, 50 training samples for each class are randomly selected. For the second image, 0.1% rate of training samples for each class are randomly selected according to the ground truth. The classification results of these two images yielded by the CNNs with some typical different iterations are shown in Figs. 7 and 8, respectively. Moreover, the maps of ground truth are also shown for better comparison.

Furthermore, to quantitatively evaluate the performance of our algorithm, the classification accuracy of each class and the overall accuracy (OA) are calculated. Figure 9 shows the curves of classification accuracy and OA versus the iteration number.

It can be observed from Figs. 7, 8 and 9 that

(1) For the first iteration where only the given training samples are used, the classification results as shown in Figs. 7b and 8b are visually bad and the OA is only 46.61% and 83.23% for the first and second images, respectively. It can be observed that some classes of the first image have high classification accuracy, such as classes 1, 3, 8, 11 and 15, while some other classes have very low classification accuracy, such as classes 9, 12, 13 and 14. These bad results

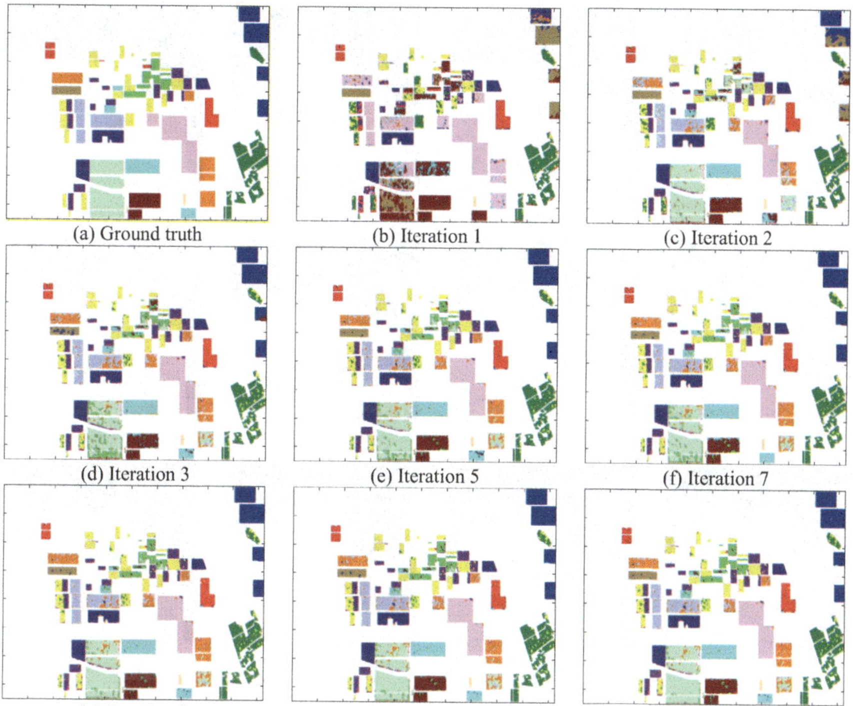
(a) Ground truth (b) Iteration 1 (c) Iteration 2 (d) Iteration 3 (e) Iteration 5 (f) Iteration 7 (g) Iteration 9 (h) Iteration 11 (i) Iteration 15

Fig. 7 **a** Ground truth and **b** (i) PolSAR image classification results yielded after 1, 2, 3, 4, 5, 7, 9, 11 and 15 iterations when 50 training samples are given for each class

can be obviously observed by comparing with Fig. 5a, b. Similarly, for the second image, obvious classification confusion occurs among the three kinds of building areas which can be observed, and the classification of the developed building area is very bad.

(2) As the number of iterations increases, the classification accuracy improves gradually, especially in the previous iterations. As observed from Fig. 9, with the increase of iteration number, for the classes with low initial classification accuracy, their accuracy increases significantly. For those with high initial classification accuracy, their accuracy maintains at high levels. Then, for the first image, the OA maintains a stable level at about 89% after 6 iterations. For the second image, the OA is over 94% after 9 iterations and reaches 95.38% after 15 iterations. For these two images, the obtained classification map becomes more and more consistent with the ground truth. These results have demonstrated the effectiveness of our algorithm by introducing a self-training scheme to the CNN-based classifier.

For further evaluating the performance of our algorithm, experiments with different initial given training samples are performed. Given the numbers of training

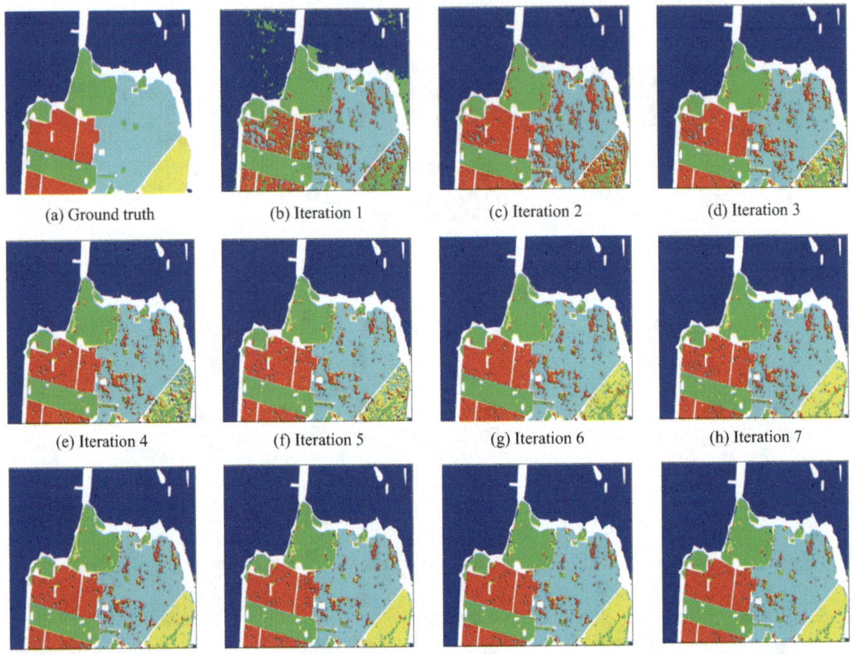

(a) Ground truth (b) Iteration 1 (c) Iteration 2 (d) Iteration 3
(e) Iteration 4 (f) Iteration 5 (g) Iteration 6 (h) Iteration 7
(i) Iteration 9 (j) Iteration 11 (k) Iteration 13 (l) Iteration 15

Fig. 8 **a** Ground truth and **b** (l) PolSAR image classification results yielded after 1–7, 9, 11, 13 and 15 iterations when 0.1% rate of training samples are given for each class

samples are set as {80, 100, 150, 300, 500} for each class. Then, the same classification experiments as before except with different numbers of initial given training samples are performed. The obtained OA curves for these two images are shown in Fig. 10.

It can be seen from Figs. 9 and 10 that

(1) The OA is generally higher when the number of given training samples is larger. Take the first image, for example, the OA is only about 46% for the first iteration when the number initial training samples of each class is 50, while that is about 94% when N_G is 500. It has validated that the amount of training samples has a great influence to the effectiveness of CNN. Similar results can be observed for the second image.

(2) Our algorithm can significantly improve the classification results especially when the number of given training samples is small. These results can be explained by the fact that more and more reliable training samples are extracted and used with the increase of iterations, which are benefit to improve the adaptability and reliability of the CNN.

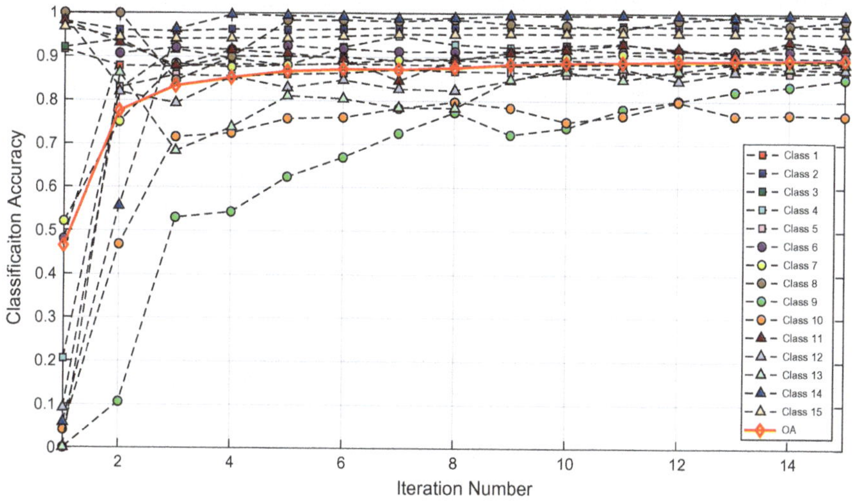

(a)

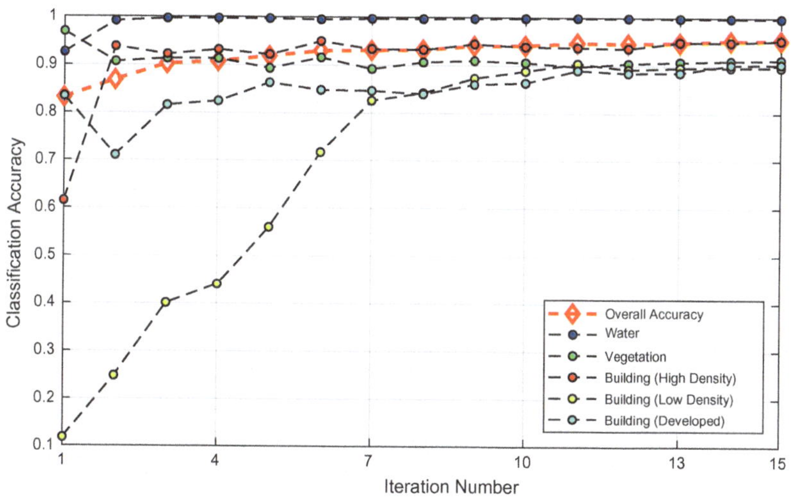

(b)

Fig. 9 Curves of classification accuracy of each class and OA versus the iteration number for **a** the first image and **b** the second image when 50 training samples are given for each class

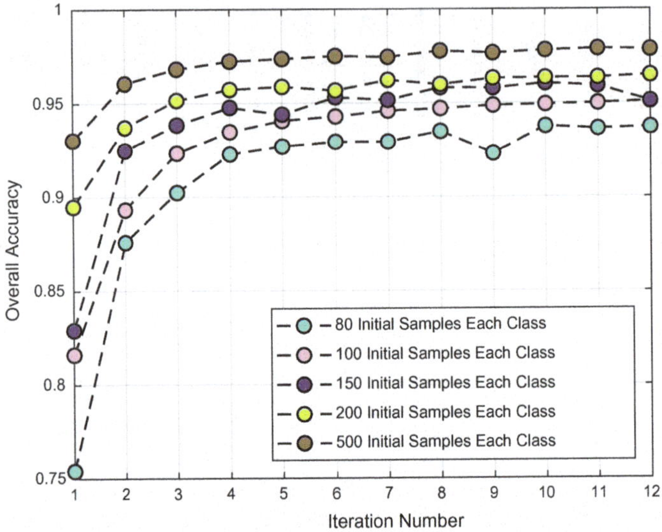

(a)

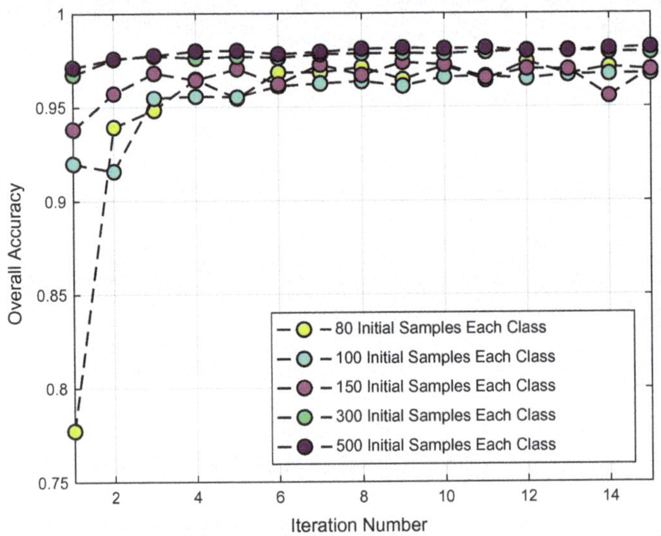

(b)

Fig. 10 Curves of OA versus iteration number with different amounts of given training samples for **a** the first image and **b** the second image

5 Summary

This paper develops a semi-supervised classification algorithm for PolSAR images using the self-training CNN. Different from conventional manners of training CNN, it selects and uses training samples from the PolSAR image to be classified itself, making the trained CNN more adaptive and suitable to the image. Experimental results have demonstrated that our algorithm can significantly improve the classification result especially when the given training samples are limited.

Acknowledgements This work is supported by National Science Foundation of China under Grant 41601436, 61403414 and 61773396 and in part by the Natural Science Basic Research Plan in Shaanxi Province of China under Grant 2018JM4029 and 2019JM-554.

References

1. Lee J-S, Pottier E (2009) Polarimetric radar imaging: from basics to applications. CRC Press, Boca Raton
2. Lee JS, Grunes MR, Ainsworth TL, Du LJ, Schuler DL, Cloude SR (1999) Unsupervised classification using polarimetric decomposition and the complex Wishart classifier. IEEE Trans Geosci Remote Sens 37(5):2249–2258
3. Wu Y, Ji K, Yu W, Su Y (2008) Region-based classification of polarimetric SAR images using Wishart MRF. IEEE Geosci Remote Sens Lett 5(4):668–672
4. Yu P, Qin AK, Clausi DA (2012) Unsupervised polarimetric SAR image segmentation and classification using region growing with edge penalty. IEEE Trans Geosci Remote Sens 50(4):1302–1317
5. Liu B, Hu H, Wang H, Wang K, Liu X, Yu W (2013) Superpixel-based classification with an adaptive number of classes for polarimetric SAR images. IEEE Trans Geosci Remote Sens 51(2):907–924
6. Hou B, Kou H, Jiao L (2016) Classification of polarimetric SAR images using multilayer autoencoders and superpixels. IEEE J Sel Topics Appl Earth Observ Remote Sens 9(7):3072–3081
7. Zhou Y, Wang H, Xu F, Jin Y-Q (2016) Polarimetric SAR image classification using deep convolutional neural networks. IEEE Geosci Remote Sens Lett 13(12):1935–1939
8. Zhang Z, Wang H, Xu F, Jin Y-Q (2017) Complex-valued convolutional neural network and its application in polarimetric SAR image classification. IEEE Trans Geosci Remote Sens 55(12):7177–7188
9. Chen S-W, Tao C-S (2018) PolSAR image classification using polarimetric-feature-driven deep convolutional neural network. IEEE Geosci Remote Sens Lett 15(4):627–631
10. Liu X, Jiao L, Tang X, Sun Q, Zhang D (2019) Polarimetric convolutional network for PolSAR image classification. IEEE Trans Geosci Remote Sens 57(5):3040–3054
11. Hu F, Xia G-S, Hu J, Zhang L (2015) Transferring deep convolutional neural networks for the scene classification of high-resolution remote sensing imagery. Remote Sens 7:14680–14707
12. Goodfellow I, Bengio Y, Courville A (2016) Deep learning. MIT Press
13. https://earth.esa.int/web/polsarpro/airborne-data-sources
14. Krizhevsky A, Sutskever I, Hinton GE (2012) ImageNet classification with deep convolutional neural networks. Proc Adv Neural Inform Process Syst 1097–1105
15. LeCun Y, Bottou L, Bengio Y, Haffner P (1998) Gradient-based learning applied to document recognition. Proc IEEE 86(11):2278–2324

A SAR Image Data Augmentation Method Based on Generative Adversarial Network

Qinglin Lu, Guojing Li, and Wei Ye

Abstract Generative adversarial network (GAN) has been increasingly used in the SAR image simulation in recent year. However, the network is difficult to keep stable, and high-quality generated samples are not available. In order to solve the problem, a model based on the Wasserstein GAN with deep convolutional structure is proposed. Training on SAR image datasets, we show evidence that our method is superior to original deep convolutional generative adversarial network (DCGAN). We also build a verification framework by YOLOv3, and tests are done to validate generated samples. The results show that our method is more stable, and the diversity of generated samples is improved.

Keywords Synthetic aperture radar · Wasserstein generative adversarial network · Image generation · YOLOv3 · SAR interpretation

1 Introduction

Synthetic aperture radar (SAR) is a key means of remote sensing reconnaissance. The research on SAR target intelligent identification has been carried out for many years. With the widespread application of deep learning methods in recent years, more and more deep learning algorithms have been applied to SAR target recognition systems. However, the target recognition network model based on the deep learning method requires a sufficient number of multi-directional and rich training sets. The small amount of data and the lack of diversity of various target images will directly affect the training of intelligent identification networks and resulting in poor recognition performance for targets in different scenarios.

The deep convolutional generative adversarial network (DCGAN) [1] combines the advantages of the deep convolutional neural network to better fit image features and generating high-quality image samples. In order to provide a sufficient number

Q. Lu · G. Li
Space Engineering University Graduate School, Beijing, China

W. Ye (✉)
Space Engineering University, Beijing, China
e-mail: yeyuhan@sina.com

L. Wang et al. (eds.), *Proceedings of the 6th China High Resolution Earth Observation Conference (CHREOC 2019)*, Lecture Notes in Electrical Engineering 657,
https://doi.org/10.1007/978-981-15-3947-3_32

of target images as the training data of the SAR target recognition system, the work of generating a simulated image against the network using the deep convolution generation method is performed. Although the resulting generated image and the real image have a large degree of similarity, the training result inevitably fails [2], such as the mode collapse.

Arjovsky et al. proposed Wasserstein generative adversarial network (WGAN) [3] in 2017, and the distance between the two distributions can be truly reflected, which basically solves the problem of unstable GAN training.

This paper attempts to solve the problem of insufficient diversity of the SAR image generated by DCGAN and the difficulty of convergence of the network. We presented a model based on the WGAN that connected by convolution layers, and it can generate more various SAR images. In order to objectively evaluate the quality and validity of the generated image, the YOLO.v3 [4] is trained by generated samples.

2 GAN-Based SAR Image Augmentation Model

Based on WGAN, the generative adversarial network for enlarging SAR image data is developed in this paper. The structure of the image augmentation model is shown in Fig. 1. We do the high-quality samples generation and build the dataset first, and then verify the validity and superiority of the dataset.

The images generated by WGAN will satisfy the lack of SAR dataset for network training. The high-quality image will be collected to build datasets with the structure like VOC2007. Finally, the superiority of the presented dataset will be tested in the detection network training.

Fig. 1 Structure of the SAR image augmentation model

2.1 Cost Function of the WGAN

Arjovsky et al. introduced the Wasserstein distance to replace the JS divergence to describe the distance relationship between two distributions. The Wasserstein distance is defined as follows:

$$W(P_r, P_g) = \inf_{\gamma \sim \prod(P_r, P_g)} E_{(x,y)\sim\gamma}[\|x - y\|]. \tag{1}$$

where $\prod(P_r, P_g)$ represents a collection of all joint distributions. A real sample x and a generated sample y can be sampled from any joint distribution γ, and $\|x - y\|$ is the distance between this pair of samples. So, $E_{(x,y)\sim\gamma}[\|x - y\|]$ is the expected value of the distance of the sample under the joint distribution γ, and the Wasserstein distance is defined as the cost of transforming the distributions P_r into the distribution P_g.

However, the original Wasserstein distance is not appropriate for cost function. In order to make the problem easy to solve, the cost function is transformed into,

$$\min_G \max_D V(D, G) = \mathbb{E}_{x\sim P_r}[D(x)] - \mathbb{E}_{z\sim P_g}[D(G(z))]. \tag{2}$$

The x represents a real image, z represents the noise that input to the G network and $G(z)$ represents the output from the G network. $D(x)$ indicates the probability that the D network judges the input image is true or not. And $D(G(z))$ is the probability that the D network determines whether the image generated by G is true.

Therefore, the loss functions of the discriminator and generator can be obtained as follows, respectively

$$L_D = \mathbb{E}_{z\sim P_g}[D(G(z))] - \mathbb{E}_{x\sim P_r}[D(x)]. \tag{3}$$

$$L_G = -\mathbb{E}_{z\sim P_g}[D(G(z))]. \tag{4}$$

2.2 Structure of the Generative Network

We build the generative model with the structure of convolutional network, and hope it can only learn the feature of SAR images but also avoid the mode collapse [5].

The generative network is designed based on a convolutional net to effectively extract spatial and spectral features for SAR images. The generator contains four fractionally strided convolutions, and the structure of the generator is depicted in Fig. 2.

The discriminator consists of four convolutions and a fully connected layer, and the structure of the discriminator is shown in Fig. 3. It can be found that the convolutional

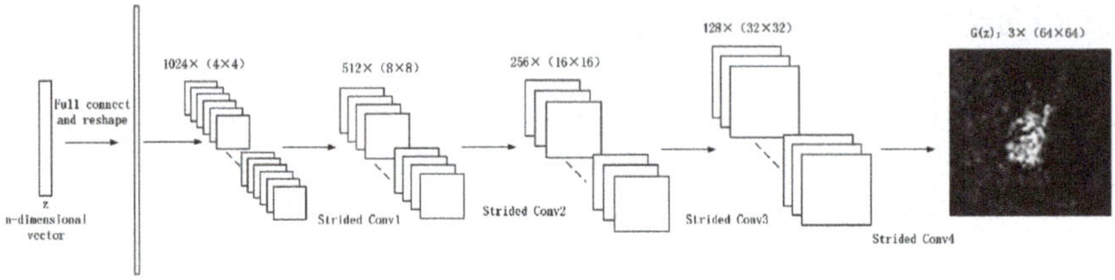

Fig. 2 Structure of the generator

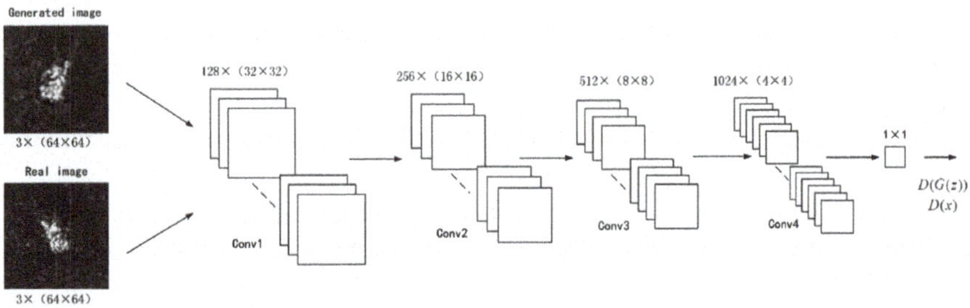

Fig. 3 Structure of the discriminator

architecture of generator and discriminator are symmetrical, which is similar to the convdeconv. The deconvolution here means upsampling operation aiming to make the size of generated samples is equal to real images.

The rectified linear unit (ReLU) activation function used in the generator replaces the Tanh function, and it kept in the output layer exceptionally. The leaky rectified linear unit (Leaky ReLU) activation function is used in the discriminator for all layers.

The convolution net replaces pooling operators with fractionally strided convolutions, and it uses images feature as an upsampling strategy. A simple example is illustrated in Fig. 4. The fractionally strided convolution is a method to enlarge the sample, and it is flexible and self-improved because the hyperparameters of convolution kernels are trainable.

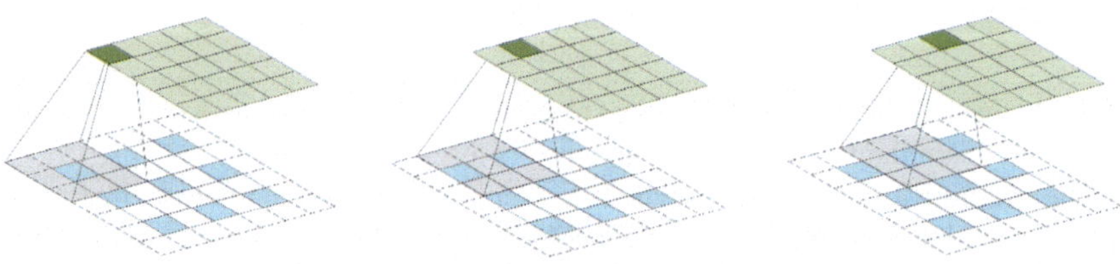

Fig. 4 Illustration to fractionally strided convolution

3 Experiment and Analysis

The experiments involved in this paper are mainly divided into two parts:

1. SAR target generation based on WGAN, using three types of target images in the MSTAR dataset as the training set.
2. The YOLO3 is trained using generated target images and real images, and the superiority of generated images is evaluated by the obtained training result.

3.1 The Generation of SAR Target Images

We tested our method in the Ubuntu 16.04 environment. The detail is shown in Table 1.

The dataset used in the experiment is the Moving and Stationary Target Acquisition and Recognition (MSTAR) image, which collects thousands of SAR images containing ground targets, including different target types, vertical and horizontal angles, depression angles, serial numbers and sharpness. The dataset is obtained by X-band spotlight SAR, the image resolution is 0.3×0.3 m and most of the target image size is 128×128 pixels. It is a common dataset for SAR target recognition research work.

The T72 main battle tank, BMP2 infantry fighting vehicle and ZSU23 air defense artillery are selected as the training set of GAN. In the MSTAR project, the measured images with radar elevation angles of 17° and 15° are provided for these three types of targets. The specific quantities and corresponding classes are shown in Table 2. In order to ensure that the number of training samples can meet the needs of network training, for each type of target, we select all the images under one of the two types of angles as the training set.

The detail of training samples shown in Fig. 5, each group from left to right is the optical image of the target, the measured SAR image, the simulated SAR image generated by DCGAN and SAR image generated by WGAN.

Table 1 Experimental configuration

Operating system	Ubuntu 16.04
CPU	Intel Core i7-6700
GPU	Nvidia GTX1070
Design language	Python 3.6
Frame	PyTorch
CUDA	Cuda 8.0

Table 2 Detailed information about the MSRAT dataset used in our experiments

Class	Serial no.	Depression	
		17°	15°
ZSU23	D_08	299	274
BMP2	SN_9563	233	195
	SN_9566	232	196
	SN_C21	233	296
T72	SN_132	232	196
	SN_812	231	195
	SN_S7	228	191

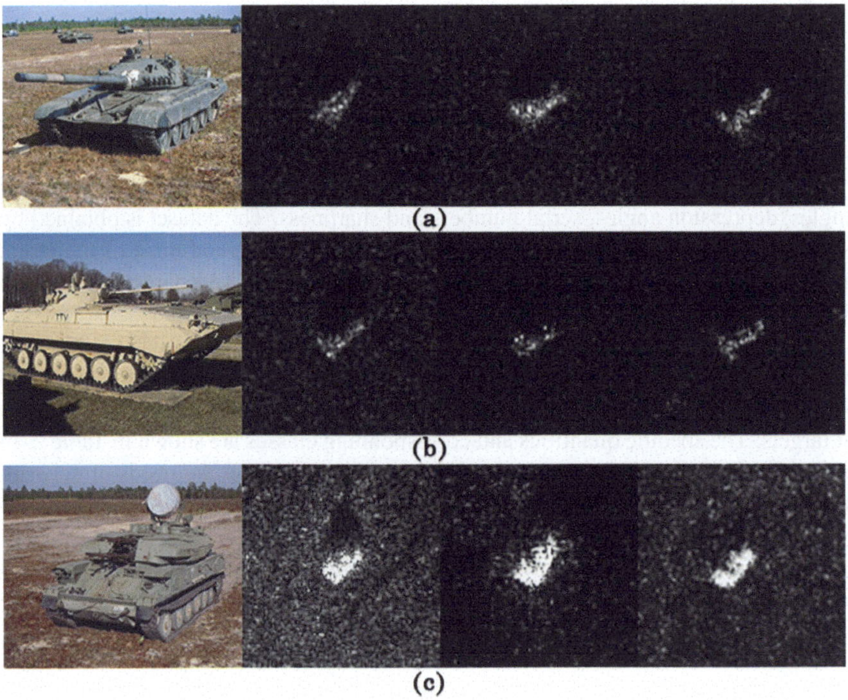

Fig. 5 Optical images, synthetic aperture radar (SAR) images, the simulated SAR image generated by DCGAN and the SAR image generated by WGAN which combined with the DCGAN structure of three classes of objects in the MSTAR database

3.1.1 SAR Image Generation by DCGAN

In the first part, we generate SAR target images based on DCGAN. In the practical training process, the convolution kernel is 4×4, the stride of the convolutional stride is $s = 2$ and the learning rate is 0.0002. We collect images that generated after the 1000th epoch and build a dataset with high-quality generated samples.

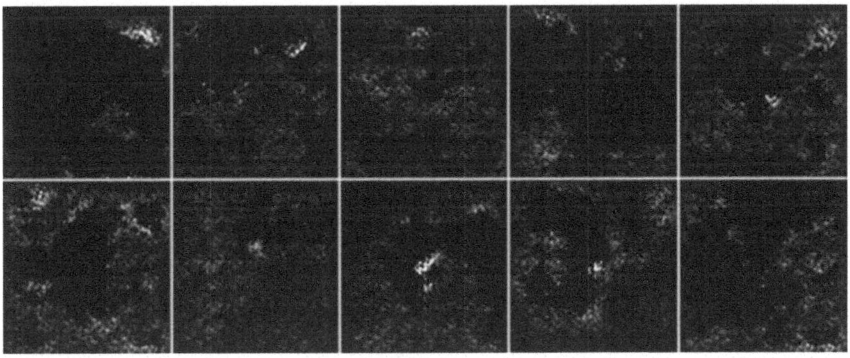

Fig. 6 Non-converged results of SAR target images generated by DCGAN

Although the DCGAN structural optimization makes it faster to converge than the original GAN and has a stronger ability to fit the image, the problem of unstable training is still hard to avoid [6]. For the generation of SAR images, the training process has become more difficult. Due to a large amount of clutter and noise interference in the SAR image, and the detailed features of the gray image are not as obvious as the ordinary optical image, the failure is easy to occur in the experiment. Figure 6 shows a generated image in which the network does not converge when the ZSU23 target image is generated. It can be seen that the generator does not effectively pass the learned target features, resulting in failure to converge and eventually generate a stray noise.

The mode collapse is another problem that needs to be overcome in training DCGAN. In the optimization process of the generator and the discriminator, the generator may concentrate on generating several kinds of patterns and lose the remaining. It will make consistency features in the same batch results. In the experiment, it is reflected as follows:

- The target under the same azimuth is repeated in the same batch. As shown in Fig. 7a, the target azimuth of the image in the white frame is almost the same, and details are approximate. It shows that DCGAN has the problem of insufficient diversity in the same batch of generated images;
- The same batch of generated results has consistency features. Figure 7b are images obtained after 2000 epoch, in the same batch of generated images there are same features which like the barrel in the middle of the vehicle. Comparing the original image, this detailed feature will only appear at a certain angle.

The reason is that the generator is trapped in a few modes in the iteration, and the output is fixed to the pattern identified by the discriminator individually, thus losing the diversity of the generated results and accompanying the consistency feature. The mode collapse happened.

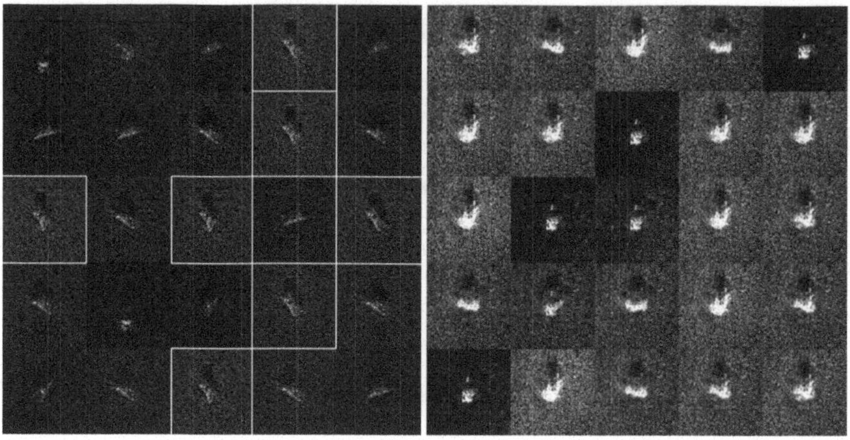

(a) (b)

Fig. 7 Images of the BMP2 in (**a**) and images of the ZSU23 in (**b**) generated by DCGAN

3.1.2 SAR Images Generation by the Optimized WGAN

In the second part of the experiment, we generate SAR images using our method. The same dataset is used in the training of the optimized WGAN. The parameters are consistent with the first part, and we still collect images after the 1000th epoch.

For the Wasserstein GAN with deep convolution structure, the surface texture and angle of the generated image have stronger diversity, and the mode collapse is basically avoided. The advantage and superiority of the images generated by the optimized WGAN are described as follows:

- In one batch of generated images, the azimuth of target cover 0–360°. Figure 8a is a collection of BMP2 images from 0° to 175° in the training sample, the angle difference between each sample is about 8° and an image of a certain specific angle (such as 180°) is missing. In Fig. 8b, generated images have the azimuth of 0–180°, and the angle difference between each sample is about 7°. The azimuth information in generated images is richer than the training set, and the angle image that does not exist in the training set can be obtained.
- Texture features of the target are more rich in the images generated by the WGAN based on the deep convolutional structure. The generated sample is more similar to the real image in visual effects, and the generated samples in the same batch have no similar features.

Comparing these two generation experiments, it can be found that the WGAN with deep convolutional structure is more superior than the original DCGAN. But the conclusion is based on the visual effects of generated samples. The verification of the superiority of recognition and detection will be carried out in the next section.

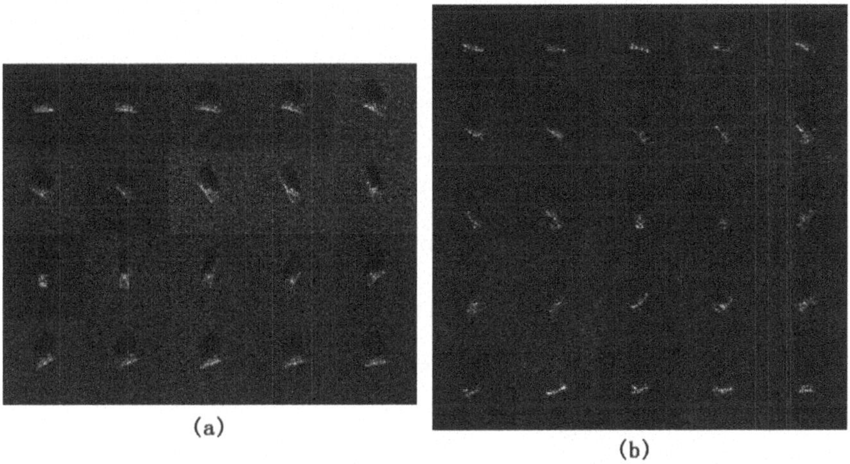

Fig. 8 Comparison between the real images and WGAN generated images: **a** the image set with the target azimuth from 0° to 175° in the training set; **b** the image collection with the target azimuth from 0° to 180° generated by WGAN

3.2 Validity Verification Based on YOLOv3

YOLOv3 is a portable model that is good at detecting small targets in the complicated background, and it costs less time in detecting process. It is appropriate for using in automatic identification of SAR targets.

We use the training set with different sample proportions to train the YOLOv3. The effect of generated samples on network training is analyzed based on the recognition accuracy of SAR target pieces.

We set up three kinds of training sets with different proportions of generated samples; details are shown in Table 3. Since it is discussed that the validity of the training set is expanded by generating samples, the total number of images in the training set could be changed.

Table 3 Setting of three types of training sets

Class	Real sample	Generated sample	The ratio of generate samples (%)
Training set 1	400	0	0
Training set 2	400	500	55.5
Training set 3	400	1000	71.4

Table 4 Results of YOLOv3 by three types of training sets

Class		Accuracy	Unrecognized rate	Maximum confidence
DCGAN	Training set 1 (0%)	0.723	0.02	0.93
	Training set 2 (55.5%)	0.741	0.01	0.97
	Training set 3 (71.4%)	0.767	0	0.97
WGAN (with conv structure)	Training set 1 (0%)	0.723	0.02	0.93
	Training set 2 (55.5%)	0.745	0.02	0.93
	Training set 3 (71.4%)	0.747	0.01	0.93

3.2.1 Recognition Performance

In order to prevent the recognition result from losing the reference meaning due to the lack of diversity of the test set, we set up a crossover experiment. From the real image of the training set, 100 images were taken as test sets multiple times, and the remaining 400 were taken as new training samples. By testing samples that are not included in different training sets, the ability of the network to identify targets in an unknown state can be derived. After all, in practical applications, it is impossible to train the network in all states of the target or in all backgrounds, so the robustness of the network needs to be considered.

The YOLO3 network is trained by using three sets of training sets with different sample proportions. The training is ended after the 50th epoch, and the batch size is 64. The test results in testing 100 pieces of three different targets, respectively, are as in Table 4.

The three types of training focused on five different test sets to test the detection performance of the target in the non-existent state of the training set and finally averaged the corresponding five results of each group. Comparing the experiment results, it can be found that the recognition accuracy of the network is improved after the training set is added to generated images; it proves that the image generated by optimized WGAN makes positive effects to the recognition ability.

3.2.2 Detection Performance

YOLOv3 has a strong detection ability for small targets in complex environments [7], and the recognition experiment of small-sized target images is limited to the ability of YOLOv3. In order to verify the effect of the generation of generated samples on the network detection capability, it is necessary to detect and identify the target in a large

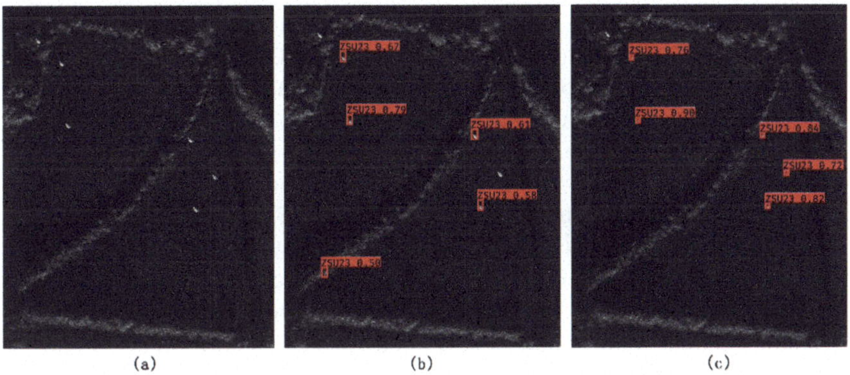

Fig. 9 Recognition results of YOLOv3 for large scene images under different training sets: **a** there are 6 measured target images embedded in the scene image as detection samples; **b** detection results from YOLOv3 trained by the original DCGAN; **c** detection results trained by our method

scene [8]. The MSTAR dataset provides SAR images of 50 different environmental scenes, corresponding to approximately 0.1 km of land in the actual scene. The measured image of the ZSU 23 is synthesized in the scene image, and the same composite image is detected using YOLOv3 corresponding to the original DCGAN generated sample and YOLOv3 corresponding to the optimized WGAN generated sample, respectively.

YOLOv3 was trained by a training set containing 1000 WGAN generated images and 400 measured images, and the total number of training was controlled at the 50th epoch. The result of detecting the same large-size scene image is as shown in Fig. 9. In Fig. 9b, YOLO trained by the DCGAN identified 4 of the 6 real targets, with an average confidence of 0.6625 and made a mistake. In Fig. 9c, the WGAN trained YOLO identified 5 of the 6 targets with an average confidence of 0.808. It can be seen that the image generated by our method is more effective in improving the detection and recognition ability.

It can be seen that after the training set added to the generated samples, the detection and recognition abilities of YOLOv3 have improved, which proves the validity and superiority of the generated samples as the training set.

4 Summary

In this paper, we proposed a WGAN-based SAR image generation model. Our method can generate SAR images with high quality, which can solve the small sample recognition problem. The new training dataset can be built by our method, and the superiority also can be verified. The following conclusions can be summarized through experiments:

1. In the image generation experiments of DCGAN and WGAN, it can be seen that the mode collapse phenomenon prevalent in DCGAN is eliminated by using Wasserstein distance. It proves that Wasserstein distance has a positive effect on the gradient transmission in training;
2. In the experiment of training YOLO by generated images, the YOLO corresponding to the training sets of 500 generated images and 1000 generated images increased the recognition accuracy by 2% and 3.4%, respectively. The validity of the generated samples in training recognition network was confirmed.
3. In the large scene target detection experiment, the accuracy of WGAN assisted training YOLO is higher, which proves that Wasserstein distance can effectively improve the quality of generated samples. It has great significance for supporting the train of deep learning networks.

The work in this paper verifies the feasibility of using GAN to generate SAR images and verifies the validity of generated images by training YOLOv3. With the gradual progress of the research on generative adversarial network, GAN has broad application prospects in the field of remote sensing image data expansion and SAR interpretation.

References

1. Radford A, Metz L, Chintala S (2015) Unsupervised representation learning with deep convolutional generative adversarial networks. J arXiv preprint arXiv:1511.06434
2. Guo J, Lei B, Ding C, Zhang Y (2017) Synthetic aperture radar image synthesis by using generative adversarial nets. J IEEE Geosci Remote Sens Lett 14(7):1111–1115
3. Arjovsky M, Chintala S, Bottou L (2017) Wasserstein gan. J arXiv preprint arXiv:1701.07875
4. Redmon J, Farhadi A (2018) Yolov3: an incremental improvement. J arXiv preprint arXiv:1804.02767
5. Guo Y, An D, Qi X et al (2019) Mode collapse and regularity of optimal transportation maps. J arXiv preprint arXiv:1902.02934
6. Arjovsky M, Bottou L (2017) Towards principled methods for training generative adversarial networks. J Stat
7. Lan W, Dang J, Wang Y, Wang S (2018) Pedestrian detection based on YOLO network model. In: 2018 IEEE international conference on mechatronics and automation (ICMA), Changchun, pp 1547–1551
8. Khan HM, Yunze C (2018) Ship detection in SAR image using YOLOv2. In: 2018 37th Chinese control conference (CCC), Wuhan, pp 9495–9499

Research on High-Resolution Synthetic Aperture Radar Image for Varied Antenna Beam Pointing

Yinghui Zhao and Xijuan Yue

Abstract The precise phase and amplitude are important for the application of the high-resolution SAR image. Airborne synthetic aperture radar antenna beam pointing varies due to the instability of platform. The effect of the varied antenna pointing on the high-resolution SAR image will be analyzed. A SAR signal model will be established based on the relationship of the beam pointing and the gain of the echo. The distortion of the phase and the amplitude of the image is discussed for two cases. Finally, the simulation results will be provided to evaluate the development in this paper.

Keywords Synthetic aperture radar · Antenna beam pointing · And impulse response function

1 Introduction

As a remote sensing system, synthetic aperture radar (SAR) transmits the coherent microwave, and the high-resolution image of the observed scene can be obtained using the data focusing techniques. Advanced high-resolution SAR system can work for many areas such as surveying and mapping, deformation monitoring, and crop measurement; soon after the accurate imaging algorithms are implemented and the precise information including the position, phase, and amplitude of the images is provided [1].

Airborne SAR systems have been used widely for its good mobility and high resolution of image. SAR imaging algorithms integrated by motion compensation are used to process SAR data since there exist trajectory deviations from the nominal rectilinear flight track for airborne SAR case due to the platform instability [2].

Besides trajectory deviations, attitude variations of aircraft are unavoidable during SAR data acquisition due to atmospheric turbulence, which affects the antenna pointing. The image quality is also degraded, which mainly presents as radiometric resolution loss, geometric, and phase distortions. Attitude variations may be reduced

Y. Zhao (✉) · X. Yue
Aerospace Information Research Institute, CAS, Beijing, China
e-mail: zhaoyh@aircas.ac.cn

L. Wang et al. (eds.), *Proceedings of the 6th China High Resolution Earth Observation Conference (CHREOC 2019)*, Lecture Notes in Electrical Engineering 657,
https://doi.org/10.1007/978-981-15-3947-3_33

by serving the real antenna to a fixed look direction and compensated by knowing the illumination characteristics of the transmitting antenna at the image calibration stage. An precise measurement of the antenna pointing angle is a pressing problem for airborne SAR, and it is important for calculating the Doppler centroid for SAR processing and performing correction of geometric displacement for ground mapping of SAR images. In the case of the X-SAR of the SIR-C/X-SAR 1994 shuttle mission, it was operated with an azimuth beam width of about 0.14° and an estimated antenna pointing uncertainty of about $\pm 1^\circ$ [3]. Besides measuring the antenna orientation angles, Doppler centroid can be estimated from the raw data. Several clutter-lock techniques have been proposed so far. They are energy balancing, correlation Doppler estimator (CDE), sign Doppler estimator (SDE), maximum-likelihood estimation (MLE), and radon transform-based ambiguity-free Doppler centroid estimation. This technique compensates the attitude variation to some degree, and the compensation performance is related to the accuracy of the Doppler centroid estimation. Nowadays, the accuracy of Doppler centroid estimation has been increased to 1% of the pulse repetition frequency (PRF) [4].

If the antenna pointing varies from its nominal direction, the processing geometry will mismatch the acquisition geometry on the SAR imaging. The difference between the processing geometry and acquisition geometry resulted from the choose of the squint angle is discussed in [5], and it shows that although Doppler centroid for data focusing varies with the change of the squint angle, the SAR image still can be correctly localized in ZD geometry on pixelscale. The main reason is that the Doppler band used for data focusing matches that of the acquisition data, where it is assumed that the amplitude of the signal keeps constant in the Doppler band. In fact, the envelope of the echo is related to the antenna pattern. If there is error in the antenna pointing, the envelope of the echo will mismatch that in the Doppler band used for data processing, which results in the distortion of the SAR image finally. In this paper, a SAR signal model will be established based on the relationship of the beam pointing and the gain of the echo, and the distortion of impulse response function (IRF) of the point target is investigated.

This paper is organized as follows. In Sect. 2, the signal echo model is presented. In Sect. 3, the data processing using an inaccurate beam pointing is derived in two-dimensional frequency domain and the IRF of the point target is discussed for two cases. In Sect. 4, the simulation results are provided, which test the developments in our paper. The conclusions are drawn in Sect. 5.

2 Signal Model

Considering the SAR imaging geometry as shown in Fig. 1, where x-axis is parallel to the trajectory of the antenna phase center(APC), z-axis is vertical, x-, y-, and z-axes form a right-handed reference system and the origin o is chosen as the projection of the APC at the time $t = 0$ onto the x-axis. The target $P(x_0, y_0, z_0)$ is located at the center of the antenna beam at the time $t = t_0$, where r is the distance from the target

Fig. 1 SAR imaging geometry

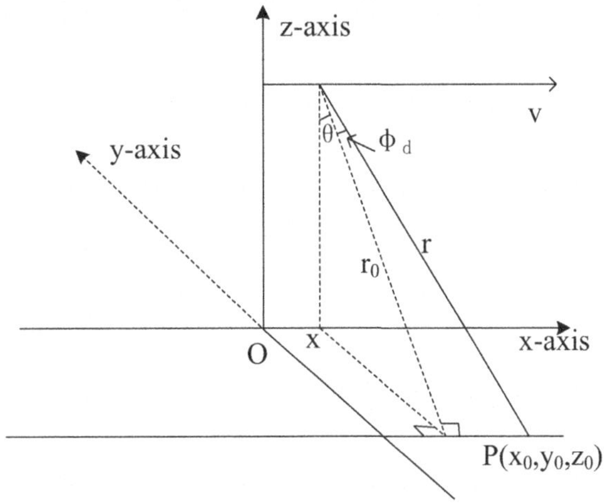

to the APC and r_0 is the range in the zero-Doppler plane, and the x-coordinate of APC is $x = vt_0$ with v being the velocity of the APC. θ and ϕ_d are the incidence angle and the squint angle, respectively.

The target location can be represented by the range r and the x-coordinate of the APC x as,

$$\begin{aligned} y_0 &= r_0 \sin\theta = r \cos\phi_d \sin\theta \\ x_0 &= x + r \sin\phi_d \end{aligned} \tag{1}$$

Analyzing (1), the location of the target P can be expressed using (x, r) in conical processing geometry [5] as well. The backscattered signal of the ground scene is received by radar in the slant range dimension, and the center of the echo of the target P is also located at (x, r) in the received data. Assuming that the radar transmits the chirped pulse, the backscattered signal of the target P can be modeled as [5]

$$\begin{aligned} ss(x', r') = {} & \sigma_P R\left(\frac{r' - R}{D}\right) \\ & \times \exp\left(-j\frac{4\pi}{\lambda}R - jk_r(r' - R)^2\right) w^2(x' - x), \end{aligned} \tag{2}$$

where σ_P is the radar cross-section of the target P, and D and k_r are the spatial counterparts of the chirp duration T and rate k, respectively, where $D = \frac{T \cdot c}{2}$ and $k_r = \frac{4\pi k}{c^2}$. λ is the wavelength corresponding to the carrier frequency f_0, $R(\cdot)$ is the rectangle envelope function, and $w(\cdot)$ is the antenna illumination pattern. R is the distance from the APC to the target.

3 Data Processing Using an Inaccurate Beam Pointing

3.1 Impulse Response Function Analysis

The signal model has been provided in Sect. 2. The expression of the backscattered signal of the target *P* is written in (2), where $w(\cdot)$ implies the gain of the backscattered signal varies when the antenna is moving in its aperture and the value of the function relates to the beam pointing directly. In this section, the impulse response function (IRF) will be used to investigate the impact of the varied beam pointing on the echo of the target *P*. After the signal is compensated by the reference function, the spectrum of the point target response can be expressed as

$$SS(\xi,\eta) = \sigma_P R\left(-\frac{\eta}{2b}\right) w'^2(\xi-\xi_d)\exp\left(-j\frac{4\pi}{\lambda}r\right) \\ \times \exp(-j\eta r)\exp(-j\xi x), \qquad (3)$$

where $b=\frac{2\pi B}{c}$ with B being the system bandwidth, and $\eta=\frac{4\pi f_r}{c}$ and $\xi=\frac{2\pi f_a}{v_s}$, and f_r and f_a are the counterparts of the fast time and slow time in the frequency domain, respectively, and the Doppler center $\xi_d=\tilde{\eta}\sin\phi_d$ with $\tilde{\eta}=\eta+\frac{4\pi}{\lambda}$ and its mean $\bar{\xi}_d=\frac{4\pi}{\lambda}\sin\phi_d$.

The antenna weighting function $w'^2(\xi-\xi_d)$ indicates the impact of the beam pointing. Ideally, $w'^2(\xi-\xi_d)$ is considered as a rectangle function theoretically, which simplifies the IRF. However, $w'^2(\xi-\xi_d)$ would derivate from the ideal rectangle function when the reference function cannot compensate the instability of the antenna beam completely. Hence, the SAR image quality would be affected by the varied beam pointing.

3.2 Ideal Case of Rectangle Antenna Weighting Function

The ideal case of rectangle antenna weighting function will be discussed in this section. Theoretically, it is assumed that the antenna pattern is rectangle, and it can be seen that the envelope of $SS(\xi,\eta)$ is a two-dimensional rectangle function. After an inverse, two-dimensional FFT is implemented on (3), and the IRF can be written as

$$yy(x',r') \approx \sigma_P \exp\left(-j\frac{4\pi}{\lambda}r\right)\exp\left(j\bar{\xi}_d(x'-x)\right) \\ \times \sin c\left(b(r'-r+\sin\phi_d(x'-x))\right)\sin c\left(a(x'-x)\right), \qquad (4)$$

where $a=\frac{2\pi}{D}$ representing the half of the azimuth spectral bandwidth. It can be seen that effect of the varied beam pointing on the IRF of the target can be neglected with the flat antenna pattern assumption. Furthermore, if the antenna beam is limited, which is often used to approximate the actual case, it means that the azimuth spectral will be truncated due to the mismatch of the Doppler center of the echo and that of the processor when the beam pointing varies. The effect on (4) presented as a new value of $\bar{\xi}_d$ and the shrink of a, which results in the change of the azimuth phase ramp, the shift of the range position for squint SAR, and the reducing azimuth resolution of IRF.

3.3 The Case of a Sinusoidal Weighting Function

A simple sinusoidal weighting function $w'^2(\xi - \xi_d)$ will be discussed. When the antenna beam swings, the amplitude of the echo of a target will be fluctuated due to the beam pointing history varying from that of the constant pointing in its synthetic aperture. A sinusoidal function $w'^2(\xi - \xi_d)$ is used to approximate the effect of the beam pointing swing history and the IRF will be investigated in this section.

With the sinusoidal assumption, $w'^2(\xi - \xi_d)$ is written as

$$w'^2(\xi - \xi_d) = \cos(\alpha(\xi - \xi_d))R\left(\frac{\xi - \xi_d}{2a}\right), \tag{5}$$

where α is used to represent the beam swing and the rectangle function $R(\frac{\xi-\xi_d}{2a})$ indicates that the bandwidth of the processor is $2a$.

Substituting (5) into (3), we have

$$\begin{aligned} SS(\xi, \eta) &= \sigma_P \cos(\alpha(\xi - \xi_d)) \\ &\times R\left(-\frac{\eta}{2b}\right)R\left(\frac{\xi - \xi_d}{2a}\right)\exp\left(-j\frac{4\pi}{\lambda}r\right) \\ &\times \exp(-j\eta r)\exp(-j\xi x). \end{aligned} \tag{6}$$

Implementing two-dimensional inverse Fourier transform, the IRF of the target can be written as

$$\begin{aligned} yy(x', r') \approx \frac{1}{2}\sigma_P \exp\left(-j\frac{4\pi}{\lambda}r\right)\exp\left(j\bar{\xi}_d(x' - x)\right) \\ \times \sin c\big(b(r' - r + \sin\phi_d(x' - x))\big) \\ \times \big(\sin c(a(x' - x - \alpha)) + \sin c(a(x' - x + \alpha))\big) \end{aligned} \tag{7}$$

It shows in (7) that there exist two peaks in the IMF of the target along the azimuth direction due to the swing of the beam. And the two peaks will separate more as the beam swings more drastically. If there are on swings in the synthetic aperture, we can see that the IRF will become the same as that in the ideal case.

The IRF of the target will be impacted by the varied antenna beam. We can see that the simple sinusoidal antenna beam swings will result in the deviation of the IRF from its normal expression. If the complicated movement of the beam pointing, it may be not easy to derive an analytic expression of the IRF and the numerical computation may be used.

4 Simulation Results

In this section, three cases of the beam pointing are simulated. The first case is a flat antenna pattern assumption with broadside looking being the nominal beam direction, which is a benchmark for the evolution of the results of another two cases. Another two cases consist of a rectangle antenna pattern assumption with a deviation of beam pointing from the nominal direction, and a sinusoidal weighting function $w'^2(\xi - \xi_d)$ assumption resulted from the varied beam pointing, respectively. The impacts of these beam pointing movements on the IRF of the target will be discussed.

The simulation results are presented in order to invalidate the proposed developments in this paper. The SAR parameters used in simulation are collected in Table 1. The radar is mounted on aircraft and operates at 6000 m altitude. The raw datasets are simulated.

Firstly, the ideal case of the flat antenna pattern with squint angle ϕ_d being zero is simulated, which can be considered as a benchmark for the other two cases. It is obvious that the movement of the beam pointing has no effect on the echo of the target. The IRF of the target is shown in Fig. 2, and it can be seen that the phase is flat in the main lobe of the IRF because the Doppler center in (3) is zero.

Secondly, the rectangle antenna pattern is assumed, and the squint angle ϕ_d is set to be zero. If the beam pointing keeps unchanged, the IRF of the target is the same as that in Fig. 2. In Sect. 3.2, it is discussed that the Doppler bandwidth would be truncated if the beam pointing deviates from its nominal direction. In order to investigate the effect of the varied beam pointing on the IRF, the deviation of 1°

Table 1 Sensor parameters used in the simulation

Carrier frequency antenna	3 GHz
Length in azimuth direction	1.0 m
Range delay	53 μs
Platform velocity	150 m/s
Pulse bandwidth	100 MHz
Pulse repetition frequency	600 Hz
Sampling frequency	200 MHz
Chirp duration	16 μs
Number of azimuth samples	6000
Number of range samples	6000

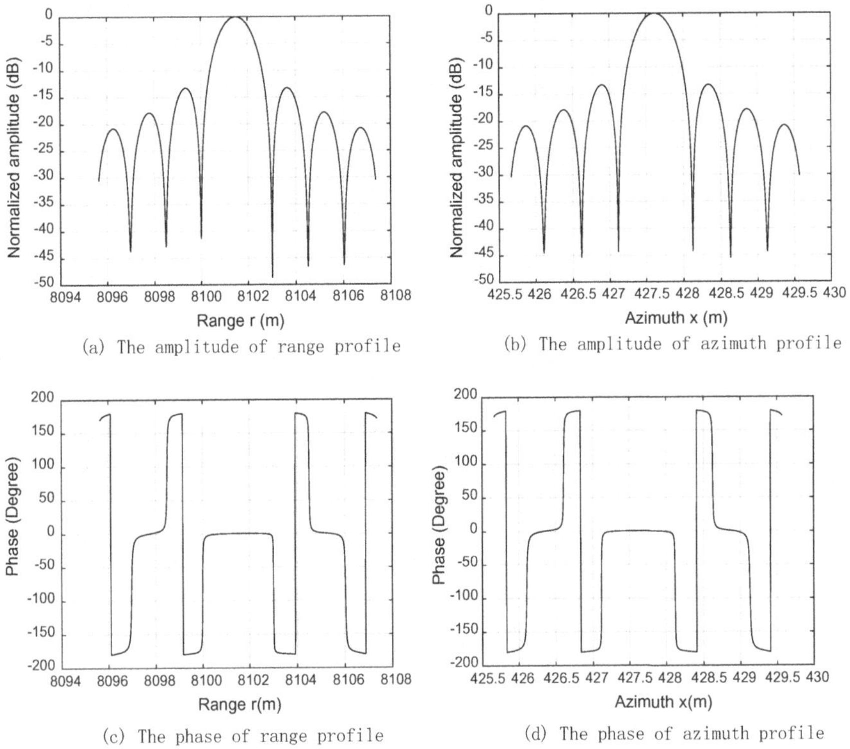

(a) The amplitude of range profile
(b) The amplitude of azimuth profile
(c) The phase of range profile
(d) The phase of azimuth profile

Fig. 2 IRF of the target in the ideal case

from the nominal broadside direction of the beam pointing is simulated. The IRF of the target is shown in Fig. 3. As can be seen, the resolution in azimuth direction is reduced in Fig. 3b since the Doppler bandwidth of the IRF is truncated, and there is a ramp in the phase in azimuth direction in Fig. 3d because the Doppler bandwidth has been shifted and the Doppler center $\bar{\xi}_d$ is shifted from zero too. If the beam pointing deviates from its nominal direction more, the azimuth resolution will be reduced more and the azimuth phase ramp becomes greater.

Thirdly, the sinusoidal weighting function resulted from the beam swinging is simulated and the nominal squint angle is set to be zero. In our simulation, the sinusoidal parameter α is selected to be 0.4. The presence of the disturbing term $\cos(\alpha(\xi - \xi_d))$ causes the deviation of the spectrum of the target from its ideal one. As can be seen from the results in Fig. 4, the envelope of the IRF of the target in range direction is almost unaffected and that in azimuth direction is affected seriously, where there are two separated peaks in Fig. 4b, which validates the analysis in Sect. 3.3. Besides the envelope, we also investigate the phases in the main lobe of IRF. It can be seen that the impact of the disturbing spectrum on the phase can be neglected. The separation of two peaks in azimuth direction results in the broadening of the azimuth main lobe actually and the flat phase has also been spread correspondingly.

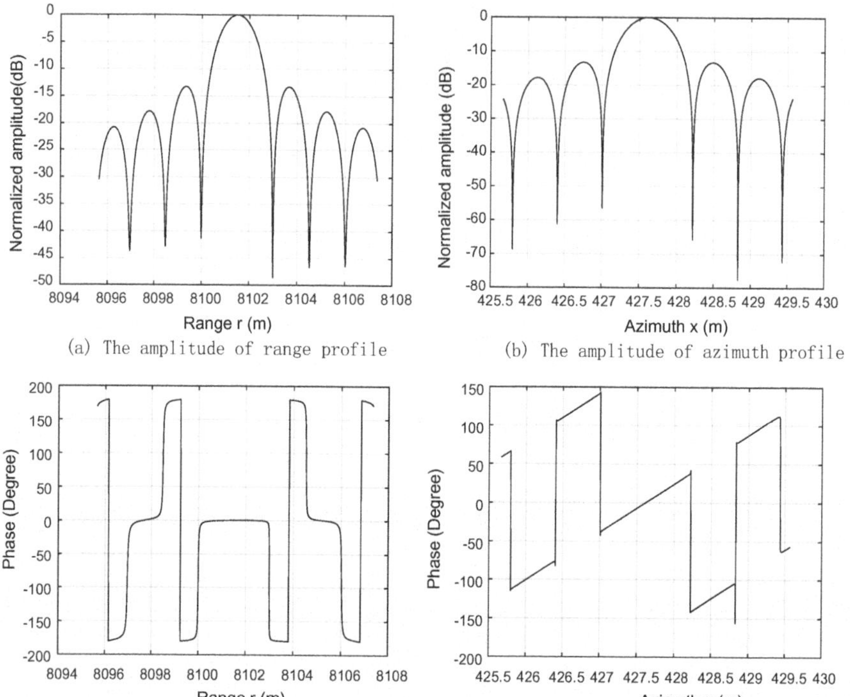

(a) The amplitude of range profile
(b) The amplitude of azimuth profile
(c) The phase of range profile
(d) The phase of azimuth profile

Fig. 3 IRF of the target in the case of the rectangle pattern assumption when there is a 1° deviation of the beam pointing from its broadside looking direction

In our simulation, we also found that it is hard to observe the separated peaks if α is less than 0.2, but the mainlobe is broadened due to the superposition of two near peaks. And the two peaks are separated totally if α is greater than 0.6. The above results illustrate that the beam swinging can result in the degradation of the SAR image, and the SAR image will be difficult to be interpreted if there exists drastic beam swinging during the data acquisition.

5 Conclusion

The varied antenna beam pointing on SAR image is studied in this paper. The echo of SAR signal is discussed and the effect of the varied beam pointing on the SAR image is analyzed in the two-dimensional frequency domain. The theoretical derivations and the simulation results illustrate that both the amplitude and the phase of the IRF will be distorted by the varied beam pointing, which is helpful for evaluating how the varied sensor attitude impacts the application of the high-resolution SAR image.

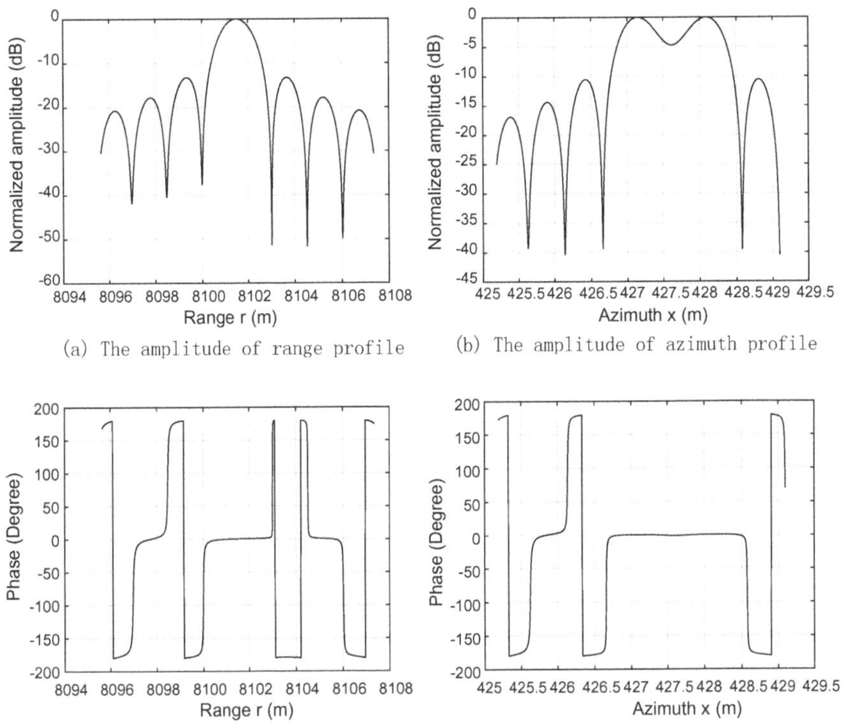

(a) The amplitude of range profile (b) The amplitude of azimuth profile

(c) The phase of range profile (d) The phase of azimuth profile

Fig. 4 IRF of the target in the case of the sinusoidal weighting function assumption with $\alpha = 0.4$

Acknowledgements This work is supported by the National Natural Science Foundation of China (Grant No. 41301487).

References

1. Oliver C, Quegan S (2004) Understanding synthetic aperture radar images, chap 11–13. SciTech Publishing, Raleigh
2. Fornaro G, Franceschetti G, Perna S (2005) Motion compensation errors: effects on the accuracy of airborne SAR images. IEEE Trans Aerosp Electron Syst 41(4):1338–1352
3. Franceschetti G, Lanari R, Marzouk E-S (1996) A new two-dimensional squint mode SAR processor. IEEE Trans Aerosp Electron Syst 32(2):854–863
4. Cumming IG, Li S (2006) Improved slope estimation for SAR Doppler ambiguity resolution. IEEE Trans Geosci Remote Sens 44(3):707–718
5. Fornaro G, Sansosti E, Lanari R, Tesauro M (2002) Role of processing geometry in SAR raw datafocusing. IEEE Trans Aerosp Electron Syst 38(2):441–454

Data-Driven Fault Detection Methods for Jilin-1 Satellite Attitude Control System

Zhi Qu, Kai Xu, Yanhao Xie, Feng Li, Mengmeng Liu, Shuangxue Han, Jiebing Liu, and Zhigang Chen

Abstract This paper studies the data processing and fault detection methods of principal component analysis (PCA) and local linear embedding (LLE) for the Jilin-1 satellite telemetry data of attitude control system (ACS). Aiming at the problem that traditional data dimensionality reduction methods cannot extract key features of nonlinear multi-dimensional telemetry data, the two methods are applied to the feature extraction of ACS telemetry data of Jilin-1 satellite. Aiming at the problem that the time-varying and multi-scale of telemetry data in the Jilin-1 satellite mission mode leads to the failure rate of fault diagnosis. In combination with the statistical SPE and T^2, design a data processing and fault detection method to obtain low-dimensional key features. Finally, two methods are verified by telemetry data in different mission modes of Jilin-1 satellite. The effectiveness of the two methods of remote sensing data mining method is compared. The results point out that the method can significantly improve the fault detection capability of ACS of Jilin-1 satellite.

Keywords Jilin-1 satellite · Telemetry data · Principal component analysis · Local linear embedding · Fault detection

1 Introduction

On October 7, 2015, Jilin-1 has been launched at the Satellite Center in Jiuquan. Up to now, Chang Guang Satellite Technology Co. LTD [1] has launched a total of ten satellites, including eight high-resolution video satellites, one high-resolution optical satellite, and one technical verification satellite. In 2018, Chang Guang Satellite

Z. Qu · K. Xu (✉)
Changchun Institute of Optics, Fine Mechanics and Physics, Chinese Academy of Sciences, Changchun 130033, China
e-mail: xukai118@126.com

University of Chinese Academy of Sciences, Beijing 100049, China

Z. Qu · K. Xu · Y. Xie · F. Li · M. Liu · S. Han · J. Liu · Z. Chen
Chang Guang Satellite Technology Co. LTD, Changchun 130102, China

L. Wang et al. (eds.), *Proceedings of the 6th China High Resolution Earth Observation Conference (CHREOC 2019)*, Lecture Notes in Electrical Engineering 657,
https://doi.org/10.1007/978-981-15-3947-3_34

Technology Co. LTD [1] will continue to launch nearly ten satellites to improve the acquisition of constellation data and improve its external service capabilities [2].

The Jilin-1 satellite attitude control system (ACS) achieves that the satellite changes its attitude slightly by the instructions, therefore, to achieve the satellite normal orbit flight. In the previous long time, the system which can control the satellite attitude will be used because it has a good reliability as the fundamental advantages for the whole satellite normal flight [3, 4]. According to the database statistics from satellite fault, highest possibility of failure can happen in attitude, the power systems, or orbit which can arrive at 23% [5]. After the orbital instructions, the Jilin-1 satellite will collect and the ground will receive a lots of telemetry information. The data can indicate the situation of operation and payload. Then extracting the lower-dimensional characteristic information from the higher-dimensional telemetry information, it will have an improvement in the capability and reliability of satellite's abnormal detection. In Reference [6], authors found that neurophysiological functions can be used to describe neural trigger rates, such as the angle and direction of the eye. In the same year, in the Journal of Science, another article published by Tenenbaum et al. In Reference [7], authors proposed the Isometric Mapping (ISOMAP) manifold learning method to shorten the dimension from nonlinear data in which the geodesic distance is used to replace Euclidean distance to obtain the neighborhood graph. In Reference [8], authors proposed a local linear embedding (LLE) manifold algorithm, which preserves the original data neighbor relationship when acquiring low-dimensional coordinates. It is the emergence of LLE and ISOMAP that the manifold learning has developed rapidly. In 2002, Silva gave the definition of manifold learning, and from then on, manifold learning really became a research direction of machine learning [9]. At the International Conference on Machine Learning held in 2004, manifold learning attracted the attention of many scholars. With the advancement of machine learning, research on manifold learning has been further deepened, and many research results have emerged at home and abroad. In foreign countries, such as the Laplace mapping method (LE) proposed by Belkin, it solves the problem of data feature extraction and dimensionality reduction when data sampling is uneven or the data itself has holes. In Reference [10], authors proposed a Hessian Eigen map (HE) method. In the application of fault diagnosis, manifold learning mostly stays in the field of signal noise reduction and data reduction, and there are few fault diagnoses documents. Although the application of manifold learning in fault diagnosis is still immature, its research value cannot be underestimated as a forward-looking research direction. In addition, the development direction after the fault diagnosis date also points the direction for the promotion of the manifold learning method, which is as follows: First, the nonlinear relationship between the observation data and the running state of the object. Second, the difficulty of data processing caused by high-dimensional data. Third, the ability of the fault diagnosis method to generalize the data. In Reference [11], authors analyze LLE-based satellite fault detection, only for fault detection in stable mode, and does not demonstrate the fault extraction effect in maneuver mode. In References [12, 13], aim to finish the fault detection in the satellite ACS, the sliding mode controller and parameter estimation identification methods are introduced, respectively. The control method [14–17] is not universal in the field of satellite fault detection and requires more accurate control system models. Parameter estimation can analyze and estimate model parameters to detect faults. However, it still has the problem of long delay time for fault detection. In Reference [18], machine learning and data extracting methods are introduced. The method can find the abnormal parts from the telemetry data, and it is used the dynamic Bayesian network to complete the deviation diagnosis. In Reference [19], authors summarized some methods such as mining and detection of satellite telemetry data. In Reference [20], firstly, the satellite telemetry data is collected according to the principal component analysis (PCA) method, and its characteristic information is extracted. The fault detection is completed by reducing the dimension and combining the vector machine (SVM). Telemetry data of the satellite ACS transmitted under orbit reflects the overall state of the control system [21]. By analyzing the telemetry data of the ACS, it is possible to determine whether a fault occurs, that is, effective fault detection. However, data-driven satellite attitude fault detection is still in its infancy, and technology development is not mature and stable, and many fault detection technologies are still needed.

With the rise of artificial intelligence and data mining, satellite telemetry data has a very important application status as big data. In this paper, the telemetry data of the satellite attitude control system is used to reduce the dimensional data, extract key features, and quickly and accurately detect the failure of the attitude control system.

2 Theoretical Derivation of Algorithms

2.1 PCA

PCA is a statistical method of multi-dimensional data, and it has very broad scientific and application values [22]. The basic idea is to linearly combine new variables through the original variables in high-dimensional data [23]. These new variables can focus on the information contained in the original variables. The algorithm can be described as follows:

Step 1: There are N sets of samples, m parameters, and the sample values are composed

$$X = \begin{bmatrix} x_{11} & x_{12} & \cdots & x_{1m} \\ x_{21} & x_{22} & \cdots & x_{2m} \\ \vdots & \vdots & \ddots & \vdots \\ x_{N1} & x_{N2} & \cdots & x_{Nm} \end{bmatrix} \tag{1}$$

Step 2: Calculate the mean $\bar{x}_i$ of each parameter and the standard deviation s_i.

$$\bar{x}_i = \sum_{j=1}^{N} x_{ji}, \quad s_i = \sqrt{\frac{\sum_{j=1}^{N} (x_{ji} - \bar{x}_i)^2}{N-1}} \tag{2}$$

Step 3: Calculate the matrix $R = [r_{ij}]$.

$$r_{ij} = \frac{\sum_{k=1}^{N} Y_{ki} Y_{kj}}{N-1}, \quad Y_{ij} = \frac{x_{ij} - \bar{x}_i}{s_i} \tag{3}$$

Step 4: Calculate the matrix R eigenvalue $\lambda_1, \lambda_2, \lambda_3, \ldots, \lambda_m$ and eigenvector $L_i = [l_{1i}, l_{2i}, \ldots, l_{pi}]^{\mathrm{T}}$.
Step 5: Looking for the main ingredient.

In the above-described characteristic value $\lambda_1, \lambda_2, \lambda_3, \ldots, \lambda_m$, an n value satisfying the following condition is calculated,

$$C_n = \frac{\sum_{i=1}^{n} \lambda_i}{\sum_{j=1}^{m} \lambda_j} \geq \sigma \tag{4}$$

where σ is the confidence level, C_n is the cumulative contribution rate, thus, n principal component components are determined, which can be expressed as follows.

$$Z_i = \sum_{j=1}^{m} l_{ji} Y_j, \; (i = 1, 2, \ldots, n) \tag{5}$$

In the actual process of satellite attitude control, it should be noted that the satellite attitude control system telemetry data parameters have different dimensions, such as pitch angle and pitch angular velocity, to eliminate noise interference and the effects of different dimensions, before applying the PCA to extract feature information, not only the pre-processing mentioned above needs to be performed on the data but also the parameters need to be standardized. The standardization criteria are as follows,

$$\tilde{X} = [X - (I, I, \ldots, I)^{\mathrm{T}} \bar{X}] D^{-1/2} \tag{6}$$

where $D = \mathrm{diag}(\delta_1^2, \delta_2^2, \ldots, \delta_m^2)$ is variance matrix and $\bar{X} = [\bar{x}_1, \bar{x}_2, \ldots, \bar{x}_m]$ is mean vector. According to the above algorithm description, it can be known that the normalized data $\tilde{X}$ can be decomposed into several principal component representations, as follows,

$$\tilde{X} = t_1 L_1 + t_2 L_2 + \cdots + t_n L_n + E = T L^{\mathrm{T}} + E \tag{7}$$

where t_i denotes the principal element score vector, L_i principal component load vector, and E model error [24, 25].

2.2 LLE

LLE is a typical nonlinear data processing method. On the one hand, it maintains the original topology of the original data, and on the other hand, it implements multi-dimensional data reduction [26]. LLE is different from other dimensionality reduction methods. It uses local linear fitting to describe the low-dimensional key features of multi-dimensional data, thereby achieving low-dimensional reconstruction of nonlinear structures. After the local linear embedded data is reduced, the reduced-dimensional data is recombined and its inherent key features are represented globally.

Let the $X = \left\{x_i \in R^D,\ i = 1, 2, \ldots, N\right\}$ be multi-dimensional data input space, the $Y = \left\{y_i \in R^d,\ i = 1, 2, \ldots, N\right\}$ be multi-dimensional data output space, and dimension of data is $d < D$.

LLE has 3 steps as follows.

Step 1: Searching for nearby points

First, the sample points in the initial data x_i are statistically listed, the Euclidean distance calculation is performed between any sample point in the list and other sample points, and then, the distance values are compared in a list. The minimum distance k from the target point is defined as the nearby point x_{ij}.

Step 2: Constructing weight matrix

The weight matrix W is constructed by calculating the distance between the nearby points of the sample points x_i, and the formula for minimizing the construction error is as follows

$$\min \varepsilon(W) = \left\| x_i - \sum_{j=1}^{k} w_{ij} x_j \right\|^2 \tag{8}$$

The minimum optimization problem has two constraints: on the one hand, the construction of the matrix must be completed by the calculation of neighboring points, $w_{ij} = 0$; on the other hand, the vectors of the weight matrix W satisfy $\sum_{j=1}^{k} w_{ij} = 1$, the weight point of the j neighborhood point $x_{i(j)}$ of the sample point x_i is w_{ij}.

From a geometric perspective, it is understood that the constraint condition $\sum_{j=1}^{k} w_{ij} = 1$ has translation invariance, that is, the arbitrary addition of a vector c to the satellite telemetry data sample point x_i and its neighboring points will not affect its reconstruction error.

$$\begin{aligned} x_i + c - \sum_{j=1}^{k} w_{ij}(x_j + c) &= x_i + c - \sum_{j=1}^{k} w_{ij}x_j - \sum_{j=1}^{k} w_{ij}c \\ &= x_i + c - \sum_{j=1}^{k} w_{ij}x_j - c \\ &= x_i - \sum_{j=1}^{k} w_{ij}x_j \end{aligned} \tag{9}$$

Solve its constraint solution $\min\varepsilon(W) = \left\| x_i - \sum_{j=1}^{k} w_{ij}x_j \right\|^2$, assuming $c = -x_i$, so

$$\begin{aligned} \left\| x_i - \sum_{j=1}^{k} w_{ij}x_j \right\|^2 &= \left\| \sum_{j=1}^{k} w_{ij}(x_j - x_i) \right\|^2 \\ &= \left\| \sum_{j=1}^{k} w_{ij}z_j \right\|^2 = w_i^{\mathrm{T}} zz^{\mathrm{T}} w_i \end{aligned} \tag{10}$$

In Eq. (10), $z_j = x_j - x_i$, zz^{T} is $k \times k$ symmetric matrix, w_i is a $k \times 1$ vector, called a Gram matrix, written as G_i. The Gram matrix contains the inner product of all neighboring points, which is transformed into solving the minimum problem,

$$\mathrm{RSS}_i = \mathbf{w}_i^{\mathrm{T}} G_i \mathbf{w}_i \tag{11}$$

According to $\sum_{j=1}^{k} w_{ij} = 1$, there is (12).

$$L(w_i, \lambda) = w_i^{\mathrm{T}} G_i w_i - \lambda\left(I^{\mathrm{T}} w_i - I\right) \tag{12}$$

Finding the partial derivative of G_i gives

$$\begin{aligned} \frac{\partial}{\partial w_i} L(w_i, \lambda) &= 2G_i w_i - \lambda I = 0 \\ \frac{\partial}{\partial \lambda} L(w_i, \lambda) &= I^{\mathrm{T}} w_i - I = 0 \end{aligned} \tag{13}$$

That is $G_i w_i = \lambda I/2$, where the construction factor λ is adjustable and $L(w_i, \lambda)$ is the minimum equation.

If $k > D$, in the case where the neighborhood point dimension k is greater than the telemetry data sample dimension D, an infinite group of solutions will appear when solving the least squares, then the reconstruction factor λ needs to be adjusted, and the optimization problem can be solved:

$$\left\| x_i - \sum_{j=1}^{k} w_{ij} x_j \right\|^2 + \alpha \sum_{j=1}^{k} w_{ij}^2 \tag{14}$$

where $\alpha > 0$, the calculation steps are the same as mentioned above.

$$L = w_i^{\mathrm{T}} G_i w_i + \alpha w_i^{\mathrm{T}} w_i - \lambda\left(I^{\mathrm{T}} w_i - I\right) \tag{15}$$

Then, there is

$$\begin{aligned} 2G_i w_i + 2\alpha w_i - \lambda I &= 0 \\ 2(G_i + \alpha I) w_i &= \lambda I \end{aligned} \tag{16}$$

We solve the weight matrix W.

Step 3: Calculating dimensionality reduction coordinates

With the weights kept constant, initial data points are constructed in the reduced-dimensional space, while minimizing construction errors.

$$\min \delta(Y) = \sum_{i=1}^{N} \left\| y_i - \sum_{j=1}^{k} w_{ij} y_j \right\|^2 \tag{17}$$

If the theory is to be true, the following constraints are required

$$\begin{aligned} &\sum_{i=1}^{N} y_i = 0 \\ &1/N \sum_{i=1}^{N} y_i y_i^{\mathrm{T}} = I \end{aligned} \tag{18}$$

where I represents the $d \times d$ unit matrix.

$$\begin{aligned} \sum_{i=1}^{N} \left\| y_i - \sum_{j=1}^{k} w_{ij} y_j \right\|^2 &= Y^{\mathrm{T}} Y - Y^{\mathrm{T}}(WY) - (WY)^{\mathrm{T}} Y + (WY)^{\mathrm{T}}(WY) \\ &= ((I - W)Y)^{\mathrm{T}}((I - W)Y) \\ &= Y^{\mathrm{T}}(I - W)^{\mathrm{T}}(I - W)Y \end{aligned} \tag{19}$$

Define the "cost matrix" $M = (I - W)^{\mathrm{T}}(I - W)$, that is $\sum_{i=1}^{N} \left\| y_i - \sum_{j=1}^{k} w_{ij} y_j \right\|^2 = Y^{\mathrm{T}} M Y$, because $1/N \sum_{i=1}^{N} y_i y_i^{\mathrm{T}} = I$, using the Lagrange multiplier factor:

$$L(Y,\mu) = Y^{\mathrm{T}}MY - \mu\left(N^{-1}Y^{\mathrm{T}}Y - I\right) \tag{20}$$

$MY = \frac{\mu}{N}Y$ can be obtained, so Y is the eigenvector of M, μ is multiplier factor, and $L(Y,\mu)$ is the minimum value of the equation containing Y and μ. In summary, the dimensionality reduction coordinates can be calculated as $M = (I-W)^{\mathrm{T}}(I-W)$, and it will obtain the feature vector Y.

3 Fault Detection Methods

The research object of this chapter is the telemetry data of Jilin-1 satellite ACS [1]. The LLE algorithm and the PCA algorithm were used to extract the dimensionality-reduced data feature information of the multi-dimensional telemetry data, and the statistics *SPE* and T^2 were introduced into the evaluation estimation scheme [1]. The abnormal changes in the statistics curve were used as the basis for detecting the fault [27, 28].

3.1 SPE *Statistic*

The calculation of *SPE* statistic is shown below.

$$SPE_{\text{new}} = \left\|x_{\text{new}} - \hat{x}_{\text{new}}\right\|^2 \tag{21}$$

where $\hat{x}_{\text{new}} = A^{\mathrm{T}}Ax_{\text{new}}$ is defined as the estimated construction weight of the new sample point x_{new}.

The theoretical formula of *SPE* statistic for the control limit is shown below.

$$SPE_{\alpha} = \theta_1\left[\frac{c_{\alpha}\sqrt{2\theta_2 h_0^2}}{\theta_1} + 1 + \frac{\theta_2 h_0(h_0-1)}{\theta_1^2}\right]^{1/h_0} \tag{22}$$

In Eq. (22), λ_j is eigenvalues of covariance matrices; $h_0 = 1 - 2\theta_1\theta_3/(3\theta_2^2)$, $\theta_i = \sum_{j=i+1}^{m}\lambda_j^i$, $i = 1, 2, 3$, c_{α} is normally distributed value of the confidence interval c. When SPE_{new} is less than the control limit SPE_{α}, there is no abnormality.

3.2 T² *Statistic*

The formula for calculating T^2 statistic for online update sample values is shown below.

$$T_{\text{new}}^2 = y_{\text{new}}^{\text{T}} \Lambda^{-1} y_{\text{new}} \tag{23}$$

where Λ is the covariance matrix in the mathematical expression of the sample training data, described as $\Lambda = \frac{YY^{\text{T}}}{N-1}$.

The mathematical expression of the control limit of T^2 statistic is

$$T_\alpha^2 = \frac{\alpha\left(N^2 - 1\right)}{N(N - \alpha)} F_\alpha(\alpha, N - \alpha) \tag{24}$$

In Eq. (24), α is the confidence level, and F distribution function of $N - \alpha$ is $F_\alpha(\alpha, N - \alpha)$; When T_{new} is less than the control limit T_α, there is no abnormality.

The LLE and PCA algorithms were used to reduce the dimensionality of the satellite ACS's telemetry data to obtain reduced-dimensional coordinates, to obtain a mapping matrix of multi-dimensional data coordinates to reduced-dimensional data coordinates, and finally to perform fault detection tasks based on whether the probability statistical results exceeded the control limit is shown in Fig. 1.

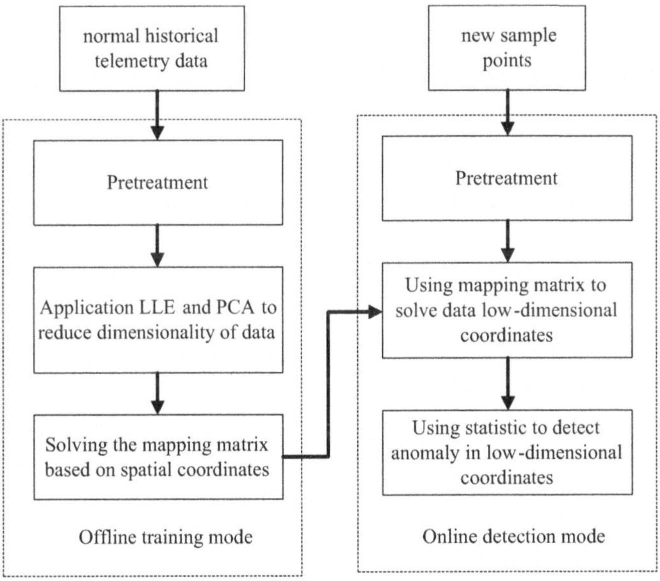

Fig. 1 Fault detection algorithm process chart

4 Fault Detection Visualization

4.1 Telemetry Data Preprocessing

Jilin-1 satellite ACS has many telemetry parameters, that is, many dimensions, and a large amount of telemetry data. In the process of receiving telemetry data, a lot of noise will be introduced, which will negatively affect the extraction and processing results of data features. Therefore, the filtering of the initial data needs to be performed before the data is reduced.

The theoretical formula of a zero-order digital filter is shown below.

$$\begin{aligned} y_1(n) &= x(n) * h(n) \\ y_2(n) &= y_1(N-1-n) \\ y_3(n) &= y_2(n) * h(n) \\ y_4(n) &= y_3(N-1-n) \end{aligned} \tag{25}$$

In Eq. (25), $x(n)$ is the input signal, $h(n)$ is the impulse response signal, and $y_4(n)$ is the output-filtered signal. In order to verify the filtering effect, a zero-phase digital filter was applied to the noise removal of the telemetry data of the satellite ACS, and the telemetry data was preprocessed: outlier removal, sequence smoothing, and uniform numerical scale [1]. The telemetry data before and after preprocessing are shown in Fig. 2.

Figure 2a is the original signal after the unified dimension, and Fig. 2b is the pre-processed data. The telemetry parameters are flywheel torque, attitude angular velocity, fiber optic gyro angular velocity and magnetic torque current.

4.2 Visualization of Algorithmic Applications

If the telemetry data of the satellite ACS has a constant deviation, it is that the attitude sensor drifts, a constant fault occurs in 530–550 points of the angular velocity in the telemetry data. The fault form is $\dot{\theta}_{\text{out}} = \dot{\theta}_0 + \Delta\dot{\theta}_f$. $\Delta\dot{\theta}_f$ is a constant value of the fault deviation, and the magnitude is about 0.26 °/s. The pre-processed data and the data obtained after applying the PCA and LLE algorithms to reduce the dimensionality are shown in Fig. 3.

In order to quantitatively describe the fault diagnosis accuracy of the two algorithms, the accuracy rate A of the fault diagnosis is defined.

$$A = \sqrt{\frac{\sigma_{T^2}^2 + \sigma_{SPE}^2}{2}} \tag{26}$$

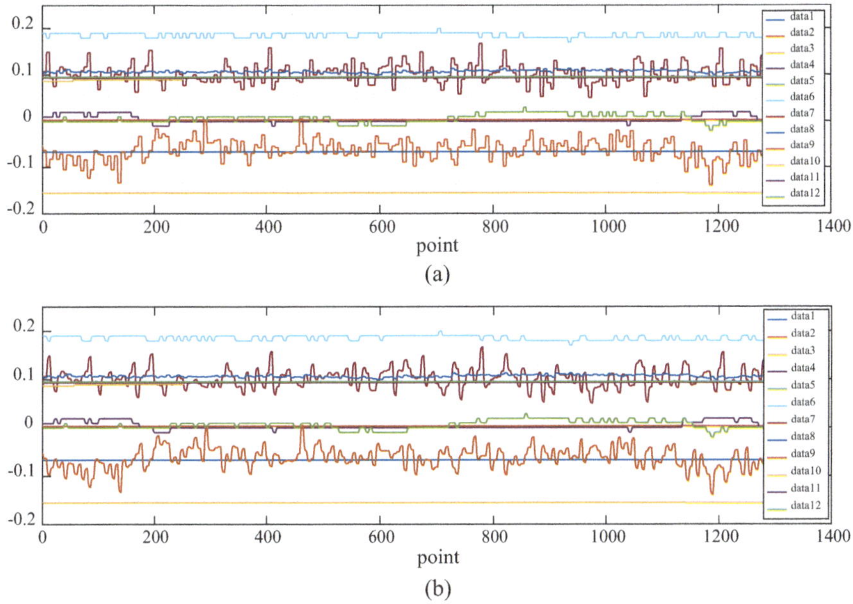

Fig. 2 High-dimensional data: **a** unfiltered telemetry: 1, 2, 3—flywheel torque; 4, 5, 6—attitude angular velocity; 7, 8, 9—fiber optic gyro angular velocity; 10, 11, 12—magnetic torque current; **b** filtered telemetry: 1, 2, 3—flywheel torque; 4, 5, 6—attitude angular velocity; 7, 8, 9—fiber optic gyro angular velocity; 10, 11, 12—magnetic torque current

In Eq. (26), σ_{T^2} and σ_{SPE}, respectively, indicate the accuracy of fault diagnosis using T^2 and *SPE* statistical monitoring charts,

$$\begin{cases} \sigma_{T^2} = 1 - (\eta_{f \cdot T^2} + \eta_{o \cdot T^2}) \\ \sigma_{SPE} = 1 - (\eta_{f \cdot SPE} + \eta_{o \cdot SPE}) \end{cases} \tag{27}$$

In Eq. (27), $\eta_{\mathrm{f}} = \frac{n}{N}$, $\eta_{\mathrm{o}} = \frac{m}{N}$, η_{f} is false alarm rate (FAR), η_{o} is missing alarm rate (MAR), n is false positive sample points, m is missing sample points, and N is total fault sample points.

Under constant deviation, the detection results of statistic T^2 and *SPE* are shown in Fig. 4. In order to quantify the effect of satellite ACS fault detection, Table 1 gives different statistics of FAR and MAR under constant deviation. The statistical results show that the quantitative statistical methods can meet the engineering application of fault detection [1].

As can be seen from Table 1, the accuracy of PCA compared to LLE dimensionality reduction data detection is slightly lower. Contrastive analysis can find out the reason: PCA is used to data covariance structure. Once the principal component

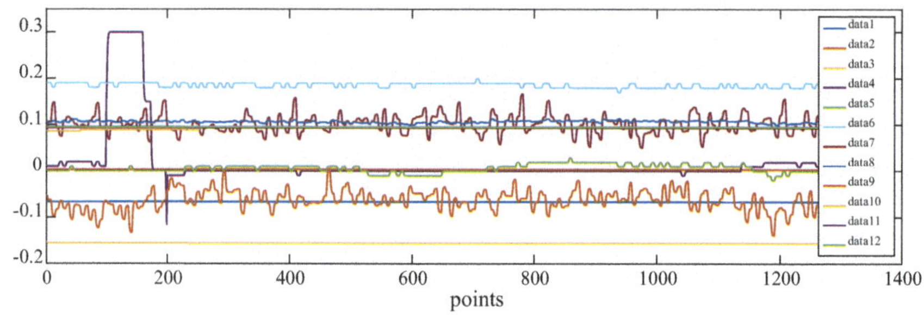

(a)

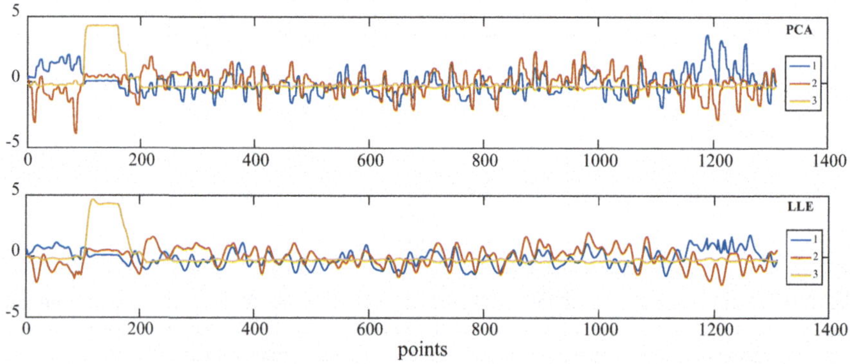

(b)

Fig. 3 Pre-processed data and reduced-dimensional results under constant deviation: **a** pre-processed data under constant deviation: 1, 2, 3—flywheel torque; 4, 5, 6—attitude angular velocity; 7, 8, 9—fiber optic gyro angular velocity; 10, 11, 12—magnetic torque current. **b** Reduced-dimensionality results under constant deviation

model is established, it will not change, but the attitude ACS telemetry data is time-varying. The method of PCA is essentially a linear transformation, and there are inherent deficiencies in dealing with nonlinear problems.

5 Summary

Aiming at the problem that it is difficult to analyze the failure detection of satellite ACS with nonlinear multi-dimensional data, the data dimensionality reduction and key features based on LLE and PCA algorithms are studied. Taking the telemetry data of Jilin-1 satellite ACS as the research object, different algorithms are applied to find out the key features of multi-dimensional data, and the probability detection *SPE* and T^2 are used to design a satellite attitude control system fault detection scheme. Finally, the simulation of the telemetry data of Jilin-1 satellite was visualized, and the

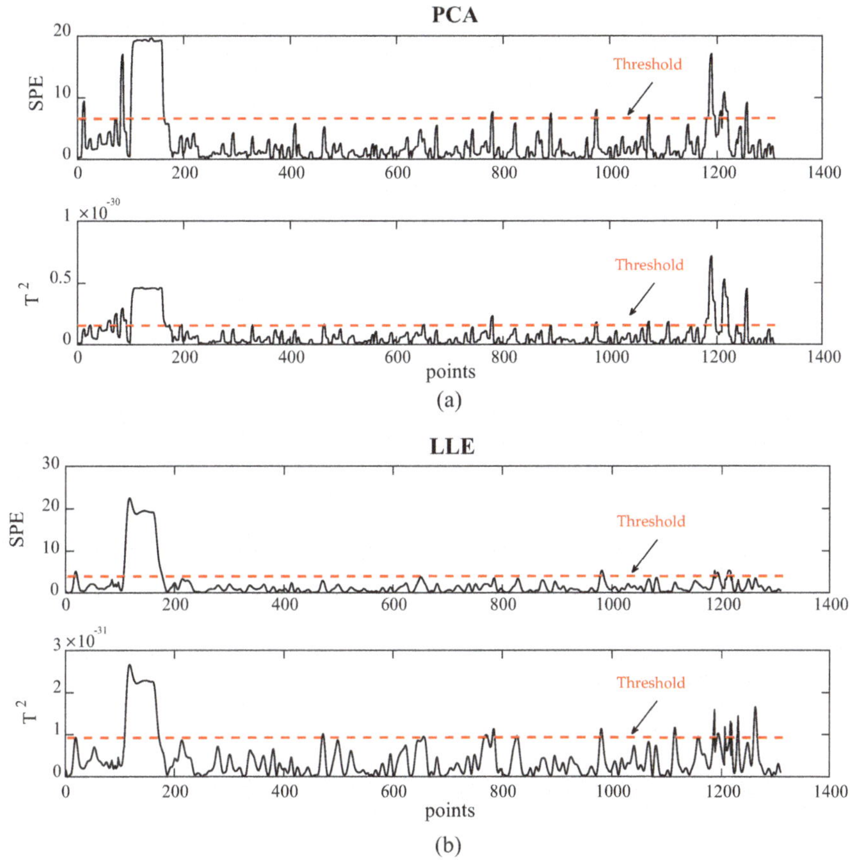

Fig. 4 Test results of T^2 and SPE under constant deviation: **a** PCA. **b** LLE

Table 1 Comparison of algorithm accuracy

Algorithm	Statistics	FAR (%)	MAR (%)	Accuracy (%)
PCA	T^2	3.28	13.11	84.43
	SPE	6.56	8.20	
LLE	T^2	4.92	6.56	93.57
	SPE	1.64	0	

results verified the effectiveness of the two schemes. The results show that the LLE algorithm improves the accuracy by 9.14% compared with the PCA fault detection method. This method effectively improves the fault detection capability of Jilin-1 satellite attitude control system, and has potential engineering application value.

Acknowledgements This work was financially supported by the Jilin Province Science and Technology Development Plan Project [29], number: 20170204069GX.

References

1. Zhi Q, Kai X, Gang CZ, Xin H, Hao XY, Meng LM, Feng L, Xue HS (2019) Fault detection method of Luojia1-01 satellite attitude control system based on supervised local linear embedding. IEEE Access
2. Information on http://www.charmingglobe.com/
3. Lei ZS, Chun ZY (2008) SVM classifier based fault diagnosis of the satellite attitude control system. In: International conference on intelligent computer technology and automation
4. Ding X, Guo L, Jeisch T (1999) A characterization of parity space and its application to robust fault detection. IEEE Trans Autom control 144(2):337–342
5. Su L, Zhao Y (2010) Fault diagnosis of navigation satellite attitude control system based on data-driven combined with artificial intelligence. In: China satellite navigation conference
6. Seung H, Lee D (2000) The manifold ways of perception. Science 290(5500):2268–2269
7. Tenenbaum JB, Silva VD, Langford JC (2000) A global geometric framework for nonlinear dimensionality reduction. Science 290(5550):2319–2323
8. Roweis S, Saul L (2000) Nonlinear dimensionality reduction by locally linear embedding. Science 290(5500):2323–2326
9. Silva VD, Tenenbaum JB (2003) Global versus local methods in nonlinear dimensionality reduction. Neural Inf Process Syst 705–712
10. Donoho D, Grimes C (2003) Hessian eigenmaps: new locally linear embedding techniques for high-dimensional data. Proc Natl Acad Sci 100(10):5591–5596
11. Geng X, Zhan DC, Zhou ZH (2005) Supervised nonlinear dimensionality reduction for visualization and classification. IEEE Trans Syst Man Cybern 35(6):1098–1107
12. Meskin N, Khorasani K (2007) Fault detection and isolation in a redundant reaction wheels configuration of a satellite. In: IEEE international conference on systems, man and cybernetics, pp 3151–3158
13. Jiang T, Khorasani K, Tafazoli S (2008) Parameter estimation-based fault detection, isolation and recovery for nonlinear satellite models. IEEE Trans Control Syst Technol 16(4):799–808
14. Tudoroiu N, Khorasani K, Tehrani ES (2006) Interactive bank of unscented Kalman filters for fault detection and isolation in reaction wheel actuators of satellite attitude control system. In: IEEE conference on industrial electronics, pp 264–269
15. Jiang B, Wang JL, Soh YC (2002) An adaptive technique for robust diagnosis of faults with independent effects on system outputs. Int J Control 75(11):792–802
16. Jiang B, Staroswiecki M, Cocquempot V (2006) Fault accommodation for nonlinear dynamic systems. IEEE Trans Autom Control 51(9):1578–1583
17. Wang T, Cheng YH, Jiang B, Qi RY (2014) Fault detection based on finite impulse response adaptive filter for satellite attitude control systems. In: Control and decision conference, pp 209–213
18. Yairi T, Kawahara Y, Fujimaki R, Sato YC, Machida K (2006) Telemetry mining: a machine learning approach to anomaly detection and fault diagnosis for space system. In: IEEE international conference on space mission challenges for information technology, pp 469–476
19. She YT, Bin C, Yu G, Hua FJ, Long ZH, Le WX (2013) Data mining-based fault detection and prediction methods for in-orbit satellite. In: International conference on measurement, information and control, pp 805–808
20. Gao Y, She YT, Nan X, Qiang XM (2012) Fault detection and diagnosis for spacecraft using principal component analysis and support vector machines. In: Industrial electronic and applications conference, pp 1984–1988

21. Fujimaki R, Yairi T, Machida K (2005) An approach to spacecraft anomaly detection problem using kernel feature space. In: Knowledge discovery and data mining conference, pp 401–410
22. Lu B, Zhao Y, Mao Z (2009) Fault diagnosis method based on moving window PCA. In: Chinese control and decision conference. IEEE Press, China, pp 185–188
23. Facco P, Bezzo F, Barolo M (2010) Nearest-neighbor method for the auto-mastic maintenance of multivariate statistical soft sensors in batch processing. Ind Eng Chem Res 49(5):2336–2347
24. Cheng CY, Hsu CC, Chen MC (2010) Adaptive kernel principal component analysis (KPCA) for monitoring small disturbances of nonlinear processes. Ind Eng Chem Res 49(5):2254–2262
25. Lee DS, Vanrolleghem PA (2004) Adaptive consensus principal component analysis for on-line batch process monitoring. Environ Monit Assess 92(1–3):119–135
26. Zheng L, Guang J, Han TS (2016) Fluctuation feature extraction of satellite telemetry data and on-orbit anomaly detection. In: 2016 prognostics and system health management conference (PHM-Chengdu). IEEE
27. Dick R, Olga K, Oleg O, Matti P, Robert D (2003) Supervised locally linear embedding. In: Proceedings of artificial neural networks and neural information processing, vol 2714, pp 333–341
28. Wang T, Cheng Y, Jiang B (2014) Feature extraction and fault detection based on telemetry data for Satellite TX-I. In: Proceedings of 2014 IEEE Chinese guidance, navigation and control conference. IEEE
29. Zhi Q, Kai X, Xin H, Chao LL (2018) Design of wind pendulum control system based on improved genetic PID algorithm. IOP Conf Ser Mater Sci Eng

A Novel Framework of Artificial Intelligent Geologic Hazards Detection Over Comprehensive Remote Sensing

Yueying Zhang, Haonan Ran, Yuexing Peng, and Yu Zheng

Abstract National reports on the total economic loss of geologic hazards demonstrate that landslides have serious negative economic impacts, especially in southwest of China. Most of geologic hazards are widely distributed, highly covert, abrupt, and devastating, which imposes more challenges on landslide detection in the early stage. Detection of slow landslides movement and prediction of the tendency of landslides in mountain terrains have great potential for crisis area targeting, hazard preventing, and people protection. Accordingly, in this study, we first propose a novel framework of artificial intelligent landslides detection over comprehensive remote sensing technique, from InSAR and high-resolution remote sensing images. The paper then discusses in detail how to construct background database with typical elements by categorizing sample sub-databases by versatile data features. In addition, a series of relevant key techniques of importance including heterogeneous data fusion, object-oriented surface coverage change detection, and intelligent comprehensive identification of hidden geohazards have been thoroughly investigated. The new framework has been successfully applied and verified in Jinsha River, China.

Keywords Landslide detection · Comprehensive remote sensing · Artificial intelligence · Heterogeneous data fusion · Sample database

1 Introduction

Geologic hazards are major adverse events resulting from natural processes of the earth, such as collapse, landslide, mudslide, ground collapse, ground subsidence and so on, which are responsible for great loss of life and destruction of property. Mountains occupy about 65% of the total land area of China with complex and active geological conditions, which makes China one of the countries in the world with

Y. Zhang · Y. Zheng (✉)
Henan Key Laboratory of Spatial Information Application on Eco-Environmental Protection, Zhengzhou, China
e-mail: yuzheng1008@yeah.net

H. Ran · Y. Peng
Beijing University of Posts and Telecommunications, Beijing, China

L. Wang et al. (eds.), *Proceedings of the 6th China High Resolution Earth Observation Conference (CHREOC 2019)*, Lecture Notes in Electrical Engineering 657,
https://doi.org/10.1007/978-981-15-3947-3_35

the most threatened vulnerable population suffering from the most severe geological disasters. In 2018, around 3000 geologic disasters occurred in China with direct economic loss reaching 1.47 billion Yuan. Most of the geologic disaster areas centralize in the west of China [1].

Areas prone to earthquake fault zones, active volcanoes, slopes in heavy rainfall can be identified as landslides risks. Currently, national authority constructed the suspecting landslides database with more than 0.3 million identified geological hazard area with high risk, where it is a great occurrence possibility in the future [2]. According to Brabb [3], at least 90% of landslide losses can be avoided if the problem is recognized before the landslide event. General Office of the State Council highlighted the need to exploit the full potential of remote sensing data to support geological hazards including disaster preventing in its various phases and aspects [1]. Obviously, the early-stage-landslides database is critical to support targeting, hazard preventing, and paying intense attention to the specific area in great crisis. However, a number of potential landslides have not been identified yet, since their highly covert, abrupt and devastating features in nature [4]. To put it another way, the identified database with potential landslides occupies only a very limited part compared to the unknown trending landslides. In general, our research that has been carried out to the known database is neither thorough or complete. Geologic hazards prevention remains a challenging and demanding task.

Generally, mainstream geological hazards detection methods can be divided into two categories: one based on expertise-oriented traditional methods and the other based on remote sensing techniques.

A majority of the research has been based on establishing the relationship between the landslide-conditioning factors and landslide occurrence through spatial or statistical data analysis. Pardeshi overviewed traditional approaches such as inventory-based mapping, heuristic approach, probabilistic assessment, deterministic approach, statistical analysis, and multi-criteria decision-making approach with rainfall threshold model and other physically-based landslide susceptibility models included [5, 6]. Many methods are applied to select the landslide-conditioning factors relative to landslide occurrence. However, these methods rely heavily on expert knowledge, which is often subjective, time-consuming, and laborious. And the extracted features are often limited to shallow information, thus unable to express the semantic information that people often understand.

Remote sensing techniques are also exploited for geohazards mapping, monitoring, and detection [7–10]. Satellite Interferometric Synthetic Aperture Radar (InSAR) can provide ground displacement estimates obtained from processing of large stacks of radar satellite images. Small baseline subsets InSAR technique is applied to calculate the time-series deformation of the Jinpingzi landslide [7]. He applies InSAR technology and surface professional monitoring method to identify geological hazards [8]. Remote sensing optical images could also be considered as a useful tool. A landslide inventory mapping framework is proposed in [9] based on the integration of the majority voting method and the multi-scale segmentation of a postevent images, making use of spatial feature of landslide. InSAR and optical EO data are exploited in landslide mapping and monitoring during all phases of emergency management: mitigation, preparedness, crisis, and recovery [10].

In recent years, thanks to the rapid development of big data and artificial intelligent technology, neural network approach has emerged and has been successfully applied in the field of landslide susceptibility assessment [11, 12], while few studies focus on applying to geohazard detection and prediction.

Unlike the previous work, considering the current situation with hazards prevention, we are trying to answer the two following critical questions in this study:

1. How can we explore and study from the identified suspecting landslide database with artificial intelligence?
2. How can we predict and detect the hidden potential landslide points with the knowledge of known data samples in an extensive and efficient way?

To figure out the answer to the questions, in this paper, our contributions mainly include the following three points:

1. We proposed a novel framework of artificial intelligent landslides detection over comprehensive remote sensing technique from InSAR and high-resolution optical remote sensing images. By integrating these two mechanisms, we find that it can effectively improve the efficiency of geological hazards detection by experiment. The proposed framework applies the knowledge of existed database and extends the simplified landslide prediction approach, which has been successfully used and verified in Baige, Jinsha River, China [12].
2. We discussed in detail how to construct sample database including heterogeneous data fusion step by step by categorizing sample sub-databases with a range of data features.
3. A comprehensive identification model to determine the type, spatial location, activity of hidden dangers of geological hazards, and to form a map of hidden dangers of geological hazards is proposed.

2 Proposed Framework

A novel framework of artificial intelligent landslides detection over comprehensive remote sensing technique is proposed through InSAR and high-resolution optical remote sensing images. High-resolution remote sensing is mainly used for surface conditions, including the acquisition of basic information such as morphology, cover, and ground type. InSAR, however, characterizes surface deformation by measuring surface displacement. The joint application of the two with the aid of deep learning method determines whether it is a landslide from the perspective of morphology and deformation, respectively. The proposed approach has the following five principal steps, as shown in Fig. 1.

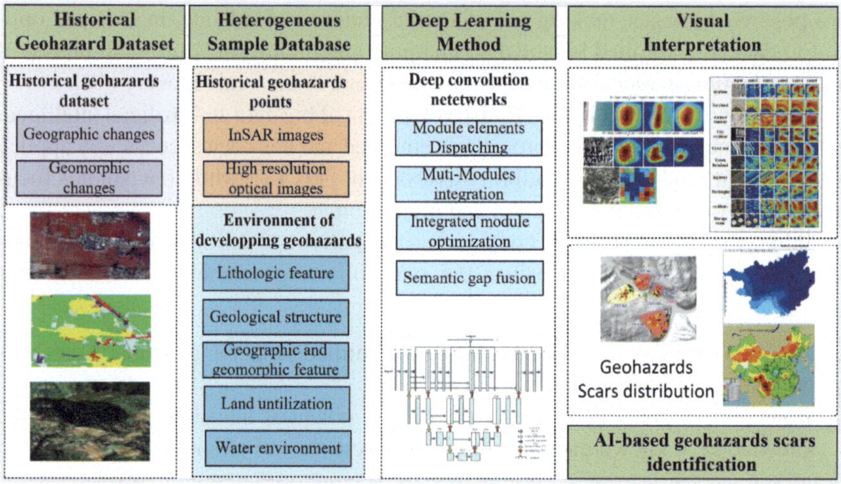

Fig. 1 Overview of proposed framework

As exploring and researching intelligent geological hazard identification methods based on deep learning and large amount of samples, comprehensively utilizing remote sensing optical image big data, InSAR products, and other geographic and geological related service data, can be employed to summarize and extract expert knowledge. Meanwhile, high-performance computing makes large-scale prediction feasible and also improves recognition efficiency. At the same time, the identification of geological hazards based on artificial intelligence can be studied and explored through changes in the wide-area spatial range under large time spans, integrating with the geological hazard sample data recorded by experts, topographic data and residential areas, roads, bridges, towns, water systems, etc. Incorporating basic geographic data like land use, land and space planning and other data, intelligently mining the characteristics of geological disasters which are relatively consistent, and further self-learning to analyze the possible slope movement direction, form and impact scope of geological disasters could improve the reliability and accuracy of early stage landslide identification overall.

Specifically, more research based on the framework in the following field is needed to further carry out:

- **The correlation model between influencing factors and geological hazards**.

Under the guidance of expert knowledge, we are able to design an intensive deep learning model that embeds expert knowledge as a priori information into a deep learning network, or fuse heterogeneous information to design different types of networks to extract a verity of characteristics and then integrate them through multi-layer perceptions. The model accelerates convergence, reduces the number of training samples, and improves the performance finally. Heterogeneous information includes information on topography, geomorphology, geology, climate, hydrology, human activities, and other factors.

- **Assessment and monitoring of geological hazard risk assessment models and key factors**.

Under various typical geological conditions, different factors have different effects on geological disasters. Based on the feature inversion technique, the impact level assessment model of geological hazards is studied, and the threshold values of key factors under various typical geological conditions are confirmed.

The traditional geohazards identification methods highly depend on subjective expert knowledge and cost lots of human work and hours to identify the possible geohazard in risk areas one by one. The proposed framework based on AI has excellent advantages over the traditional ones with the following three aspects:

1. **Summary and highlight of expert knowledge**.

Under the guidance of expert knowledge, the model of geological disaster area discovery and risk assessment can be obtained through the study of historical geohazard samples. When there are few samples or no samples at all, a semantic information-based model is established to accomplish the explicit grammatical representation of expert knowledge.

2. **Promotion and application of expert knowledge**.

The comprehensive integration of expert knowledge and the overall consideration of both general rules and special scenarios are employed in the proposed framework to improve the applicability of the model; In addition, the objectivity of the model and the superior ability of big data processing improve the application value of the framework.

3. **Discovery of new knowledge and update of expert knowledge**.

Machine learning/deep learning establishes a correlation model between influencing factors and output results. Compared with the expert causal analysis model, it is possible to discover new rules that experts have not mastered yet. After the expert assessment/field survey, the new knowledge will further enrich the expert knowledge base with original cognition and can feed back the optimization of the machine learning/deep learning model in the other way.

3 Heterogeneous Database Fusion and Sample Database Build

3.1 Sample Database Content

The construction of heterogeneous sample database mainly includes two aspects. One is the element database related to geological disasters, and the other is the construction of the algorithm database. The element database is the basis and premise of the algorithm database, and the algorithm database is the value embodiment of the element database.

The two databases need to be gradually accumulated and improved, and the number and types of basic elements are increased through the accumulation, targeted enhancement of the model, the universality of expert knowledge, and the discovery of new knowledge.

1. The contents of the element database include the following types:
   - Landform elements: mountains, slopes, ridges, rivers, lakes, etc.
   - Ground cover elements: snow, vegetation (forests, pastures), clouds, etc.
   - Human activity elements: roads, bridges, houses, etc.
2. The algorithm database is designed to achieve the goals of remote sensing big data for various tasks. Taking geological hazard identification as an example, the construction of the algorithm database mainly includes:
   - Land subsidence clustered area identification model based on InSAR geological deformation data;
   - Ground element recognition model based on high-resolution remote sensing image, including edge segmentation and target recognition model;
   - Based on element segmentation and recognition, the high-resolution image calibration model realizes image inconsistency caused by differences in viewing angle, illumination, season, and surface cover;
   - A time variation identification model of the geomorphic elements;
   - Geological hazard identification model based on multi-source data fusion;
   - Geological hazard risk assessment model and early warning model of key factors.

3.2 Sample Database Category

Specifically, the contents of the sample database include the following four categories:

1. **Generate a library of geological remote sensing elements**.

The input material library includes two types: One is a public service dataset of remote sensing images like GF-1, GF-2, ZY-3, etc. This type of database contains various types of geological elements but needs algorithms such as edge segmentation, object recognition, automatic labeling to realize the identification, and labeling of geological elements, as well as standardized pre-processing processes such as digitization and coding of feature descriptions, and finally, form an element library. The second category is a single-element database, but it may not be a remote-sensing image. It needs to be digitized and encoded by various image processing such as scale

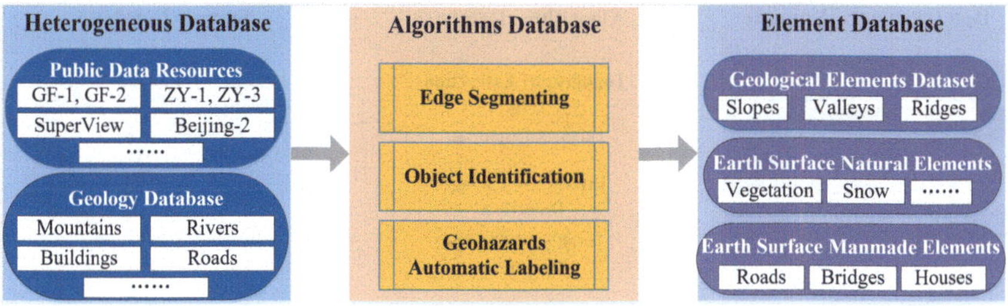

Fig. 2 Database of geological remote sensing elements

transformation and perspective transformation, and finally constitutes an element library (Fig. 2).

2. **Construction of a geological hazard identification sample library based on InSAR data**.

The typical feature of geological disasters is obvious land subsidence deformation, which is closely related to geological features such as slope and stratum, and the geological characteristics and deformation characteristics of historical samples provide reference value for hidden dangers. Therefore, the corresponding algorithm model is designed, and finally, a typical sample of geological disasters based on InSAR ground deformation clusters under various typical geological conditions is obtained (Fig. 3).

3. **Construction of geological disaster identification sample database based on remote sensing imagery**.

Based on multi-temporal remote sensing images, after image registration, through the edge segmentation, target recognition and time variation of the same target with respect to surface cover, combined with gradient calculation, and slope recognition of digital elevation model (DEM) data, the positive and negative samples of geological disasters are constructed accordingly. The negative samples include the changes in earthwork accumulation caused by artificial excavation of mountains such as road construction, house construction, and the temporal changes of mountain volume

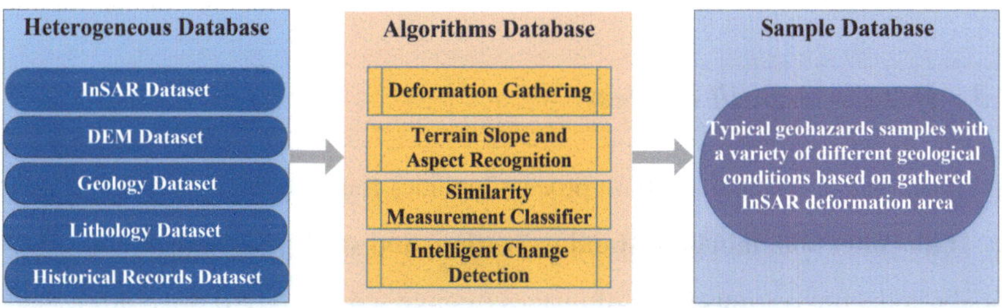

Fig. 3 Construction of a geological hazard identification sample library based on InSAR data

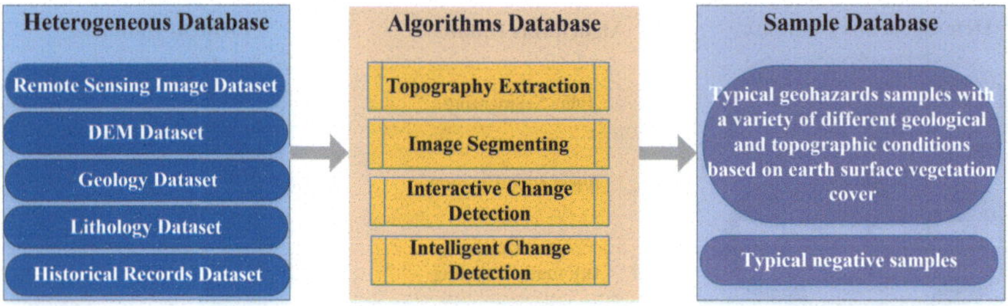

Fig. 4 Geological disaster identification sample database based on remote sensing imagery

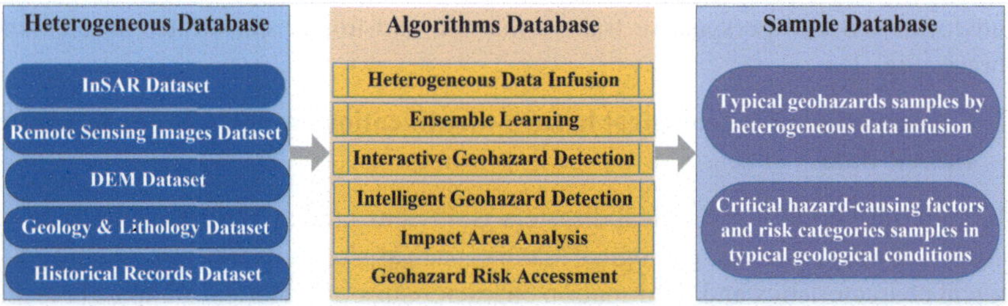

Fig. 5 Geological disaster identification sample database with multi-sources heterogeneous information fusion

snow as the seasons. We could identify and label geological hazard levels with the aid of geological data and historical samples (Fig. 4).

4. **Construction of a geological hazard identification sample library with multi-sources heterogeneous information fusion**.

On the basis of more types of data, through data fusion technology, not only the typical sample construction of geological disaster types, grades, and their influencing factors but also the key influencing factors and thresholds for triggering geological disasters under various typical geological conditions are constructed. A sample of values provides a basis for the identification and key monitoring of key elements (Fig. 5).

3.3 Database-Built Process

As shown in Fig. 6, database-built process includes four phases.

1. **Data collection, labeling, standardization, and storage management**.

Firstly, collecting various types of images (remote sensing types) of geology, landforms, and ground coverings (roads, bridges, houses, snow, etc.) processed by remote

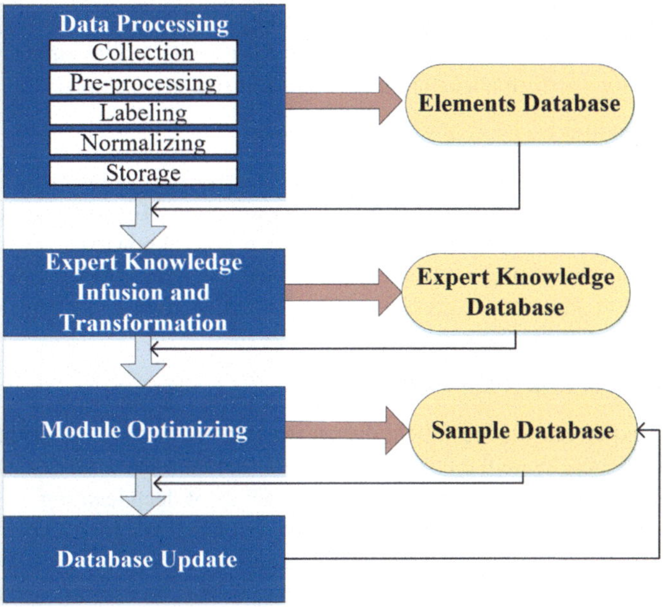

Fig. 6 Flowchart of database-built processing

sensing images, and normalizing them through viewing angles, illuminance, deformation, etc., are needed. And then through edge detection, target recognition, classification/clustering, automatic labeling, etc., the generation of basic element samples can be achieved. Finally, file storage of sample element datasets can be constructed by the storage and retrieval processing by unified coding, abstract generation, etc.

2. **Regularization of expert knowledge**.

In the early stage of the establishment of the sample library, expert knowledge has extremely important guiding value, and it is necessary to translate expert knowledge into rules that can be utilized by machine learning. Specifically, it includes four aspects of design:

- **Determination of the influencing factors**.

When the sample dataset is small, it is difficult to quantitatively analyze the relativity between the influencing factors and the results. At the same time, the convergence of the machine learning/deep learning model is difficult to guarantee. Therefore, it is necessary to draw on the expert knowledge to select the influencing factors and perform appropriate preprocessing and digitization to make the model applicable with fast iteration.

- **Performance metric design and labeling accuracy review**.

The model performance of machine learning/deep learning depends largely on the design of the cost function, and the measure of performance depends on the needs of the application. Thus, the judgment and grading of the expert knowledge on the result are transformed into the design of the cost function. As the sample increases, the cost function, performance metric, etc., are adjusted accordingly.

- **Integration and revision of expert knowledge**.

Due to the diversity of knowledge with different field, the complexity of geological remote sensing problems, the difficulty of evaluation of results, and the diversity of modeling methods, inconsistencies between expert knowledge are inevitable. Research committee voting, multi-mechanism weighting, and other methods should be considered to adjust the expert knowledge, and the expert knowledge should be gradually revised as the sample increases.

- **Embedding of expert knowledge**.

Embedding expert knowledge into the deep learning network through data processing can effectively accelerate the convergence of the model and reduce the number of training samples, thus greatly improves the efficiency of deep learning.

In this method, by referring to the effectiveness of expert knowledge on the data processing method, the hidden layer of the data processing form is added at the bottom of the deep learning network, and the specific data processing method can be utilized efficiently. In addition, integrated learning can be used to process the specific processing of heterogeneous data with a suitable network, and then integrate it with the end-to-end learning network through a fully connected layer to achieve a larger model capacity at a smaller network scale. Moreover, model training and feature extraction can further accelerate network convergence.

3. **Model training and rule learning on the sample set to build and optimize a model library**.

On the different datasets, the appropriate model and training algorithm are designed for construction of the algorithm library. For example, on the InSAR dataset, the surface deformation group is identified by the two-dimensional filtering model; the Bayesian classifier is designed and trained by combining DEM elevation data, stratigraphic/geological data, and historical sample data to realize the assessment of geological disaster risk level. On the remote sensing image dataset, through the models of edge segmentation, target recognition, time change rules identification, geological hazard identification, etc., the positive and negative sample algorithm library and sample library of geological disasters are established. On the basis of mastering more types of heterogeneous data, such as InSAR images dataset, multi-temporal remote sensing image dataset, geological/stratigraphic dataset, and DEM dataset, we need to design integrated learning and deep learning models of multi-source heterogeneous fusion to implement a sample library build that includes critical factors and algorithm models.

4. **The updating and improvement of expert knowledge**.

As the project progresses, more and more datasets and samples can be obtained, so that the algorithm library and the sample library can be gradually improved through incremental learning to enhance the versatility, applicability, performance, and efficiency of the model. At the same time, according to the specific application, the active learning model is designed, and the model is quickly converged by actively selecting the most similar and effective samples.

4 Object-Oriented Surface Coverage Change Detection and Automatic Information Extraction

As shown in Fig. 7, the technical process includes four phases of data acquisition, change detection, object-oriented classification, and slope surface change extraction. According to the differences in the formation, development, and image characteristics of geological hazards in different regions and types, the surface slope change detection model of subsurface slopes is established to increase the detection accuracy.

The regional surface cover change detection is carried out for the slope, and the high-resolution remote sensing data of repeated observations is used to find out the

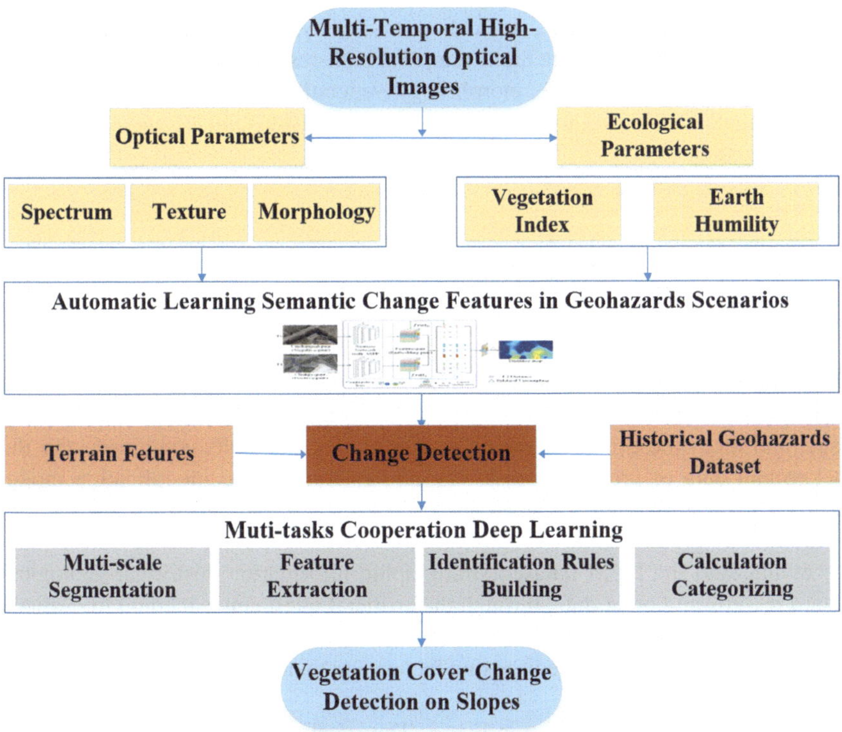

Fig. 7 Technical process of object-oriented surface coverage change detection and automatic information extraction

state change process related to geological disasters such as landslides and mudslides. The object-oriented high-resolution remote sensing image change detection method is used to segment the ground object. The spectrum, texture, shape, and related ecological element information of the object are comprehensively considered to improve the accuracy and integrity of the change detection.

Multi-temporal high-resolution optical remote sensing data, the image spectrum, texture, morphological characteristic parameters, and vegetation parameters such as vegetation index and soil moisture obtained through image inversion are taken as the main indicators, and image registrations and transformation methods are also used. The change detection method of the element performs change detection to retain more detailed information of the image while quickly acquiring regional surface changes.

Based on the above-mentioned indicators of change detection results, combined with topographic information such as elevation, slope, and aspect, we use existing geological hazards as samples, and specific information on the effective slope surface changes can be obtained by object-oriented classification techniques, multi-scale segmentation, feature extraction, recognition rule creation, and classification extraction.

Finally, landslides and debris flows information on slope surface cover changes associated with geological hazards can be extracted by object-oriented classification change detection method using multi-temporal high-resolution optical satellite data, taking image spectrum, texture, morphology, vegetation index, soil moisture as the main parameters, referring to existing geological disaster samples and topographic data all together.

5 Intelligent Comprehensive Identification of Hidden Dangers of Geological Hazards

The comprehensive identification of hidden dangers of geological disasters uses the information of surface deformation and coverage change obtained in the early stage, and takes the existing geological disasters as a sample, combined with the background data of pregnancy disasters, and establishes the identification model based on artificial intelligence technology. Geological hazard identification depends on the InSAR ground deformation dataset, the remote sensing image dataset data processing, and the DEM dataset, stratigraphic information, historical geological hazard records and other data, through the artificial intelligence method to evaluate the risk level of the geological hazard area, risk prediction based on time variation rules, determination of key predisposing factors and their thresholds, expert systems, etc.

The technical route is shown in Fig. 8. The specific process is: under the comprehensive guidance of expert knowledge, firstly, based on the InSAR ground deformation map, determine the ground deformation and accumulation area, obtain the

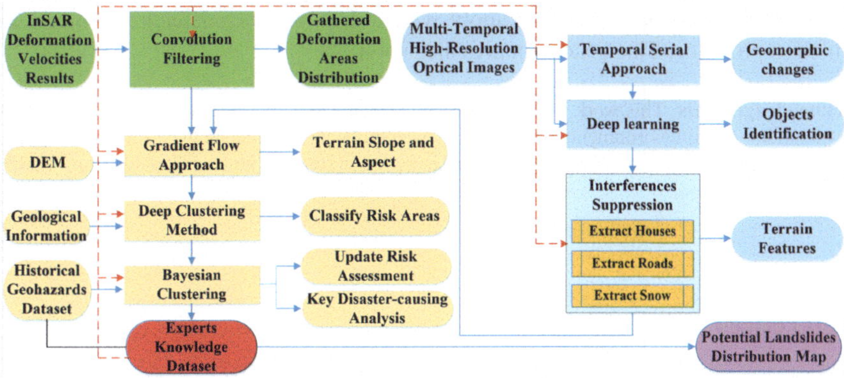

Fig. 8 Intelligent comprehensive identification of hidden dangers of geological hazards

suspected area distribution of hidden dangers of geological disasters; meanwhile, based on multi-period remote sensing image dataset through the image processing, the change law of the landform cover is identified, and the target area is recognized to eliminate the hidden area of the non-geological hazard. Combining the results of the above two types of geological hazard suspected areas, with the aid of DEM, stratigraphic information and historical geological hazard area information, artificial intelligence methods such as machine learning, deep learning, and integrated learning are employed to accomplish the identification and risk of geological hazard areas. We have proved this solution feasible and efficient in case study in Jinsha River, China [12].

6 Summary

This paper proposed a novel artificial intelligent framework of geological hazards detection over comprehensive remote sensing. This approach investigates morphology of the landslide, surface cover changing, studies the formation, development of disasters, and reveal the potential disaster situation by the high-resolution remote sensing image change detection. InSAR monitoring is used to obtain the surface deformation of the disaster body to characterize the movement and deformation state of the slope, and to determine the slip size, activity stage, and development trend of the slope; The embedded deep learning algorithms take a thorough judgment from the disaster situation, the current deformation situation, and the potential disaster situation; the early identification before the disaster happens is realized automatically and efficiently. The main procedures in the framework, such as sample database building, object-oriented surface coverage change detection, and intelligent comprehensive identification methods are also fully investigated in our study. The proposed framework is going to apply with hidden disaster detection projects in Guizhou province, Tibet, Sichuan province soon.

Acknowledgements This work was supported by Henan Key Laboratory of Spatial Information Application on Eco-environmental Protection. The authors also appreciate the efforts and suggestions from Liqiang Tong, Yi Wang and Jiancun Li, experts in Remote Sensing Geology from AGRS, China.

References

1. Ministry of Natural Resources (2019) National geologic disasters in 2018 and prediction of geologic disasters trends in 2019
2. Xu Q, Dong X, Li W (2019) Integrated space-air-ground early detection, monitoring and warning system for potential catastrophic geohazards. Geomatics Inf Sci Wuhan Univ 44:957–966
3. Brabb E (1993) Proposal for worldwide landslide hazard maps. In: Proceedings of 7th international conference and field workshop on landslide in Czech and Slovak Republics, pp 15–27
4. Ge D, Dai K et al (2019) Early identification of serious geological hazards with integrated remote sensing technologies: thoughts and recommendations. Geomatics Inf Sci Wuhan Univ 44:949–956
5. Pardeshi SD, Autade SE, Pardeshi SS (2013) Landslide hazard assessment: recent trends and techniques. SpringerPlus 2:1–10
6. Musaev A, Wang D, Cho C, Pu C (2014) Landslide detection service based on composition of physical and social information services. In: Proceedings of IEEE ICWS, pp 1–6
7. Zhao C, Kang Y, Zhang Q et al (2019) Landslide identification and monitoring along the Jinsha River Catchment, China using the InSAR method. Remote Sens 10:1–6
8. He C, Ju N, Xie M (2019) Application of InSAR technology in early recognition of geohazards. J Xihua Univ (Nat Sci Ed) 38:32–39
9. Lv Z, Shi W, Zhang X et al (2018) Landslide inventory mapping from bitemporal high-resolution remote sensing images using change detection and multiscale segmentation. IEEE J Sel Topics Appl Earth Obs Remote Sens 11:1520–1532
10. Casagli N, Cigna F et al (2016) Landslide mapping and monitoring by using radar and optical remote sensing: examples from the EC-FP7 project SAFER. Remote Sens Appl Soc Environ 4:92–108
11. Dou J, Yamagishi H, Pourghasem HR et al (2015) An integrated artificial neural network model for the landslide susceptibility assessment of Osado Island, Japan. Nat Hazards 78:1749–1776
12. Zhang Y, Ran H et al (2019) Automatic prediction of landslides over InSAR techniques and differential detection using high-resolution remote sensing images: application to Jinsha River. In: Submitted to China high resolution earth observation conference

Automatic Prediction of Landslides Over InSAR Techniques and Differential Detection Using High-Resolution Remote Sensing Images: Application to Jinsha River

Yueying Zhang, Haonan Ran, Yuexing Peng, and Yu Zheng

Abstract With a rapidly increasing population on or near steep terrain in southwest of China, landslides have become one of the most significant natural hazards. Thus, quick detection, prediction, and identification of early signals of landslide occurrence are critical for prompt emergency information, rescue efforts, and mitigation of further damage such as collapse of a landslide dam. Accordingly, in this paper, a prototype of early-stage-landslide detection is introduced. We construct a new automatic approach to extract landslides from remote sensing imagery, both optical and radar for quick prediction and detection before disasters, achieved by combining interferometric synthetic aperture radar (InSAR) with differential detection method using multi-temporal high-resolution optical images. The idea behind the novel approach is to identify potential landslides by typical distribution features, deformation, and tendency of landslides, including the result of expert second opinion. To verify the feasibility of this landslide prediction system, a case study was performed in Jinsha River area. It is found that the use of automatic detection of the potential landslides over satellite imagery allows the identification and characterization of the affected areas based on landslide features. It is expected that the prototype of landslide prediction can provide early recognitions before severe landslides occur. Total of eight active small-scaled landslides in early edge covering study region are automatically detected.

Keywords Landslides · Early recognition · Remote sensing · InSAR

1 Introduction

Landslides are a kind of serious natural geological hazards caused by masses of soil, rock, and debris flowing down steep slopes under the influence of gravity during

Y. Zhang · Y. Zheng (✉)
Henan Key Laboratory of Spatial Information Application on Eco-Environmental Protection, Zhengzhou, China
e-mail: yuzheng1008@yeah.net

H. Ran · Y. Peng
Beijing University of Posts and Telecommunications, Beijing, China

L. Wang et al. (eds.), *Proceedings of the 6th China High Resolution Earth Observation Conference (CHREOC 2019)*, Lecture Notes in Electrical Engineering 657,
https://doi.org/10.1007/978-981-15-3947-3_36

periods of heavy rainfall, earthquake, rapid snow melt, etc. It is reported in 2017 and 2018 in China that landslides happened more repeatedly than before, and their damages are massive and destructive with significant national impact, including loss of life, and damage to communication routes, human properties, agricultural fields and forest lands. In 2018, around 3000 geologic disasters occurred in China with direct economic loss reaching 1.47 billion Yuan [1]. Most of the landslides occurred in the southwest of China. In these mountainous regions, full of rocks with steep slopes are highly likely to collapse in active geological events, which makes Sichuan, Tibet, etc., the most threatened vulnerable provinces suffering from the most severe geological disasters, such as landslides.

National and local disaster mitigation initiatives have setup research and development agenda to propose new approaches for landslides prevention in advance and more in general to improve the efficacy detection provided for crisis response and mitigation. For example, General Secretary Xi Jinping presided over the third meeting of Central Financial and Economic Affairs Commission. It was clearly stated: Constructing natural disaster prevention and control system to improve natural disaster prevention ability in the whole society. Meanwhile, it was proposed to promote the construction of the "Nine Major Projects," including: Implement disaster risk investigation and key hidden danger investigation projects, master the risk hidden danger base, implement natural disaster monitoring, and early warning project [2]. Thus, quick detection, prediction, and identification of early signals of landslide occurrence are critical for prompt emergency information, rescue efforts, and mitigation of further damages and arouse lots of attention from the academic field and government.

This paper constructs an automatic method to extract landslides from comprehensive remote sensing techniques for quick prediction and detection before disasters. The remainder of the paper is organized as follows.

Section 2 describes related work in landslide movement monitoring, and other methods for landslide detection and prediction. The contributions of our work have been introduced. In Sect. 3, we detail the methodology including geographic changes detection over interferometric synthetic aperture radar (InSAR) products and object-oriented surface coverage change detection over differential detection from high-resolution remote sensing imagery. In addition, we demonstrate this concept through a simulated example and practical case studies in the Jinsha River in Sect. 4. Performance with respect to InSAR deformation area, edge segmentation, and interference suppression has been analyzed and recognition of hidden landslide map in the study area has been automatically demonstrated and categorized by the risk level. Finally, we conclude in Sect. 5 and in the same section, future work is also discussed.

2 Related Work and Contributions

In this context, various mapping techniques have been implemented for the landslide movement and prevention. These include landslide inventories, landslide displacement inventories, landslide hazard assessments, and landslide vulnerability assessments [3–10]. Among all of these studies, landslide prediction and detection are fundamental for the prevention, assessment, and reduction of landslide hazards and risks.

Detection and real-time monitoring after landslides occurred have been thoroughly focused in the academic and industrial field. Wireless sensor networks (WSN) are one of the major technologies that have contributed to real-time monitoring [3, 4]. However, WSN has a group of limitations, such as low memory, power, and bandwidth. Additionally, it is practically difficult to deploy WSN in a largescale in order to explore the whole area at risk in view of high expense with great number. It is more acceptable for use in a specific limited area locally. Furthermore, implement in hostile environment where no one could get access to is still impractical, since landslides are widely sparsely distributed.

Besides WSN, remote sensing imagery, both optical and radar, has been used extensively for observation of the geomorphologic changes of landslides [5–10], creating new opportunity for scientific community. Optical remote sensing images with high spatial resolution offer the possibility for landslides recognition and objects classification on earth surface. Nichol utilized stereo satellite images from IKONOS very high-resolution sensors to detail the landslide hazard assessment for larger areas [5]. A high-resolution digital terrain models based on the detection of thresholds derived by the statistical analysis of variability of landform curvature is proposed in [6] to automatically detect and highlight the location of shallow slope failures and bank erosion. The use of automatic detection of the landslides over satellite imagery also allowed the identification and characterization of the affected areas, the mapping of the landslide features and the calculation of the displaced sediment volume [7].

Compared that optical remote sensing could be impacted by certain weather condition like fog or cloud, the InSAR technique shows major advantages due to its broad coverage and its high spatial resolution under all weather conditions. InSAR is an imaging technique for measuring Earth's surface topography and deformation, especially where standard interferometry is problematic [8–10]. InSAR images can be obtained by directly measuring the radar phase difference between two separated antennas taken from the same view in the same surface area. These features provide landslide detection possible solutions with InSAR. Authors of [8] quantify how the choice of independent digital elevation model (DEM) influences InSAR results for the purposes of resolving the movement of active landslides. TerraSAR-X images have been analyzed in [9] and a good combination of filters and enhancement technique for the automatic landslide detection is described. InSAR deformation images, InSAR coherence maps, SAR backscattering intensity images, and a DEM gradient map are combined in [10] to detect active landslides by setting individual thresholds.

As mentioned above, although many efforts have been made and approaches to landslide monitoring and detection, fewer researches carried out with respect to landslide prediction. It is of great significance to improve accuracy, efficiency, and automation for early landslide detection and prediction before the landslide occurs. The existing methods have more or less performance limitations and many approaches are sometimes limited to certain fixed scenarios. In addition, as the most common techniques to landslide prediction, visual interpretation, and geomorphological field surveys highly depend on expert knowledge and are also quite time-consuming and labor-intensive. It is impossible to predict the potential landslides in largescale efficiently in the traditional way. Therefore, this paper aims at improving the performance of landslide prediction in relation to automation and accuracy by incorporating InSAR techniques with high-resolution remote sensing imagery to identify morphology, deformation, and situation of the active area.

In this study, our contributions mainly include the following four points:

1. We first introduce a comprehensive remote sensing method for automatic detection and prediction of slowly moving landslides using InSAR products and high-resolution images, including deformation maps, DEM, and GF-2 images.
2. We then apply this method to map active landslides in a large region in Baige, Jinsha River, Southwestern China. The spatial segmentation for a specific landslide is analyzed in detail, the temporal time-series deformation results are achieved using differential detection technique, and the accuracy of deformation measurements is verified using independent InSAR observations.
3. We were successfully able to suppress noise in the image, identify landslide signals, and classify potential landslides from other surface changes with expert knowledge.
4. The detected potential landslides are finally analyzed and verified by independent expert survey. Our approach proposed here will contribute to the rapid assessment of landslide hazards, prompt emergency information, rescue efforts, and further hazard mitigation.

3 Methodology

The comprehensive identification of hidden dangers of hidden landslides uses the information of surface deformation and coverage change obtained in the early stage, and combines with the geomorphic features of hidden disasters, and establishes the identification model based on comprehensive remote sensing technology. Landslide identification depends on the InSAR ground deformation map, the multi-temporal remote sensing image processing, and the DEM dataset, stratigraphic information, historical geological hazard records and other data to perform the hidden landslide identification map, and category the risk level of the geological hazard.

3.1 General Procedures

The technical procedures of proposed landslide prediction method are shown in Fig. 1. According to the comprehensive guidance of expert knowledge and the specific geomorphic features of hidden disasters, firstly based on the InSAR ground deformation map, we determine the gathered ground deformation and accumulation area which is coherent with the characteristics of hidden landslide physically collapsing together with the whole slope in a steep mountain cliff. Then, we obtain an overall distribution of suspected area with hidden dangers. Meanwhile, we need to refine the rough version of hidden landslide map in detail by recognizing the developing and continuing slide tendency based on multi-temporal remote sensing image change detection. Furthermore, the deformation in the target area by man-made interference which is non-geological hazards like building houses or roads should be eliminated

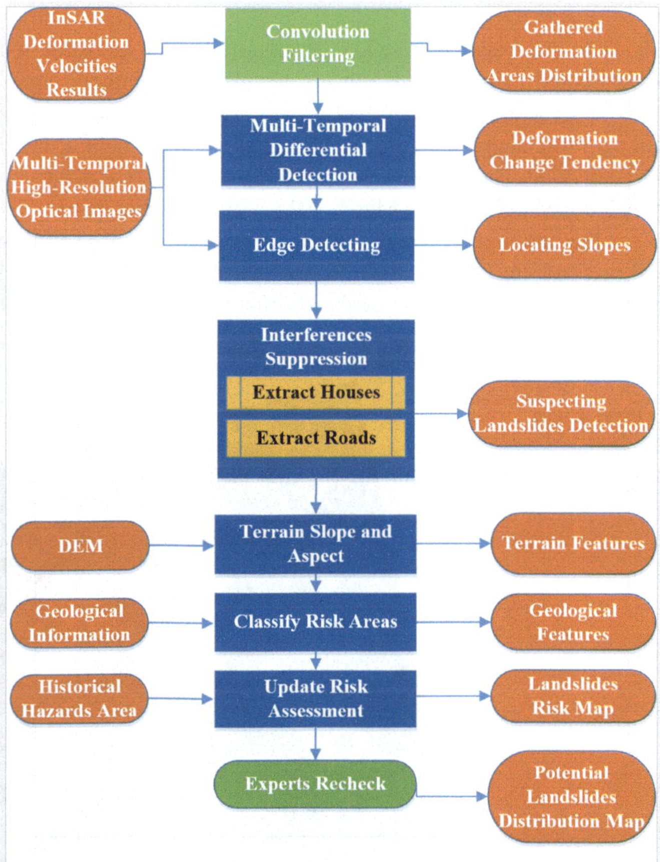

Fig. 1 General procedure of landslide prediction methodology with proposed approach

and deleted from the first version of the hidden danger distribution map. Comparing the results of the above two types of geological hazard suspected areas, with the aid of DEM, stratigraphic information, and historical geological hazard area information, comprehensive remote sensing techniques are used to accomplish the identification and grade the risk of geological hazard areas.

3.2 Study Area and Input Data Selections

The Jinsha River is the upper course of the Yangtze River. Study area Baige is located around the border of Sichuan Province, Tibet, southwestern China. Because of frequent geological activities, the structure of the mountain body is relatively fragmented and is easily affected by rainfall and earthquakes [11]. This fact is probably responsible for large-scale landslide that historically occurred in Baige twice in 2018 (Fig. 2).

The analysis of this paper is based on automatic hidden landslide recognition, which was conducted using remote sensing techniques. Synthetic aperture radar (SAR) images were acquired by Sentinel-1, which provides continuity of C-Band SAR data for operational applications and to contribute to Global Earth Observation System of Systems [12]. The 30 SAR images we utilize in this research from Sentinel-S1A spanning from March 2017 to March 2018 with ascending track.

Imagery for the automatic geomorphic change detections used in this research was acquired by GaoFen-2 (GF2), with high spatial resolution (4 m/pixel for bands R, G, B and NIR and 1 m/pixel for the panchromatic band). The width of imagery

Fig. 2 Study area located in Baige, Jinsha River, China (GF-2 satellite data)

is 45 km and coverage period is 69 days. We preprocessed three remote sensing images, which were acquired on April 16, 2016, February 9, 2017, and March 6, 2018, covering 45 km^2.

3.3 Suspicious Landslide Recognition Over InSAR

With the capability of surveying large areas efficiently under all kinds of weather conditions both during day and night, InSAR techniques have proven very useful for detecting and mapping landslides. InSAR techniques have been described in detail in [8–10]. We employ it to show the time-series deformation of Baige, Jinsha River area, and identify suspicious landslide signals from the InSAR deformation map shown in Fig. 3.

For the automatic detection of landslides from an InSAR deformation map, we need to suppress noise in the image, identify signals, refine, and classify landslides from other surface changes. In this section, we explain a method to identify landslide signal from the InSAR products efficiently.

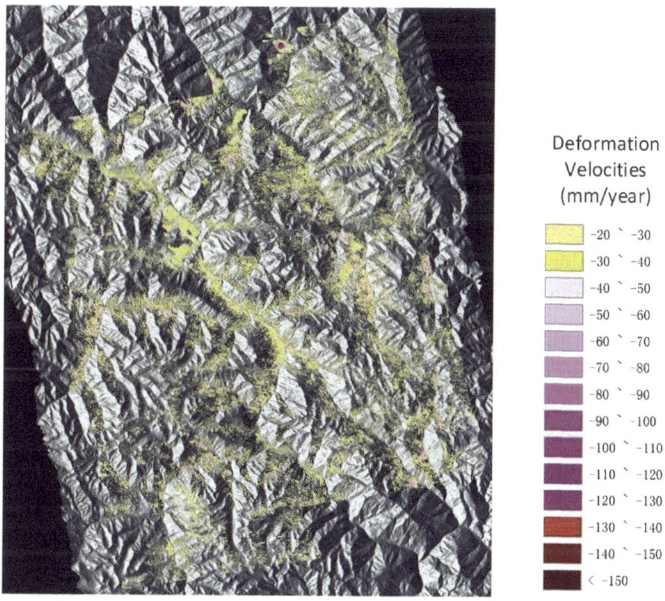

Fig. 3 InSAR deformation map of study area

3.3.1 Suppress Noise in the Image

A landslide is considered a relatively rapid mass wasting process that causes the down slope movement of mass of rock, debris or earth triggered by variety of external stimulus. From Fig. 3, we can see the deformation velocities per year in the study area roughly between 20 and 100 mm. An appropriate threshold is set to determine if the area is prone to landslides, because a place with low deformation velocities is unlikely to be a geohazard threat.

Additionally, from other characteristics of hidden landslides which a small area of earth moves together down the same slope, it is coherent to the gathered distribution cluster over the InSAR deformation map. We applied multiple filters and enhancement process to suppress noises and enhance signals. The original images include random noises and foreshortening zones, which are usually in form with scattered points. We tried to suppress the random noise and mask foreshortening zones.

A two-dimensional 5×5 sized convolution filter was applied to keep the accumulated clusters and reduce the random noise pattern. The standard deviation of the bilateral filter, threshold of intensity for masking, and window size of the median filter are adapted to control the effects of the filtering.

3.3.2 Identify Signals of Hidden Landslides in the Image

Landslides change the vegetation of the ground surface and topography, which affects the distribution of deformation velocities map. The region of the suspicious landslides should clearly be shown in the clustered points from the deformation velocities map. Pixels with intensity larger than a threshold were extracted and a binary image was created. And small-isolated objects were filtered out by applying 2D convolution filter on the image. To identify small objects close each other which is prone to continue movement down slope, we stored the selected 5×5 pixel frames as identified signals of suspicious landslide scars.

3.4 Object-Oriented Surface Differential Coverage Change Detection

Based on obtained suspicious landslide scars mentioned above, we have a guidance to examine and investigate further if the suspicious landslide scars are potentially risky or not. High-resolution remote sensing is a great aid to exclude the interferences and accomplish the prediction of hidden landslide distribution. As Fig. 4 described, through the time difference processing of remote sensing images taken at different times in the same suspicious landslide area framed before, the trend graph of geomorphic features changing with time is obtained. Since this trend is expected to continue and may increase in the future with the influence of gravity and geological causes.

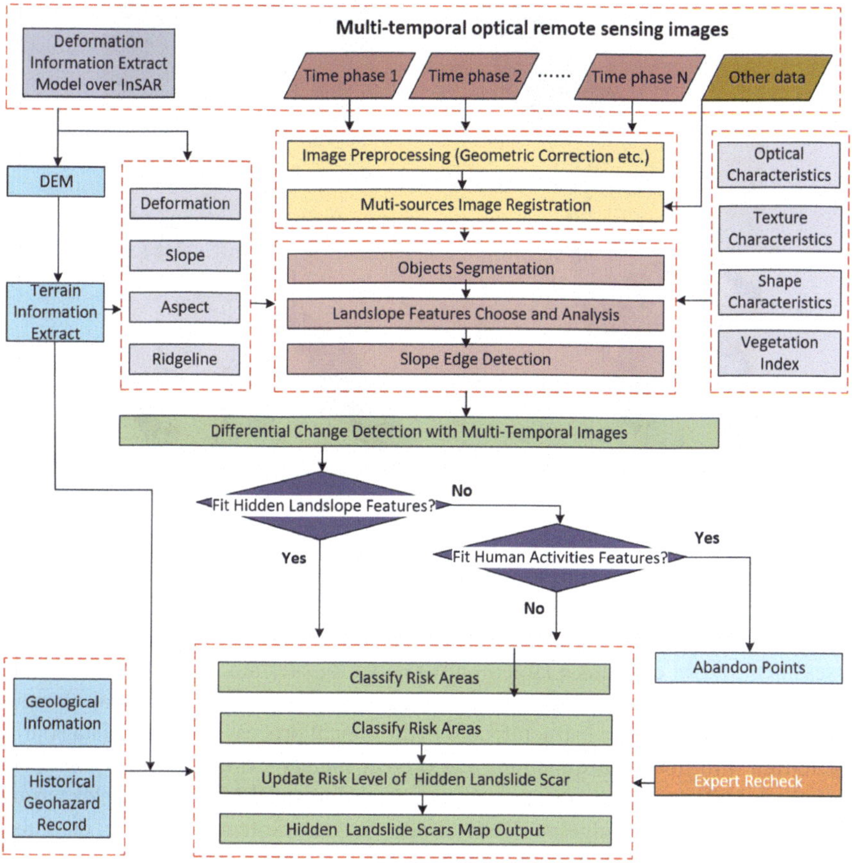

Fig. 4 Object-oriented surface differential coverage change detection

Based on the knowledge of experts, the features extraction and prediction model of geographic feature change detection are established for specific regions whose changes exceed the predetermined threshold. Specific network details and design ideas are shown in the following sub-sections.

3.4.1 Landslide Geomorphic Features

The geomorphic variation characteristics of hidden landslide include:

1. Morphological characteristics: long strip changes with down slope direction;
2. Size: usually in smallscale;
3. Positional features: extension along slope with long axis;
4. Dynamic change characteristics: gradually move down to the bottom of the slope, etc.

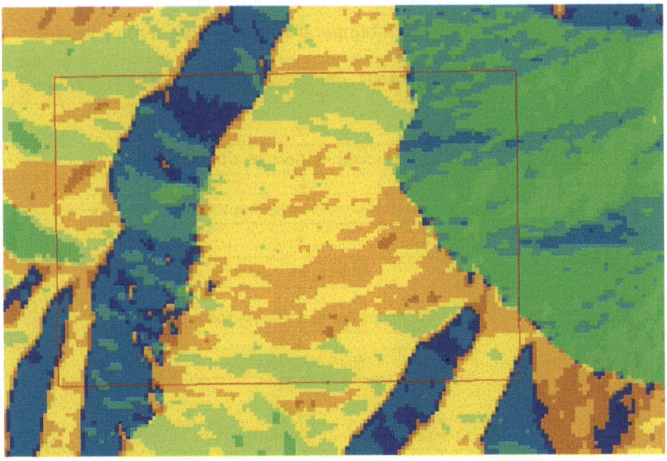

Fig. 5 Terrain slope and aspect of study area

3.4.2 Terrain Information Extraction

Use of digital elevation model (DEM) is of immense importance in landslide hazard assessment. DEM contains a large number of various terrain structures and feature information, which is the basic data for quantitatively describing spatial changes such as geomorphological structure and hydrological process and is the data source for terrain information acquisition. The project mainly extracts basic topographic information, slope, aspect, surface area, etc., according to DEM data. Complex terrain information including ridgeline, valley line, topographic feature surface, etc., also can be extracted from DEM with good resolution. Figure 5 shows the terrain slope and aspect of study area that we obtained from DEM.

3.4.3 Multi-temporal Differential Change Detection of Developing Hidden Landslides

Image preprocessing includes geometric correction, radiation correction, and multi-sources registration over several high-resolution GF-2 remote sensing data shooting for the same area in separate years. Image registration is the very important process of transforming different sets of data into one coordinate system. Data applied here are multiple optical images, from different times, depths, or viewpoints. Registration is necessary in order to be able to compare the data obtained from these different measurements.

After image preprocessing, object-oriented remote sensing image segmentation classification is introduced. According to the color, shape, texture, and other characteristics of the image pixel, the adjacent pixels with the same feature or satisfying certain similarity conditions are composed into one image object, and then according to the image object and the various features of the object classify it. We use edge segmentation method to detect the slopes which are suspicious area shown in processed deformation distribution map over InSAR. Differential detection method in [13] is employed here to compare the change difference. Especially, we extend the method to investigate the dynamic change characteristics of the hidden landslide area over years.

Clustered deformation area shown over InSAR can be triggered by gradual processes or by external mechanisms including human activities, such as road building or construction activity loading on upper slopes, or a combination of these and other factors. In the absence of geological survey data, we use the machine learning methods such as classification/clustering to distinguish the geographical changes of geological hazards and non-geological hazards. Construction of buildings and roads/rails/bridges are identified and excluded to achieve more reliable geological hazard areas identification and prediction.

Finally, after comparing the investigating area with radar and optical remote sensing techniques carefully and exclusion of the non-geological hazards-caused clustered deformation, we utilize DEM geological information and historical geohazard record to update the risk level. Different rules are taken into consideration, for example, slopes less than a threshold (15° is set as the threshold here) obtained from DEM are excluded, since slopes are not steep enough and it is unlikely for landslide failures soon. The hidden landslide distribution map with different risk levels is achieved with experts recheck.

4 Experimental Results and Analyses

We focus on the same framed out 45 km^2 area with that of optical remote sensing image we chose from GF-2, so that the clustered deformation patterns are supposed to be consistent with the landslide distribution investigated in the later section. Before noise suppresses process, we found out 249 areas depicted in Fig. 6a where the average deformation higher than the threshold we set. After noise suppress process, 22 clustered area pointed in Fig. 6b are found as the suspicious landslide deformation area where we need to further investigate in high-resolution remote sensing images.

We investigated these 22 small-scale areas one by one after object-oriented surface differential coverage change detection shown in Fig. 7, we successfully found out four hidden landslide scars fitting with geomorphic features described in Sect. 3.4, and another four hidden landslide scars with clustered deformation from InSAR product but without obvious geomorphic changes detected over years from multi-temporal optical images. Fourteen other landslide scars have been detected as non-geohazards, caused by human activities with six houses building in blue and eight road constructions in green, depicted in Fig. 7, respectively.

Comparing the landslide deformation map over InSAR with high-resolution images for the specific area considered, the clustered deformation patterns are verified to be consistent with the landslide distribution in Fig. 8. The landslide scars shown

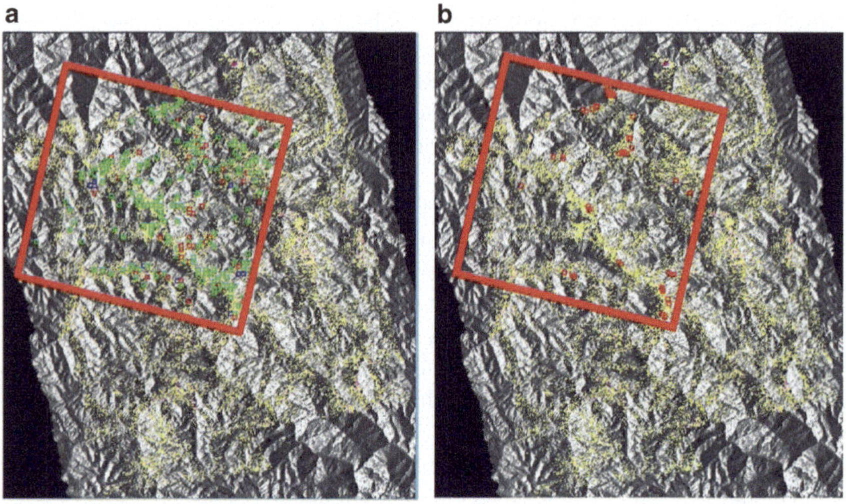

Fig. 6 Landslide deformation map over InSAR (before and after noise suppress as **a**, **b**, respectively)

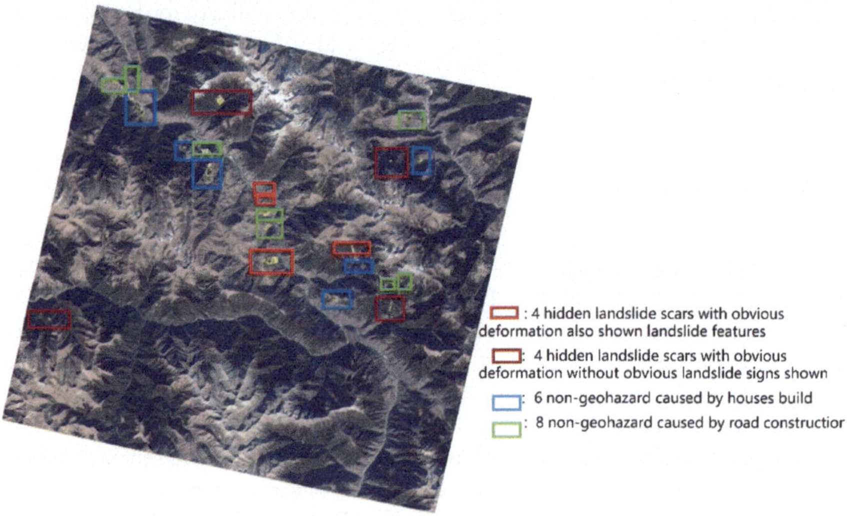

Fig. 7 Hidden landslide scars distribution map (before excluding non-geological hazards)

change signals in both remote sensing methods will be considered as high-risk hidden landslide spots.

Figure 9 depicts six non-geohazard places caused by houses building, explaining the deformation signals shown in landslide deformation map over InSAR but turning out to be human activities caused deformation. The deformation or landslides caused by human activities such as road construction and houses build should not

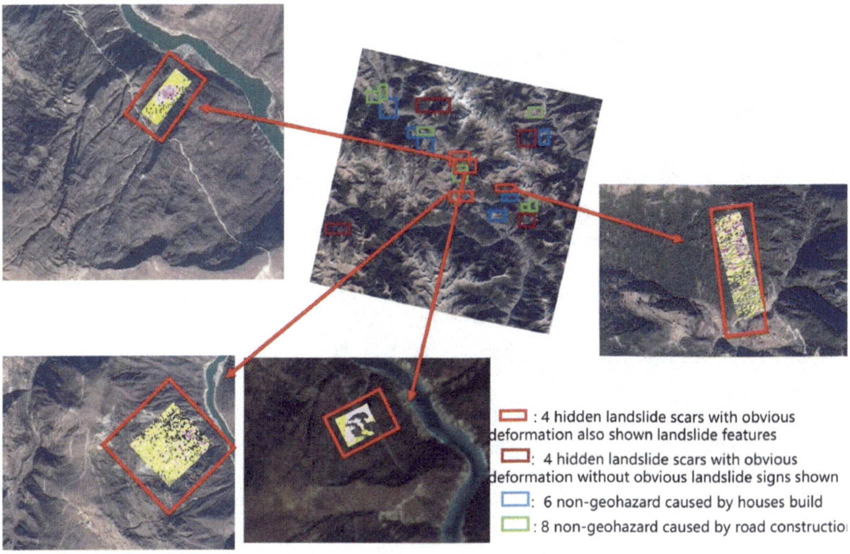

Fig. 8 Hidden landslide scars analysis combining with deformation map over InSAR

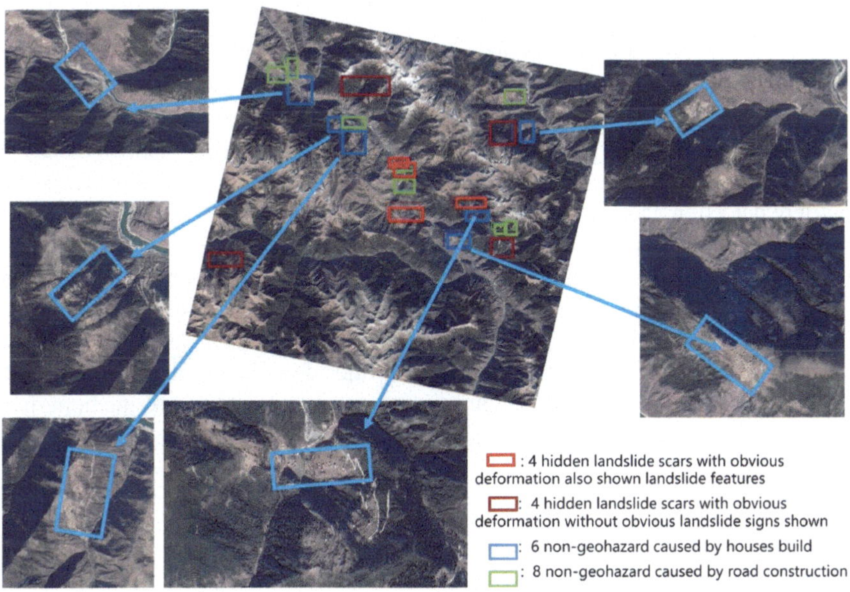

Fig. 9 Non-geological hazards analysis in hidden landslide scars distribution

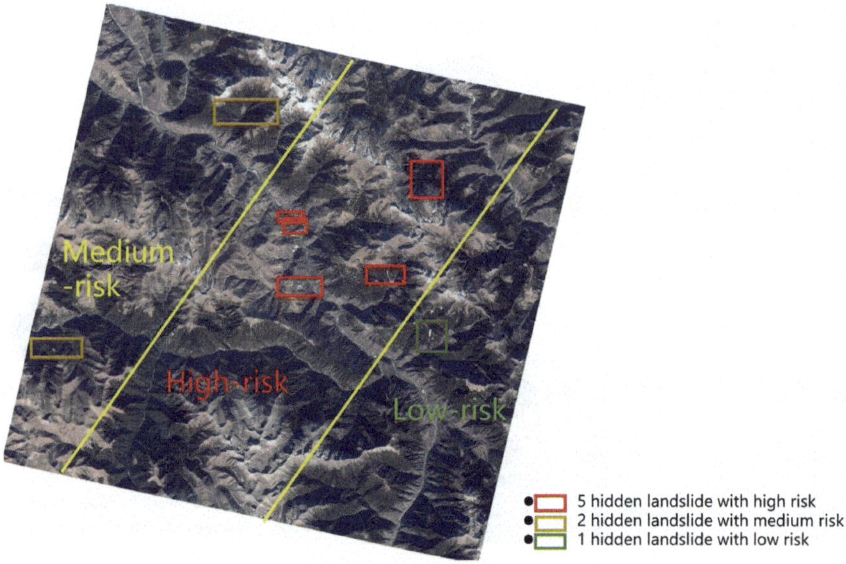

Fig. 10 Hidden landslide scars distribution map with risk-level assessment

be considered as threats since geomorphic features changing with time is unlikely to continue and increase in the future compared to geohazards caused by continuous geological factors. Therefore, we discard the detected non-geohazards scars as inference samples. In this way, the performance of the proposed automatic prediction of landslide in early stage in view of efficiency has been significantly improved by interfere detection and automatically discard.

Figure 10 shows the final early-stage hidden landslides distribution and risk assessment map. We were successfully able to automatically suppress noise and human interference in the image identified hidden landslide signals and classify landslides in the early stage from other surface changes, DEM information, historical geohazards records according to the risk-level assessment rules described in Sect. 3.4. Total of eight detected hidden landslides in this area is categorized as high risk, medium risk, and low risk shown in red, yellow, and green frames in Fig. 10.

5 Conclusion and Future Work

We attempted to propose an automatic prediction of landslides over InSAR techniques and differential detection using high-resolution remote sensing images and apply to Baige, Jinsha River, China. Over 50 active potentials are identified based on features extraction, human activities exclusion, and then updated by terrain slope and aspect detection and historical records. Finally, eight hidden landslide scars distribution maps with risk-level assessment are achieved automatically. The proposed approach is proved to be feasible and efficient after experts verifying. However, this research is currently based on geologist knowledge with the aid of comprehensive remote sensing techniques, which may not thoroughly and completely consider enough geohazard-causing factors. In this result, artificial intelligence should be added in the future work, such as learning/deep learning establishing a correlation model between various influencing factors and output landslide recognition results. Authors also proposed an artificial intelligent framework for hidden landslide detection in [14] as a counterpart with this study. Compared with the expert causal analysis model, it is possible to discover new rules that experts have not mastered yet and improve the overall performance.

Acknowledgements This work was supported by Henan Key Laboratory of Spatial Information Application on Eco-environmental Protection. The authors also appreciate the efforts and suggestions from Liqiang Tong, an expert in Remote Sensing Geology from AGRS, China.

References

1. Ministry of Natural Resources (2019) National geologic disasters in 2018 and prediction of geologic disasters trends in 2019
2. Information on http://www.gov.cn/xinwen/2018-10/11/content_5329688.htm
3. Yu Z, Xu S, Zhang S et al (2012) Distributed detection in landslide prediction based on wireless sensor networks. In: World automation congress 2012, pp 235–238
4. Liu Z, Tsuda T, Watanabe H et al (2019) Data driven cyber-physical system for landslide detection. Mob Netw Appl 24:991–1002
5. Nichol JE, Shaker A, Wong MS (2016) Application of high-resolution stereo satellite images to detailed landslide hazard assessment. Geomorphology 76:68–75
6. Tarolli P, Sofia G, Fontana GD (2012) Geomorphic features extraction from high-resolution topography: landslide crowns and bank erosion. Nat Hazard 61:65–83
7. Lira C, Lousada M, Falcão AP et al (2011) Automatic detection of landslide features with remote sensing techniques: application to Madera Island. In: Geoscience & remote sensing symposium 2011, pp 1997–2000
8. Bayer B, Schmidt D, Simoni A (2017) The influence of external digital elevation models on PS-InSAR and SBAS results: implications for the analysis of deformation signals caused by slow moving landslides in the northern Apennines (Italy). IEEE Trans Geosci Remote Sens 55:2618–2630
9. Yamada M, Hashimoto M, Fukushima Y (2013) Automatic detection of landslides from SAR images: application to the 2011 Kii landslides. In: Synthetic aperture radar, pp 245–248
10. Zhao C, Lu Z, Zhang Q et al (2012) Large-area landslide detection and monitoring with ALOS/PALSAR imagery data over Northern California and Southern Oregon, USA. Remote Sens Environ 124:348–359
11. Li Y, Jiao Q, Hu X et al (2020) Detecting the slope movement after the 2018 Baige landslides based on ground-based and space-borne radar observations. Int J Appl Earth Obs Geoinf 84:101949
12. Information on http://spaceflight101.com/copernicus/sentinel-1/
13. Yamaguchi Y, Tanaka S, Odajima T et al (2002) Detection of a landslide movement as geometric misregistration in image matching of SPOT HRV data of two different dates. Int J Remote Sens 1–12

14. Zhang Y, Ran H et al (2019) A novel framework of artificial intelligent geologic hazards detection over comprehensive remote sensing. In: Submitted to China high resolution earth observation conference

Internal Calibration and Range Replica Extraction Scheme for Ultrahigh-Resolution Spaceborne SAR

Fan Feng, Axin Jin, Jia Sun, Ruohan Hou, and Hongxing Dang

Abstract In order to guarantee the image quality of ultrahigh-resolution spaceborne SAR, highly precise phase and amplitude responses of system are needed via calibration. Since this kind of SAR employs multiple subbands and be equipped with multiple transmit and receive channels to acquire the wider range bandwidth and antenna footprint, respectively, it becomes much more complex as compared with conventional SAR. In this paper, we propose a novel internal calibration and range replica extraction scheme to deal with this problem. The calibration loop, module, and workflow are provided. And effectiveness of this new scheme is verified by range compression results achieved with an experimental SAR system.

Keywords Spaceborne SAR · Multiple subbands · Multiple channels · Internal calibration · Range replica

1 Introduction

Spatial resolution is a very important figure of merit for spaceborne SAR [1]. The finer the resolution, the more detailed information we can acquire from SAR images. Nevertheless, ultra-high resolution is quite different from the conventional SAR on the system level as well as the processing level. Firstly, it should be capable to transmit and receive signals within multiple adjacent subbands so as to acquire the wider range bandwidth and the resultant higher resolution [2–4]. Secondly, the imaging processing needs more precise phase and amplitude information to construct the range compression replica as the residual errors will cause the distortion of images [5–7]. As a consequence, the conventional calibration approach is not valid for ultra-high resolution SAR any more [8, 9].

According to the characteristics of high-resolution spaceborne SAR, a new internal calibration scheme and its associated range replica construction algorithm are suggested to deal with the above-mentioned problems. By introducing (1) two electronics switches, (2) waveguide couplers, and (3) a specifically designed calibration

F. Feng (✉) · A. Jin · J. Sun · R. Hou · H. Dang
China Academy of Space Technology (Xi'an Branch), Xi'an, China
e-mail: sailingvon@qq.com

L. Wang et al. (eds.), *Proceedings of the 6th China High Resolution Earth Observation Conference (CHREOC 2019)*, Lecture Notes in Electrical Engineering 657,
https://doi.org/10.1007/978-981-15-3947-3_37

module, we can effectively acquire the properties of each transmit–receive channel for each subbands. And then, range compression replica can be deduced using the calibration loop signals via corresponding algorithm to realize the precise matching in range dimension.

This paper is organized as follows. In Sect. 2, the calibration scheme is illustrated in detail, including the calibration loop, module unit, and the workflow to ensure all calibration signals would be acquired without interference. Section 3 gives the replica construction algorithm to extract range compression function for each subband, thus ensuring the precise image focusing. Section 4 gives the calibration and focusing results using an experimental system to verify the effectiveness of the scheme.

2 Internal Calibration Loop and Workflow

2.1 Internal Calibration Loop

Figure 1 shows the block diagram of ultrahigh-resolution spaceborne SAR with internal calibration module (ICM). The yellow blocks are added to the conventional SAR system to acquire the calibration signals.

Internal calibration involves three calibration modes, namely transmission calibration, receiving calibration, and reference calibration. The first two modes will obtain the phase and amplitude characteristics of power amplifiers and LNAs, respectively, and the last mode will compensate the common parts of the first two so as to make the final replica to be identical to the real radar echoes as much as possible.

The signal routes for SAR imaging mode and three calibration modes are listed in Table 1. It can be seen that signal routes of all working modes will cover the

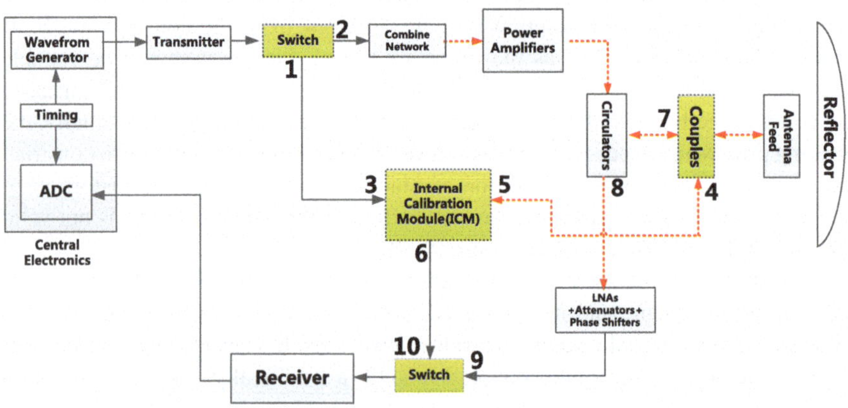

Fig. 1 Block diagram of ultrahigh-spaceborne SAR with ICM, the red-dashed line indicates multiple parallel channels

Table 1 Signal routes for SAR imaging and three calibration modes

Working modes	Signal routes
Imaging	Transmitter → 2 → 7 → Antenna → 7 → 8 → 9 → Receiver
Ref Cal	Transmitter → 1 → 3 → 6 → 10 → Receiver
Transmit Cal	Transmitter → 2 → 7 → 4 → 5 → 6 → 10 → Receiver
Receive Cal	Transmitter → 1 → 3 → 5 → 4 → 7 → 8 → 9 → Receiver

transmitter receiver, and central electronics. Specifically, imaging mode will cover (1) combine network, (2) power amplifiers, (3) circulators, (4) waveguide couplers and (5) LNAs; transmit calibration mode includes (1), (2), (3), (4), internal calibration (Port5 → Port6); reference calibration covers internal calibration (Port3 → Port6), whereas receive calibration internal calibration (Port3 → Port5), (3), (4), (6) LNAs.

2.2 Calibration Module

Assuming the reflector SAR antenna employs four feeds to transmit and receive signals, the system will thus be equipped with four transmit as well as four receive channels. Under this assumption, Fig. 2 illustrates the interior of internal calibration

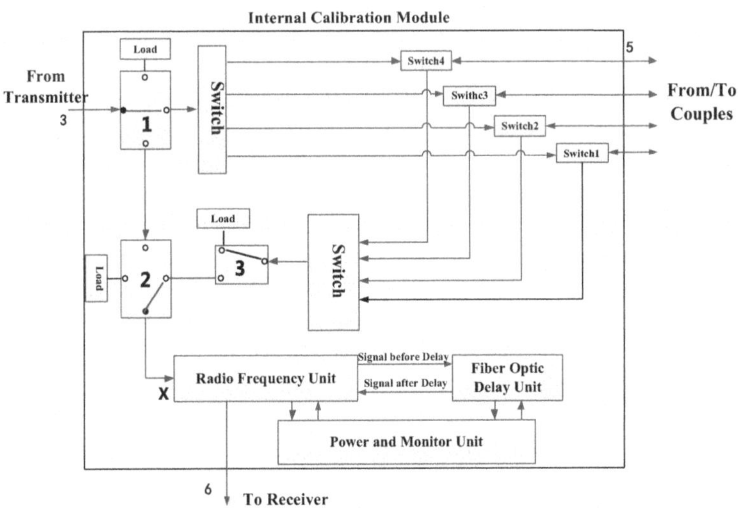

Fig. 2 Interior units of ICM

module. It includes several switches to adjust the signal flow directions to accommodate different calibration modes. Fiber optic delay unit (FODU) is employed to ensure that calibration signal receiving can be separated from the high power transmission temporally during the transmit calibration, so as to avoid the power leakage and the contamination of calibration signals. The radio frequency unit (RFU) is employed to fine-tune the signal power levels to meet the input requirements of FODU as well as that of receiver. Power and monitor units are equipped to control the calibration module and acquire its state information.

2.3 Calibration Workflow

Since SAR system will use multiple channels to transmit and receive signals, phase and amplitude imbalance among them should be removed firstly to make sure that signals from different channels could be combined coherently. Thus, this adjustment procedure should be performed first before the internal calibration. For each subband, three single frequency (lowest, central, and highest frequency within the spectrum) subpulses will be selected to obtain the phase and amplitude imbalance between different channels, and their results will then be averaged to evaluate the channel imbalance within the whole bandwidth. During this procedure, routes of sub-pulse will be the same as that of the internal calibration as shown in Table 1.

Based on imbalance results obtained by real-time evaluation onboard, attenuators and phase shifters in both transmit and receive links will be tuned to compensate them. Then, the internal calibration can be implemented to acquire the range replica. For each subband, one complete calibration period consists of one reference calibration, four transmit and receive calibrations, respectively. And the whole process will be repeated over many times to reduce the impact of additional Gaussian white noise (AWGN) and improve the signal-to-noise ratio (SNR). In one single period, the calibration sequence will be one reference calibration, then four transmit calibration, and four receive calibration at last. The reason for this is that both reference and transmit calibration pulses will go through the FODU, whose phase will be linearly drifted with time. Thus, the time interval between reference and transmit calibration should be as short as possible to make the extra phase of FODU on them to be same as much as possible. And thus their impacts could be considered as common factors in the range replica extraction, thereby simplifying the algorithm.

Figure 3 gives the calibration workflow. For each subband, it will include the channel imbalance adjustment and the internal calibration posterior to it. In the end, the range replica will be extracted with the calibration signals. The algorithm of extraction will be elaborated in next section.

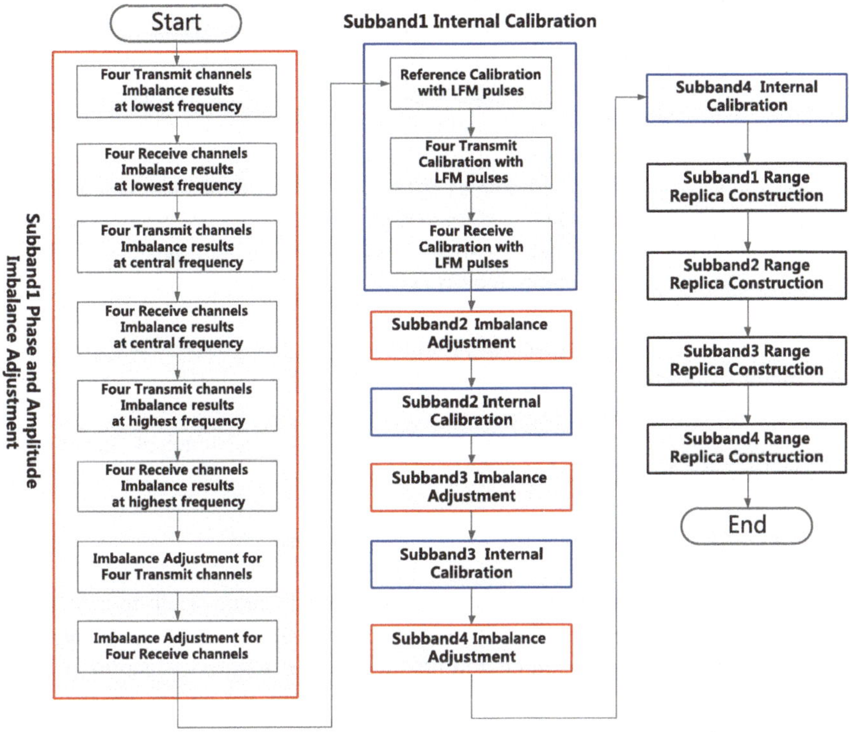

Fig. 3 Flowchart of channel imbalance adjustment and internal calibration procedure for high-resolution spaceborne SAR

3 Algorithm of Range Replica Extraction

3.1 Comparison Between Calibration and Imaging Signals

References are cited in the text just by square brackets [1]. (If square brackets are not available, slashes may be used instead, e.g. /2/.) Two or more references at a time may be put in one set of brackets [3, 4]. The references are to be numbered in the order in which they are cited in the text and are to be listed at the end of the contribution under a heading References, see our example below.

According to signal flow routes of three calibration modes, their transfer functions could be explicitly expressed as

$$\mathrm{H}k_{\mathrm{T}} = \mathrm{H}_{\mathrm{trans}} \cdot \mathrm{H}_{\mathrm{switch11}} \cdot \mathrm{H}k_{\mathrm{power}} \cdot \mathrm{H}k_{\mathrm{cir1\text{-}2}} \cdot \mathrm{H}k_{\mathrm{coup1\text{-}3}} \cdot \mathrm{H}k_{\mathrm{c}\rightarrow\mathrm{r}} \cdot \mathrm{H}_{\mathrm{switch22}} \cdot \mathrm{H}_{\mathrm{rec}} \quad (1)$$

$$\mathrm{H}_{\mathrm{ref}} = \mathrm{H}_{\mathrm{trans}} \cdot \mathrm{H}_{\mathrm{switch12}} \cdot \mathrm{H}_{\mathrm{t}\rightarrow\mathrm{r}} \cdot \mathrm{H}_{\mathrm{switch22}} \cdot \mathrm{H}_{\mathrm{rec}} \quad (2)$$

$$\mathrm{H}k_{\mathrm{R}} = \mathrm{H}_{\mathrm{trans}} \cdot \mathrm{H}_{\mathrm{switch12}} \cdot \mathrm{H}k_{\mathrm{t}\to\mathrm{c}} \cdot \mathrm{H}k_{\mathrm{coup3\text{-}1}} \cdot \mathrm{H}k_{\mathrm{cir2\text{-}3}} \cdot \mathrm{H}k_{\mathrm{LNA}} \cdot \mathrm{H}_{\mathrm{switch21}} \cdot \mathrm{H}_{\mathrm{rec}} \quad (3)$$

where $\mathrm{H}k_{\mathrm{T}}$ is the overall transfer function of k_{th} ($k = 1, 2, 3, 4$) transmit calibration channel, including that of transmitter $\mathrm{H}_{\mathrm{trans}}$, switch(from transmitter to combine network) $\mathrm{H}_{\mathrm{switch11}}$, the k_{th} power amplifier $\mathrm{H}k_{\mathrm{power}}$, Port1 $\to$ Port2 of k_{th} circulator $\mathrm{H}k_{\mathrm{cir1\text{-}2}}$, Port1 $\to$ Port3 of k_{th} coupler $\mathrm{H}k_{\mathrm{coup1\text{-}3}}$, ICM(from coupler to receiver) $\mathrm{H}k_{\mathrm{c\text{-}r}}$, switch(from ICM to receiver) $\mathrm{H}_{\mathrm{switch22}}$, and receiver $\mathrm{H}_{\mathrm{rec}}$; $\mathrm{H}_{\mathrm{ref}}$ is the transfer function of reference calibration channel, consisting $\mathrm{H}_{\mathrm{trans}}$, switch(from transmitter to ICM) $\mathrm{H}_{\mathrm{switch12}}$, ICM(from transmitter to receiver) $\mathrm{H}_{\mathrm{t\text{-}r}}$, $\mathrm{H}_{\mathrm{switch22}}$, and $\mathrm{H}_{\mathrm{rec}}$; $\mathrm{H}k_{\mathrm{R}}$ is the overall transfer function of k_{th} receive calibration channel, comprising $\mathrm{H}_{\mathrm{trans}}$, $\mathrm{H}_{\mathrm{switch12}}$, ICM(from transmitter to coupler) $\mathrm{H}_{\mathrm{t\text{-}c}}$, Port3 $\to$ Port1 of k_{th} coupler $\mathrm{H}k_{\mathrm{coup3\text{-}1}}$, Port2 $\to$ Port3 of k_{th} circulator $\mathrm{H}k_{\mathrm{cir2\text{-}3}}$, the k_{th} channel of LNA Unit $\mathrm{H}k_{\mathrm{LNA}}$, switch(from LNA to receiver) $\mathrm{H}_{\mathrm{switch21}}$, and $\mathrm{H}_{\mathrm{rec}}$.

As to the imaging mode, the transfer function of kth channel of transmitter–receiver loop (excluding antenna) is given by

$$\mathrm{H}k_{\mathrm{sig}} = \mathrm{H}_{\mathrm{trans}}\mathrm{H}_{\mathrm{switch11}} \cdot \left(\mathrm{H}k_{\mathrm{power}}\mathrm{H}k_{\mathrm{cir1\text{-}2}}\mathrm{H}k_{\mathrm{coup1\text{-}2}}\mathrm{H}k_{\mathrm{coup2\text{-}1}}\mathrm{H}k_{\mathrm{cir2\text{-}3}}\mathrm{H}k_{\mathrm{LNA}}\right) \cdot \mathrm{H}_{\mathrm{switch21}}\mathrm{H}_{\mathrm{rec}} \quad (4)$$

By comparing (1)–(3), and (4), we can clearly found that most sub-transfer functions of the imaging mode are already embodied within the (1)–(3), except for a few ones. And this fact lays the foundation for the range of replica extraction given in the next subsection.

3.2 Algorithm for Range Replica Extraction

If you follow the "checklist" your paper will conform to the requirements of the publisher and facilitate a problem-free publication process.

On the basis of expressions of (1)–(4), we can observe that $\mathrm{H}k_{\mathrm{sig}}$ can be expressed in terms of $\mathrm{H}k_{\mathrm{T}}$, $\mathrm{H}_{\mathrm{ref}}$, and $\mathrm{H}k_{\mathrm{R}}$ as

$$\mathrm{H}k_{\mathrm{sig}} = \frac{\mathrm{H}k_{\mathrm{T}}\mathrm{H}k_{\mathrm{R}}}{\mathrm{H}_{\mathrm{ref}}} \cdot \frac{\mathrm{H}_{\mathrm{t}\to\mathrm{r}}}{\mathrm{H}k_{\mathrm{c}\to\mathrm{r}}} \cdot \frac{1}{\mathrm{H}k_{\mathrm{t}\to\mathrm{c}}} \cdot \frac{\mathrm{H}k_{\mathrm{coup1\text{-}2}} \cdot \mathrm{H}k_{\mathrm{coup2\text{-}1}}}{\mathrm{H}k_{\mathrm{coup1\text{-}3}} \cdot \mathrm{H}k_{\mathrm{coup3\text{-}1}}} \quad (5)$$

And since all four transmit and receive channels have been used for transmission and receiving, respectively, the range replica could be given by the combination of four transmit–receive channels as

$$\mathrm{H}_{\mathrm{sig}} = \mathrm{H1}_{\mathrm{sig}} + \mathrm{H2}_{\mathrm{sig}} + \mathrm{H3}_{\mathrm{sig}} + \mathrm{H4}_{\mathrm{sig}} \quad (6)$$

From (5) and (6), it can be observed that besides $\mathrm{H}k_{\mathrm{T}}$, $\mathrm{H}_{\mathrm{ref}}$, and $\mathrm{H}k_{\mathrm{R}}$, transfer functions of both coupler and ICM are needed to derive the final range replica. Due to the passive microwave properties, couplers are comparatively constant and will not change significantly. Thus, their transfer functions can be acquired with the aid of vector network analyzer (VNA) on land. As regards to ICM properties, $Hk_{t\text{-}c}$ only consists of switches and will also perform stably onboard. This term could also be determined in advance. Nevertheless, properties of $Hk_{c\text{-}r}$ and $H_{t\text{-}r}$ will vary with operation circumstance and time due to the RFU and FODU in their signal routes. And this problem must be treated to guarantee the range replica effectiveness. In the following, we will show the corresponding approach to deal with it.

The term $H_{t\rightarrow r}/Hk_{c\rightarrow r}$ in (5) can be decomposed as

$$\frac{H_{t\rightarrow r}}{Hk_{t\rightarrow c}} = \frac{H_{switch} \cdot H_{RFU+FODU}}{Hk_{switch} \cdot H_{RFU+FODU}} \tag{7}$$

where H_{switch} and Hk_{switch} represent transfer functions of Port3 and Port5 to the input port of RFU (as shown in the Capital letter X in Fig. 2) in the ICM, respectively, and $H_{RFU+FODU}$ the combined transfer functions of RFU and FODU.

As illustrated above, the reference and transmit calibration are adjacent in time. And such a short time interval can promise very little changes in the $H_{RFU+FODU}$ that this term in both numerator and denominator of (7) can be eliminated. In consequence, $H_{t\rightarrow r}/Hk_{c\rightarrow r}$ could be rewritten as

$$\frac{H_{t\rightarrow r}}{Hk_{c\rightarrow r}} = \frac{H_{switch}}{Hk_{switch}} \tag{8}$$

And since both terms of H_{switch} and Hk_{switch} only include switches, their properties will have stable performance and can be measured before the integration of ICM.

In conclusion, the range replica will be obtained from calibration signals of three calibration modes, properties of couplers, and that of ICM without RDU and FODU.

4 Verification Results

4.1 Experimental System Parameters

In order to verify the effectiveness of suggested internal calibration approach, we have taken advantage of an existing experimental SAR system and made necessary modifications so that it can be equipped with multi-channels to transmit and receive signals and can accommodate multiple subpulses located within adjacent subbands. Furthermore, the ICM has been embedded within the SAR system to realize the transmission, and receive as well as reference calibration. Table 2 lists parameters of SAR system, making sure that SAR system has met the requirement of ultrahigh-resolution standard in terms of bandwidth.

Table 2 Ultrahigh-resolution spaceborne SAR parameters

Working modes	Value
Subband1 carrier frequency	8.7 GHz
Subband2 carrier frequency	9.45 GHz
Subband3 carrier frequency	10.15 GHz
Subband4 carrier frequency	10.9 GHz
Bandwidth of each subband	830 MHz
Bandwidth of combined spectrum	3.03 GHz

4.2 Range Replica Extraction

Taking subpulse1 as an example, we give the amplitude–frequency response of $\mathrm{H}k_{\mathrm{T}} \cdot \mathrm{H}k_{\mathrm{R}} / \mathrm{H}_{\mathrm{ref}}$ $(k = 1, 2, 3, 4)$ associated with four transmit–receive loops in Fig. 4, where the blue line indicates the result of single cycle and the red one the average result of 1024 repetitions. It can be seen that differences among them are obvious

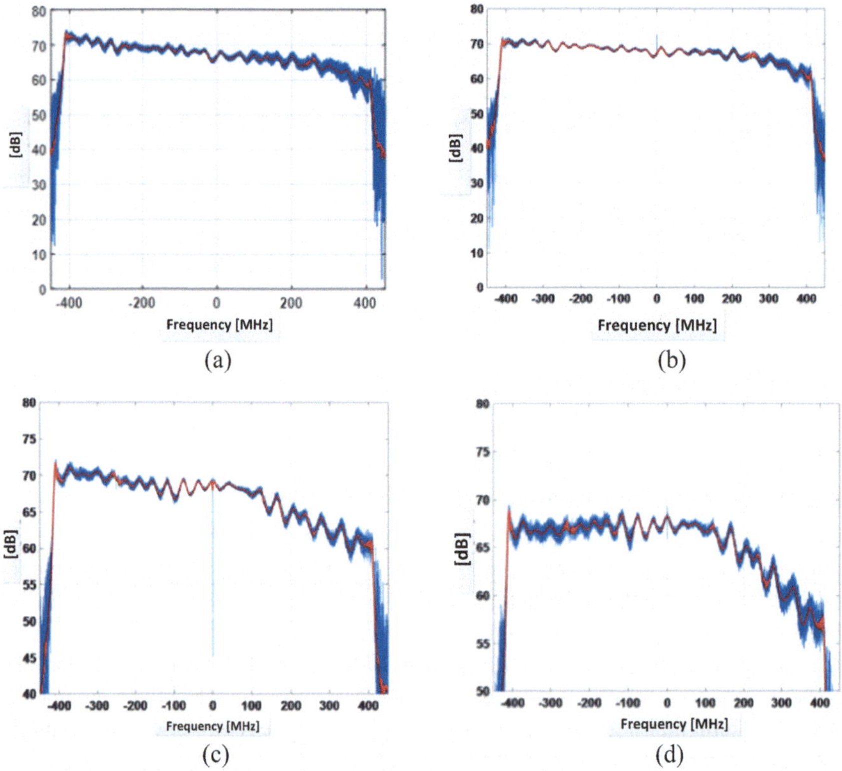

Fig. 4 Amplitude–frequency response of four transmit–receive channel loops associated with subpulse1. **a**–**d** represents that of channel 1–4, respectively

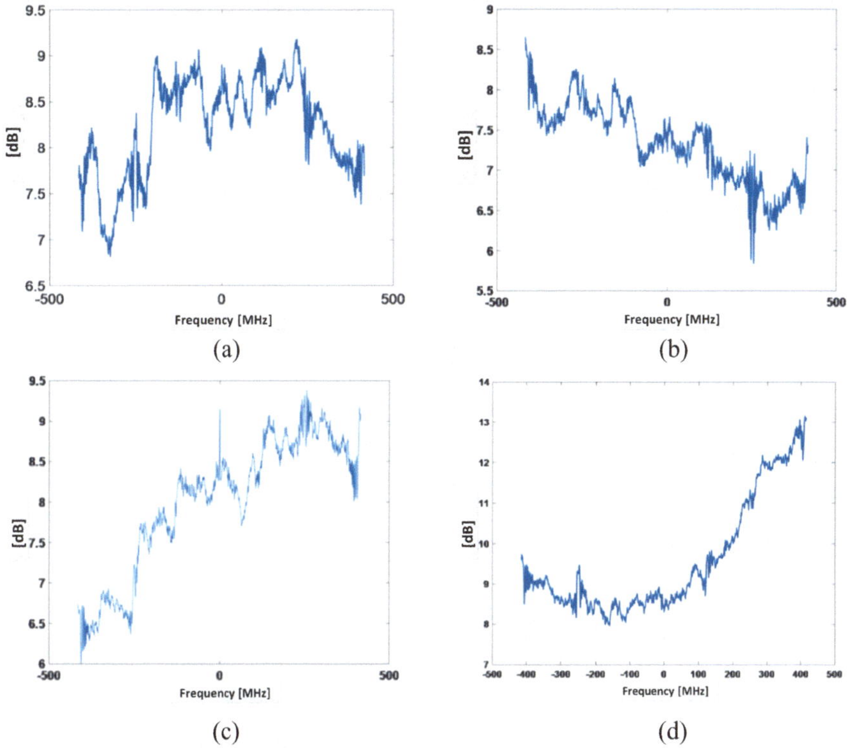

Fig. 5 Amplitude–frequency response of four ICM channels associated with subpulse1. **a**–**d** represents that of channel 1–4, respectively

due to the combined influence of transmit–receive loops and ICM. Thus, in order to remove the effect of ICM, the term $\mathrm{H}_{t\to r}\big/\mathrm{H}k_{c\to r}\cdot 1\big/\mathrm{H}k_{t\to c}$ in Eq. (5) must be taken into account. Figure 5 shows the corresponding results associated with four internal calibration channels.

By combining results shown in Figs. 4 and 5, we will attain the final range replica for subpulse1. And the same procedure could be applied to attain the range replica for other three subpulses.

4.3 Processing Results

In Fig. 6, we give the range compression results with the range replica before and after the ICM term compensation. It can be clearly found that much better performance will be achieved after the residual term caused by ICM being compensated.

In the end, range spectra of four subpulses should be combined in frequency domain to achieve wider bandwidth and consequently ultrahigh resolution. Figure 7

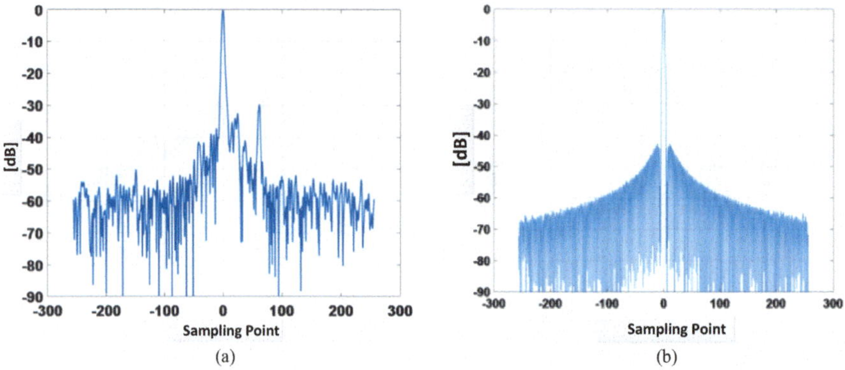

Fig. 6 Range compression results with range replica before and after the ICM term compensation

Fig. 7 Range compression results after the combination of spectra of four subpulses

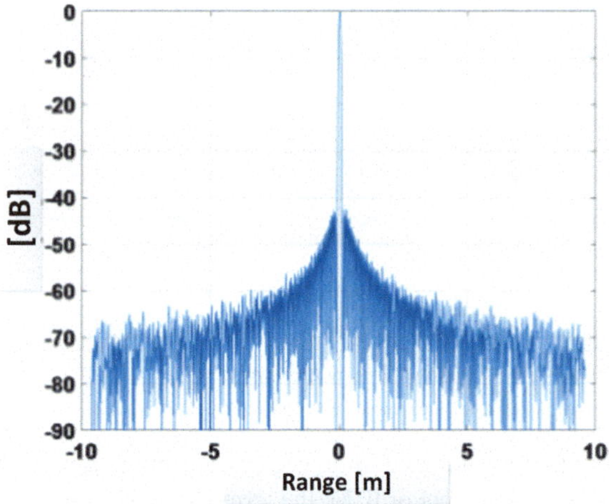

gives this result which shows better performance in terms of resolution as compared to Fig. 6, and the impulse response width, peak-sidelobe ratio (PSLR), and integrated-sidelobe ratio (ISLR) are listed in Table 3.

Table 3 Compression parameter of combination results

IRW	PSLR	ISLR
1.48	−42 dB	−26.5 dB

5 Conclusion

To realize the goal of high image quality for ultrahigh-resolution SAR, this paper proposes a novel internal calibration scheme and module to attain the characteristics of each transmit–receive channel loop for each subpulse. Moreover, the flowchart and algorithm are provided to acquire the range replica for precise compression using calibration signals. By using a modified SAR system, we have finally validated the effectiveness of this calibration approach for high-resolution SAR.

References

1. Gebert N (2009) Multi-channel azimuth processing for high-resolution wide-swath SAR imaging. Ph.D., University of Karlsruhe, Karlsruhe, Germany
2. Ender JHG, Brenner AR (2003) PAMIR—a wideband phased array SAR/MTI system. In: IEE proceedings—radar, sonar and navigation, vol 150, pp 165–172
3. Brenner AR, Ender JHG (2006) Demonstration of advanced reconnaissance techniques with the airborne SAR/GMTI sensor PAMIR. In: IEE proceedings—radar, sonar and navigation, vol 153, pp 152–162
4. Gebert N, Krieger G, Moreira A (2009) Digital beamforming on receive: techniques and optimization strategies for high-resolution wide-swath SAR imaging. IEEE Trans Aerosp Electron Syst 45:564–592
5. Rippler J (2012) Ultrahigh resolution X-band SAR images with SmartRadar. In: Proceedings of European conference on synthetic aperture radar, Nuremberg, Germany, pp 426–428
6. Freeman A (1992) SAR calibration: an overview. IEEE Trans Geosci Remote Sens 30:1107–1121
7. Zink M, Torres R, Buck CH (2003) Calibration and early results of the ASAR on ENVISAT. In: Proceedings of IEEE 25th international geoscience and remote sensing symposium, Toulouse, France, pp 3096–3098
8. Brautigam B, Schwerdt M, Bachmann M (2007) In-flight monitoring of TerraSAR-X radar instrument stability. In: Proceedings of IRS, Cologne, Germany
9. Brautigam B, Schwerdt M, Bachmann M (2010) TerraSAR-X instrument calibration results and extension for TanDEM-X. IEEE Trans Geosci Remote Sens 48:702

Monitoring and Analysis of Surface Deformation and Glacier Motion Along the Sichuan–Tibet Railway: A Case Study of the Lhasa–Nyingchi Railway Section

Jinghui Luo, Qing Ding, Gong Zhang, Wei Zhang, Xiaoxia Wang, and Changli Zheng

Abstract The Sichuan–Tibet Railway which stretches as far as 1600 km starts from Chengdu in Sichuan Province and west to Lhasa in the Tibet Autonomous Region. The terrain along the railway is undulating, the ecological environment is fragile, and the railway passes through high-intensity active seismic belts and geological fault zones, facing a variety of geological disaster risks. Keeping abreast of changes in surface morphology along the railway and further monitoring and early warning of disasters can provide important technical support for the smooth construction and safe operation of the Sichuan–Tibet Railway. Taking the Lhasa to Nyingchi Railway section as an example, using the spaceborne SAR data, the surface deformation information along the railway was extracted by using the interferometry point target analysis (IPTA) technology, and the glacier motion information was derived by using the pixel offset-tracking (POT) technology. The observation results showed that the annual subsidence velocity was small in most areas of our study area, except the local areas along the LinMao Highway, whose subsidence velocity was more than 3.5 cm/year, which can be served as a key monitoring area. Another observation was that the velocity of glacier motion had a great relationship with the temperature change in our study area. As a whole, the velocity of glacier motion increases with the increase of temperature, and the glacier motion velocity in summer is significantly higher than that in winter. In addition, local topographic conditions such as slope and aspect also had a great influence on the glacier motion velocity of the glacier. Therefore, it is necessary to analyze the glacier motion in combination with local topographic.

Keywords Sichuan-Tibet railway · Deformation monitoring · Glacier motion · IPTA · POT

J. Luo (✉) · Q. Ding · G. Zhang · W. Zhang · X. Wang · C. Zheng
Science and Technology on Electronic Information Control Laboratory, Chengdu, China
e-mail: wsljh555@163.com

L. Wang et al. (eds.), *Proceedings of the 6th China High Resolution Earth Observation Conference (CHREOC 2019)*, Lecture Notes in Electrical Engineering 657,
https://doi.org/10.1007/978-981-15-3947-3_38

1 Introduction

The Sichuan–Tibet Railway starts from Chengdu in Sichuan Province in the east and Lhasa in the Tibet Autonomous Region in the west. It is a key line for the country to implement the "One Belt, One Road" development strategy. The planning and construction of the Sichuan–Tibet Railway have great and far-reaching significance for the long-term stability of China and the economic construction and development of Tibet. The construction of the Sichuan–Tibet Railway faces many challenges such as large terrain differences, strong seismic activity along the route and frequent mountain disasters. And it will face enormous challenges to ensure the smooth construction and future safe operation of the Sichuan–Tibet Railway in the future.

Interferometric Synthetic Aperture Radar (InSAR) technology provides the capability to monitor precise surface displacements over time with a wide coverage in a time-efficient manner [1]. It has been used successfully in various applications and fields of research as surface deformation monitoring [2], earthquake and plate movement [3, 4], infrastructure deformation [5], glacial drift [6, 7], landslide [8] and other fields. This paper takes the Lhasa–Nyingchi Railway section as an example to carry out monitoring and analysis of geological disasters along the railway, which is under construction and is also an important part of the Sichuan–Tibet Railway. We focused on surface time series deformation monitoring and glacier motion monitoring along the railway, which can provide a scientific and accurate basis for quantitative evaluation of potential geological disasters along the railway.

2 Methods

2.1 Surface Deformation Monitoring

Persistent Scatters InSAR (PS-InSAR) technology was proposed by Italian scholar Ferretti et al. [9, 10], extracting deformation information by using various ground object targets with strong backscattering of radar waves and stable timing. It inherits the advantages including wide range and high spatial resolution of differential SAR interferometry (D-InSAR) and overcomes the time and baseline miscorrelation of traditional D-InSAR. To solve the problem that the traditional PS-InSAR method can only select a small number of stable phase points (Persistent Scatters, PS) in low coherence regions, a method called interferometry point target analysis (IPTA) was developed on the basis of PS-InSAR. IPTA method only performs time dimension and space dimension analysis on extracted points to obtain large-scale surface deformation of long time series. The processing flow of IPTA is shown in Fig. 1. Taking there is only a single reference image as an example, the main technical ideas are as follows.

It is assumed that there are N SAR images of different phases in the study area, and one of them is selected as the reference image, and the remaining N-1 images are the

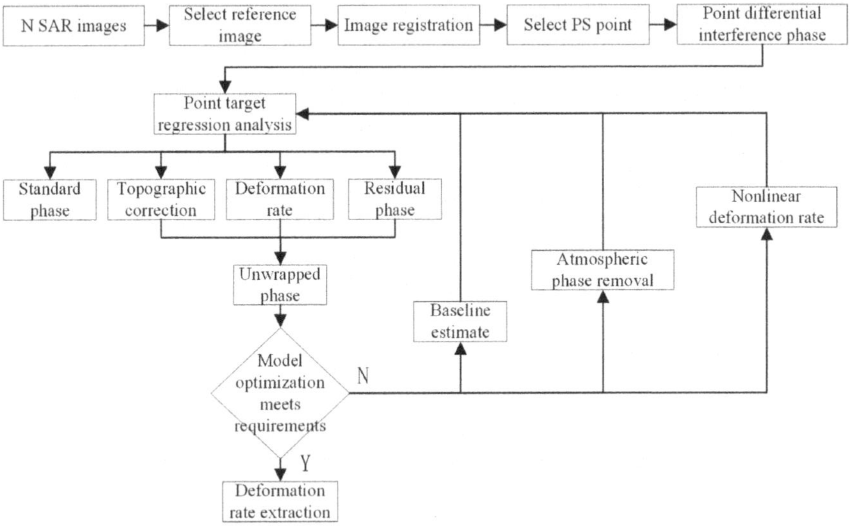

Fig. 1 Flowchart of IPTA method

slave images, which are, respectively, registered with the reference image to obtain N-1 interference pairs. By using the external DEM data and performing differential interference processing, N-1 differential interferograms can be obtained, thereby obtaining N-1 time series differential interference phases at each PS point. Then, using the regression analysis method, the elevation error and surface deformation of the PS point are obtained. The phase model of the IPTA method can be expressed as follows:

$$\varphi_{\text{unw}} = \varphi_{\text{topo}} + \varphi_{\text{def}} + \varphi_{\text{atmos}} + \varphi_{\text{noise}} \tag{1}$$

where φ_{unw}, φ_{topo}, φ_{def}, φ_{atmos} and φ_{noise} denote unwrapped interference phase of PS point, topographic error, the phase components along the radar line of sight (LOS) due to surface deformation, atmospheric artifacts and decorrelation/thermal noise, respectively. Equation (1) can also be expressed as follows using the observed geometric parameters:

$$\varphi_{\text{unw}} = \frac{4\pi}{\lambda R} \cdot \frac{B_{\perp}}{\sin\theta} \cdot \Delta h + \frac{4\pi}{\lambda} \cdot v \cdot T_i + \varphi_{\text{res}} \tag{2}$$

where λ, R, θ, $B_{\perp}$, T_i, Δh denote radar wavelength, the slant distance between radar and the ground target, radar incident angle, time baseline of interference pairs, vertical baseline of interference pairs, topographic correction, respectively. v is linear deformation rate in the LOS direction. And φ_{res} is the residual phase of the PS point including the atmospheric delay phase, the nonlinear deformation phase and the noise phase.

2.2 Glacier Motion Monitoring

The pixel offset-tracking technique uses the intensity tracking method [11] or the coherence tracking method [12] to register the two SAR images pixel by pixel and then estimates the surface displacement from the registration offset. This paper monitors the glacier motion based on the intensity tracking method, considering the POT method is more suitable for areas with low coherence and obvious surface features. The processing flow of the POT technology is shown in Fig. 2. The speckle noise information was used to perform normalized cross-correlation calculation on the intensity information of SAR images to obtain offsets between pixels, where the correlation coefficient was calculated as follows:

$$cc(x,y) = \frac{\sum_{x,y}(r(x,y)-u_r)(s(x-u,y-v)-u_s)}{\sqrt{\sum_{x,y}(r(x,y)-u_r)^2\sum_{x,y}(s(x-u,y-v)-u_s)^2}} \tag{3}$$

where $cc(x,y)$ is normalized cross-correlation. (x,y) and $(x-u,y-v)$ represent pixel locations in reference image and slave image, respectively. r and s denote pixel values in reference image and slave image, respectively. u_r and u_s are the average pixel values in reference window and search window, respectively.

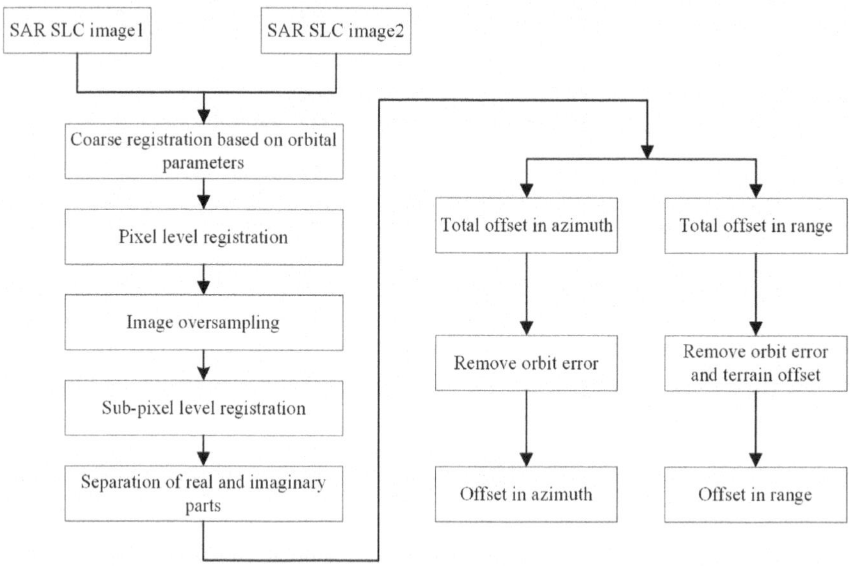

Fig. 2 Flowchart of POT method

3 Study Sites and Data

The research area of this paper is along the line of Lhasa–Nyingchi Railway and its surrounding area, N29°0′–N29°55′, E93°42′–E95°18′, mainly involving the part of Bayi district and Mainling County of Nyingchi City (Fig. 3). The red color box in Fig. 3 shows the surface deformation research area, covering a range of about 32 km × 50 km, mainly including the Lhasa–Nyingchi Railway construction section, Nyingchi Mainling Airport and other areas. The blue color box in Fig. 3 shows the glacier motion research area, covering a range of about 28 km × 30 km, with glaciers and frozen soils in the area.

The image data used to extract the surface deformation of the study area is the Sentinel-1 interference wide-mode SAR image. All images are in single-view complex format data, the incident angle is about 44°, the polarization mode is VV, the azimuth sampling interval is 20 m, and the sampling interval is 5 m. The PS candidate point targets were extracted according to the spectral characteristics and backscattering characteristics of the coherence point in the study area. When generating the differential interferogram, we selected multiple reference images, set the time baseline (Delta_T) threshold to 100 days, and the perpendicular baseline (Bperp) threshold to 200 m. Then, we got a total of 87 interference image pairs. After eliminating the poor quality interference pairs, we finally got 61 high-quality interference pairs (Fig. 4). In Fig. 4, the axis of abscissa represents the image acquisition time, and the axis of ordinate represents perpendicular baseline. The number 1–31 in Fig. 4 corresponds to the 31 images at 12 days interval between September 5, 2017 and August 31, 2018.

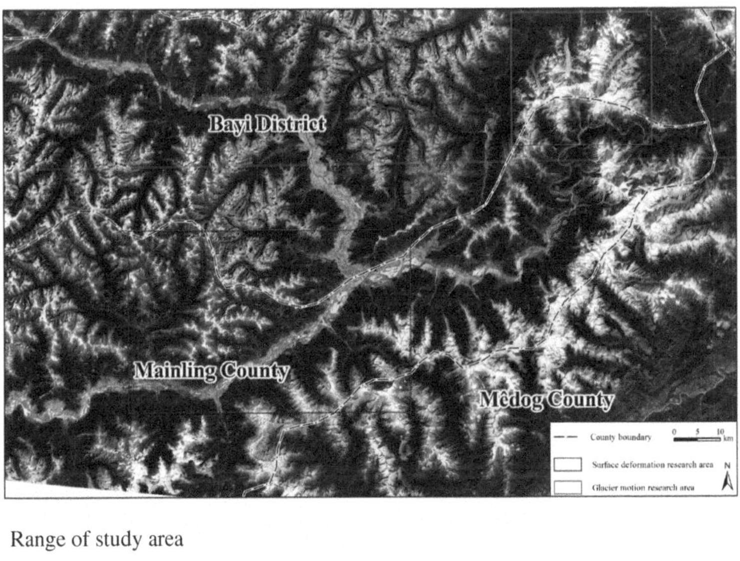

Fig. 3 Range of study area

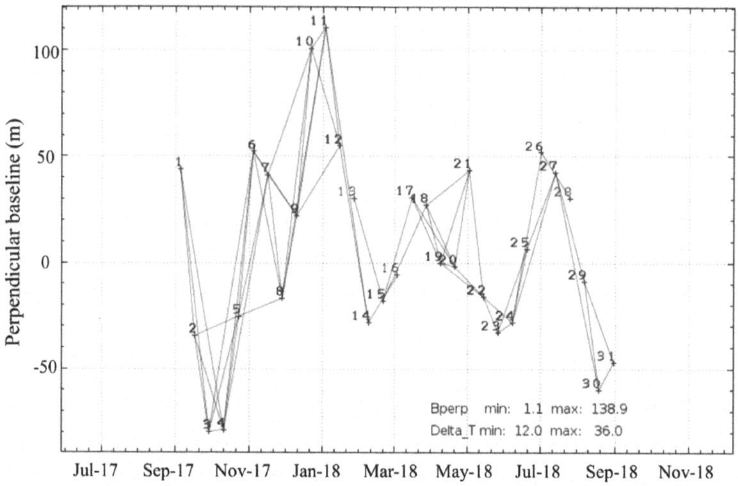

Fig. 4 High-quality interferometry pairs

Table 1 Details of image pairs

Orbit	No.	Reference image	Slave image	Delta_T/d	Bperp/m	Mode
1	1	20170614	20171101	140	−85.3	Fine
	2	20171101	20180307	126	484.2	Fine
2	3	20170706	20171109	126	332.8	Fine
	4	20171109	20180510	182	−128.0	Fine

Image data used for dynamic monitoring of glacier motion is ALOS-2 strip mode data. The image acquisition time is from 2017 to 2018, with a total of six scene images, distributed in two orbits (satellites in orbital 1 and orbit 2 have opposite directions of flight, the orbital 1 has an incident angle of about 41°, and the orbital 2 has an incident angle of about 31°). The image has a width of 70 km and an azimuth sampling interval of 3.2 m. Image information and combinations are shown in Table 1.

4 Results and Discussion

4.1 IPTA

The PS-InSAR results along the Lhasa–Nyingchi Railway section and its surrounding areas are shown in Fig. 5. The name of the county, major roads and some locations are marked in figure. The deformation velocity range is set to −3.5 to 3.5 cm/y. In

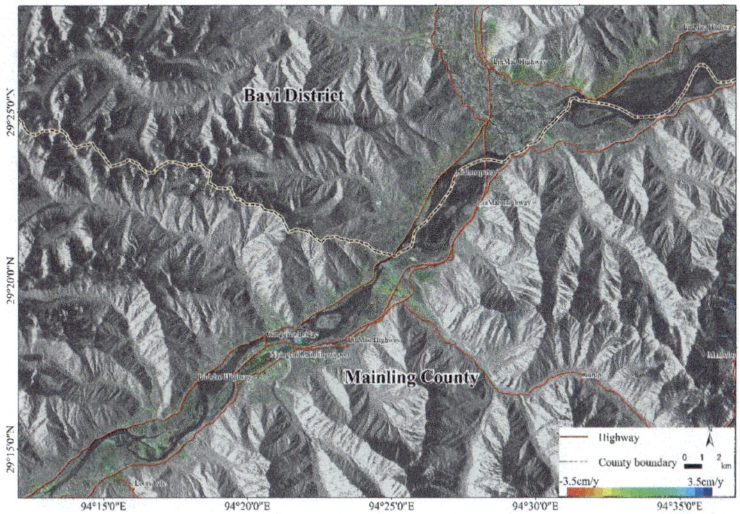

Fig. 5 Deformation velocity map

this study, we only care about the deformation information along the railway and the surrounding area. Therefore, most of the mountainous areas are masked, and the deformation velocity maps of relevant areas are not shown the figure.

It can be seen from Fig. 5 that during the period from September 5, 2017 to August 31, 2018, the annual deformation of most areas in the study area was small, and only a small part area has a large deformation. The ground subsidence was mainly concentrated along the LinMao Highway in the northeast of velocity map. The local deformation was more than 3.5 cm/year, which can be served as a key monitoring area.

Ground uplift mainly occurs in the northeastern part of Nyingchi Mainling Airport, where there was an overall uptrend. This condition may be related to the changes of flood season and dry season of the Brahmaputra River. In the summer (July, August and September), the Brahmaputra River was in its flood season, the water volume increased and flooded some of the tidal flats and coastal areas. In other seasons, river water volume decreased, and some flooded areas appeared.

4.2 POT

The dynamic monitoring results of glacier motion are shown in Figs. 6, 7, 8 and 9. The slope and aspect distributions of the corresponding study area are shown in Fig. 10, which were extracted from the SRTM DEM data of 30 m resolution. Figures 6, 7, 8 and 9 represent the motion velocity maps of different time periods derived from ALOS-2 data. Each color cycle in the figures represents a motion rate of 60 cm/d.

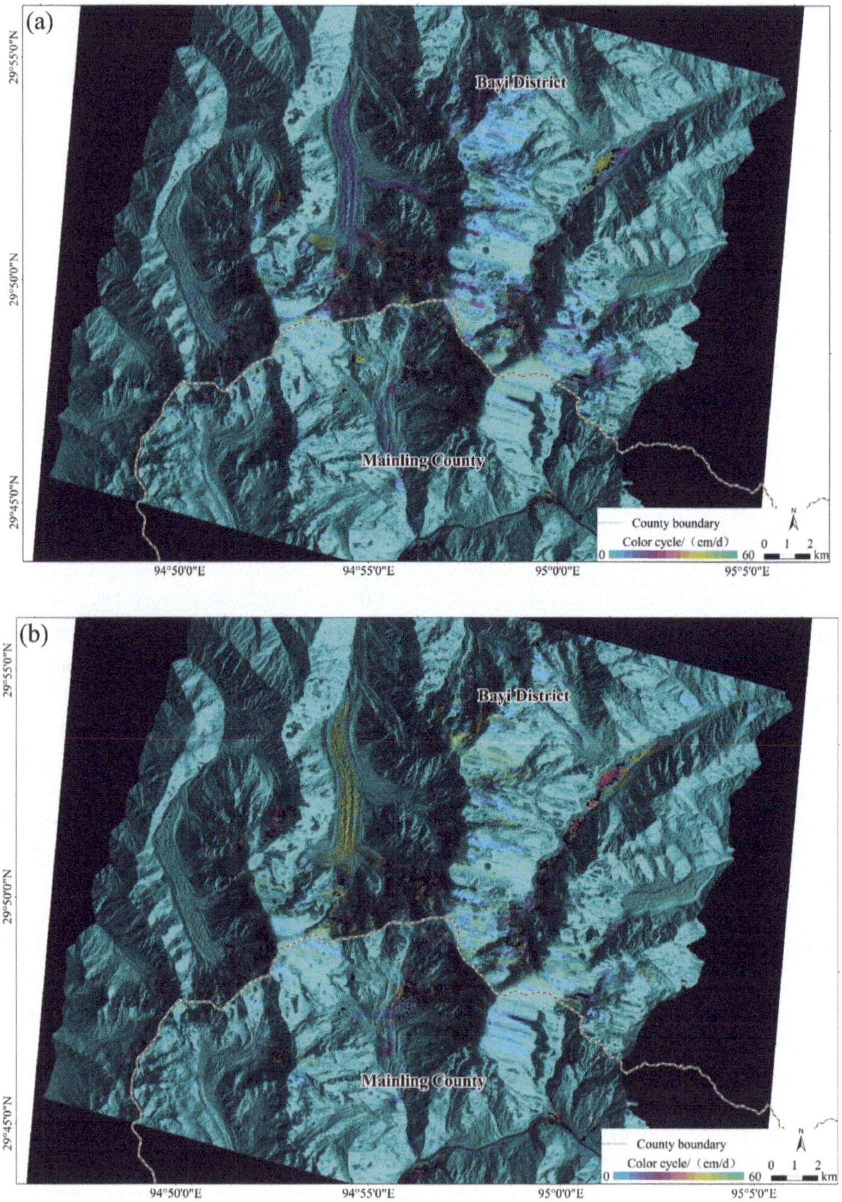

Fig. 6 Motion velocity maps of glacier from image pair 20170614–20171101. **a** In range; **b** in azimuth

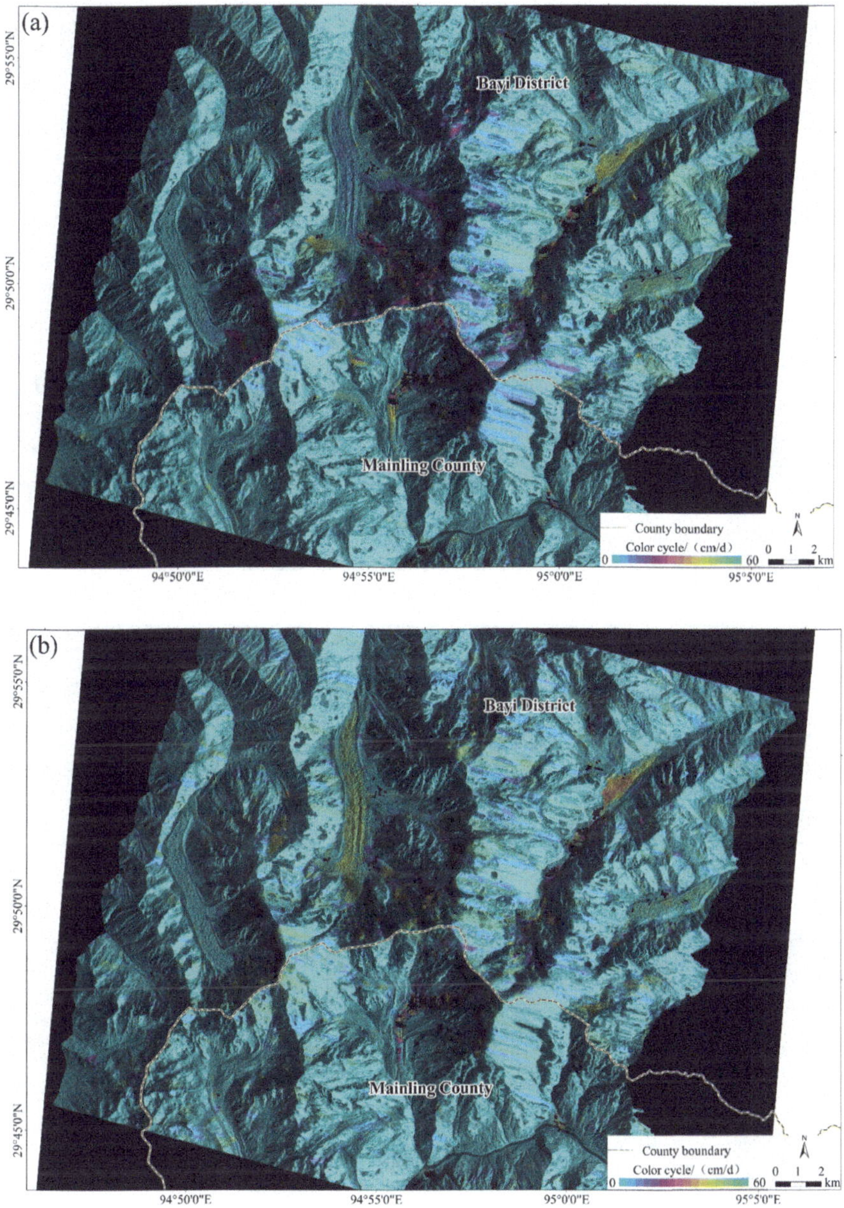

Fig. 7 Motion velocity maps of glacier from image pair 20171101–20180307. **a** In range; **b** in azimuth

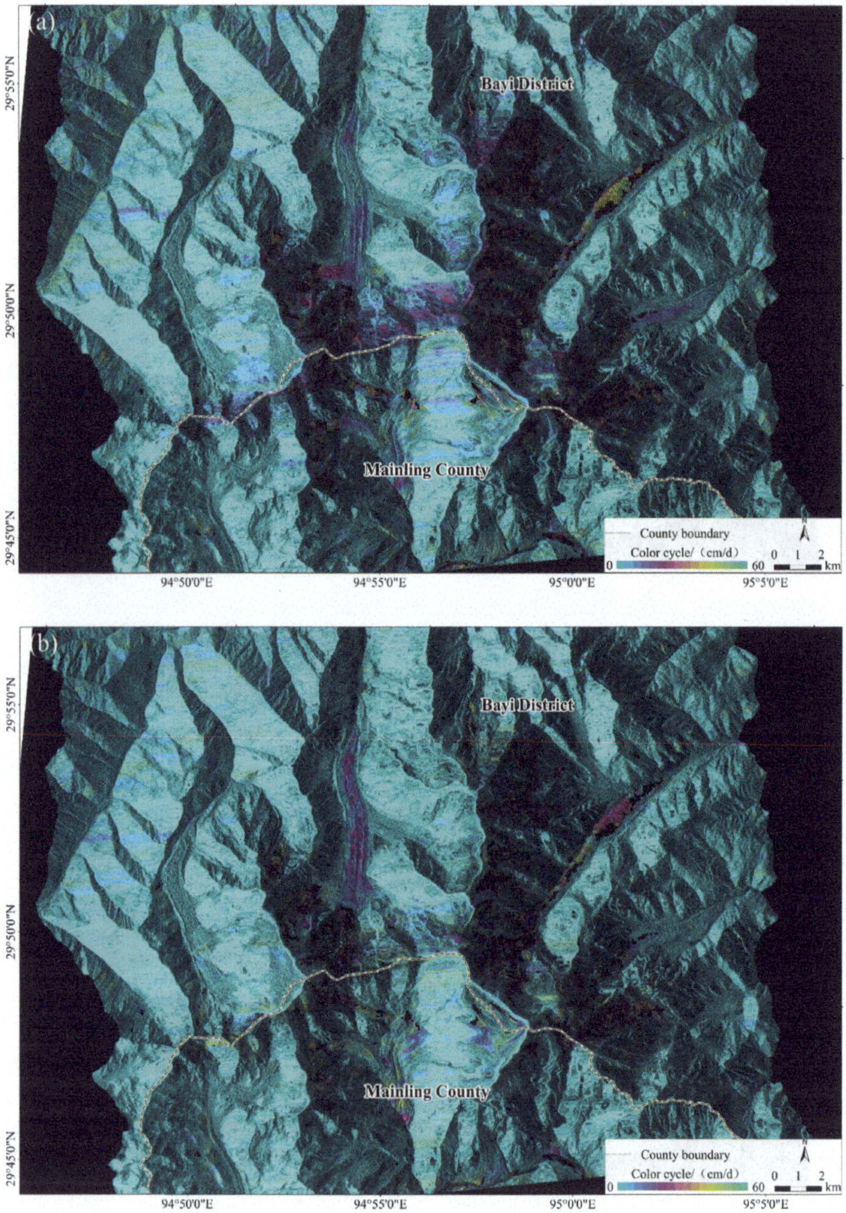

Fig. 8 Motion velocity maps of glacier from image pair 20170706–20171109. **a** In range; **b** in azimuth

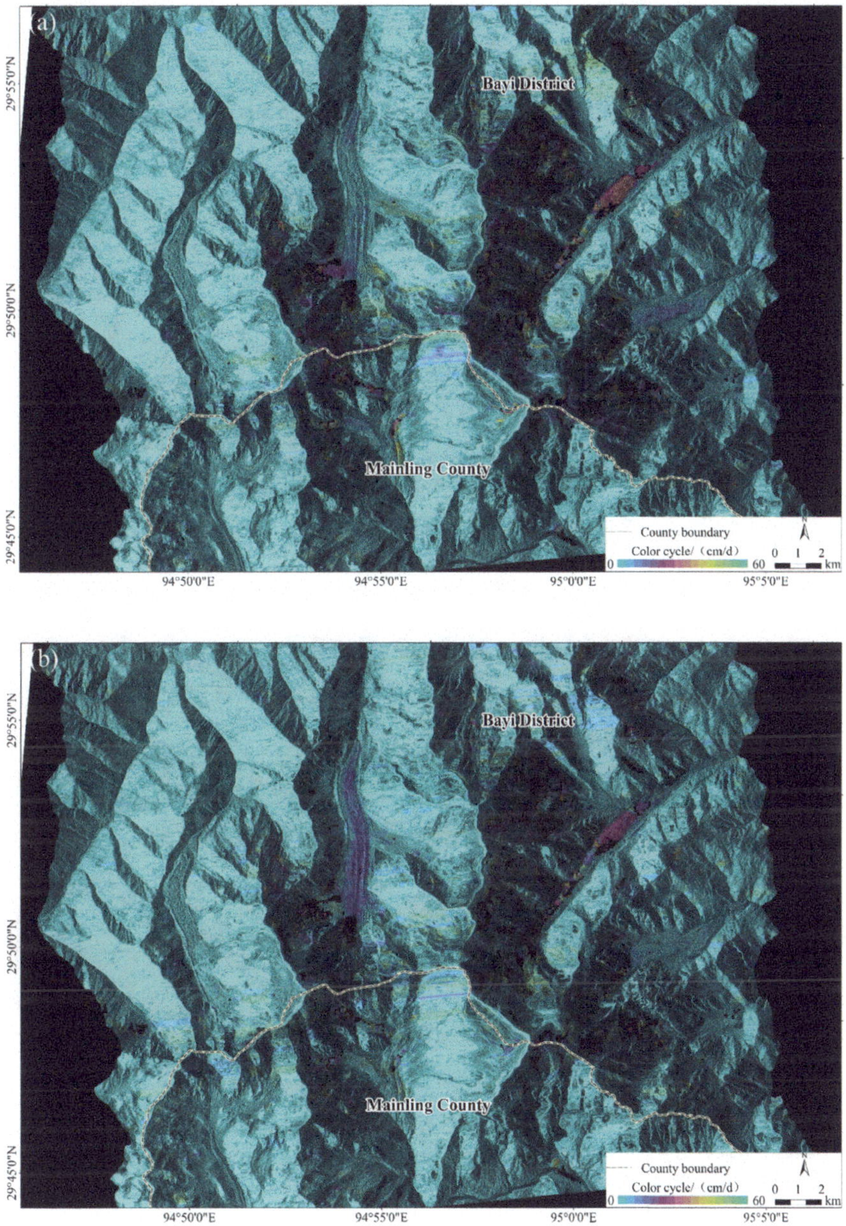

Fig. 9 Motion velocity maps of glacier from image pair 20171109–20180510. **a** In range; **b** in azimuth

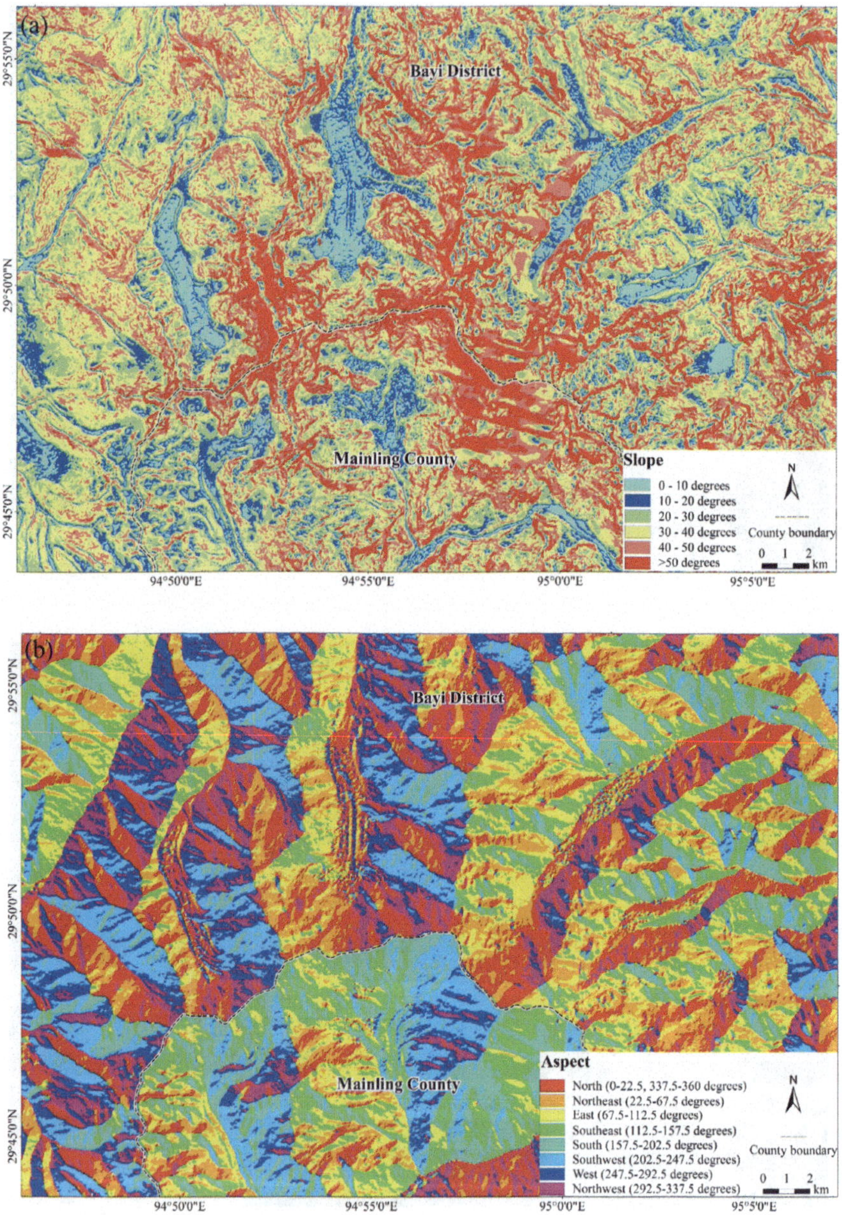

Fig. 10 Slope and aspect distribution maps of study area. **a** Slope; **b** aspect

There were some black areas in the effective data range of the study area, which are low correlation areas and SAR imaging overlap shadow areas. In addition, due to natural phenomena such as wind, rain and snow, the scattering characteristics of the target may have changed, resulting in registration accuracy in some areas. It also caused empty values and motion rate mutation points in some areas.

The results show that the motion velocity of the glacier has a great relationship with the temperature change. As the summer temperature rises, the speed of glacier movement increases, and the velocity of glaciers in summer is significantly faster than that in winter. Reflected in the figure, the velocity of glacier motion in 20170614–20171101 period (Fig. 6) is faster than that in 20171101–20180307 period (Fig. 7) of orbit 1; the velocity of glacier motion in 20170706–20171109 period (Fig. 8) is faster than that in 20171109–20180510 period (Fig. 9) of orbit 2; the flow rate of the glaciers in 20171109–20180510 period of orbit 1 is significantly increased compared with 20171101–20180307 period of orbit 2, which is consistent with the increase of glacier flow rate with increasing temperature. At the same time, we also noticed that the derived glacial motion velocity defers by using SAR data acquired in different orbits at the same time period. It is obviously that the glacier motion velocity in Fig. 6 is about 10 cm/d faster than that in Fig. 8. The main difference is that the satellite heading and incidence angle of the two orbits are different. In complex mountainous terrain conditions, the range of data obtained by the two orbits varies.

Local topographic also has a large impact on the flow rate of the glacier. The satellite flies in the north–south direction in both orbit 1 and orbit 2, but orbit 1 is an ascending orbit, and orbit 2 is the opposite. As shown in Figs. 6, 7, 8 and 9, the glacier motion velocity varies greatly in range and in azimuth. In the figures, the north–south flowing glaciers have a smaller motion velocity in range than that in azimuth. And the east–west flowing glaciers have a smaller velocity of motion in azimuth than that in range. Take the north–south flowing glacier that in the middle of the figures as an example, considering the data of which is relatively complete. We can find that from the upper part of the glacier to the end of the ice tongue, the motion velocity of the glacier increases first and then decreases gradually. The flow velocity of the glacier is less than the middle on both sides, and the velocity at the end of the ice tongue is obviously slowed down. From Fig. 10, we can find that the glacier slope is generally northward, the glacier slope is mostly between 0° and 20°, and the glacier slope in middle is slightly higher than that in the two sides. This also confirms that the glacier velocity we found is affected by local topographic.

In addition, the glacier surface and bottom dam will also affect the motion velocity of glaciers. Even in different glaciers or different glaciers in the same area, the glacier motion velocity may be different. Therefore, specific analysis is needed for specific study areas to quantitatively evaluate their potential disaster risks and hazards.

5 Conclusions

Taking the Lhasa–Nyingchi Railway section as an example, this paper used PS-InSAR technology to extract the surface time series deformation information along the railway on the basis of spaceborne SAR data. In addition, considering that there are many glacial groups along the Sichuan–Tibet Railway, with global warming, glaciers melting accelerated, and there is a potential risk of geological disasters. Therefore, POT technology was used to extract the glacier motion velocity and dynamically monitor the evolution of glaciers. The results showed that SAR interferometry technology can be used for surface deformation and glacier motion monitoring, thus can provide important technical support for disaster monitoring and early warning. It should be pointed out that due to the limitation of data, the monitoring results were not verified in this paper. Next, we will apply for the leveling surveying data in similar periods from relevant departments to calibrate and verify our PS-InSAR and POT monitoring results.

Acknowledgements The ALOS-2 data used for glacier motion monitoring comes from the Surveying and Mapping Emergency Support Center of the Sichuan Surveying and Mapping Geographic Information Bureau.

References

1. Tazio S, Frank P, Andreas W et al (2017) Circum-Arctic changes in the flow of glaciers and ice caps from satellite SAR data between the 1990s and 2017. Remote Sens 9(9):947
2. Zhang Y, Zhang J, Gong W et al (2009) Monitoring urban subsidence based on SAR interferometric point target analysis. Acta Geod Cartogr Sin 38(6):482–487
3. Ji L, Liu C, Xu J et al (2017) InSAR observation and inversion of the seismogenic fault for the 2017 Jiuzhaigou Ms 7.0 earthquake in China. Chin J Geophys 10:4069–4082
4. Jiang M, Ding X, Li Z et al (2009) Study on coseismic deformation of WenChuan earthquake by use of L and C wavebands of SAR data. J Geod Geodyn 29(1):21–26
5. Qin X, Yang M, Wang H et al (2016) Application of high-resolution PS-InSAR in deformation characteristics probe of urban rail transit. Acta Geod Cartogr Sin 45(6):713–721
6. Li J, Li Z, Wang C et al (2013) Using SAR offset-tracking approach to estimate surface motion of the South Inylchek Glacier in Tianshan. Chin J Geophys 56(4):1226–1236
7. Wang X, Liu Q, Jiang L et al (2015) Characteristics and influence factors of glacier surface flow velocity in the Everest region, the Himalayas derived from ALOS/PALSAR images. J Glaciol Geocryol 37(3):570–579
8. Li Z, Song C, Yu C et al (2019) Application of satellite radar remote sensing to landslide detection and monitoring: challenges an solutions. Geomatics Inf Sci Wuhan Univ 44(7):967–979
9. Ferretti A, Prati C, Rocca F (2001) Permanent scatterers in SAR interferometry. IEEE Trans Geosci Remote Sens 39(1):1–20
10. Ferretti A, Fumagalli A, Novali F et al (2011) A new algorithm for processing interferometric data-stacks: SqueeSAR. IEEE Trans Geosci Remote Sens 49(9):3460–3470

11. Rott H, Stuefer M, Siegel A et al (1998) Mass fluxes and dynamics of Moreno Glacier, Southern Patagonia Icefield. Geophys Res Lett 25(9):1407–1410
12. Pattyn F, Derauw D (2002) Ice-dynamic conditions of Shirase Glacier, Antarctica, inferred from ERS SAR interferometry. J Glaciol 48(163):559–565

Automatic Detection of Ship Based on Rotation Invariant RetinaNet

Luyang Zan, Kaixuan Lu, and Zhengchao Chen

Abstract Ship detection in the aerial image is an active yet challenging in remote sensing image processing. Due to the difficulties of locating the arbitrarily rotated ships and the complexity of the background around the ship, the existing deep learning-based object detection algorithm is difficult to accurately identify the ship target. For multi-angle ships, the horizontal bounding box contains not only ship objects but also a lot of irrelevant backgrounds, they are learned by the deep convolutional network as a learning object, and this leads to the common misalignment between the final classification confidence and localization accuracy. In this paper, we propose a new module called rotation RetinaNet (RRNet) to handle this problem. Firstly, we start from the popular one-stage RetinaNet approach, with ResNet50 as a basic network. Then, we apply a rotation head to RetinaNet to guarantee the rotation invariance of the model. Finally, we add an angular loss to the original loss so that the model can learn the angular offset of the bounding box. As a consequence, the proposed RRNet achieves high performance on the open-source DOTA datasets.

Keywords Ship detection · Convolutional neural network · Rotation RetinaNet

1 Introduction

As a comprehensive technology of earth observation, remote sensing has a large range of high-resolution imaging capabilities. With the rapid growth of the number of high-resolution satellites in orbit, the acquisition speed of remote sensing image data has been accelerated, and the data volume has increased significantly. As a result, the comprehensive observation capability of human beings to the earth has reached an unprecedented level. Ship detection is the focus of research in remote sensing object

L. Zan · K. Lu · Z. Chen (✉)
Airborne Remote Sensing Center, Institute of Remote Sensing and Digital Earth, Chinese Academy of Sciences, Beijing 100094, China
e-mail: chenzc@radi.ac.cn

L. Zan
College of Resources and Environment, University of Chinese Academy of Sciences, Beijing 100049, China

L. Wang et al. (eds.), *Proceedings of the 6th China High Resolution Earth Observation Conference (CHREOC 2019)*, Lecture Notes in Electrical Engineering 657,
https://doi.org/10.1007/978-981-15-3947-3_39

detection. It is very important in the fields of military defense and port monitoring. Over the past years, experts and scholars have proposed many methods to detect ships [1–3], these works have achieved remarkable results. These methods only focus on the position information of ships but ignore the direction information, which is also of great significance. The ships on the sea are easy to detect because of the simple background; however, the inshore ships are more difficult to be detected than other objection detections [4]. The inshore ships are often of large aspect ratio, tightly assigned, and arbitrary rotated, other than this, the harbors and ships are similar in gray level and texture. Those factors will bring great challenges to rotated ship detection.

In recent years, with the development of deep learning technology, the feature learning algorithm based on deep neural network has made a breakthrough in many research fields such as computer vision and natural language processing. In essence, deep learning is a deep neural network containing multiple hidden layer nodes. Through feature extraction step by step, the optimal representation of data is obtained. In the context of remote sensing big data, the intelligent deep learning algorithm has significant advantages in performance and precision in the information extraction of complex high-resolution remote sensing images. As an efficient target feature extraction network, convolutional neural network has been widely used in target detection model. In recent years, Pascal VOC [5], COCO [6], DOTA [7], and DIOR [8], the dataset of object detection in the field of large-scale computer vision or remote sensing, have made a breakthrough in the object detection algorithm based on deep learning. According to whether candidate regions are generated or not, the existing deep learning-based object detection model can be divided into two categories: candidate region-based method and location-based regression method. Since R-CNN was proposed, target detection models based on candidate regions have achieved great success in the field of natural images, including R-CNN [9], fast R-CNN [10], faster CNN [11], FPN [12], HRNet [13], etc. This method divides object detection into two phases: the first phase focuses on generating a set of candidate regions that may contain objects; the second phase aims to further classify the candidate regions obtained in the first phase and adjust the coordinates of the boundary boxes. The method based on position regression is to use a single-stage position detector for object instance prediction to simplify the detection to a regression problem. Compared with the method based on candidate regions, this method is simpler and faster, including YOLO [14], SSD [15], RetinaNet [16], etc. Among them, RetinaNet is the state-of-the-art detector, which can guarantee speed and accuracy at the same time. These methods can be applied to ship detection of remote sensing images and can significantly improve the efficiency and automation level of ship recognition. However, these methods cannot inference the angle information of the ship.

Based on the framework of target detection network RetinaNet, we proposed an arbitrary rotated network named as rotated RetinaNet (RRNet), which is more suitable for ship detection.

The rest of the paper is organized as follows. Section 2 introduces the details of the dataset and the proposed method. Section 3 describes an experiment conducted on

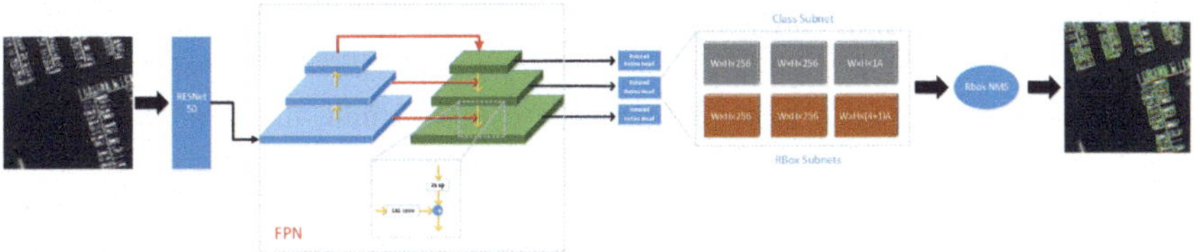

Fig. 1 Pipeline of RR, including FPN and Rotated Retina Head

DOTA datasets to evaluate the performance of the method. Finally, Sect. 4 concludes the results of the method.

2 Methodology

2.1 *Pipeline*

Our rotation RetinaNet's overall structure is illustrated in Fig. 1. Our network is a one-stage method based on RetinaNet. By adding a Rotated Retina Head structure to make the network learn target's direction information, the final detection result changed from the horizontal bounding box (HBB) to the oriented bounding box (OBB). We use ResNet50 as our backbone network to generate a multi-scale convolution feature pyramid, then we use an FPN module to fusion different scale feature maps, on each FPN level we add two additional small FCN subnet, one for classes prediction the other for RBox regression. After the five-parameter regression and the rotation non-maximum suppression (R-NMS) operation for each RBox in the RBox regression subnet, we get the final detection results.

2.2 *Network Design*

2.2.1 Data Augmentation and Multi-scale Training

To ensure the robustness of the model and avoid overfitting of the network, the training set images are randomly flipped and randomly mirrored in the pre-processing stage. Because the proportion of ships in the image is too small and the distribution is too dense, to ensure the accuracy of the ship detection, we will expand the image from 640 × 640 to 1024 × 1024 during the training process. Small ships are zoomed into make them easier to detect. The experimental results show that the model can reduce missed detection after the image is enlarged.

2.2.2 Feature Pyramid Networks

On the one hand, because the ship's target is usually small, the shallow network responsible for detecting small targets usually contains less semantic information, so the detection network cannot detect small ship targets well. On the other hand, because the image contains a large number of ships of different scales, it is difficult for the network to cope with the problem of multiple different scales of the same object. Therefore, in this paper, the FPN module is added to the detection model, and the FPN is used to make the detection network overcome the differences mentioned above in scale and enhance the semantic information of the shallow network.

The basic idea of FPN: Using different scales of feature information to predict targets of different scales, the specific structure in FPN is mainly divided into two processes: bottom-up and top-down. Bottom-up process: The process by which a common convolutional network extracts features through feedforward calculations. Top-down process: Contains two steps of upsampling and horizontal join. Firstly, the features extracted from the bottom-up process are sampled so that the scale of the high-level features can satisfy the scale of the horizontal connection with the features of the lower layer, and then, the obtained sampling features and the underlying features are added to the pixels, that is, the horizontal connection gets a multi-level feature map.

2.2.3 RBox Parameterized Description and RBox Regression

For the object detection method using the HBB to locate the target position, the HBB is usually parameterized into four variables consisting of the coordinates of the center point and the width and height, for example, box $= (x_c, y_c, w, h)$. The objects with arbitrary direction on the remote sensing image, such as the ship target in the image with a tilt angle of 45°, may account for less than 40% of the minimum external bounding box. The learning target of the network leads to a decrease in the detection accuracy of the network. Therefore, this paper uses the arbitrary quadrilateral to mark the target object. The arbitrary quadrilateral is a more accurate definition method of the target position. The labeling of the arbitrary quadrilateral is usually represented by a total of eight parameters of its four endpoints coordinates $(x_1, y_1, x_2, y_2, x_3, y_3, x_4, y_4)$, and then, in the bounding box regression, we simplify the above-mentioned arbitrary quadrilateral to a OBB with the following five parameters (x_c, y_c, w, h, θ). As show in Fig. 2, where the box Contains the coordinates of the center point (x_c, y_c), the width w and the height h and the direction angle θ.

For RBox regression [13], the following five coordinates are used for parameterization:

$$t_x = (x - x_a)/w_a, \ \ t_y = (y - y_a)/h_a \tag{1}$$

$$t_w = \log(w/w_a), \ \ t_h = \log(h/h_a) \tag{2}$$

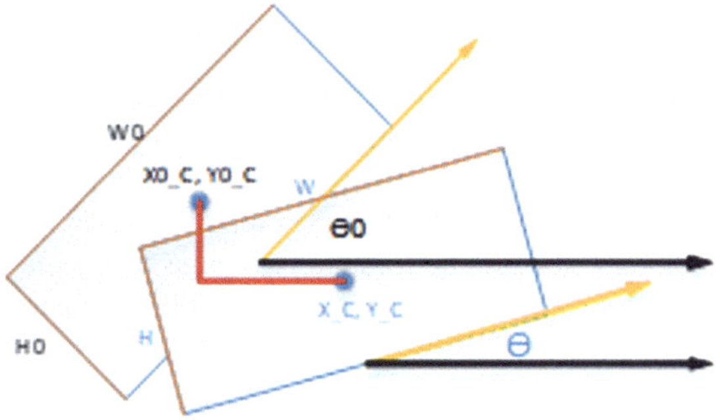

Fig. 2 RBox parameterization

$$t_\theta = \sin(\theta - \theta_a) \tag{3}$$

$$t_x^* = \left(x^* - x_a\right)/w_a,\ t_y^* = \left(y^* - y_a\right)/h_a \tag{4}$$

$$t_w^* = \log\left(w^*/w_a\right),\ t_h^* = \log\left(h^*/h_a\right) \tag{5}$$

$$t_\theta^* = \sin\left(\theta^* - \theta_a\right) \tag{6}$$

where the variables x, x_a, x^* are for the prediction box, default box, and ground truth box (as for x, y, w, h, θ).

2.2.4 Rotated Retina Head

Rotated Retina Head consists of two similarly structured sub-networks, one for classification and the other for RBox regression. We add this Rotated Retina Head module on each FPN level output layer to generate a series of candidate rboxes and classes.

Classification Subnet

The class prediction sub-network discriminates the objects in each spatial position in the feature map. The subnet attaches a small FCN to each FPN layer. The parameters of this subnet are shared at all pyramid levels. The input feature map comes from each FPN layer and has 256 channels. Then, four 3 × 3 conv layers are applied. Each layer has 256 filters. After each filter, ReLU is activated, and then, a 3 × 3 conv layer has two filters. Finally, the softmax activation function is used to calculate a set of category confidence for each region to complete the classification task.

RBox Regression Subnet

Similar to the class prediction sub-network, the position regression sub-network adds a small FCN structure to each FPN layer to return the offset between the anchor box and the nearest ground truth object. Each position has a 5A output. Relative to each a anchor of each spatial position, there are 5 offsets calculate by the prediction anchor and the ground truth box to achieve accurate positioning. Classification and regression have the same structure and different parameters.

2.2.5 Multi-task Loss Function

In this paper, the multi-task loss function with angle penalty is used for the end-to-end training of the network, and the loss function is shown as follows:

$$L = \frac{1}{N_{\text{cls}}}\sum_{i} L_{\text{cls}}(p_i, l_i) + \frac{1}{N_{\text{reg}}}\sum_{i} p_j\left(L_{\text{reg}}(v_j^*, v_j)\right) \tag{7}$$

Here, l_i represents the object category, p_i represents the probability distribution of different categories calculated by softmax function, v_j represents the coordinate vector deviation of model prediction output, and v_j^* represents the real vector deviation. Where classification loss L_{cls} is the cross entropy loss, regression loss L_{reg} is smooth L1 loss.

3 Experiment

3.1 Dataset and Setting

Object detection dataset of aerial remote sensing image. It contains 2806 expert interpreting aerial images from different sensors and platforms. Each image ranges in size from 800 × 800 to 4000 × 4000 pixels and contains a wide range of scales, directions, and shapes. These DOTA images are then annotated by aviation image interpretation specialists with an arbitrary quadrilateral. We selected the images containing ship targets in the DOTA dataset, obtained a total of 1572 + 4902 samples, divided the training set and test set, and obtained 4902 training set samples and 1572 test set samples. We divide the images into 640 × 640 sub-images with an overlap of 100.

All the experimental contents in this paper were completed under the PyTorch deep learning framework. The hardware environment is a server with four NVIDIA GeForce Titan XP GPU (12 GB memory) and Intel Xeon E5 CPUs. We use ResNet50 imagenet pre-training model to initialize the network. For the DOTA dataset, we trained 60 epochs in total.

3.2 Performance Evaluation

We evaluate the performance of RRNet by the ship's validate set of DOTA. The performance evaluation of neural network includes mean average accuracy, precision, and recall. The classification results of each assessment can be divided into four categories:

- True positives (TP): the number of ships correctly classified as ships.
- False positives (FP): the number of ships incorrectly classified as ships.
- True negatives (TN): the number of ships correctly classified as non-ships.
- False negatives (FN): the number of ships incorrectly classified as non-ships.

Precision is often used to measure the ability of a model to distinguish binary classes (that is, ship or non ship). In this study, combined with TP and FP calculation:

$$\mathrm{Precison} = \frac{\mathrm{TP}}{\mathrm{TP} + \mathrm{FP}} \tag{8}$$

Recall is a commonly used method to quantify the performance of algorithms in machine learning. It measures the proportion of correctly detected ships in all positive samples, i.e., a missing metric.

$$\mathrm{Recall} = \frac{\mathrm{TP}}{\mathrm{TP} + \mathrm{FN}} \tag{9}$$

mAP is a specific value that can more intuitively describe the performance of the detector, and it is the area of the precision–recall curve and the horizontal axis. The definition is described as follow:

$$\mathrm{mAP} = \int_{0}^{1} P(R)\mathrm{d}R \tag{10}$$

3.3 Evaluation of RRNet

We use 4375 training sets and 1357 validation sets to train RRNet, where resize is set to 640 * 640. We use the pre-training model ResNet50 to initialize the network and train the model by using the optimizer SGD with 0.9 momentum, 0.0001 weight decay, and batch size 8, and maximum number of epochs is set to be 50. The initial learning rate at the beginning of training is set to be 0.01, we use the learning policy "warm up" to adjust the learning rate, the warm-up step is set to be [10, 30, 50], and the warm-up ratio is 1/3.

On one hand, for the small dense ships because of its small volume quantity, in the process of feature extraction network constantly down sampling of the figure map will be lost in a large number of target information, thus detection model needs to be done the ship detection task using the characteristics of the previous figure map, but due to the characteristics of the previous figure map layer contained in the semantic information is too shallow, it is easy to make the detection model can't locate the ship object precisely. On the other hand, for large ships with characteristics of strong semantic, due to the complexity of its semantic information, so the model need contain deep characteristic figure of high-level semantic information to complete judgment, and high-level semantic feature maps cannot very accurately locate the position information of large ships. The RRNet is a multi-scale detection network, and it chooses five different feature maps to combine, the lowest layer is mainly used for small ship detection, and the highest layer is mainly used for large ship detection. When training RRNet on the ship datasets, the trend curve of mAP is shown as Fig. 3, and RRNet achieves 66.8% mAP after 50 epochs. In order to prove our proposed method is more outstanding, we compare the RRNet with other methods, and the results are shown in Table 1. In Table 1, the SSD, YOLOv2, and R-FCN [15] are horizontal region detectors, and the R-DFPN [16] is rotation region detector. The result shows that the mAP of our RRNet is better.

In addition to using mAP to evaluate the detection accuracy of the model, we also calculate the accuracy and recall rate of the model to evaluate the performance of the model and record the accuracy and recall rate under different thresholds in the ship validation set. Figure 4 shows the precision–recall curve, it can be seen that the best precision rate and recall rate of our proposed method are 93.9% and 72.1%, respectively.

Figure 5 shows some detection results in different scenes, including tightly arranged ships and arbitrary oriented ships, ships in the harbor and on the sea. These

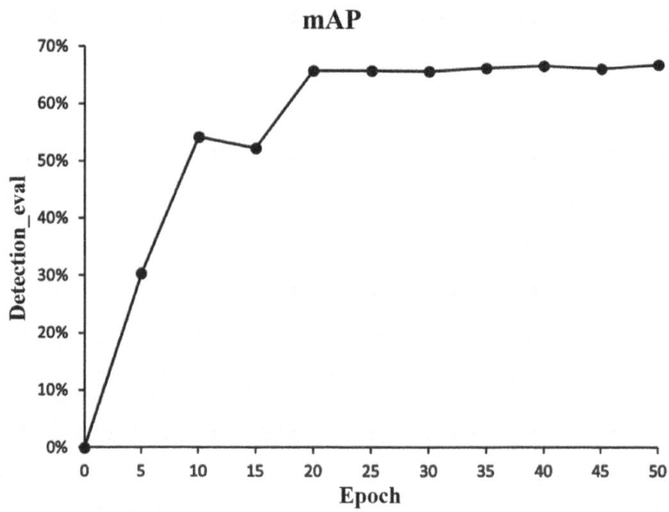

Fig. 3 mAP of RRNet

Table 1 Detection performance comparisons of different models

Methods	AP of ship
SSD	13.21
YOLOv2	7.37
R-FCN	7.45
R-DFPN	54.78
Ours	66.80

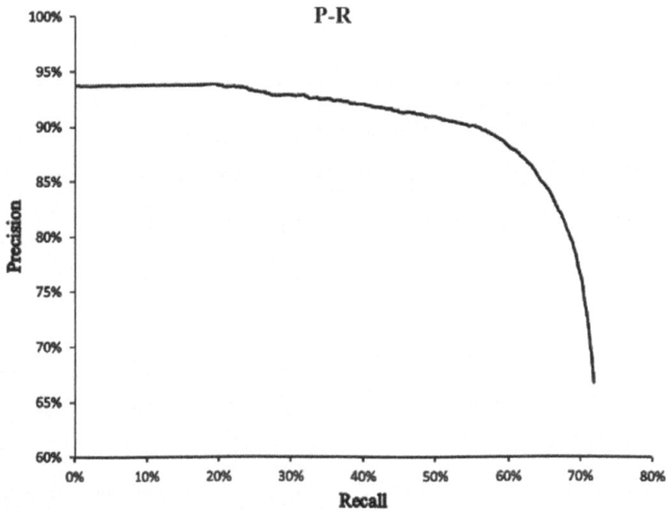

Fig. 4 Precision–recall curve of RRNet

complex scenes and various ships have a great influence on the performance of RRNet.

4 Conclusion

In this paper, we propose a rotated neural network named RRNet to detect arbitrary-oriented ships. We apply a rotation head to guarantee the rotation invariance of the model and add an angular loss to the original loss so that the model can learn the angular offset of the bounding box. The results of the experiment based on the DOTA dataset show that our proposed RRNet has a great performance on rotated ship detection. However, due to the difficulty of features in ship detection, our method has a relatively low recall rate, there still exists some missed inspections, and we need to explore how to effectively increase recall rate in the future.

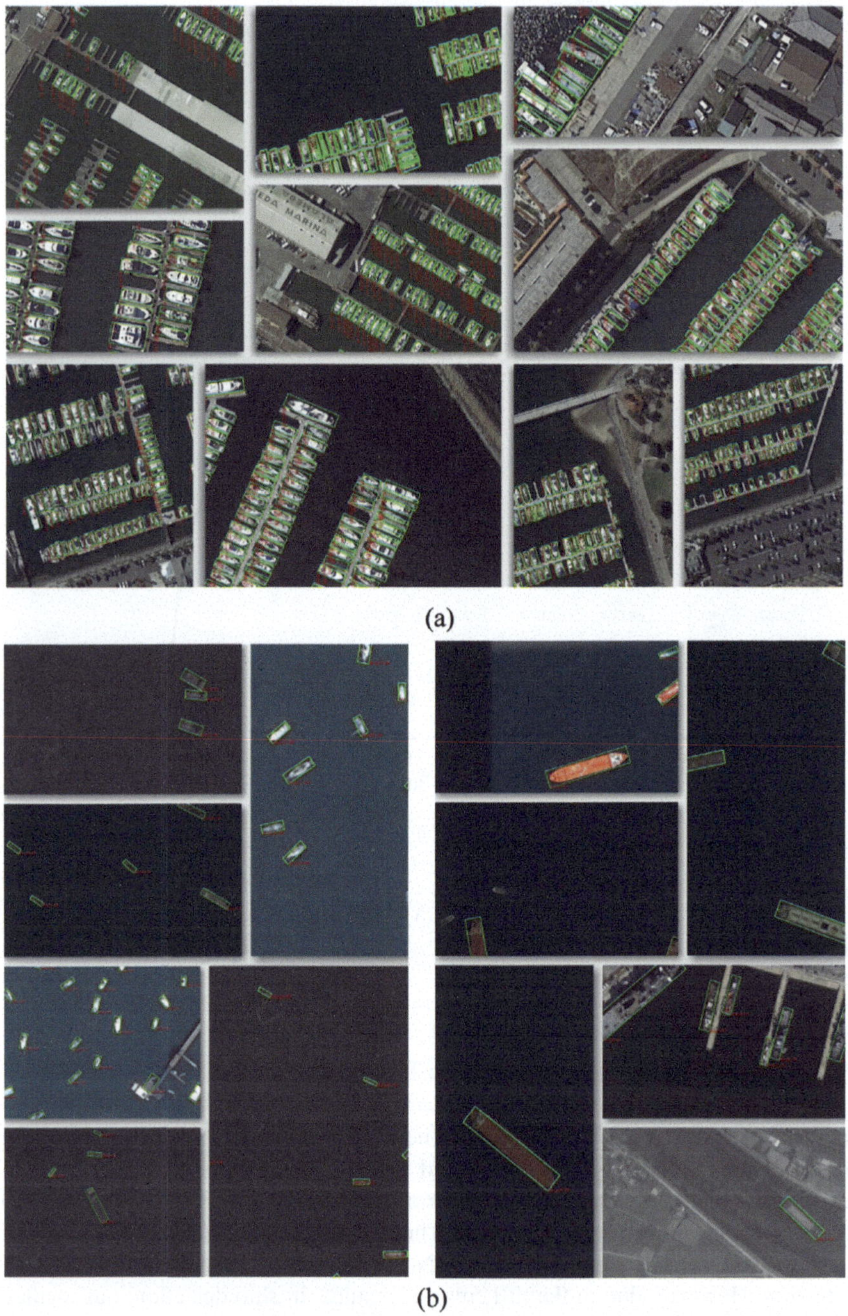

Fig. 5 Detection results of ships in different scenes. **a** Detection results of arbitrary oriented and tightly arranged ships; **b** detection results of ships in the harbor and on the sea

Acknowledgements This work was jointly supported by the Strategic Priority Research Program of the Chinese Academy of Sciences (Grant No. XDA19080302 and No. 23100304).

References

1. Zhang R, Yao J, Zhang K, Feng C, Zhang J (2016) S-CNN ship detection from high-resolution remote sensing images. ISPRS Int Arch Photogramm Remote Sens Spatial Inf Sci XLI-B7:423–430
2. Kang M, Ji K, Leng X, Lin Z (2017) Contextual region-based convolutional neural network with multilayer fusion for SAR ship detection. Remote Sens 9(8):860
3. Tang J, Deng C, Huang GB, Zhang B (2014) Compressed-domain ship detection on spaceborne optical image using deep neural network and extreme learning machine. IEEE Trans Geosci Remote Sens 53(3):1174–1185
4. Chen Z, Lu K, Gao L, Li B, Gao J, Yang X, Yao M, Zhang B (2019) Automatic detection of track and fields in China from high-resolution satellite images using multi-scale-fused single shot multibox detector. Remote Sens 11:1377
5. Everingham M, Van Gool L, Williams CKI et al (2010) The pascal visual object classes (voc) challenge. Int J Comput Vision 88(2):303–338
6. Lin TY, Maire M, Belongie S et al (2014) Microsoft coco: common objects in context. In: European conference on computer vision. Springer, Cham, pp 740–755
7. Xia G-S, Bai X, Ding J, Zhu Z, Belongie S, Luo J, Datcu M, Pelillo M, Zhang L (2018) DOTA: a large-scale dataset for object detection in aerial images. In: Proceedings of the IEEE conference on computer vision and pattern recognition, Salt Lake City, 18–22 June 2018
8. Li K, Wan G, Cheng G, Meng L, Han J (2019) Object detection in optical remote sensing images: a survey and a new benchmark
9. Girshick R, Donahue J, Darrell T et al (2014) Rich feature hierarchies for accurate object detection and semantic segmentation. In: Proceedings of the IEEE conference on computer vision and pattern recognition, pp 580–587
10. Girshick R (2015) Fast R-CNN. In: Proceedings of the IEEE international conference on computer vision, pp 1440–1448
11. Ren S, He K, Girshick R et al (2015) Faster R-CNN: towards real-time object detection with region proposal networks. In: Advances in neural information processing systems, pp 91–99
12. Lin TY, Dollár P, Girshick R et al (2017) Feature pyramid networks for object detection. In: Proceedings of the IEEE conference on computer vision and pattern recognition, pp 2117–2125
13. Huang J, Zhu Z, Huang G (2019) Multi-stage HRNet: multiple stage high-resolution network for human pose estimation. arXiv preprint arXiv:1910.05901
14. Redmon J, Divvala S, Girshick R et al (2016) You only look once: unified, real-time object detection. In: Proceedings of the IEEE conference on computer vision and pattern recognition, pp 779–788
15. Liu W, Anguelov D, Erhan D et al (2016) SSD: single shot multibox detector. In: European conference on computer vision. Springer, Cham, pp 21–37
16. Lin T-Y, Goyal P, Girshick RB, He K, Dollár P (2017) Focal loss for dense object detection. In: ICCV

A High-Resolution Remote Sensing Images Segmentation Algorithm Based on PCA and Fuzzy C-Means

Chenchen Jiang, Hongtao Huo, and Qi Feng

Abstract Aiming at optimal segmentation scale for different surface features with different features in high-resolution remote sensing images takes a lot of experiments and exists subjectivity. This paper proposes an optimal segmentation algorithm, a method that combines principal component analysis (PCA) with fuzzy c-means (FCM). In this method, the initial clustering centers of FCM are generated by sorting values after dimension reduction by PCA on high-resolution remote sensing images. Then using fuzzy c-means algorithm merges the homogenous image units into one object, and thus, we can gain the segmentation results which rule out influence of subjectivity and uncertainty of initial clustering centers and segmentation scale. Our final result, visual evaluation and clustering internal evaluation indicators and segmentation evaluation indicators show that the high-resolution remote sensing images segmentation algorithm based on PCA and FCM is better than original FCM, and other traditional image segmentation methods mentioned in the paper.

Keywords High resolution remote sensing images · Image segmentation · PCA · Fuzzy c-means

1 Introduction

Image segmentation is an important step in high-resolution remote sensing images classification and is a key factor of accuracy of images classification results, and

C. Jiang · H. Huo (✉)
Institute of Police Information Engineering and Cyber Security, People's Public Security University of China, Beijing 100038, China
e-mail: Huohongtao@ppsuc.edu.cn

Q. Feng
Remote Sensing Center of Public Security, People's Public Security University of China, Beijing 100038, China

Civil-Military Integration Center for Public Security, People's Public Security University of China, Beijing 100038, China

L. Wang et al. (eds.), *Proceedings of the 6th China High Resolution Earth Observation Conference (CHREOC 2019)*, Lecture Notes in Electrical Engineering 657,
https://doi.org/10.1007/978-981-15-3947-3_40

thus, researching optimization segmentation scale is of great significance to high-resolution remote sensing images object-oriented classification. Object-oriented segmentation means merging neighboring pixels into an object based on specific similarity detection (such as spatial characteristic [1], textural features and spectral information et al.) [2]. Recently, researching on selection of optimal segmentation scale in high-resolution remote sensing images focuses on three aspects including supervised segmentation with samples, unsupervised segmentation without samples and semi-supervised segmentation with few samples.

Previous researches [3–7] have shown that object-oriented supervised segmentation of remote sensing images depended on the segmentation scale parameters of the trial-and-error approaches, which had a great impact on the subsequent classification. This trial-and-error method which based on repeated experiments and the experience of researchers is instructive and inadvisable [8]. Therefore, many researchers have proposed unsupervised segmentation and semi-supervised segmentation. At present, image segmentation based on clustering is widely recognized [9], among them, FCM algorithm is very important and widely used segmentation method [10]. Zhang et al. [11] propose an adaptive spatially constrained fuzzy c-means algorithm (ASCFCM) which does not require any parameter adjustment realizing unsupervised image segmentation, but there are some shortcomings in the selection of initial clustering center and the number of clustering. Fan and Wang [12] present a two-phase fuzzy clustering algorithm based on neuro-dynamic optimization. In the first stage, the method uses the linear assignment initialization and eliminates the instability of the results of random initialization clustering for polarimetric synthetic aperture radar (PolSAR) remote sensing image segmentation.

Although the original FCM algorithm does not need to set samples compared with supervised segmentation in the process of image segmentation, it also has some shortcomings that the determination of initial clustering center and the number of clustering.

The segmentation results of high-resolution remote sensing images are greatly influenced by the quality of original images, the number of images bands and the resolution of remote sensing images [13, 14], and evaluating the segmentation results is not only an indispensable key technology in the segmentation process but also one of difficulties in the technical field [15]. G. Meinel and M. Neubert [16] presented a method for evaluating segmentation quality and compared the qualities in eight different image segmentation results between presently available segmentation programs (*eCognition 2.1*, *eCognition 3.0*, *Data Dissection Tools*, *CAESAR 3.1*, *InfoPACK 1.0*, *Image segmentation for Erdas Imagine*, *Minimum Entropy Approach to Adaptive Image Polygonization* and *SPRING 4.0*). In this paper, we combine qualitative and quantitative indicators to evaluate to the segmentation results.

Aiming at above problem, we propose a high-resolution remote sensing images segmentation algorithm based on PCA and fuzzy c-means which is an incremental approach to clustering segmentation. The sorting result of PCA is used as the initial value of the original FCM, and it does not depend on any initial clustering centers conditions and has the more stable segmentation results. In the article, we take GF-2 remote sensing images database, based on MATLAB2017a platform, through visual evaluation and quantitative evaluation evaluating the new method segmentation results.

2 Methodology

In this paper, the methods include the machine learning such as PCA and FCM, introducing PCA dimension reduction analysis into FCM clustering algorithm. Thus, it is eliminating the sensitivity of the initial value of FCM.

2.1 Technical Route

This study is based on preprocessing GF-2 remote sensing images ROI using a new object multi-scale image segmentation which is first taking PCA to reduce the dimension of multi-band high-resolution remote sensing images and then sorting one-dimensional array. In the second stage, the sorted PCA remote sensing images are used for FCM clustering to get the results of object-oriented multi-scale images segmentation. Finally, the images segmentation results are analyzed by using clustering internal index and segmentation evaluation indictors. The technical route is summarized in Fig. 1.

2.2 PCA Unsupervised Dimension Reduction

PCA [17] is generally used to replace the original more variables with fewer new variables, and these new variables can keep all the information in the original data as much as possible. Let $\boldsymbol{X} \in \mathbb{R}^{m \times N}$ denote a set of normalized data with m samples and N variables.

$$\boldsymbol{X} = \begin{bmatrix} x_{11} & x_{12} & \cdots & x_{1N} \\ x_{21} & x_{22} & \cdots & x_{2N} \\ \vdots & \vdots & \ddots & \vdots \\ x_{m1} & x_{m2} & \cdots & x_{mN} \end{bmatrix} \tag{1}$$

The covariance matrix $\text{cov}(\boldsymbol{X})$ of $\boldsymbol{X}$ derives

$$\text{cov}(\boldsymbol{X}) = \frac{1}{m-1}\boldsymbol{X}^{\mathrm{T}}\boldsymbol{X} \tag{2}$$

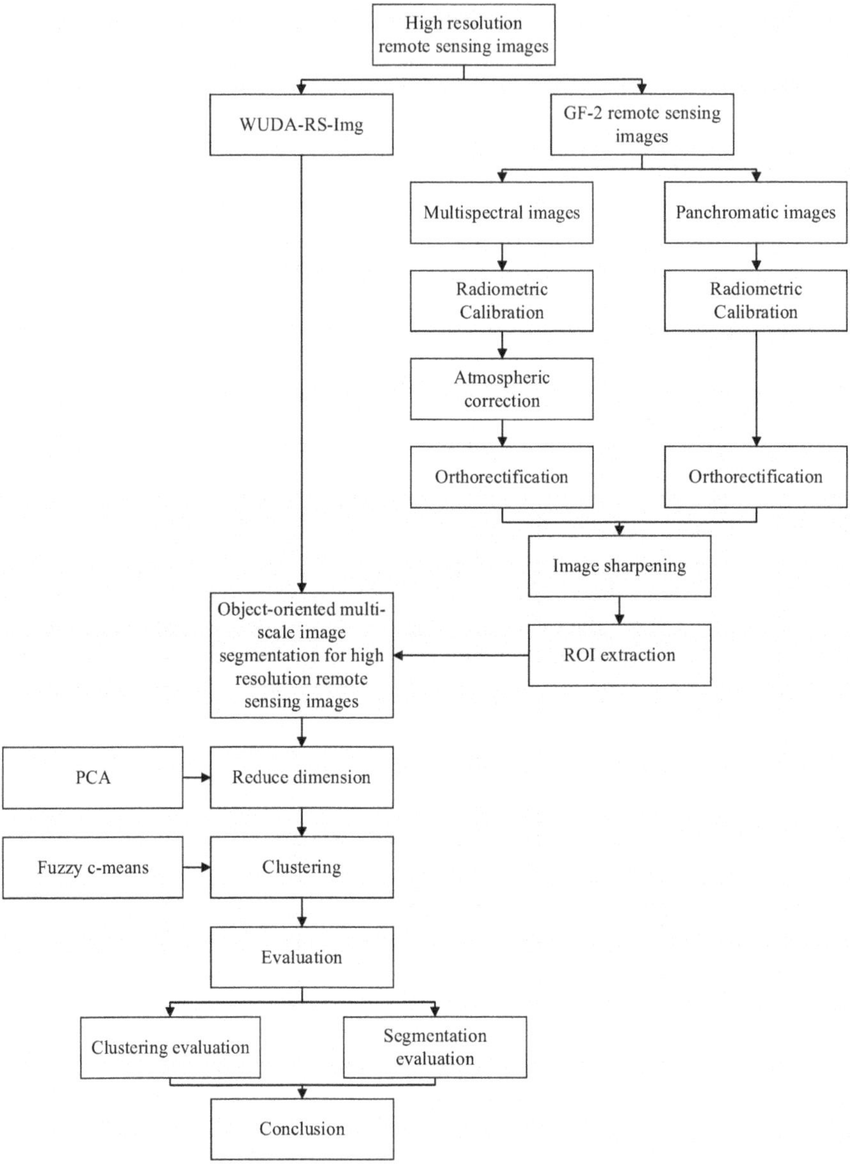

Fig. 1 Technical route

where $\mathbf{\Lambda} = \mathrm{diag}(\lambda_1, \lambda_2, \ldots, \lambda_N)$ and $\lambda_1 \geq \lambda_2 \geq \ldots \geq \lambda_N \geq 0$. In PCA, calculation of principal component contribution rate $\frac{\lambda_i}{\sum_{k=1}^{N} \lambda_k} (i = 1, 2, \ldots, N)$ and cumulative contribution rate $\frac{\sum_{k=1}^{i} \lambda_k}{\sum_{k=1}^{N} \lambda_k}$ $(i = 1, 2, \ldots, N)$ are taken to extract the corresponding principal components $z_1, z_2, \ldots, z_n$ $(n < N)$.

2.3 FCM Clustering Algorithm

FCM [18] was proposed by Dunn [19] in 1973 and improved by Bezdek [20] in 1981. Let matrix $\boldsymbol{X} = \{x_i\},\ i = 1, 2, \ldots, n$ denote a dataset with n samples and d dimensions. In FCM, the matrix is divided as k clusters denote the number of groups. Represent k clusters as:

$$C = \{c_i,\ i = 1, 2, \ldots, k\} \tag{3}$$

where μ_k is the center of mass of c_k. The objective function of FCM is

$$J_{\mathrm{f}}(X\colon U, V) = \sum_{j=1}^{k}\sum_{i=1}^{n}\left(u_{ij}\right)^{w}\left\|x_i - \mu_j\right\|^2,\ 1 < w < \infty \tag{4}$$

where $u_{ij} \in [0, 1]$ is a degree to which data $\{x_i\},\ i = 1, 2, \ldots, n$ belong to cluster. u_{ij} is desirable to have any real number between 0 and 1, but the sum of the partition matrix has constraint $\sum_{i=1}^{k} u_{ij} = 1$. And $w \in (1, +\infty)$ is a fuzzification parameter. $\left\|x_i - \mu_j\right\|^2$ is a Euclidean distance.

2.4 PCA and FCM Clustering Algorithm

Let initialization clustering centers of FCM be defined as the sorting the first principal component after PCA:

$$V_j = \lceil \frac{1 + 2l}{2k} m \rceil \left(l = 0, 1, \ldots, \frac{m}{k} - 1\right) \tag{5}$$

where V_j denotes the initial clustering centers, k represents the number of clustering (Fig. 2).

2.5 Evaluation Indicators

Evaluation indicators include clustering evaluation sum of square error (SSE) [21], segmentation accuracy [22], over-segmentation rate [22] and under-segmentation rate [22].

As the objective function of clustering, SSE is expressed as following:

$$\mathrm{SSE} = \sum_{i=1}^{k}\sum_{x \in c_i} \mathrm{dist}(x, v_i)^2 \tag{6}$$

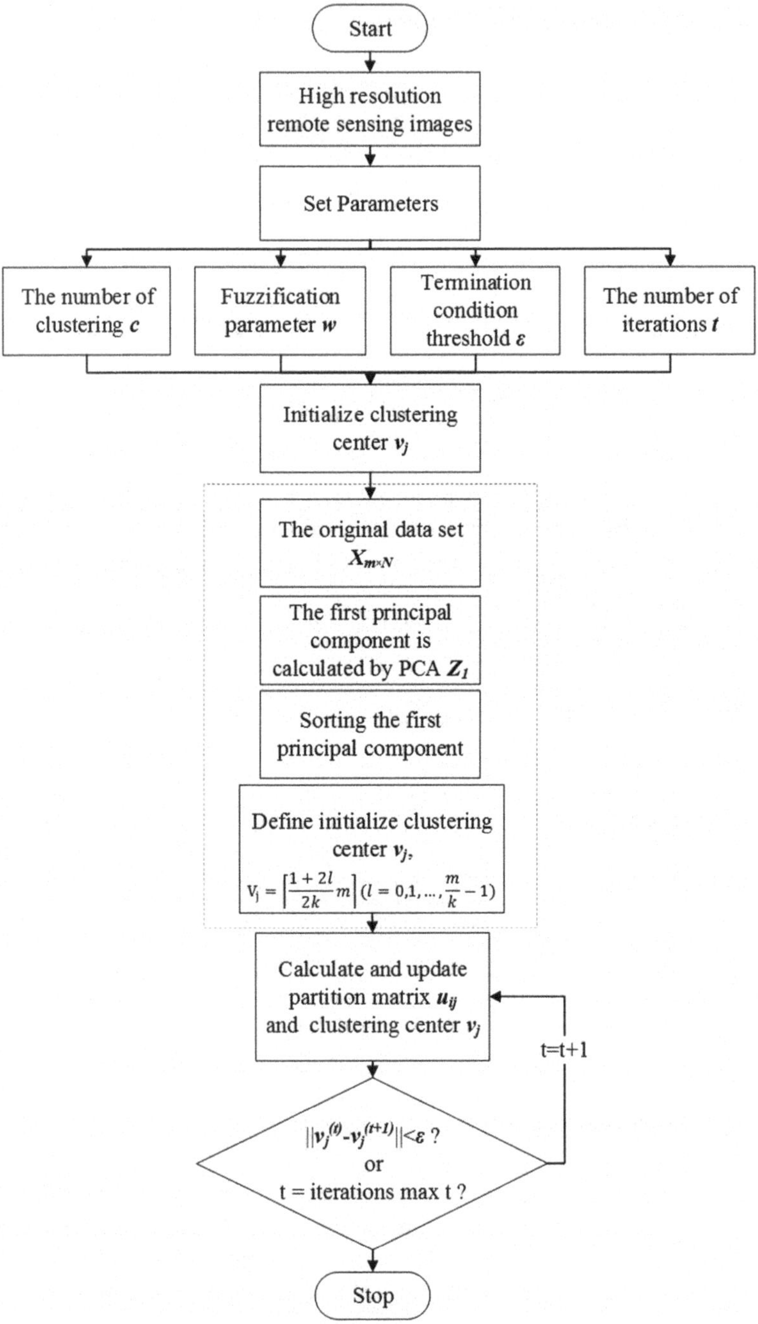

Fig. 2 Flowchart of FCM algorithm based on PCA

where k is the number of clustering, and $\text{dist}(x, v_i)^2$ means the Euclidean distance between the sample point x which is in the cluster and its corresponding clustering center v_i.

Segmentation evaluation indicators include overall segmentation accuracy, over-segmentation rate and under-segmentation rate. The overall segmentation accuracy (SA) is defined as following:

$$\text{SA} = \left(1 - \frac{|R_s - T_s|}{R_s}\right) \times 100\% \tag{7}$$

where R_s is the number of pixels of original high-resolution remote sensing images, T_s represents the number of pixels of segmentation images. $|R_s - T_s|$ denotes the number of pixels of incorrect segmentation.

Over-segmentation is defined as

$$\text{OS} = \frac{O_s}{R_s + O_s} \tag{8}$$

where O_s denotes the number of pixels that are not actually in the segmented image but are segmented in the segmented results.

Under-segmentation rate is defined as

$$\text{US} = \frac{U_s}{R_s + O_s} \tag{9}$$

where U_s denotes the number of pixels in a segmented image that is not in the segmented image but is supposed to be in the segmented images.

Figure 3 shows the three indicators above all. Figure 3a divides the image into the three clusters such as 80, 100 and 150. In Fig. 3b, pixels with a value of 100 are not actually in the value of 80 segmented image but are segmented in the value of 80 segmented results. And Fig. 3c denotes the under-segmentation rate, and pixels with a value of 80 are not in the 80 segmented images but are supposed to be in the value of 80 segmented images.

3 Experiments

In this paper, the hardware environment of the experiment is ASUS FX59 Intel(R) Core(TM) i7-9750H CPU @ 2.60 GHz 2.59 GHz RAM 8.00 GB, operation system: Windows 10 x64, and the software environment of the experiment is MATLAB2017a and ENVI 5.3.1.

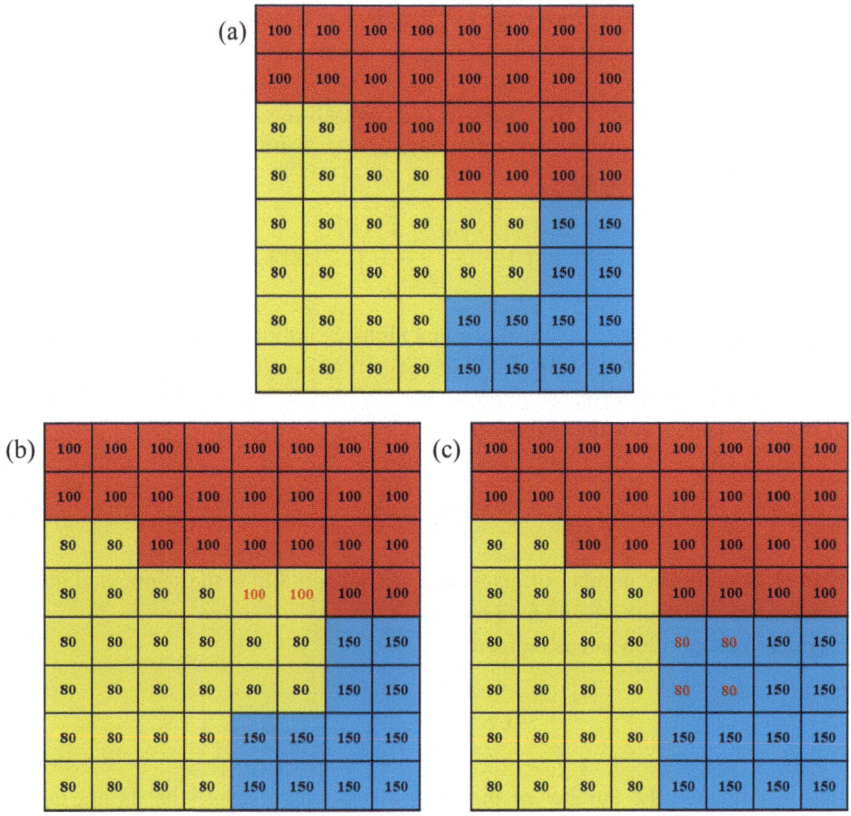

Fig. 3 Schematic diagram of evaluation of segmentation results **a** segmentation accuracy, **b** over-segmentation rate, **c** under-segmentation rate

3.1 Data Description

The time phase and scene number of GF-2 high-resolution remote sensing image are September 2, 2015 and GF2_PMS1_E116.3_N39.7_20150902_L1A0001015327. Preprocessing includes radiometric calibration, atmospheric correction, orthorectification, image sharpening and ROI extraction (Fig. 4).

Wuhan University GaoFen image dataset (GID) [23] includes 150 high-resolution GF-2 images acquired from more than 60 different cities in China for land-use classification. In GID label, five categories are labeled with five different colors: built-up (red), farmland (green), forest (cyan), meadow (yellow) and waters (blue) (Fig. 5).

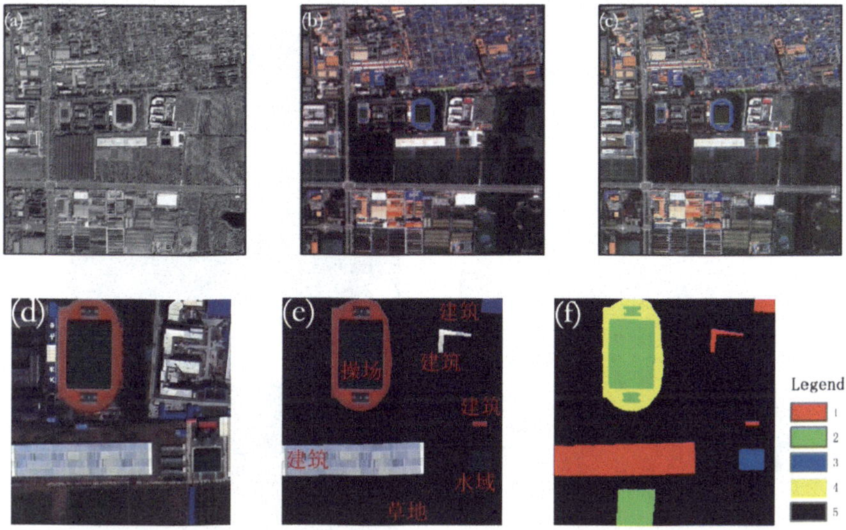

Fig. 4 GF-2 high-resolution remote sensing images **a** panchromatic image with the resolution of 1 m, **b** multispectral image with the resolution of 4 m, **c** the fusion image after Nearest Neighbor Diffusion pan sharpening, **d** ROI extract from the fusion image, **e** ROI after removing background value, **f** ROI classification label

3.2 Segmentation Results

As shown in Fig. 6, the trial-and-error approaches experiment the segmentation and merging scales with the values of 25, 50 and 75. The method depends on repeated experiments which also rely on the experience of the researchers.

The methods in this paper are taking PCA and FCM algorithms for avoiding subjective experience and sensitive initial value. The results of PCA are shown in Fig. 7 as following:

Under the condition that other parameters are the same, it can be found in Fig. 8 by visual interpretation that the segmentation result by PCA and FCM is better than the original method. But the segmentation of water area is under-segmentation in both methods. Thus, the later experiment is testing with the large water area image, as shown in Fig. 9.

The segmentation effect of the water area is poor in both methods, and algae are mistakenly classified into other categories. The segmentation of farmland is more obvious in PCA and FCM method than the original FCM method.

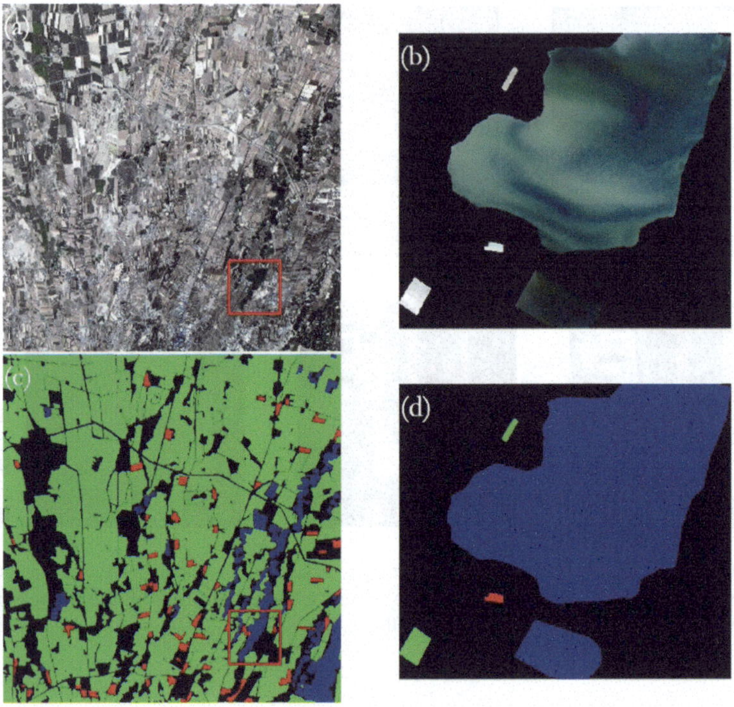

Fig. 5 Wuhan University Gaofen image dataset (GID) **a** 20160912_L1A0001821711 data source, **b** ROI extraction, **c** data source label, **d** ROI label

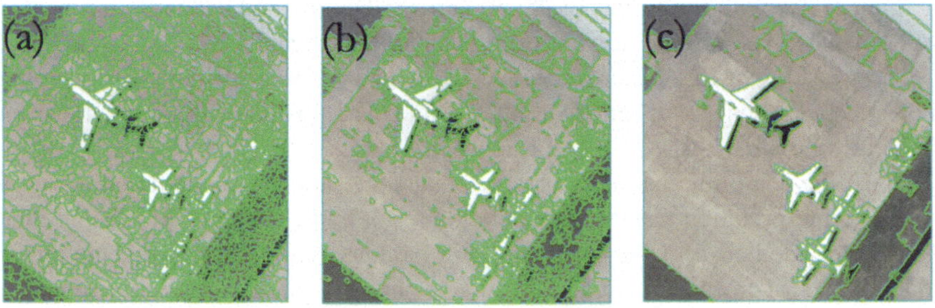

Fig. 6 Multi-scale images segmentation **a** the scales of segmentation and merging are 25, **b** the scales of segmentation and merging are 50, **c** the scales of segmentation and merging are 75

3.3 Evaluation Indicators of Segmentation Results

For evaluation of segmentation results, we take four indicators that sum of square error (SSE), segmentation accuracy (SA), over-segmentation rate (OS) and under-segmentation rate (US).

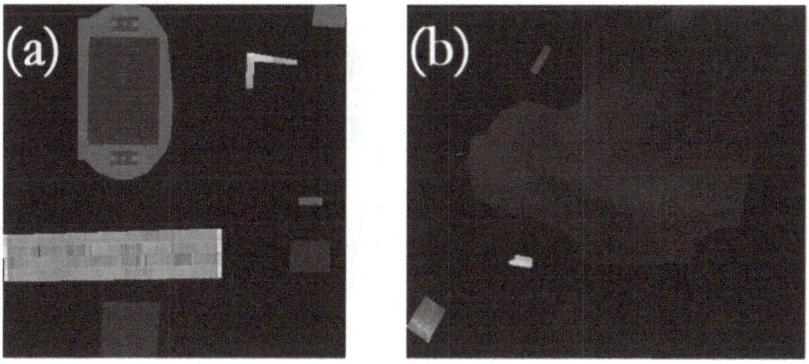

Fig. 7 The results of PCA **a** GF-2 remote sensing image, **b** GID

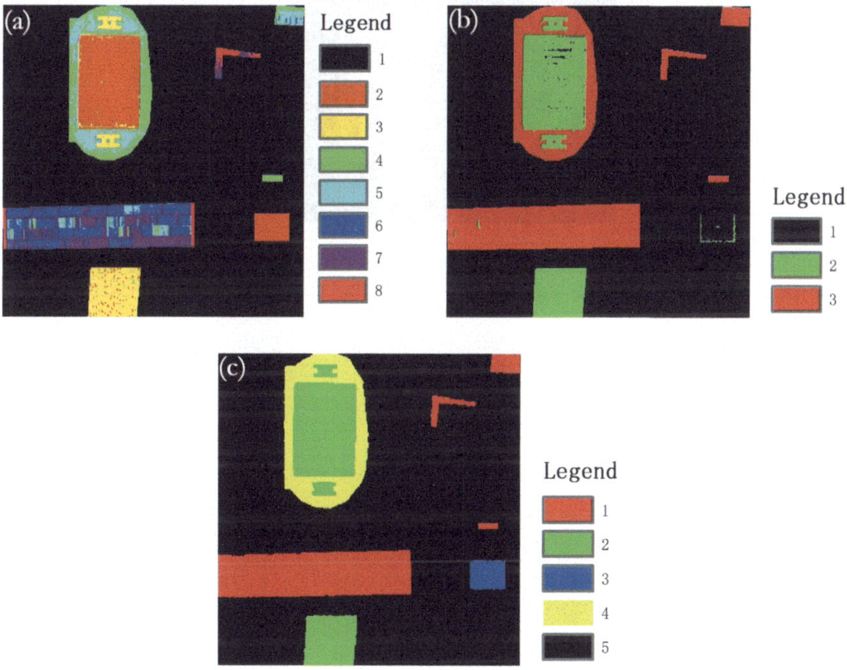

Fig. 8 The GF-2 remote sensing image segmentation result with PCA and FCM **a** the segmentation result with the original FCM method, **b** the segmentation result with PCA and FCM, **c** the label of remote sensing image

Part of GF-2 remote sensing images segmentation evaluation in Table 1 and the class 1 of buildings are better segmented, although the runway in class 4 is in the class 1, it is also well segmented and distinguished from adjacent pixels. The segmentation accuracy of class 2 which represents the farmland (green) reaches 99.4%. And the segmentation accuracy of class 5 is also as high as 98.6%.

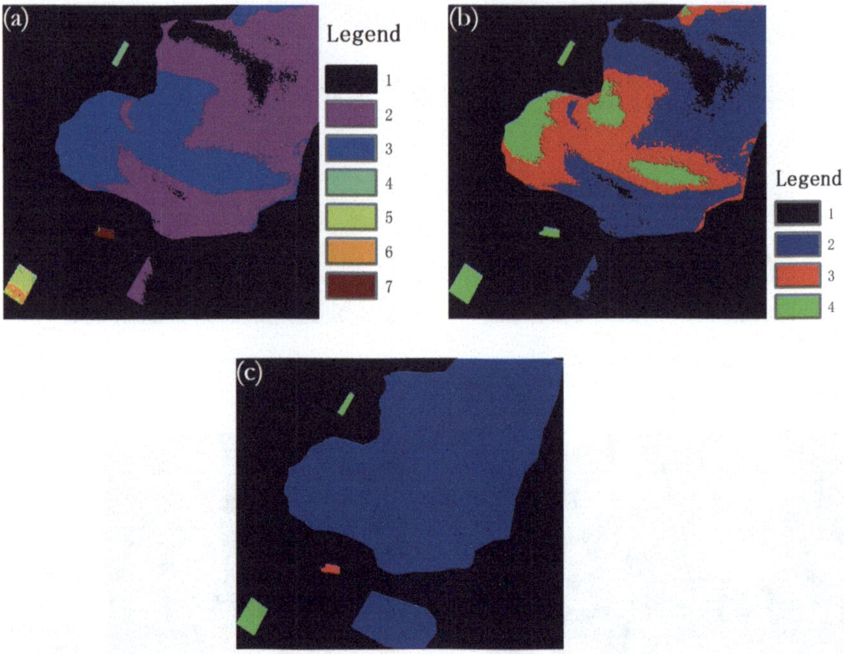

Fig. 9 The segmentation result of GID GF-2 remote sensing image **a** the segmentation result with the original FCM method, **b** the segmentation result with PCA and FCM, **c** the label of remote sensing image

Table 1 Evaluation of segmentation results

Data set		Indicators			
		SSE	SA	OS	US
GF-2	Class 1	1.84e+07	0.994	0.018	0.024
	Class 2		0.994	0.029	0.023
	Class 3		0.126	0.018	0.876
	Class 4		0.977	0.027	0.050
	Class 5		0.986	0.018	0.004

4 Summary

In this paper, we have presented a new remote sensing image segmentation method to eliminate the sensitivity of FCM initial clustering centers and avoid the selection of segmentation scale by trial-and-error method with PCA. We gain the following results.

1. After PCA analysis and sorting, it is found that the mean center eliminating initial value sensitivity of FCM.

2. The traditional method takes trial-and-error method for multi-scale segmentation, the PCA and FCM method avoid the shortcoming.
3. In the terms of segmentation accuracy, GF-2 remote sensing image has a better segmentation effect especially in class 1 and class 2.

References

1. Ma L, Li M, Ma X et al (2017) A review of supervised object-based land-cover image classification. ISPRS J Photogramm Remote Sensing 130:277–293
2. Deng SB (2014) ENVI remote sensing image processing methods, 2nd ed. Higher Education Press
3. Arbiol R, Zhang Y, Palà V (2006) Advanced classification techniques: A review. In: ISPRS Commission VII Mid-term Symposium Remote Sensing: From Pixels to Processes, Enschede, the Netherlands 292–296
4. Smith A (2010) Image segmentation scale parameter optimization and land cover classification using the Random Forest algorithm. Spat Sci 55(1):69–79
5. Kim M, Warner TA, Madden M et al (2011) Multi-scale GEOBIA with very high spatial resolution digital aerial imagery: scale, texture and image objects. Int J Remote Sens 32(10):2825–2850
6. Myint SW, Gober P, Brazel A et al (2011) Per-pixel vs. object-based classification of urban land cover extraction using high spatial resolution imagery. Remote Sens Environ 115(5):1145–1161
7. Hussain M, Chen D, Cheng A et al (2013) Change detection from remotely sensed images: from pixel-based to object-based approaches. ISPRS J Photogramm Remote Sens 80:91–106
8. Gao RQ, Ou YJ, Chen LX et al (2019) Research on high resolution image classification of Hedi reservoir based on object-oriented method. Sci Surv Mapp 1–14. http://kns.cnki.net/kcms/detail/11.4415.P.20181115.1629.002.html
9. Chen K, Chen XH (2019) Fuzzy C-means clustering image segmentation algorithm with local spatial information based on ELM. J Data Acquis Process 34(1):100–110
10. Wu YJ (2018) Modified fuzzy c-means clustering algorithm of image segmentation. Northwest Normal University
11. Zhang H, Shi W, Hao M et al (2018) An adaptive spatially constrained fuzzy c-means algorithm for multispectral remotely sensed imagery clustering. Int J Remote Sens 39(8):2207–2237
12. Fan J, Wang J (2018) A two-phase fuzzy clustering algorithm based on neurodynamic optimization with its application for PolSAR image segmentation. IEEE Trans Fuzzy Syst 26(1):72–83
13. Drăguţ L, Tiede D, Levick SR (2010) ESP: a tool to estimate scale parameter for multiresolution image segmentation of remotely sensed data. Int J Geogr Inf Sci 24(6):859–871
14. Fortin MJ, Olson RJ, Ferson S et al (2000) Issues related to the detection of boundaries. Landsc Ecol 15(5):453–466
15. Yangyang C, Dongping M, Lu X et al (2017) An overview of quantitative experimental methods for segmentation evaluation of high spatial remote sensing images. J Geo-Inf Sci 19(6):818–830
16. Meinel G, Neubert M (2004) A comparison of segmentation programs for high resolution remote sensing data. Int Arch Photogrammetry Remote Sens 35:1097–1105
17. Qingchao J, Xuefeng Y (2018) Parallel PCA–KPCA for nonlinear process monitoring. Control Eng Pract 80:17–25
18. Arora J, Khatter K et al (2019) Fuzzy c-means clustering strategies: a review of distance measures. Softw Eng
19. Dunn JC (1973) A fuzzy relative of the ISODATA process and its use in detecting compact well-separated clusters. J Cybern 3(3):32–57

20. Bezdek JC (1981) Pattern recognition with fuzzy objective function algorithms. Adv Appl Pattern Recogn 22(1171):203–239
21. Zhou BJ, Tao YZ et al (2018) Optimizing k-means initial clustering centers by minimizing. Comput Eng Appl 54(910)(15):53–57
22. Jiang WB, Liu LP, Sun XH (2019) K-means model based on adaptive weight method for remote sensing image segmentation. Comput Appl Softw 5:231–234+261
23. Tong X-Y et al (2018) Learning transferable deep models for land-use classification with high-resolution remote sensing images. [Online]. Available http://arxiv.org/abs/1807.05713

Structural Design and Analysis of a Belly Radome

Liuqing Xu

Abstract The aircraft modification is a research hot spot in the field of general aviation. It has played an important role in disaster relief, meteorological detection, ocean monitoring, scientific experiments, remote sensing mapping and other application fields. Radome is one of the typical aircraft modifications. In this paper, we propose an aircraft belly radome and focus on its structural design and analysis. Firstly, a 3D model of the aircraft belly radome is constructed, and then, the aerodynamic performances of the radome in four kinds of typical flight conditions are simulated by ANSYS software. Next, based on the fluid-structure interaction platform, the flight aerodynamic simulation data in various flight conditions are extracted as the high-altitude environmental parameters to provide more realistic static force and deformation simulations of the radome. In addition, a pressure exchanged device is proposed to clear the pressure difference of the inside and outside of the radome. Moreover, the static stress of the bolts fastened the radome to the aircraft is analyzed to support the fastener material selection of the radome. The research can be as a strong engineering reference for the structural research of the aircraft belly radome.

Keywords Belly radome · Fluid-structure interaction · Aircraft modification

1 Introduction

In order to meet the developed requirements of aerial remote sensing research, such as surveying, mapping, exploration and so on, more and more aircrafts have been modified as aerial remote sensing experimental platforms, loading various sensors to carry out scientific experiments. The belly radome is one of the typical aircraft modifications. The mainly function is to improve the aerodynamic shape of the abdominal hanging equipment with generally thin shell type structure. Its application can be traced back to the 1940s. With the development of computer technology, the design

L. Xu (✉)
Aerospace Information Research Institute, Institute of Remote Sensing and Digital Earth, Chinese Academy of Sciences, Beijing, China
e-mail: xulq@aircas.ac.cn

L. Wang et al. (eds.), *Proceedings of the 6th China High Resolution Earth Observation Conference (CHREOC 2019)*, Lecture Notes in Electrical Engineering 657,
https://doi.org/10.1007/978-981-15-3947-3_41

of radome has changed from the cumbersome process of experience estimation, theoretical validation and testing to computer aided design, manufacturing and testing, which greatly reduces the designation cycle. The research of airborne radome began in 1950s in our country. It also has been developed promptly in recent years with the advance of computing performance [1, 2].

We design one aircraft belly radome and discuss its structural performance in the paper. Firstly, a 3D model of an aircraft with belly radome is constructed by software CATIA, and the aerodynamic simulation and static analysis are comprehensively analyzed by software ANSYS. Based on the fluid-structure interaction platform, the aerodynamic simulation data under various flight conditions are extracted as the environmental parameters to reach a more realistic steady-state simulation of belly radome in flight. We also discuss the deformation and pressure relief of the radome. In addition, the stress distribution of the bolts is simulated to assist the material selection of fasteners. This study provides data and methods as strong structural references for the belly radome modification.

2 Modeling

Firstly, both the geometric original model of the aircraft (Model A) and that with the belly radome (Model B) are established. Model B is shown in Fig. 1. Considering on the symmetry of the model, only right side of the geometric model is constructed, and then, the spatial flow field is meshed. The length of the aircraft model is 14 m, and the wingspan is 15 m. The shape of radome is 3.5 m in length, 0.9 m in width and 28 kg in weight separately.

In the aspect of mesh generation in fluid field, we simulate the flight environment of the aircraft belly radome by software ANSYS. The structural mesh generation is applied on the flight flow field model in Fig. 2. The mesh quality of the second-order Jacobian matrix meets the engineering requirements. For boundary conditions, the surface of the aircraft and the radome is setting as wall. The plane of the aircraft's

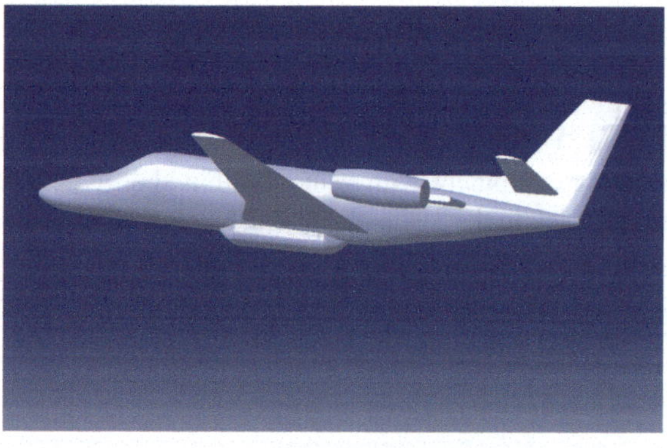

Fig. 1 Geometrical model of Model B

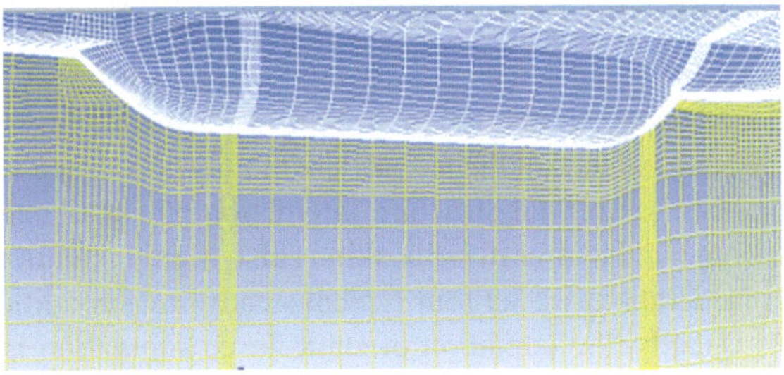

Fig. 2 Structural meshing generation of model B

central axis is setting as symmetry, and the others are setting according to the working conditions [3–5]. Solution method is SIMPLEC. Under-relaxation factor is pre-0.1. Momentum is 0.5.

In the fluid-structure interaction design of Workbench platform, the static stress analysis of the bolts fastened the radome to the aircraft belly skin, adopts the method of fluid-structure interaction to increase the simulating accuracy. In the meantime, it really increases the computing complexity. The research framework of the fluid-structure interaction of radome is presented in Fig. 3. The tetrahedral mesh model for static research of radome is presented in Fig. 4. The material of the radome is glass steel. The density is 1800 kg/m^3. The tensile strength is 290 MPa, and the Poisson's ratio is 0.22.

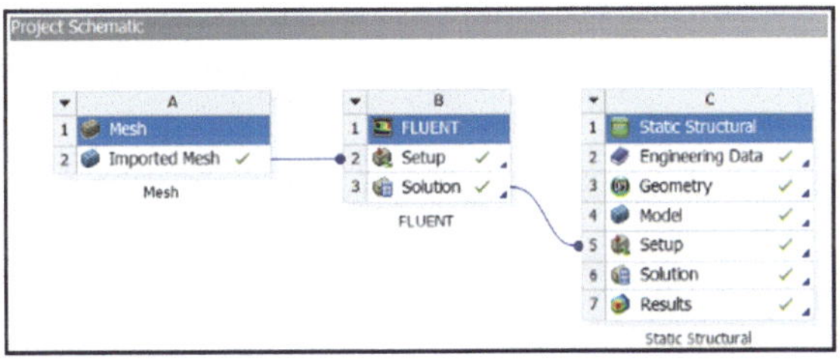

Fig. 3 Fluid-structure interaction platform

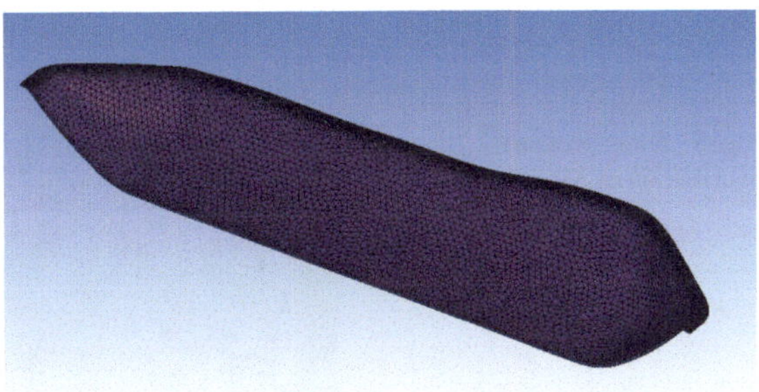

Fig. 4 Tetrahedral mesh model of the belly radome

3 Simulation Results and Discussions

Based on the simulation of aircraft in various flight conditions, the deformation and pressure balance of the radome are mainly concerned in the paper. On the Workbench platform with fluid-structure interaction, the static stress analysis of bolts fastened radome to the aircraft belly skin is also highly attentioned. According to the scheme consideration, we set four typical steady-state conditions as Table 1.

Usually, the condition of static force simulation is simulated on ground, which generally ignores the flight conditions, easily causing the deviation of simulation results. Thus, we focus on the various flight conditions with high altitude and high speed. Although it increases the computing loads, it can provide more realistic belly radome simulation results.

3.1 Static Pressure Effect of the Belly Radome

The static pressure distribution of Model A and Model B in the condition 3 is shown in Figs. 5 and 6, respectively. Compared to Model A in Fig. 5, the front and rear static pressure near the fuselage increase significantly in Fig. 6. The front-end pressure increases due to the influence of wind resistance. The back-end pressure increases because the air flow has been disturbed obviously in this section, which forms lots of eddy currents. The airflow velocity decreasing on the back-end of the radome leads to pressure rising quickly. In addition, the undersurface of the radome is in lower

Table 1 Four kinds of aircraft flight conditions

No	Model	Height	Temperature	Pressure	Velocity	Mach
1	A.B	11000 m	216.7 K	22,632pa	680 km/h	0.64
2	A,B	7500 m	239.4 K	38,251pa	680 km/h	0.61
3	A,B	7500 m	239.4 K	38,251pa	580 km/h	0.61
4	A,B	5000 m	255.7 K	47,305pa	680 km/h	0.59

Fig. 5 Static pressure distribution of Model A

Fig. 6 Static pressure distribution of Model B

pressure than surroundings, which indicate that is subjected to greater external tension during flight. Hence, the radome installation causes the obvious aerodynamical effect on flight.

3.2 Pressure Difference Discussions of the Belly Radome

The static pressure difference on the radome is obvious in Fig. 6. It can lead to structural rip. Thus, we need to consider a pressure exchange device to clear the pressure difference between inside and outside of the radome. It is not suitable for pressure exchanging on the windward side because of the high wind resistance. The undersurface is also not a good position. It will affect the radome aerodynamical shape apparently if it is installed. Therefore, the optimum position of the pressure relief device is the central axis area of the leeward side, which is used to keep the pressure balance to reduce the deformation of the radome by pressure difference. Maximum steady-state pressure difference value of 7 MPa is obtained from above four kinds of flight condition simulations, which can be used as the pressure exchanged capability index of the device in steady-state flight.

3.3 The Deformation of the Belly Radome

The deformations of the belly radome are discussed by Figs. 7, 8, 9 and 10 separately. At the same flight speed, the maximum deformation of the radome is about 9 mm in condition 1 of Fig 7. The maximum deformation is about 13 mm in condition 2 of Fig 8. The maximum deformation is about 14 mm in condition 4 of Fig 10. It indicates that the deformation of radome decreases with the flight altitude increasing. The flying altitude is higher, and the ambient pressure is lower. It causes that the pressure difference is fewer between the inside and outside of radome with higher altitude. In the same flight altitude, the maximum deformations of radome are about 13 mm and 8 mm in the condition 2 and 3, respectively, which indicates that the deformations of radome increase with the flight speed increasing. Therefore, the elastic design of radome shall refer to the maximum flight speed and mainly flight altitude. The

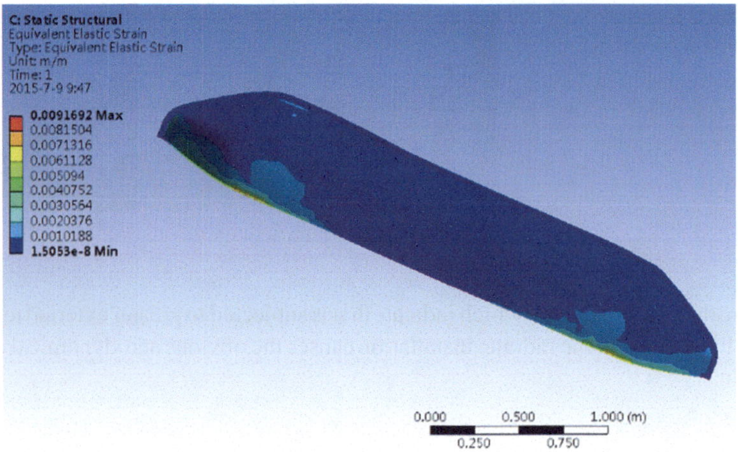

Fig. 7 Reformation of radome in condition 1

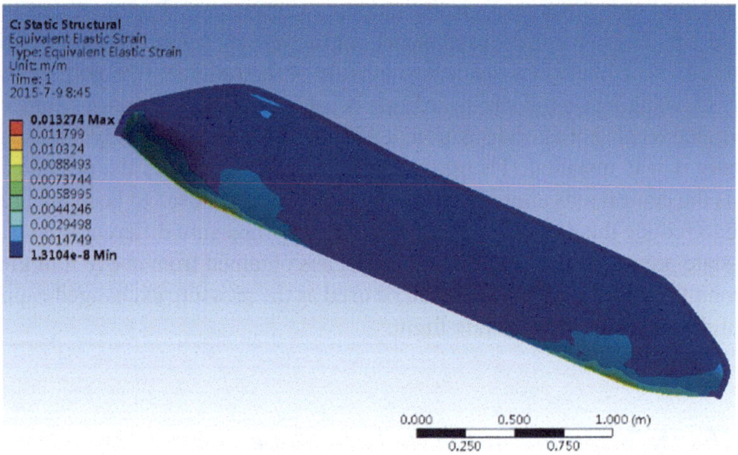

Fig. 8 Reformation of radome in condition 2

maximum deformation data of radome are on the back of radome's mid-axle in four flight conditions, which also prove the rationality of the central axis area of the leeward side as the pressure exchanged device position.

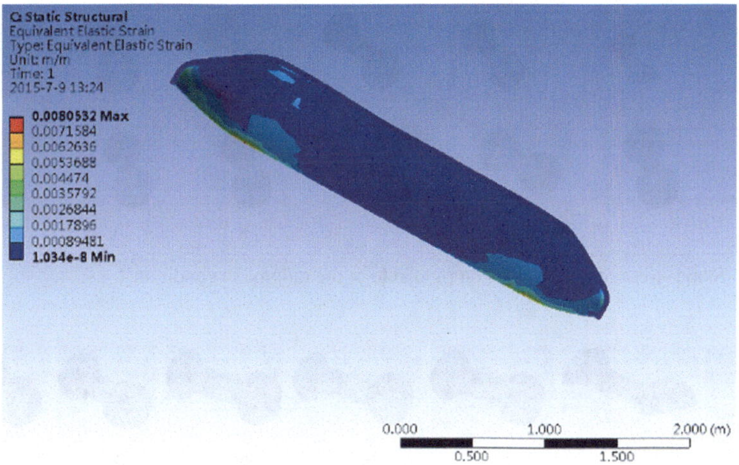

Fig. 9 Reformation of radome in condition 3

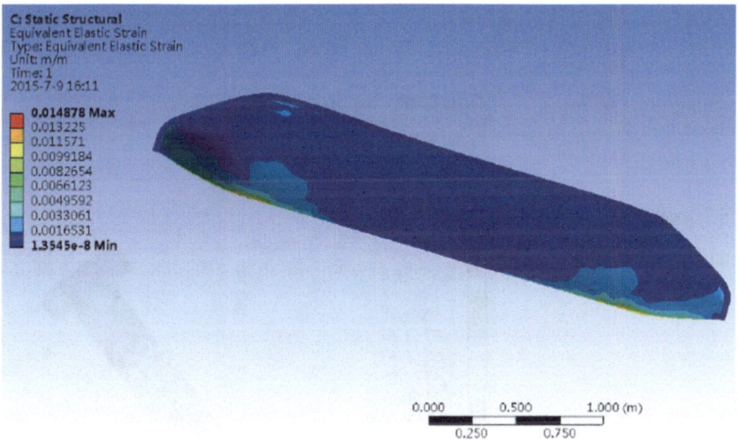

Fig. 10 Reformation of radome in condition 4

3.4 *The Static Stress Anlysis of the Bolts Fastened to Radome*

Considering on configuration of the radome, a total of 23pcs bolts are designed as fasteners connected the radome to the aircraft belly. Because the model is symmetrical, we only discuss the static stress of 12pcs bolts on right side. The sequence number of the bolts is defined as 1–12 along the radome longitudinal from forward to backward. The stress on the fastening bolts of the radome is shown in Figs. 11, 12, 13 and 14. The stresses of 12pcs bolts are similar generally shown in Figs. 11 and 12. It is presented that the stresses of No. 1–6 bolt are slightly greater than that of No.

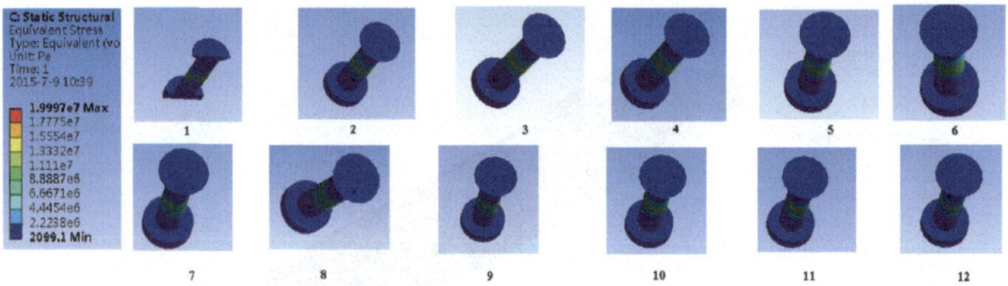

Fig. 11 Static stress of 12 pcs bolts on right side of the radome in condition 1

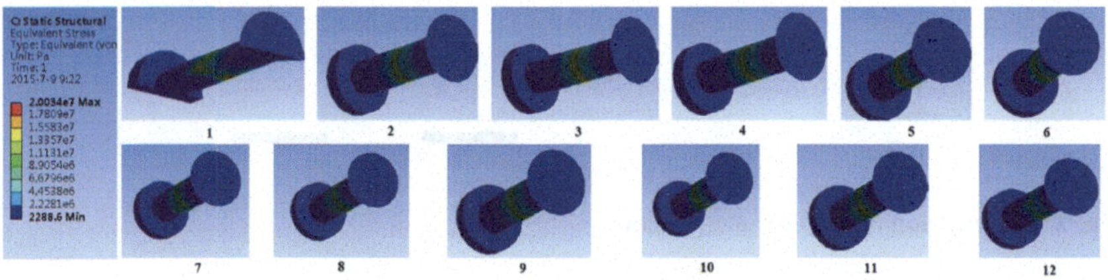

Fig. 12 Static stress of 12 pcs bolts on right side of the radome in condition 2

Fig. 13 Static stress of No. 1 bolt in condition 3

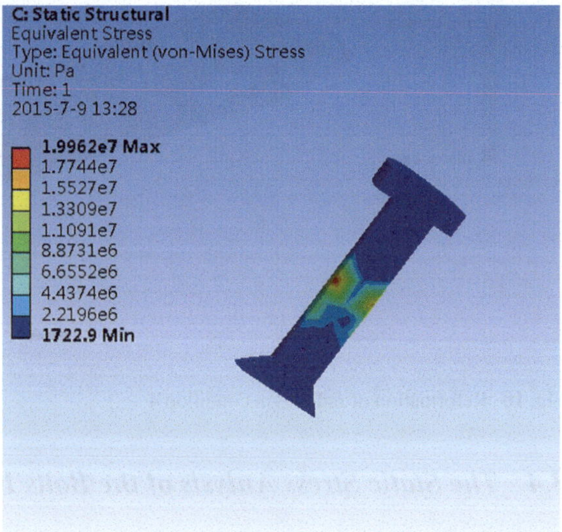

7–12 bolt. The stress changes of bolts have a similar tendency to the deformations of radome in various flight conditions shown in Figs. 12 and 13. According to Fig. 13, the maximum average cross-sectional pressure of the bolt is about 6 MPa, and the maximum force of No. 1 bolt is about 100 N. The material of bolt is recommended to use aluminum alloy. It not only satisfies the force requirement but also has the preferred advantage of lightweight. Next, we will have another important study to

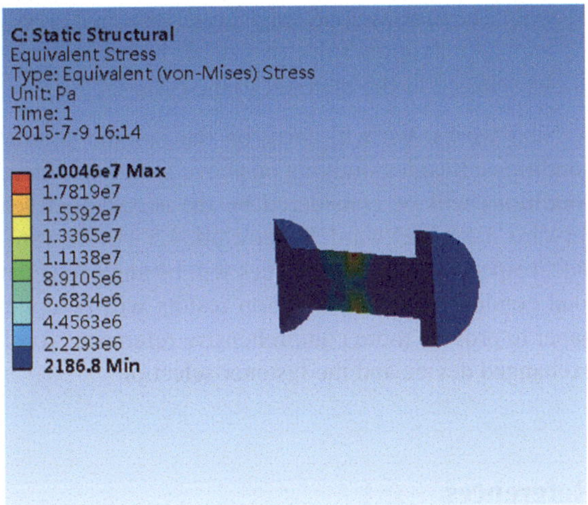

Fig. 14 Static stress of No. 1 bolt in condition 4

carry on the aircraft flight simulation in the various limited conditions with different accelerated velocity.

4 Conclusions and Next Works

We design a belly radome according to the scheme requirements. We propose one aircraft belly radome and focus on its structural design and analysis. Firstly, a 3D model of the aircraft belly radome is constructed, and then, the aerodynamic performances of the radome in four kinds of typical flight conditions are simulated by ANSYS software. Next, based on the fluid-structure interaction platform, the flight aerodynamic simulation data in various flight conditions are extracted as the high-altitude environmental parameters to provide the static force and deformation analysis of the radome. It presents the simulating results more realistically compared to that with the ground simulating condition. From the above study, we can acquire conclusions as:

1. The radome installation strongly effects the aircraft aerodynamics. The static pressure has obviously changed on the belly surroundings of Model B compared to that of Model A.
2. One pressure exchanged device is proposed. To keep the pressure balance both inside and outside of the radome, we discuss the central axis area of its leeward surface as the optimum position for the device installation. 7 MPa is designed as the pressure exchanged capability index of the device in steady flight condition.
3. Total of 23 bolts fastened the radome to the aircraft skin are designed and discussed in the four flight conditions. Though the static stress generally changes a little at high-altitude steady-state flight, the maximum stress is applied on No.1

bolt. The maximum average cross-sectional pressure of the bolt is about 6 MPa, and the maximum force of a single bolt is about 100 N. The aluminum alloy can be selected as the material of the bolts.

Next works, we will carry on the radome simulation in various limited flight conditions. Besides studying on above-mentioned steady flight condition, the limited conditions will be considered by the accelerated velocity load of aircraft on 3 g forward, 1.5 g backward, 2 g upward, 4.5 g downward, 1.5 g left side and 1.5 g right side, respectively [6]. The forces will be analyzed continually in the various limited load conditions. The simulation results will be combined with the results of this paper to provide more comprehensive references for the designation of the pressure exchanged device and the fastener selection.

References

1. Zhu LL (2015) An Introduction to US airborne IR measurement systems and their test aircraft. Electron Opt Control 22(6):103–108
2. Wang ZY, Zhong CF, Zhao K (2016) Structural design and mechanical analysis of an airborne electronic pod. Technol Econ Guide 17:1–3
3. Song P, Jia XP, Zhang MC (2014) Research on fast estimation method of climbing performance for aircraft mounting a pod. Aeronaut Sci Technol 25(1):20–24
4. Liang B, Shen HJ, Meng H (2012) Aerodynamic analysis of an airborne equipment pod. Telecommun Eng 52(5):808–811
5. Xu LQ, Yang J, Yan ZX (2018) Aerodynamic analysis of an airborne belly shell. Aircr Des 38(1):34–37
6. Tan GH, Zhang ZY, Song ZH, Cheng L (2010) The structural Design Technology for Airborne Radar Pod. Radar Sci Technol 8(3):284–288

Cross-comparison and Analysis of SJ-9A, SPOT5, and THEOS Based on a Satellite Network

Fangyan Yuan, Guoqing Li, Zhengli Zuo, Quan Ran, LiDong Guo, and Guangbin Ma

Abstract Although a number of inter-sensor comparisons have been conducted, no comparison between the SJ-9A, SPOT5, and THEOS sensors has been conducted. This study provides such a comparison, with the aim of evaluating the SJ-9A image quality. A synchronized satellite network technical system was proposed, and a comparison of SJ-9A, SPOT5, and THEOS based on synchronous data was successfully completed. Compared with previous methods, proposed method is an active mode for quality assessment and shown to be practical and operable. The study had two outcomes; first, a new comparison method for new sensor quality assessment was proposed and verified, and second, the new SJ-9A sensor was evaluated.

Keywords Satellite network · Satellite synchronous observation · SJ-9A · SPOT5 · THEOS · Cross-comparison

1 Introduction

Optic satellite sensor imaging quality is affected by sensor design, installation, and orbit degradation. The remote imaging process is very complex and can be affected by the atmosphere, sun angle, cloud, terrain, and other factors. Image quality is an important topic, especially when new satellites are launched. In the initial phase of imaging quality research, studies focused on evaluating the quality of satellite images [1–4]. With many earth observation satellites being launched, image quality assessment methods have changed, and a popular area of research is the cross-comparison of different satellites with similar operating parameters [5–10]. To date, studies have used three comparison methods to cross-compare different satellites. The first method is considered a per-band-based comparison [11, 12] or a multiband-combination-based comparison, such as the thematic-indicator-based comparison [13, 14]. The second method is the context difference comparison. Some comparisons are made based only on the quality of the sensor imaging process itself [15–17], including its

F. Yuan (✉) · G. Li · Z. Zuo · Q. Ran · L. Guo · G. Ma
Aerospace Information Research Institute, Chinese Academy of Sciences, No. 9 Dengzhuang South Road, Haidian District, Beijing, China
e-mail: yuanfy@aircas.ac.cn

L. Wang et al. (eds.), *Proceedings of the 6th China High Resolution Earth Observation Conference (CHREOC 2019)*, Lecture Notes in Electrical Engineering 657,
https://doi.org/10.1007/978-981-15-3947-3_42

attention to geometric accuracy, radiation accuracy, amount of information produced, contrast, and other factors. Some studies have focused on satellite applications, such as the comparison of vegetation indicators [9, 18], agricultural crops [19, 20], burnt areas, including the severity of forest fires, and bare soil [21, 22]. The third method is the comparison of the sensor system style. Cross-comparisons among satellite sensor systems can be used for almost all the existing airborne and spaceborne sensors, including optical and synthetic aperture radar (SAR) [23, 24].

Among the various cross-comparisons mentioned above, the synchronous image pair-based comparison is one of the most frequently used methods and is the basis of objective comparisons. According to the literature, simultaneous cross-comparison images are mainly used for historical data; however, there are some limitations to this method, due to its passive mode. This method is suitable for satellites that produce a wide range of images and that have many applications because they have a large amount of historical data; however, it is not suitable for more recently launched sensors because new satellites have no historical data. To solve this problem, we propose an active synchronized satellite network method that will provide an effective way to assess the quality of new satellite images.

2 Methodology

2.1 Cross-comparison of Satellites

A similar spectral band design and spatial resolution was the basis for the cross-comparisons of SJ-9A, THEOS, and SPOT5. Table 1 shows the detailed satellite parameters.

2.2 Test Area

To successfully collect synchronous images, the test areas used in this study were selected based on the following criteria.

1. Test area must have low rainfall and little cloud cover. Accordingly, Zhangye City in Gansu Province, China, was selected, and it contains regions of forest, grassland, built-up areas, and desert.
2. Cross-comparison image pairs have as little sensor view angle as possible and a similar solar elevation angle.
3. An abundance of different land cover types is important for good comparison results. Summer is the best time for synchronous image collection because ground vegetation is in the growing stage. Accordingly, we implemented a satellite network synchronous test in June.

Table 1 Cross-comparison satellites parameters

	Orbit type	Spatial resolution		Orbital inclination	Spectral type	Spectral interval
SJ-9A	Sun synchronous	Pan	2 m	97.982°	Pan	0.45–0.89
					Blue	0.45–0.52
		Multiband	15 m		Green	0.52–0.59
					Red	0.63–0.69
					Infrared	0.77–0.89
SPOT5	Sun synchronous	Pan	2.5 m	98.72°	Pan	0.50–0.73
					green	0.50–0.59
		Multiband	10 m		red	0.61–0.68
					infrared	0.78–0.89
					Shortwave infrared	1.58–1.75
THEOS	Sun synchronous	Pan	2 m	98.7°	Pan	0.45–0.90
					Blue	0.45–0.52
		Multiband	15 m		Green	0.53–0.60
					Red	0.62–0.69
					infrared	0.77–0.90

By following these rules, comparison image pairs were collected in Zhangye City, Gansu Province, including panchromatic and multiband images, which were cross-compared to obtain more objective comparison results. We used preliminary product data for each satellite.

2.3 *Satellite Network*

A satellite network is a virtual constellation, which organizes different kinds of earth observation satellites to serve specific purposes, after consideration of the satellite resources, the ground system capability and its various constraints, overall time planning, attitude control, and the selection of the ground target for observation.

Depending on the exact application required, there are three types of satellite networks (see Fig. 1).

1. Simultaneous imaging of different satellites at the same time and place. This type of network is the most demanding and the most basic. It is generally difficult to achieve complete synchronization, but it can achieve near synchronization. This type of cooperative data is mainly used for the comparison of the performance and image quality of different satellite sensors. This type of network was used in this study.

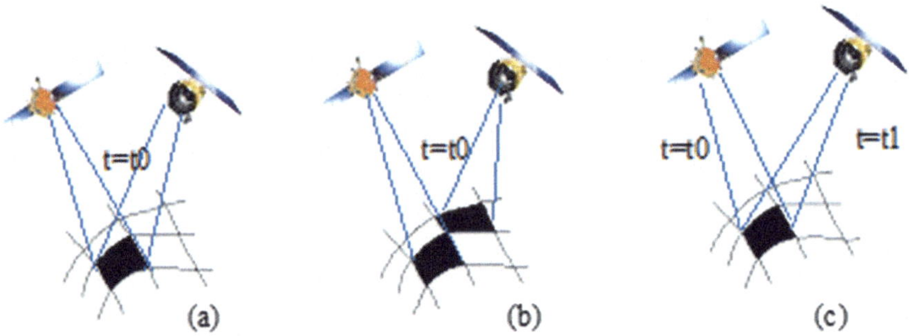

Fig. 1 Satellite network type

2. Imaging of different satellites at the same (or similar) time but in a different place. This type of satellite networking is primarily used to acquire full area coverage as soon as possible and can be used for images of the country or even the whole world.
3. Imaging of different satellites at different times but in the same place. Compared with the previous two methods, this network is easier to achieve. This type of data is primarily from data sequences with a high time frequency of observations. Data can be acquired for specific locations from many satellites at different times, providing a wealth of data sources for the detection.

2.4 Synchronous Imaging

The establishment of a satellite network is restricted by many factors and is sensitive to certain constraints, planning measures, and strategies. The network constraints include the satellite number, number of satellite orbits, payload number, time range, target area characteristics, and the cloud distribution within the optical satellite image. Correct planning measures and strategies should provide solutions for meeting the coverage requirements for the target area. To improve the coverage of the target area, strict satellite conditions were planned in advance, with highly mobile satellites used to fill vacancies.

To ensure the successful execution of synchronous imaging, before determining the imaging area and imaging time, we used the systems tool kit (STK) to simulate multiple simultaneous satellite trajectories. According to the orbit simulation results, SJ-9A, SPOT5, and THEOS were synchronized in Zhangye City, Gansu Province on June 13, 2013. The study area included Zhangye City and the ground observation site of the Cold and Arid Regions Environmental and Engineering Research Institute, Chinese Academy of Science (Fig. 2). SJ-9A, SPOT5, and THEOS synchronization tests were performed successfully (Table 2). One scene image of a panchromatic and multiband image of SPOT5 and THEOS, respectively, was collected, and many scenes

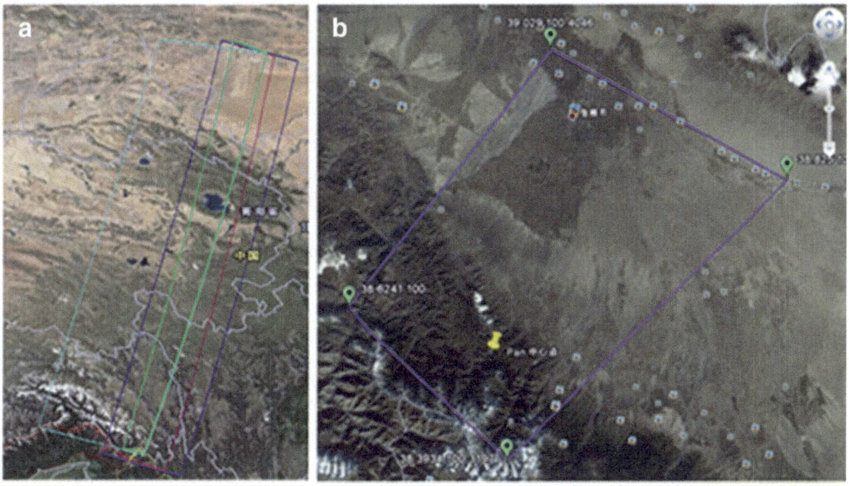

Fig. 2 **a** SJ-9A, SPOT5, and THEOS synchronous orbit simulation. **b** Location of the study area

Table 2 Synchronous images

Sensor	Date	Time (GMT)	Incidence angle	Image type	Sun Azimuth	Sun Elevation
SJ-9A	6/13/2013	03:58:03	1°	Pan/MS	122.5°	66.5°
SPOT5	6/13/2013	03:57:19	9.77°	Pan/MS	126.3°	66.9°
THEOS	6/13/2013	03:42:29	10°	Pan/MS	121.62°	64.93°

of SJ-9A panchromatic and multiband images were collected. Panchromatic (Fig. 3) and multiband image (Fig. 4) pairs were used to conduct the cross-comparison.

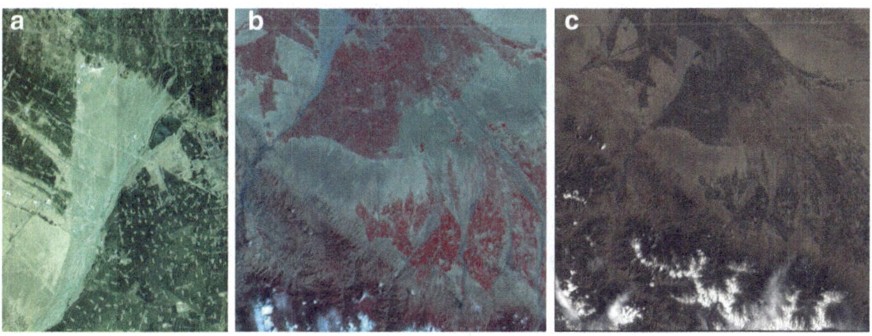

Fig. 3 **a** One of the SJ-9A multispectral images. **b** SPOT5 multispectral images. **c** THEOS multispectral image

Fig. 4 **a** SJ-9A Panchromatic image. **b** SPOT5 Panchromatic image. **c** THEOS Panchromatic image

2.5 Assessment Indicator

1. Information content

Information content is an indicator that reflects the rich degree of information contained in images. It is always measured by the information entropy. The more complicated the texture is in an image, the larger the amount of information it contains. The entropy is large in a complicated image. It is calculated using the following equation.

$$H = -\sum_{i=1}^{n} p_i \log\ p_i \tag{1}$$

where i is the value of pixels, n is the pixel number, and Pi is the probability of value i.

2. Radiometric accuracy

Radiometric accuracy reflects the radiation status, and its parameters are mainly mean (m), variance (d), skewness (s), and steepness (k). They are defined as follows:

$$m = \frac{1}{M \times N} \sum_{i=1}^{M \times N} g(i) \tag{2}$$

$$d = \frac{1}{M \times N} \sum_{i=1}^{M \times N} [g(i) - m]^2 \tag{3}$$

$$s = \frac{1}{M \times N} \sum_{i=1}^{M \times N} [g(i) - m]^3 / d^3 \tag{4}$$

$$k = \frac{1}{M \times N} \sum_{i=1}^{M \times N} [g(i) - m]^4 / d^4 \tag{5}$$

where M represents the line number of the image, N is the row number of the image, i is the number of pixels, and $g(i)$ is the gray value of pixels in the test image.

3. Sharpness

The sharpness reflects the sensitivity of changes in the image edge. The more distinctive the color contrast of the image junction, the clearer the image. An improved point sharpness algorithm was used to calculate the image sharpness. It was calculated using the following formula:

$$p = \frac{\sum_{i=1}^{m \times n} \sum_{\alpha=1}^{8} |df/dx|}{m \times n} \tag{6}$$

where p is the sharpness, m and n are the line and row number of pixels, respectively, df is the amplitude of the intensity change, and dx is the distance increment between pixels.

4. Signal-to-noise ratio

The signal-to-noise ratio (SNR) is the ratio of signal-to-noise in the image. A variance method can determine the SNR level of sensors. It is calculated using the following formula:

$$SNR = m/d \tag{7}$$

$$m = \frac{1}{M \times N} \sum_{i=1}^{M \times N} g(i) \tag{8}$$

$$d = \frac{1}{M \times N} \sum_{i=1}^{M \times N} [g(i) - m]^2 \tag{9}$$

where m is the mean, d is variance, M represents the line number of the image, N is the row number of the image, and $g(i)$ is the gray value of pixels in the test image.

5. Modulation Transfer Function

MTF is a function of spatial frequency, reflecting the ability of the optical system to respond to different spatial frequencies; therefore, it is also known as the spatial contrast transfer function. The MTF has been widely used to characterize the performance of optical systems because it contains almost all of the optical information in a system. A straightforward approach to measuring the MTF involves imaging a known input, such as a point or a line source, to produce the point spread function (PSF) and the line spread function (LSF), respectively. However, the use of a point or a line source requires precise fabrication and high exposure. Thus, an edge source is imaged to yield the edge spread function (ESF) instead, which can then be differentiated to obtain the LSF. The detailed steps are as follows:

1. According to the edge imaging, the gray distribution is used to extract the sub-pixel edge location.

2. The sample data is extracted based on sub-pixel edge location and is then fitted to the ESF curve.
3. A derivation of the ESF curve can obtain the line spread function curve.
4. The MTF curve can be obtained, with the curve of the linear diffusion function obtained by Fourier transform.
5. The MTF curve is corrected by the image width of slit imaging.

2.6 Data Processing

There were three data processing stages for an image:

1. Image registration

A SPOT-5 image was the reference image, with a no-cloud image selected as the region of interest through the ERDAS IMAGINE 9.1 AOI (Area of Interest) tool. Figure 5a shows a multispectral image of THEOS, SPOT-5, and SJ-9A satellite images. Here, the THEOS and SJ-9A images were synthesized by 4-, 3-, and 2-bands; SPOT-5 satellite images were synthesized by the 3-, 2-, and 1-bands. Figure 5b shows the panchromatic image of the study area for the three satellites.

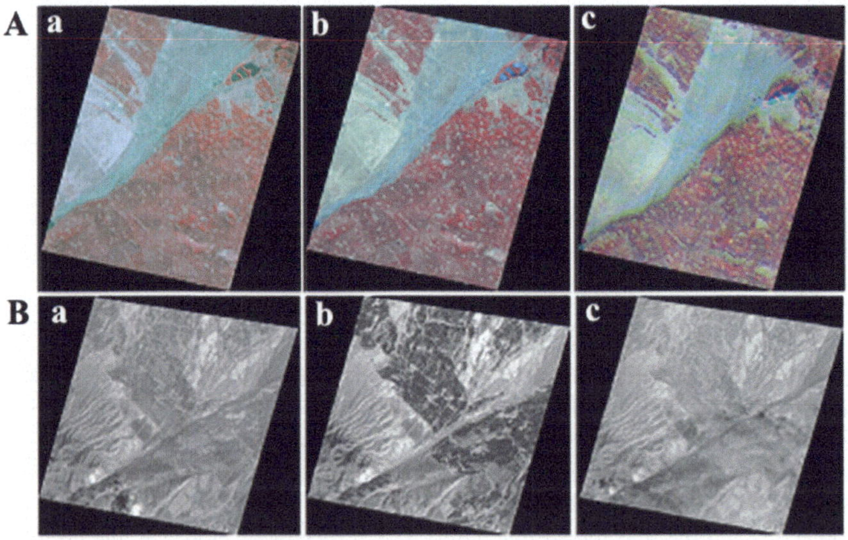

Fig. 5 Multispectral and Panchromatic study area. **A** Multispectral study area (a) THEOS image (b) SPOT-5 image (c) SJ-9A image. **B** Panchromatic study area (a) THEOS image (b) SPOT-5 image (c) SJ-9A image

2. Quantitative level normalization

The quantization levels of SJ-9A, SPOT5, and THEOS are 10, 8, and 8, respectively. The quantization level was normalized to maintain them at the same level.

3. Indicator result normalization

The indicator values cannot directly indicate if an indicator is good or bad. The judging standard is compared with the quality value of the ideal state. To ensure that the indicator value and the ideal value can be compared with each other, the indicator values were normalized.

3 Results and Discussion

To obtain more objective and accurate comparison results, panchromatic images of sandy land, urban areas, and mountains were selected as interest areas. For multi-spectral images, sandy land, urban areas, and water were selected as interest areas. To increase the reliability of the experimental results, for each feature object, two sample areas were selected for information quantity, radiation quantity, and sharpness of indicator comparisons. However, the SNR and MTF curves mainly reflected the ground information, in which the result was mainly influenced by the imaging quality of the sensor. Therefore, the SNR and MTF curves were calculated based on the whole image, not just the interest area.

3.1 Comparison of Panchromatic Images

The SJ-9A panchromatic image information entropy indicated that the maximum amount of information and a consistent performance were obtained for sandy land, urban areas, and mountains. The mean of the radiation quantity was the smallest of the parameters investigated, which means that the SJ-9A image was the darkest. The variance was the largest of the parameters investigated, which was in good agreement with the amount of information. The skewness of the different features was different, and the gray value of SJ-9A was generally closer to the mean value. In the sandy land and mountains, the steepness of SJ-9A was the largest, indicating that the gray value distribution was concentrated in a relatively small dynamic range. In urban areas, the steepness was the largest in THEOS and SPOT-5. In the three interest areas, the SJ-9A image sharpness value was the largest, and the edge changes in images were most sensitive in THEOS and SPOT-5 images. The SNR was the lowest in SJ-9A, indicating that the noise level was large. The MTF results show that the spatial frequency response of SJ-9A was worse than that of SPOT-5 but better than THEOS. The detailed results are shown in Table 3. Figure 6 shows the results of a comparison of the MTF.

Table 3 Results of the comparison of all indicators for panchromatic images

	Indicators		Interest areas	Comparison in the interest area	Conclusion
Panchromatic Images	information content		Sandy area	The SJ-9A entropy was the largest	Compared with SPOT5 and THEOS, SJ-9A had the most information
			Urban area	The SJ-9A entropy was the largest	
			Mountains	The SJ-9A entropy was the largest	
	SNR		Whole image		The lowest SJ-9A SNR indicates large noise
	Radiation accuracy	Mean	Sandy area	The mean was the smallest	Compared with SPOT5 and THEOS, the SJ-9A image was the darkest
			Urban area	The mean was the smallest	
			Mountains	The mean was the smallest	
		Variance	Sandy area	SJ-9A has the largest variance with gray-level dispersion	Overall, SJ9-A variance was the largest, reflecting a large amount of information
			Urban area	SJ-9A variance values were between SPOT5 and THEOS	
			Mountains	The SJ-9A has the largest variance with gray-level dispersion	
		Skewness	Sandy area	The SJ-9A skewness absolute value was the smallest, and the histogram gray value distribution was more concentrated around mean value, with good symmetry	In different regions, the skewness value was not completely consistent, and not very stable
			Urban area	SJ-9A skewness absolute value was the largest, and the symmetry was not good	
			Mountains	SJ-9A skewness absolute value was larger in area 1 and smaller in area 2	

(continued)

Table 3 (continued)

	Indicators		Interest areas	Comparison in the interest area	Conclusion
		Steepness	Sandy area	The SJ-9A steepness value was the largest	Gray value distribution of SJ-9A in the sandy area and mountains was relatively concentrated, and the dynamic range was the smallest; in urban areas, the value was between THEOS and SPOT-5, indicating relatively concentrated gray value distribution and small dynamic range
			Urban area	SJ-9A steepness value is in between THEOS and SPOT-5, and the concentration of gray value distribution is lower than SPOT-5, higher than THEOS	
			Mountains	In region 1, the steepness values of SJ-9A and SPOT-5 were similar, and higher than THEOS, while they were larger in region 2	
	MTF		Whole image		The spatial frequency response of SJ-9A was better than THEOS but worse than SPOT-5
	sharpness		Sandy area	Compared with SPOT-5 and THEOS, the SJ-9A image was the most sensitive with changed edge details	Compared with SPOT-5 and THEOS, the edge details of the SJ-9A image changed, and the SJ-9A image was the most sensitive and had the best sharpness in the three interest areas
			Urban area	The change in the image edge details of the SJ-9A was more sensitive than SPOT-5 but slightly worse than THEOS	
			Mountains	Same as that in sandy area	

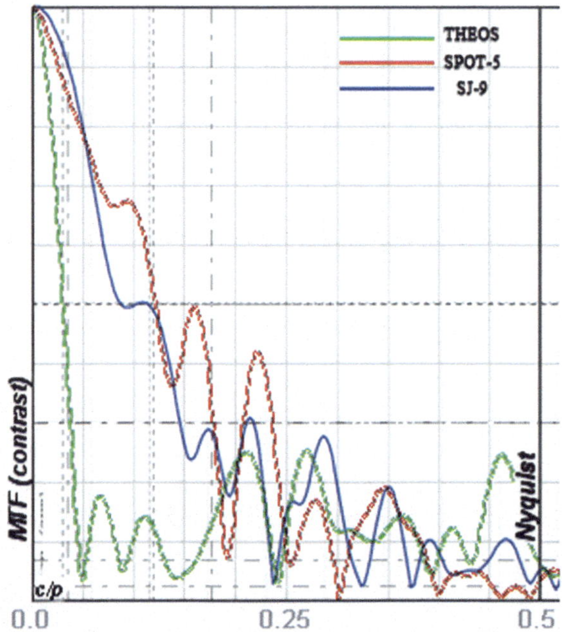

Fig. 6 Comparison of MTF results for panchromatic images

3.2 Comparison of Multispectral Images

The SJ-9A multispectral image information entropy indicated that the maximum amount of information and a consistent performance were obtained for sandy land, urban areas, and mountains. The mean of the radiation quantity was the smallest of the parameters investigated, which means that the SJ-9A image was the darkest. The variance was the largest parameter, and this is in good agreement with the amount of information acquired. Overall, the skewness was smallest for SJ9-A, and the symmetry of the histogram was the best. In different regions, steepness was not completely the same for the different bands and was not very stable, but overall, the SJ9-A gradient was the largest, and the histogram distribution was more concentrated. For all three interest areas, the SJ-9A image sharpness value was largest, and the details of the edge changes in the images were more sensitive than THEOS and SPOT-5. The SJ-9A SNR was the lowest, indicating that noise was large. The MTF results show that the spatial frequency response of SJ-9A was worse than that of SPOT-5 and THEOS. The detailed results are shown in Table 4. Figure 7 shows the results of a comparison of the MTF.

Table 4 Results of the comparison of all indicators for multispectral images

	Indicator		Area type	Comparison result of interest area	Comparison result
Multiband image	information content		Sandy land	Compared with SPOT5 and THEOS, SJ-9A entropy was the largest, which indicates that information was most abundant	SJ-9A information is most abundant
			Urban area	Compared with SPOT5 and THEOS, SJ-9A entropy was the largest, which indicates that information was most abundant	
			Water	Compared with SPOT5 and THEOS, SJ-9A entropy was the largest, which indicates that information was most abundant	
	SNR		Whole image		SJ-9A SNR was the lowest, indicating that noise was large
	Radiation accuracy	Mean	Sandy land	The mean is the smallest, which means the SJ-9A image is the darkest	SJ-9A image was the darkest
			Urban area	The mean was the smallest, which means the SJ-9A image was the darkest	
			Water	In the blue (B1) and green (B2) bands, the mean of SJ-9A was the smallest. In the red (B3) and near infrared (B4), the mean of SJ-9A was similar to THEOS but larger than SPOT5	
		Variance	Sandy land	The gray value distribution of the SJ-9A multispectral image was relatively dispersed	SJ-9A variance value is the largest, and the information is the most abundant

(continued)

Table 4 (continued)

	Indicator		Area type	Comparison result of interest area	Comparison result
			Urban area	The gray value distribution of the SJ-9A multispectral image was relatively dispersed	
			Water	The gray value distribution of the SJ-9A multispectral image was relatively dispersed	
		Skewness	Sandy land	In the green (B2), red (B3), and near infrared (B4) bands, the SJ-9A skewness absolute value was the smallest, and the histogram gray value distribution was more concentrated around the mean value, with good symmetry	On the whole, the skewness of SJ9-A is relatively small, and the symmetry of histogram is the best
			Urban area	In the green (B2), red (B3), and near infrared (B4) bands, the SJ-9A skewness absolute value was the smallest, and the histogram gray value distribution was more concentrated around the mean value, with good symmetry	
			Water area	In the green (B2), red (B3), and near infrared (B4) bands, the SJ-9A skewness absolute value was the smallest, and the histogram gray value distribution was more concentrated around the mean value, with good symmetry	

(continued)

Table 4 (continued)

	Indicator		Area type	Comparison result of interest area	Comparison result
		Steepness	Sand area	In the green (B2) and red (B3) bands, the steepness value of SJ-9A was larger than that of SPOT5 and THEOS, and the gray value distribution was relatively concentrated. However, in the blue (B1) and near infrared (B4)	In different regions, steepness was not completely the same for the different bands and was also not
				bands, the steepness value was smaller, and the gray value distribution is dispersive	very stable. Overall, the SJ9-A gradient was relatively large, and the histogram distribution was more concentrated
			Urban area	The steepness value of SJ-9A in the blue (B1) band was smaller than in THEOS in the two interest area, but in area 1, the SJ-9A steepness value for the green (B2) and red (B3) bands was larger, and in area 2, the SJ-9A green (B2) band steepness value was larger. For the remaining bands, the steepness value of SJ-9A was in between that of THEOS and SPOT-5	
			Water	In the green (B2) and red (B3) bands, the steepness value of SJ-9A was larger than that of SPOT5 and THEOS, and the gray value distribution was relatively concentrated, and the dynamic range was small	

(continued)

Table 4 (continued)

	Indicator	Area type	Comparison result of interest area	Comparison result
	MTF	Whole image		The spatial frequency response of SJ-9A is worse than that of SPOT-5 and THEOS
	sharpness	Sand area	Compared with SPOT-5 and THEOS,the image edge details change of the SJ-9A image is the most sensitive	Compared with SPOT-5 and THEOS, the change in the details of the image edge of the SJ-9A image was the most sensitive and provided most sharpness in all three interest areas
		Urban area	Compared with SPOT-5 and THEOS, the change in the details of the image edge of the SJ-9A image was the most sensitive	
		Water area	Compared with SPOT-5 and THEOS, the change in the details of the image edge of the SJ-9A image was the most sensitive	

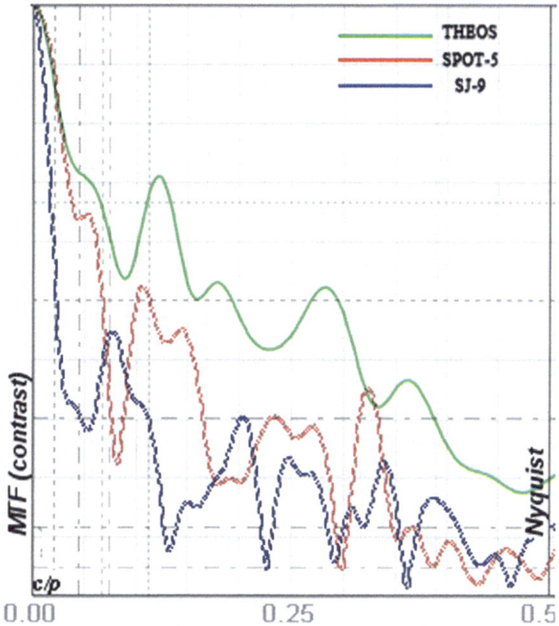

Fig. 7 Comparison of MFT results for multispectral images

4 Conclusions

In this paper, an image quality evaluation method is presented which is based on synchronous observations by multisatellites. The technology of the method is described in detail, and a comparison and analysis of the THEOS, SPOT5, and SJ-9A satellites were conducted using this method. The method was shown to be practical and operable.

Through cross-comparison, the overall quality of SJ-9A was found to be better than SPOT5 and THEOS, in terms of the amount of information and sharpness. This study can only make a reference to the quality of SJ-9A. As there were no ground observations, the comparison can only be relative.

It is clear that image quality evaluation based on synchronous observation by multisatellites is restricted by many factors, such as synchronous data acquisition, the data product level, and the data quantization digit. This study largely solved the problem of synchronous data acquisition, but the impact of the product level and quantization digit was not considered. The original data received from the ground station was the best choice for satellite comparison; however, because most original data is not accessible for the public, a preprocessing image was used, and the impact of such data on the results needs to be further studied. The quantization digit is the digital of the amplitude axis of the analog audio signal and determines the dynamic range of the analog signal to digital. To compare with the same level, the data level should be normalized, but the impact of the normalization on the comparison of results needs further verification. In addition, due to the high cost of satellite networks, only a synchronized test was performed. In follow-up studies, different study areas should be selected, and pairs of multiple simultaneous images should be used for comparison.

References

1. Ran Q et al (2009) Noise assessment of Beijing 1 small satellite images. J Remote Sens 10(3):554–558
2. Ryan R et al (2003) IKONOS spatial resolution and image interpretability characterization. Remote Sens Environ 88(1):37–52
3. Zhang X, Zhang B, Zhao Y (2002) Image Quality Assessment for the Infrared Multi Spectral Scanner of the Chinese-Brazil Earth Resources Satellite. J Image Graph 7(06):63–68 + 108
4. Elshehaby AR, Taha LGE-D (2011) Assessment of cartographic potential of EgyptSat-1 satellite image (case study in flat areas). Appl Geomatics 3(3):159–169
5. Chander G, Markham BL, Helder DL (2009) Summary of current radiometric calibration coefficients for Landsat MSS, TM, ETM+ , and EO-1 ALI sensors. Remote Sens Environ 113(5):893–903
6. Chander G et al (2010) Monitoring on-orbit calibration stability of the Terra MODIS and Landsat 7 ETM+sensors using pseudo-invariant test sites. Remote Sens Environ 114(4):925–939
7. Lin L, Heng L, Xinming T, Zhen LI (2014) Characteristic analysis and quality assessment of ZY-3 multi-spectral image. Remote Sens Land Resour 26(01):17–24
8. Li P, Jiang L, Feng Z (2013) Cross-comparison of vegetation indices derived from Landsat-7 enhanced thematic mapper plus (ETM+) and Landsat-8 operational land imager (OLI) sensors. Remote Sens 6(1):310–329
9. Chen X et al (2013) Cross-sensor comparisons between Landsat 5 TM and IRS-P6 AWiFS and disturbance detection using integrated Landsat and AWiFS time-series images. Int J Remote Sens 34(7):2432–2453
10. Ke S, Shen X, Lei Y (2012) Quality evaluation of the CBERS-02B high spatial resolution image. Geosci Remote Sens Symp (IGARSS), 2012 IEEE Int 2312–2315
11. Thome KJ, Biggar SF, Wisniewski W (2003) Cross comparison of EO-1 sensors and other earth resources sensors to Landsat-7 ETM+ using Railroad Valley Playa. Geosci Remote Sens IEEE Trans 41(6):1180–1188
12. Chander G, Meyer DJ, Helder DL (2004) Cross calibration of the Landsat-7 ETM+ and EO-1 ALI sensor. Geosci Remote Sens IEEE Trans 42(12):2821–2831
13. Miura T et al (2008) Inter-comparison of ASTER and MODIS surface reflectance and vegetation index products for synergistic applications to natural resource monitoring. Sensors 8(4):2480–2499
14. Soudani K et al (2006) Comparative analysis of IKONOS, SPOT, and ETM+ data for leaf area index estimation in temperate coniferous and deciduous forest stands. Remote Sens Environ 102(1):161–175
15. Deng H et al (2014) Geometric accuracy assessment and correction of imagery from Chinese earth observation satellites (HJ-1 A/B, CBERS-02C and ZY-3). Int Arch Photogram Remote Sens Spat Inf Sci 2(71–78)
16. Hongwei W, Qingjiu T (2012) Quality evaluation and analysis f HJ1B-CCD images. Remote Sens Inf 27(05):31–36
17. Wu M et al (2012) Data quality evaluation of ZY-102C satellite. IFIP Int Fed Inf Process 9
18. Teillet PM, Ren X (2008) Spectral band difference effects on vegetation indices derived from multiple satellite sensor data. Can J Remote Sens 34(3):159–173
19. Oguro Y, Tsuchiya K, Suga Y (1999) Comparison of land cover features observed with different satellite sensors over a semi-arid land in central Australia. Adv Space Res 23(8):1401–1404

20. Oguro Y et al (2003) Monitoring of a rice field using Landsat-5 TM and Landsat-7 ETM+data. Adv Space Res 32(11):2223–2228
21. Van Wagtendonk JW, Root RR, Key CH (2004) Comparison of AVIRIS and Landsat ETM + detection capabilities for burn severity. Remote Sens Environ 92(3):397–408
22. Vazquez A, Cuevas J, Gonzalez-Alonso F (2001) Comparison of the use of WiFS and LISS images to estimate the area burned in a large forest fire. Int J Remote Sens 22(5):901–907
23. Teillet P et al (2007) Impacts of spectral band difference effects on radiometric cross-calibration between satellite sensors in the solar-reflective spectral domain. Remote Sens Environ 110(3):393–409
24. Oguro Y et al (2001) Comparison of SAR and optical sensor data for monitoring of rice plant around Hiroshima. Adv Space Res 28(1):195–200

Water Extraction of Airborne Polarimetric SAR by Introducing Eigenvalue Relative Difference

Zheng Changli, Zhang Wei, Ding Qing, Wang Xiaoxia, and Luo Jinghui

Abstract In high-resolution airborne SAR images, water bodies, paddy fields, roads, bare soils and other ground objects exhibit similar weak scattering characteristics. At present, the water body extraction algorithm based on the weak echo characteristics cannot eliminate the interference of other ground objects, resulting in unsatisfactory water extraction results. Aiming at the KU-band airborne polarimetric SAR image, this paper proposes a fine extraction method of airborne polarimetric SAR water with the relative difference of single-reflection eigenvalues. Firstly, the H-α Wishart classification algorithm is used to classify the weak scattering features, and then, the weakly scattered features are re-segmented based on the pre-classification results to construct the object unit. Finally, based on the single-reflex eigenvalue relative difference feature, the threshold method is used to realize the water fine extraction. Experiments show that the relative difference of single-reflection eigenvalues can effectively separate the water body from the rest of the weakly scattered features, and the object unit constructed by the fractal network evolution algorithm can obtain better water body fineness than the simple pixel-level analysis.

Keywords Polarimetric SAR · Water extraction · Object-oriented analysis · Speckle noise · Scattering characteristics

1 Introduction

Synthetic aperture radar (SAR) has the ability of all-day and all-weather surface observation. It can quickly respond to the severe weather conditions of rain, snow and fog and obtain the surface information of the target area. Water information plays an important role in flood assessment, urban management and river management. In recent years, with the development of airborne SAR system, UAV SAR with full polarization earth observation capability has played an important role in disaster prevention and mitigation. Compared with single-pol SAR system, full-pol SAR reveals the scattering difference of ground objects and provides a new technical

Z. Changli (✉) · Z. Wei · D. Qing · W. Xiaoxia · L. Jinghui
Science and Technology on Electronic Information Control Laboratory, 610036 Chengdu, China
e-mail: zs_zcl@163.com

L. Wang et al. (eds.), *Proceedings of the 6th China High Resolution Earth Observation Conference (CHREOC 2019)*, Lecture Notes in Electrical Engineering 657,
https://doi.org/10.1007/978-981-15-3947-3_43

means for the interpretation, classification and information extraction of ground objects.

At present, scholars at home and abroad mainly focus on four aspects [1–9]. (1) In view of the weak backscattering intensity of calm water surface and the characteristics of dark pixels in the image, water extraction research was carried out; (2) Research on water extraction based on the smooth texture characteristics of water body; (3) Study on water extraction based on the polarization characteristics of Bragg scattering; (4) Study on water extraction by mixing the bove characteristics. The above algorithms have achieved remarkable results in water extraction of spaceborne SAR images, but the results are not ideal for airborne SAR images. Compared with spaceborne SAR, airborne SAR has a larger incidence angle, which makes the water, paddy field, road and bare soil indistinguishable from airborne SAR images, and all of them have similar weak backscattering intensity characteristics. In addition, they all have smooth texture characteristics and odd scattering polarization characteristics, which make use of the above spaceborne SAR algorithm. The results of water extraction from airborne polarimetric SAR images are not ideal.

In order to eliminate the interference of low scattering intensity targets on the water extraction of airborne polarimetric SAR, this paper proposes a method of fine water extraction incorporating the relative difference of eigenvalues. Firstly, the weak scattering objects are extracted based on H-α Wishart [10] classification algorithm, and then, the weak scattering objects are segmented based on the pre-classification results. Then, we construct the single bounce eigenvalue relative difference [11, 12] (SERD) feature of the image and use the threshold method to screen out the water target from the weak scattering object unit. In order to verify the validity of the proposed algorithm, the full polarization SAR images of Nanchong City obtained from UAV SAR system were tested.

2 Construction of Weak Scattering Object Unit

In airborne SAR images, the weak scattering objects are seriously affected by speckle noise. If only processed at the pixel level, the results will be fragmented, and the image points will be isolated. Therefore, based on H-α Wishart pre-classification results, this paper constructs objects based on fractal network evolution algorithm.

The fractal network evolutionary segmentation algorithm is a bottom-up segmentation strategy. The basic idea is to merge pixels or objects in the direction of decreasing heterogeneity until the increment of heterogeneity after merging exceeds the threshold, and then, the merging terminates. The traditional fractal network evolutionary segmentation method uses spectral information and shape information to describe the heterogeneity of objects. Spectral information and shape information of various polarization features are integrated to describe the heterogeneity of objects and are used to segment polarization SAR buildings. The principle is as follows:

Suppose that before merging, object 1 contains n_1 pixels, heterogeneity is h_1, object 2 contains n_2 pixels, heterogeneity is h_2, and merged object heterogeneity is

h_m. The variation of heterogeneity before and after merger is as follows:

$$h_{\text{diff}} = (n_1 + n_2)h_m - (n_1 h_1 + n_2 h_2) = n_1(h_m - h_1) + n_2(h_m - h_2) \tag{1}$$

A K-dimension feature participates in the segmentation, assigns a weight w_k to each dimension feature and describes the heterogeneity with the standard deviation of the eigenvalue in the object, and then the variation of the heterogeneity of the object's feature becomes as follows:

$$h_{\text{pol}} = \sum_k w_k(n_1(\sigma_m - \sigma_1) + n_2(\sigma_m - \sigma_2)) \tag{2}$$

Shape heterogeneity is generally defined in terms of compactness and smoothness. If the compactness and smoothness of the segmented object are h_{cmpt} and h_{smooth}, respectively, the shape heterogeneity h_{shape} can be expressed as follows:

$$h_{\text{shape}} = w_{\text{cmpt}} h_{\text{cmpt}} + \left(1 - w_{\text{cmpt}}\right) h_{\text{smooth}} \tag{3}$$

Let the weight of feature heterogeneity be w_{pol}. Then, the similarity F of adjacent image objects can be expressed as:

$$f = w_{\text{pol}} h_{\text{pol}} + \left(1 - w_{\text{pol}}\right) h_{\text{shape}} \tag{4}$$

3 Fine Extraction of Water Body Integrating SERD

SERD is calculated from the average coherence matrix satisfying the assumption of reflection symmetry. In addition to reflection symmetry, it is stipulated that for natural media, the correlation between the same polarization channel and the cross polarization channel can be assumed to be zero. Based on the S matrix satisfying the reflection symmetry, the expression of the corresponding unsorted eigenvalues is derived as follows:

$$\begin{aligned}
\lambda_1 &= \frac{1}{2}\left\{ |S_{HH}|^2 + |S_{VV}|^2 + \sqrt{\left(|S_{HH}|^2 - |S_{VV}|^2 + 4\left|S_{HH}S_{VV}^*\right|^2\right)} \right\} \\
\lambda_2 &= \frac{1}{2}\left\{ |S_{HH}|^2 + |S_{VV}|^2 - \sqrt{\left(|S_{HH}|^2 - |S_{VV}|^2 + 4\left|S_{HH}S_{VV}^*\right|^2\right)} \right\} \\
\lambda_3 &= 2|S_{HV}|^2 \\
\text{SERD} &= \frac{\lambda_s - \lambda_3}{\lambda_s + \lambda_3}
\end{aligned} \tag{5}$$

In the formula, λ_s corresponds to the eigenvalue of single scattering, and α_i is the parameter α extracted from the eigenvector corresponding to λ_i. When $\alpha_1 \leq \frac{\pi}{4}$ or $\alpha_2 \geq \frac{\pi}{4}$, $\lambda_s = \lambda_1$; when $\alpha_1 \geq \frac{\pi}{4}$or $\alpha_2 \leq \frac{\pi}{4}$, $\lambda_s = \lambda_2$.

Influenced by surface roughness, the soil is characterized by low-entropy scattering in long-wavelength SAR images, while in Ku-band polarimetric SAR images, the soil is not dominated by Bragg scattering but by medium–low entropy. For water body, paddy field and road, because the scattering intensity of each channel is relatively low and close and affected by coherent speckle noise of the system, SAR images in Ku-band always show high entropy scattering characteristics, while SERD parameters are very useful for medium with high polarization entropy, which can determine the characteristics and magnitude of different scattering mechanisms.

4 Experimental Results and Analysis

4.1 Experimental Data

This paper makes use of the full polarization SAR data acquired by the unmanned aerial vehicle (UAV) airborne SAR system of Southwest China Research Institute of Electronic Equipment to carry out experiments on the fine extraction method of water body.

The UAV airborne SAR system operates in Ku-band with an incident angle ranging from 20 to 60 degrees. The working mode is strip imaging mode. The resolution of range and azimuth is 0.3 m. The maximum relative flight altitude is 3 km. The UAV-borne SAR system can support flight at altitude of 5 km. The maximum operating distance is 6 km, the maximum mapping bandwidth is 3 km, and it has *HH*. Full polarization measurement capability of *HV*, *VH* and *VV* is shown in Fig. 1a.

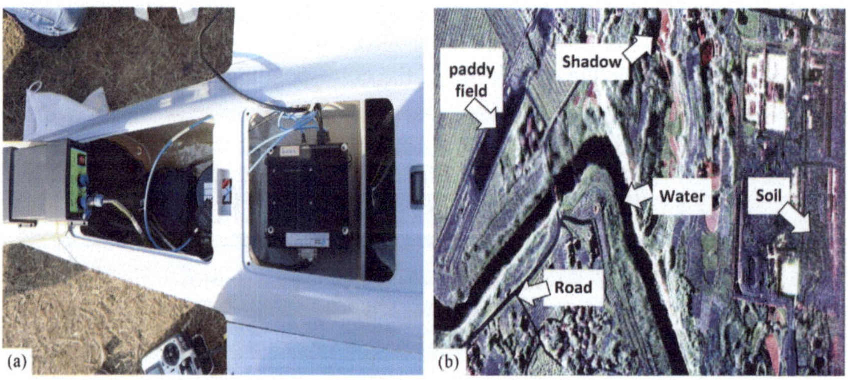

Fig. 1 Experimental data profiles. **a** UAV airborne SAR system. **b** Pauli RGB image

The full polarization SAR data obtained from the flight test of the UAV-borne SAR system in Nanchong in January 2018 were used to intercept a 6200 × 5400 pixel size area in the airstrip ID5, as shown in Fig. 1b Pauli RGB image. In Fig. 1b, in addition to water, weak scattering objects include shadows caused by houses and vegetation during crop growth, paddy fields, suburban roads and soil.

4.2 Pre-classification and Construction of Weak Scattering Objects

In order to satisfy the better interpretation effect, refined Lee filtering (filter window is 7*7) is applied to the image. Then, *H*-*α* Wishart classification is carried out in the experimental area, and the classification results are shown in Fig. 3. Even if the anisotropy parameter is added, it is still impossible to distinguish the weak scattering objects, and there are still a lot of weak scattering interference in the water extraction results. This is due to the following two reasons: (1) due to speckle noise interference, pixel-level results are difficult to distinguish weak scattering objects, but due to the impact of echo noise, the object based on the initial image is greatly disturbed. (2) In SAR images acquired under UAV SAR system, due to the large incidence angle of radar wave, the weak scattering objects show similar polarization scattering characteristics and are easy to be confused.

On the basis of the pre-classification results, this paper extracts the seventh kind of confused weak scattering objects from Fig. 2. Because of the large incidence angle, objects such as buildings, slopes and vegetation which are higher than the ground

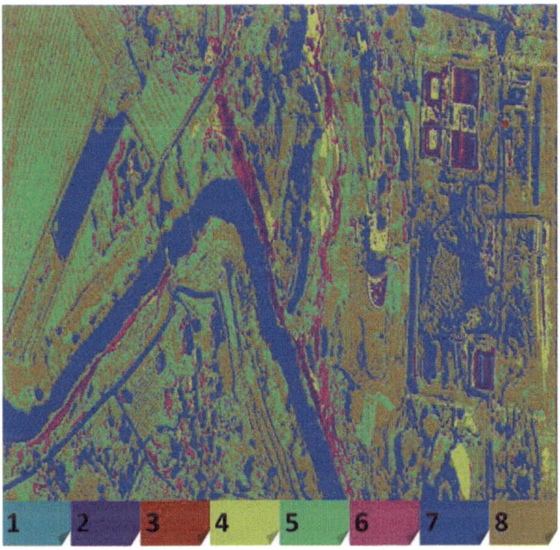

Fig. 2 *H*-*α* Wishart classification results in experimental areas

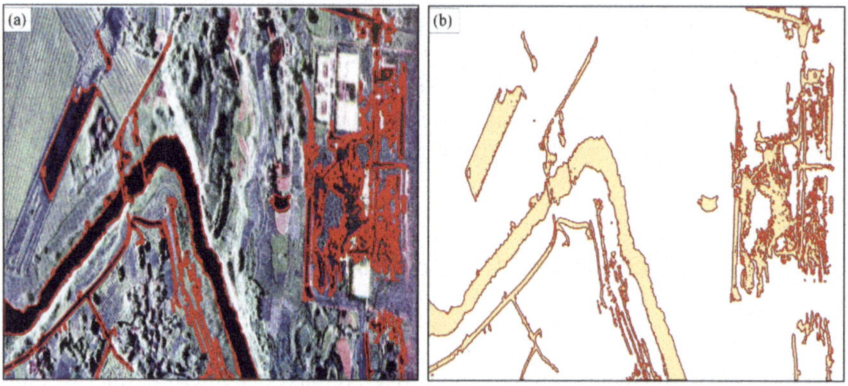

Fig. 3 Construction of weak scattering objects. **a** Overlapping Pauli RGB image of weak scattering objects. **b** Vector of weak scattering object

are easily affected by shadows. Because shadows do not contain or contain a small amount of information, shadows are often removed directly from the classification results without considering the impact of shadows. Airborne SAR images can get ground height by interferometry, so as to recognize and remove the shadow effects caused by buildings and other objects. This paper does not focus on this, so the shadow can be removed directly from seven categories. Using the extracted weak scattering objects to construct a mask, the polarization data are segmented by fractal network evolution under the restriction of the pre-segmentation class, and the object unit of weak scattering objects analysis is constructed, as shown in Fig. 3.

4.3 Fine Water Extraction by Introducing SERD

There are two steps in water body fine extraction: (1) initial determination of threshold based on histogram of SERD characteristics; (2) fine adjustment of threshold based on actual distinguishing results of objects to achieve fine extraction of water body.

The dynamic range of SERD is −1 to +1. It can be seen from Fig. 4a that the SERD value of water body is at a lower level than other low scattering objects. Using JET imaging, it can be seen that the color of water body region transits between yellow and red, while the color of other roads, paddy fields and bare soil is yellow–green to blue. There is a large SERD difference between water and non-water bodies during the transition. In order to distinguish water from non-water from weak scattering objects, SERD is masked by weak scattering classes, and the image shown in Fig. 4a is obtained. The distribution of SRED eigenvalues in Fig. 4a is counted, and the gray distribution in Fig. 4b is obtained. In Fig. 4b, there is an obvious double peak in the gray level distribution. The SERD value at the boundary between the two peaks is −0.3 as the initial threshold of judgment. In order to obtain a better separation effect,

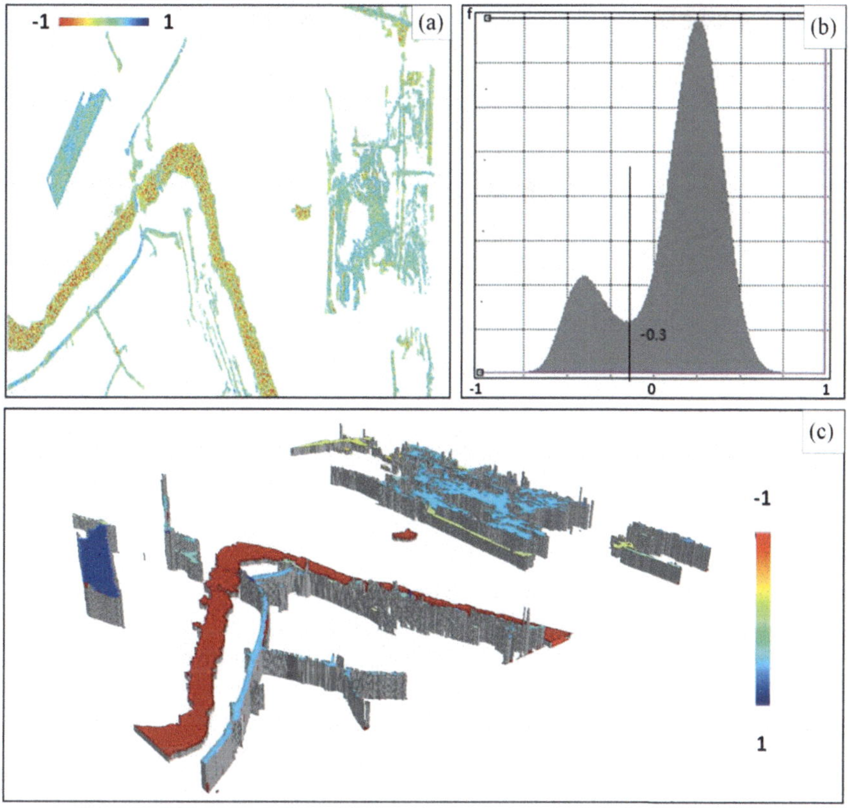

Fig. 4 SERD characteristic image. **a** SERD characteristics under pre-classification mask. **b** SERD histogram under pre-classification mask. **c** Object-level SERD height image

the SERD threshold was finally determined to be = 0.35 by fine-tuning the threshold according to the distinguishing effect between water and non-water. In Fig. 4c, the object is imaged to the JET color bar according to the object mean of SERD, and the object mean of SERD is also highly reflected.

On the premise of the same segmentation threshold, the precise extraction results of water body are obtained by using the pixel as the analysis unit and the object as the analysis unit, as shown in Fig. 5. As can be seen from the figure, the result of pixel level is not ideal because of speckle noise. There are not only isolated and misclassified pixel points, but also a considerable part of water body is not separated from non-water body, which makes the boundary of water body extraction result fragmented and confused. The results of the object as an analysis unit are not affected by speckle noise. The classification of water body and non-water body is realized. The boundary of the water body results is accurate and clear.

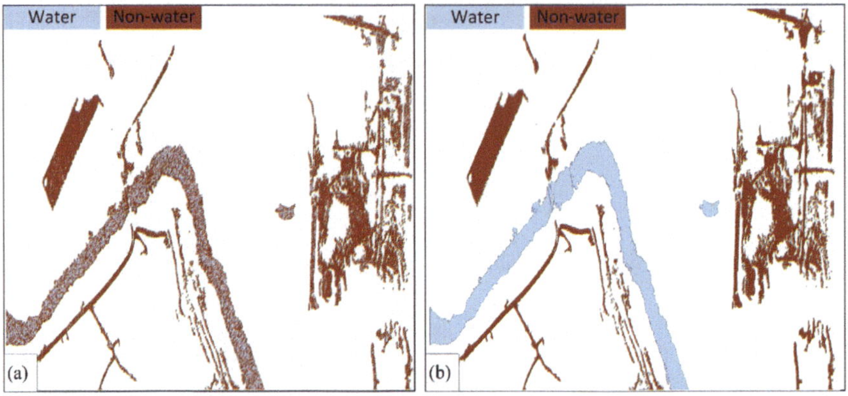

Fig. 5 Fine water extraction results. **a** Pixel-level water results. **b** Object-level water results

5 Conclusion

In this paper, the problem of object aliasing in high-resolution Ku-band airborne SAR system in water extraction is studied and discussed. In view of the fact that conventional spaceborne polarimetric SAR water extraction algorithm cannot distinguish water from weak scattering objects, this paper proposes a new method of fine water extraction incorporating SERD. Axis is pre-classified, fractal network evolution is used to segment weak scattering objects in pre-classification to obtain object-level analysis units; SERD eigenvalues of polarimetric SAR are calculated under pre-classification masks, and objects are taken as analysis units, and water body is fine extracted by threshold of SERD gray analysis, and other weak scattering objects are eliminated. In the future, we intend to do in-depth research and analysis on the adaptability of the algorithm with more data from different sources.

References

1. Fulong C, Hong Z, Chao W (2007) The Art in SAR Chang detection—a systematic review[J]. Remote Sens Technol Appl 22(1):109–115
2. Fan W, Chao W, Hang Z (2005) Residential area extraction from high resolution SAR images based on texture features[J]. Remote Sens Technol Appl 20(1):148–152
3. Changli Zheng, Guoman Huang, Zheng Zhao et al (2018) Water segmentation using multiscale level set method based on joint distribution of G0 and Gamma [J]. Remote Sens Inf 33(1):13–20
4. Chuan Xu, Feng Hua, Haigang Sui et al (2014) Automatic segmentation of water from multiscale level set SAR images [J]. J Wuhan Univ (Inf Sci Ed.) 39(1):27–31
5. Xiao C (2012) Research on extraction algorithms of flood and snow disaster area based on SAR image [D]. Univ Electron Sci Technol

6. Refice A, Capolongo D, Pasquariello G et al (2014) SAR and InSAR for flood monitoring: examples with COSMO-SkyMed data[J]. IEEE J Sel Top Appl Earth Obs Remote Sens 7(7):2711–2722
7. Lu J, Li J, Chen G et al (2017) Improving pixel-based change detection accuracy using an object-based approach in multitemporal SAR flood images[J]. IEEE J Sel Top Appl Earth Obs Remote Sens 8(7):3486–3496
8. Cossu R, Schoepfer E, Bally P et al (2009) Near real-time SAR-based processing to support flood monitoring[J]. J Real-Time Image Proc 4(3):205–218
9. Auynirundronkool K, Chen N, Peng C et al (2012) Flood detection and mapping of the Thailand central plain using RADARSAT and MODIS under a sensor web environment[J]. Int J Appl Earth Obs Geoinf 14(1):245–255
10. Ferro-Famil L, Pottier E, Lee J-S (2001) Unsupervised classification of multifrequency and fully polarimetric SAR images based on the H/A/A-wishart classifier[J]. IEEE Trans Geosci Remote Sens 39(11):2332–2342
11. Allain S, Ferrofamil L, Pottier E (2004) Two novel surface model based inversion algorithms using multi-frequency polSAR data[C]. IEEE Int Geosci Remote Sens Symp
12. Allain S, Lopez-Martinez C, Ferro-Famil L et al (2005) New eigenvalue-based parameter for natural media characterization[J]

Research on Dimension Reduction Detection Technology of Space-Borne GPS Receiver Fault

Zhang Xiaopeng, Wang Xiaochen, Zhuang Haixiao, Guo Yongfu, Chen Xi, Zuo Miao, Zhang Xiangyan, Wang Weiwei, and Yang Ping

Abstract In order to realize dimension reduction detection of satellite-borne GPS receivers, the working mode and on-orbit historical fault mode of GPS receivers are analyzed, and a dimension reduction detection method of satellite-borne GPS receivers based on principal component analysis is proposed. The main components of the fault alarm variables are extracted from the on-orbit fault data samples of a satellite GPS receiver, and the fault features are effectively reduced in dimension, which is verified by historical fault simulation. The dimension reduction detection of satellite-borne GPS receivers is of great significance for ensuring the safe and stable operation of satellites in orbit and effectively realizing their mission.

Keywords GPS receiver · Fault detection · Principal component · In-orbit

1 Introduction

With the increasing number of on-orbit satellites, the large amount of fault monitoring and alarm and high false alarm rate become an important factor affecting the efficiency and accuracy of on-orbit satellite fault identification. The historical fault phenomenon of satellite-borne GPS receiver has the characteristics of large alarm volume and high repetition rate. At present, the fault detection of satellite-borne GPS receivers is mainly realized by threshold judgment or expert knowledge method. Threshold judgment method can detect on-orbit faults timely and accurately, but the amount of fault alarm is large. Expert knowledge method has less alarm volume, but it needs to rely on prior knowledge input, and fault knowledge is difficult to achieve full coverage. In this paper, the working mode and historical fault mode of the satellite-borne GPS receiver are analyzed, and the telemetry parameters of the GPS receiver are analyzed by principal component analysis. The dimensionality reduction of the fault detection variables of the satellite-borne GPS receiver is realized, and the redundant information of fault identification is removed.

Z. Xiaopeng (✉) · W. Xiaochen (✉) · Z. Haixiao (✉) · G. Yongfu · C. Xi · Z. Miao · Z. Xiangyan · W. Weiwei · Y. Ping (✉)
China Academy of Space Technology, Beijing, China
e-mail: 18613805915@163.com

L. Wang et al. (eds.), *Proceedings of the 6th China High Resolution Earth Observation Conference (CHREOC 2019)*, Lecture Notes in Electrical Engineering 657,
https://doi.org/10.1007/978-981-15-3947-3_44

2 Failure Mode Analysis of Space-Borne GPS Receiver

2.1 Working Mode of GPS Receiver

Space-borne GPS receiver receives signals from GPS navigation satellite system through GPS filter and GPS receiving antenna. GPS receiver demodulates and calculates the signal, outputs positioning information, orbit determination information, time information, original measurement data information, speed-to-height ratio and integration time information to satellite-related subsystems. In this way, the functions of satellite positioning, orbit determination, time service and image aided analysis are completed. GPS receiver system is mainly composed of channel board, data processing interface board and power board [1].

The space-borne GPS receiver has two working modes: real-time positioning and extrapolation.

In the real-time positioning mode, the GPS receiver is turned on after power-on, and the acquisition begins after self-check is normal. After blind acquisition, the channel board starts to position normally, and the measurement signal is output to the satellite-related subsystem [2]. At the same time, the channel board outputs the original measurement data and navigation positioning data to the orbit interface board, and the orbit determination interface board uses the filtering algorithm based on the orbit dynamics model to output the positioning data and the orbit determination data after the solution. After the filtering solution for a period of time, the higher precision navigation positioning information and orbit determination information can be outputted.

Extrapolation mode means that after blind acquisition and positioning by GPS receiver, if the channel board status changes from positioning to non-positioning, the orbit determination interface board extrapolates and outputs positioning data and orbit determination data based on orbit dynamics model.

2.2 Historical Failure Mode of GPS Receiver

Satellite-borne GPS receivers can achieve precise positioning, time calibration and provide precise time reference under normal working conditions. By analyzing the historical fault data of a LEO satellite platform, it is found that the historical fault modes of GPS receivers can be summarized into three categories: non-positioning of GPS receivers, autonomous reset of GPS receivers and non-updating of GPS receivers.

Under normal working conditions, GPS receivers of on-orbit satellites can achieve precise positioning, time calibration and provide precise time reference. By summarizing the historical fault data of a LEO platform satellite in orbit, it is found that the historical fault modes of GPS receiver can be summarized into three categories: non-positioning of GPS receiver, autonomous reset of GPS receiver and non-updating of GPS receiver data.

Non-positioning and autonomous reset anomalies of satellite-borne GPS receivers are usually related to satellite attitude maneuver or space environment impact. This kind of phenomenon lasts for a short time, and the non-positioning lasts for a certain period of time can trigger the automatic restoration measures of autonomous reset. After the autonomous reset occurs, the GPS receiver can be positioned again within a few minutes without ground processing, and the amount of fault alarm is small, which does not affect the normal operation of the satellite.

Space-borne GPS receivers do not update on-orbit abnormal data, mainly due to the impact of space environment, resulting in single event latch-up. Generally, ground commands are required to power off and re-energize the GPS receivers, which affect the normal operation of the GPS receivers for a long time, and the alarm amount is large when the failure occurs.

3 Principal Component Analysis

Principal component analysis (PCA) is a multivariate statistical analysis method that transforms multiple indicators into a few comprehensive indicators. It is a method to find independent comprehensive indicators reflecting multiple indicators. After examining the irrelevance of these few comprehensive indicators, it is easy to grasp the main contradictions according to the information contained. Principal component analysis (PCA) is used to extract features of multi-telemetry parameters for fault dimension reduction detection. The steps are as follows [3, 4]:

(1) Establishing the original telemetry parameter detection matrix.

There are p variables to represent the original index, which are recorded as $X1, X2$, …, Xp, and the number of samples is n. The matrix $X = \left(X_{ij}\right)_{n \times p}$ is obtained from the data of the original telemetry parameters. Among them, $i = 1, 2, \ldots n$ represents n samples; $j = 1, 2, \ldots, p$, X_{ij} represents the j index value of the sample i.

(2) Standardization of original telemetry parameter data.

In order to make the principal component analysis treat each original telemetry parameter variable equally, the original telemetry parameter data are standardized.

In order to make the principal component analysis treat each original telemetry parameter variable equally, the original telemetry parameter data are standardized, that is to say, changed $X = \left(X_{ij}\right)_{n \times p}$ into $X^* = \left(X^*{}_{ij}\right)_{n \times p}$.

(3) The sample co-variance matrix is calculated.

Find the sample co-variance matrix S (or S^*, or R) of the original evaluation index.

$$S = (s_{ij})_{p \times p} = \frac{1}{n-1}\sum_{m=1}^{n}(X_{mi} - \overline{X_j})(X_{mj} - \overline{X_j}) \tag{1}$$

$$R = (r_{ij})_{p \times p} = \frac{s_{ij}}{\sqrt{s_{ii}}\sqrt{s_{jj}}} \tag{2}$$

$$\overline{X_j} = \frac{1}{n}\sum_{i=1}^{n} X_{ij}; i = 1, 2, \cdots p; j = 1, 2, \cdots p \tag{3}$$

(4) Computing eigenvectors.

The eigenvalues and eigenvectors of the sample co-variance matrix S (or correlation matrix R) are calculated to determine the principal components. Find the eigenvalue $\lambda_1 \geq \lambda_2 \geq \lambda_3 \cdots \geq \lambda_p > 0$ of the sample co-variance matrix S, the default eigenvector $t_1, t_2, \cdots t_p$, $t_i = (t_{i1}, t_{i2}, \cdots t_{ip}), i = 1, 2, \cdots, p$. The size of the eigenvalue reflects the influence of each indicator. Principal components: $F_i = t_{i1}X_1 + t_{i2}X_2 + \cdots t_{ip}X_p$. Among them, X_p are the original index variables, $i = 1, 2, \cdots p$.

The eigenvalue of the first principal component is the variance of the principal component. The greater the variance, the greater the contribution to the total variance.

Its contribution rate:

$$\alpha_i = \frac{\lambda_i}{\sum_{i=1}^{p} \lambda_i} \tag{4}$$

It reflects the percentage of the information of the first principal component synthesizing the original variable.

(5) Extract the principal component.

The contribution rate of variance was calculated, and the number of principal components K was determined. The K principal component ($K < p$) is selected, and the K value is determined by the cumulative contribution rate of variance. Generally, the cumulative contribution rate is more than 85% as the basis for determining the K value.

(6) Fault detection model F is established.

The comprehensive parameter model for calculating the final fault detection of K principal component can be used in the formula:

$$F = \frac{\lambda_i}{\sum_{i=1}^{p} \lambda_i} F_i, i = 1, 2, \cdots K. \tag{5}$$

4 Dimension Reduction Detection of GPS Receiver Faults Based on Principal Component Analysis

In order to reduce dimension and detect the operation status of GPS receivers of satellites in orbit, the remote measurement data of GPS receivers in orbit of a satellite are selected for data analysis. The design life of the satellite is three years, and it has been operated beyond its lifetime. One day, when a satellite passes through a circle, the satellite GPS receiver telemetry parameters appear abnormal data not updated, and there are 27 alarms. Fifty remote sensing sample data of 27 parameters from normal period to fault period and to normal period were selected to carry out fault dimension reduction test (Table 1).

The original variable curve of the satellite-borne GPS receiver is shown in Fig. 1. The curve after standardizing the original variable data is shown in Fig. 2:

Through principal component analysis of 50 sample data of 27 variables, it was found that the cumulative contribution rate of the first three principal components was 99.9995% (>85%). Therefore, the first three principal components should be retained. The characteristic values and contribution rates of the three principal components are as follows (Table 2):

The calculation results are as follows:

The above results show that the original 27 alarm variables are reduced to three variables ($m = 3$) by principal component analysis. Figure 3 shows that the principal component samples are abnormal at a certain time. Figure 1 shows that the original variables should be abnormal at the same time. The curve of the original variables in Fig. 1 verifies this inference.

Fig. 1 Original variable curve of a GPS receiver

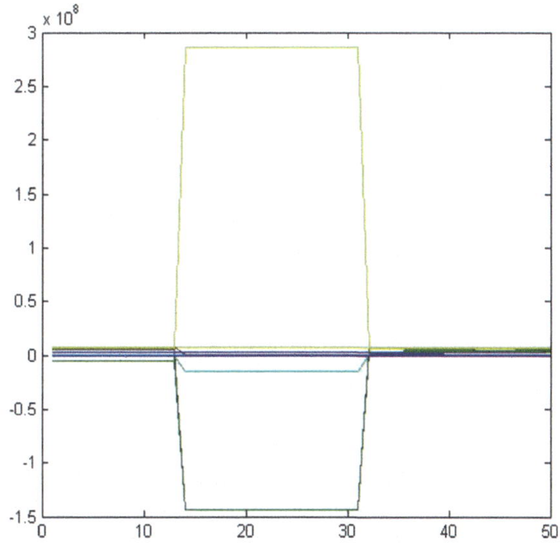

Table 1 Original alarm variables for GPS receivers

Serial Number	**Variable Name**
1	*X*-axis position
2	*Y*-axis position
3	*Z*-axis position
4	*X*-axis velocity
5	*Y*-axis velocity
6	*Z*-axis velocity
7	*X*-axis position
8	*Y*-axis position
9	*Z*-axis position
10	*X*-axis velocity
11	*Y*-axis velocity
12	*Z*-axis velocity
13	Track semi-length axis
14	Orbital eccentricity
15	Orbital inclination
16	Ascending intersection red meridian
17	Perigee angular distance
18	Flat and near point angle
19	Greenwich stellar time angle
20	Half-length axis variation rate
21	Rail eccentricity variability
22	Quadratic term of amplitude and angle change time
23	Longitude
24	Latitude
25	Satellite height
26	GDOP
27	TIC count

5 Conclusion

Principal component analysis (PCA) deals with multi-variable parameter matrices for complex system problems without changing the sample data structure. The principal component obtained is a linear combination of the original variables and is not related to each other. It can reflect the information contained in the original variables to the greatest extent while greatly reducing the redundancy of the system variables. Good selection of low-dimensional feature sets from the initial high-dimensional feature sets of on-orbit satellites is an effective and simplified analysis method for on-orbit satellite state features and fault detection.

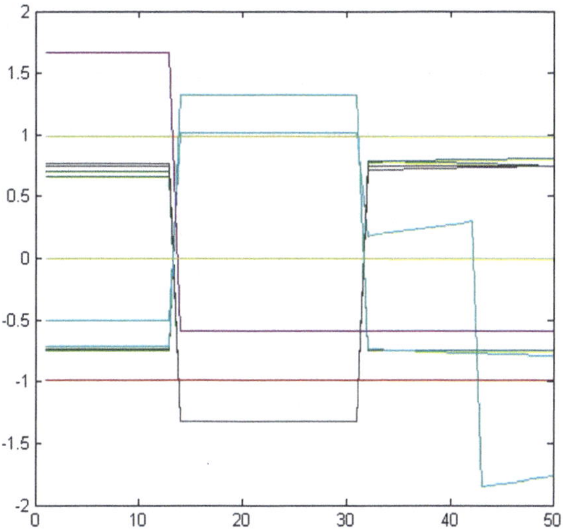

Fig. 2 Standardized curve of original variables for GPS receivers

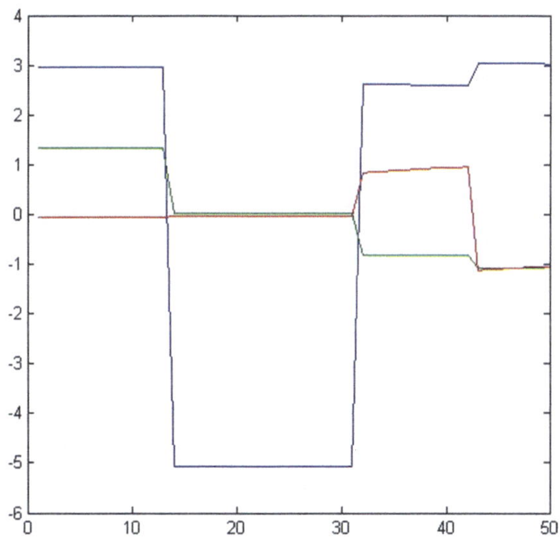

Fig. 3 Variable principal component curve of space-borne GPS receiver

Table 2 Principal component eigenvalues and contribution rates of GPS receivers

Serial Number	Principal Component	Eigenvalue	Contribution Rate	Cumulative Contribution Rat
1	A1	14.8152	92.5953%	92.5953%
2	A2	0.8075	5.0469%	97.6422%
3	A3	0.3772	2.3573%	99.9995%

References

1. Weichi T, Jingang H (2009) Spacecraft Systems Engineering, ISBN978-7-5046-5417-5, China Science and Technology Press, 4
2. Zhai Q (2004) Satellite-borne GPS receiver and its application, Papers Collection of 2004 National Academic Conference on Surveying and Mapping Instruments. 2004: 145–147
3. Jun Zhang (2014) Performance analysis of mobile phone software based on principal component analysis. Inf Technol 3:45–47
4. Huafei F, Wenjun W (2017) Flight load measurement technology based on principal component analysis. Value Eng 5:101–103

Establishing the Downscaling Model of NDVI Based on the Iterated Function System

Haijun Luan, Meng Zhang, Yunya Wan, Yuanrong He, Qin Nie, and Xinxin Zhang

Abstract The moderate resolution imaging spectroradiometer (MODIS) normalized difference vegetation index (NDVI) product plays an important role in eco-environmental monitoring of natural disasters. However, its validation has long been an important scientific problem that has yet to be resolved. In this study, the downscaling model of NDVI was established based on the fractal Iterated Function System (IFS), and the MOD13 Q1 product was then validated based on the model. Xiamen was selected as the core study area, and utilizing the 30 m resolution Landsat 8 operational land imager (OLI) images as the validation data, the validation of MOD13 Q1 was implemented. The results showed the following. (1) The overall quality of the MOD13 Q1 product is good. While in the NDVI range of 0.2 to 0.6, the MOD13Q1 has an overestimation and the difference recognition of the NDVI is low, which should be paid attention to in practical applications; (2) The experiment proved that the fractal IFS was an effective methodology to establish downscaling models for RS land surface parameters such as NDVI. The inherent physical meaning and dynamic process expression advantages of this method make it have great application potential, which needs further digging.

Keywords NDVI · Downscaling · Iterated Function System · MOD13 Q1 · OLI

1 Introduction

As the best indicator of vegetation growth status and vegetation coverage, NDVI is widely used in environmental (climate) changes, crop yield estimation, and other fields. Among the existing vegetation index products, MODIS vegetation index products are highly valued for their unfettered use, stable supply, global coverage, and

H. Luan (✉) · M. Zhang · Y. Wan · Y. He · Q. Nie · X. Zhang
College of Computer and Information Engineering, Xiamen University of Technology, 361024 Xiamen, Fujian, China
e-mail: luanhaijun@xmut.edu.cn

H. Luan · Y. He · X. Zhang
Digital Fujian Institute of Big Data for Natural Hazards Monitoring, Xiamen University of Technology, 361024 Xiamen, Fujian, China

L. Wang et al. (eds.), *Proceedings of the 6th China High Resolution Earth Observation Conference (CHREOC 2019)*, Lecture Notes in Electrical Engineering 657,
https://doi.org/10.1007/978-981-15-3947-3_45

continuous phase. They are widely used in forest fires [1, 2], grassland vegetation growth [3, 4], drought [5, 6], land desertification [7], and other ecological environment monitoring research. However, the maximum spatial resolution of MODIS vegetation index products is 250 m. The validation of this RS land surface parameter is an important issue that cannot be avoided [8–10]. The validation of MODIS vegetation index products needs to be carried out by means of scale conversion. The most representative MODIS NDVI product MOD13 Q1 will be studied in the paper. This paper will also focus on establishing the downscaling model of NDVI and validating the MOD13 Q1 product based on the model.

Liang [11] has reviewed some current downscaling methods, including linear decomposition methods and nonlinear statistical decomposition methods, methods for generating continuous regions, NDVI time series decomposition, multi-resolution data fusion, and statistical downscaling method of Global Climate Model products (GCM), etc. Further, Gao [12], Zhu [13], and Huang [14, 15] have done systematic and effective work in the spatio-temporal fusion downscaling of land surface reflectance, which becomes a research hotspot. The spectral-spatial features fusion by Wang [16–19] and Shi [20] also achieved good results in subpixel mapping. However, in these studies, the scale conversion process is little considered from the perspective of dynamics, and the research on surface parameter downscaling based on the fractal Iterated Function System (IFS) has paid attention to it.

The use of fractals for surface parameter scale conversion modeling usually contains two important research contents: (1) the performance of fractal features, that is, fractal metrics, also the fractal dimension of the research object. For example, Zhang [21, 22] used the information dimension method to describe the fractal dimension of LAI scale conversion. Luan [23, 24] and Wu [25] used the similar dimension method to measure the fractal dimension of NDVI and LAI scale-up conversion, respectively. Conduct a series of studies: (2) the intrinsic nature of the fractal phenomenon, that is, the dynamics produced, which is the combined effect of multi-factor effects on the surface. The mathematical basis of fractal generation is the Iterated Function System (IFS); Kim [26] constructed the *r* function by kinetic factors (soil sediment content, vegetation water content) of soil moisture scale conversion, and then established the description. The soil moisture downscaling IFS has a good conversion effect. The established model can describe the dynamic process of soil moisture scale conversion, which has physical significance and demonstrates the advantages of downscaling of surface parameters based on fractal IFS. In general, there are currently few reasons for the dynamics of fractals. The fractal IFS in mathematics is continuously iterative calculation based on the whole research object [27], and the RS land surface parameter image is performed in units of local pixels. This determines that the IFS vertical conversion factor (*r* function) in mathematics is usually constant [28], while the vertical conversion factor of RS land surface parameters (such as soil moisture) is based on the physical elements of each pixel (such as soil sandy). The amount of space and vegetation water content varies dynamically and temporally [26]. This is why the IFS function can describe the scale-switching dynamics of surface parameters and the model has certain physical meanings. The vertical conversion factor is used to describe the inter-scale conversion of surface parameter values and is the key to determine the IFS function. Different surface parameters are different due to the spatial distribution and scale conversion factors (or dynamic factors), and the vertical conversion factor (*r* function) contains different types of variables and function forms. How to determine the *r* function is difficult in determining the IFS function, which is also an important reason why the IFS function is less applied in quantitative RS land surface parameter scale conversion description. Therefore, the NDVI downscaling model based on the fractal IFS function can be considered to describe the dynamics process of scale conversion. This research has a large research space and is of great significance. The following is a preliminary implementation.

2 Study Area and Data Sources

2.1 Study Area

A rectangular area surrounding Xiamen in Fujian Province, China, was selected as the study area. The rectangle shape facilitated the upscaling of land surface parameters. Xiamen is located in the southeastern part of Fujian Province and west of Taiwan Strait. Its geographic coordinates are 24°23’-24°54’N latitude and 117°53’-118°26’E longitude. The land area and sea area comprise 1699.39 km^2 and over 300 km^2, respectively. The terrain is dominated by plains, highlands, and hills. The land inclines from the northwest to the southeast. Xiamen has a pleasant climate and flourishing production activities. There are various ground object categories in remote sensing images. The study area also included parts of Quanzhou, Zhangzhou, and Kinmen islands, although these areas overall comprise a small portion. Because of the social and natural conditions in the study area, the underlying surface exhibits significant spatial heterogeneity, which will result in significant scale effect of the remote sensing parameters (such as NDVI). As such, this provided a great opportunity for the implementation of this study.

2.2 Experimental Data and Its Preprocessing

The images used in this study include Landsat 8 OLI image (track number: 119, line number: 43) and MODIS NDVI product MOD13 Q1 (row and column number: H28, V06), collected on January 3, 2017.

For the purpose of atmospheric correction of the OLI image, basic data including elevation and meteorological parameters were obtained from the Xiamen government website (http://www.xm.gov.cn) and meteorological data website.

During the study, field sampling was conducted with a uniform distribution method. A total of 700 sampling points were obtained. Some of the sampling points in Jimei District are shown in Table 1. On the one hand, structural features in the

Table 1 Examples of field sampling points

Number	Sampling time	Sampling location	Object category of the sample point	Latitude (°.'.")/longitude (°.'.")/elevation (m)	Location of sample point	Field image
1	2017/1/3 13:26:40	Jimei District, Xiamen	Road (construction land)	24.37.23/ 118.05.25/ 126.511	Intersection of Tianma road and Xinglinwan road	
2	2017/1/3 14:28:17	Jimei District, Xiamen	Forestland	24.36.33/ 118.05.07/ 22.400	Intersection of South Sunban road and Xiasha Highway ramp	
3	2017/1/7 15:09:25	Jimei District, Xiamen	Residential building (construction land)	24.35.19/ 118.06.23/ 6.452	Between North Xujiang road and Zonger road, Quanshun Quanshuiwan residential community	
4	2017/1/7 15:40:31	Jimei District, Xiamen	Athletic field (construction land)	24.35.51/ 118.06.35/ 4.441	Intersection of Letian road and Fenglinsan road, Jimei middle school	
5	2017/1/7 16:30:42	Jimei District, Xiamen	Forestland	24.37.06/ 118.05.01/ 4.912	Across Yingcun Elementary school on Ligong road	
6	2016/1/7 12:31:34	Jimei District, Xiamen	Road (construction land)	24.37.10/ 118.3.11/ 3.655	Intersection of Jimei road and North Jimei road	
7	2016/1/7 13:06:43	Jimei District, Xiamen	Bareland	24.38.33/ 118.3.55/ 20.364	Yongtongchang Square on Yantong road	
8	2017/1/7 15:07:13	Jimei District, Xiamen	Training field (construction land)	24.35.14/ 118.06.25/ 13.458	Intersection of Lehai road and North Xunjiang road, Lianhuang Driver Training school	

Fig. 1 Preprocessed OLI image (R:4, G:3, B:2)

sampling points (such as a road intersection) can be used for geometric correction of OLI images. On the other hand, field measurement points can be used for accuracy assessment of the OLI image classification.

2.2.1 Preprocessing of the OLI Image

With Xiamen as the main study area, the ENVI software was used to crop a rectangular area from the original image. Atmospheric correction and anomaly processing (using the ENVI BandMath module to modify pixels with a negative or >1 land surface reflectance) were then carried out. In addition, based on field measurement control points, geometric correction of the image was carried out following the atmospheric correction. A total of 85 control points were selected. The accuracy of geometric correction was 0.45 pixels, which met the application requirement. The OLI image obtained after preprocessing is shown in Fig. 1. The NDVI and histogram of the image were then calculated, as shown in Figs. 2 and 3.

2.2.2 Preprocessing of the MOD13 Q1 Product

The MOD13 product has a different coordinate system and projection method from the OLI image. Therefore, it is necessary to carry out re-projection prior to MOD13 validation, for which the MRT software was used in this study. By referencing the OLI image, the MOD13 product was then cropped following re-projection. For the post-preprocessing 240 m MOD13 Q1 product, the sea area pixels were set to −3000. These pixels should be processed before validation. The abnormal pixels were replaced with the corresponded ones of 240 m re-sampled image of OLI NDVI.And the 240 m MOD13 Q1 image and its histogram were obtained as shown in Fig. 4.

Fig. 2 30 m OLI NDVI image

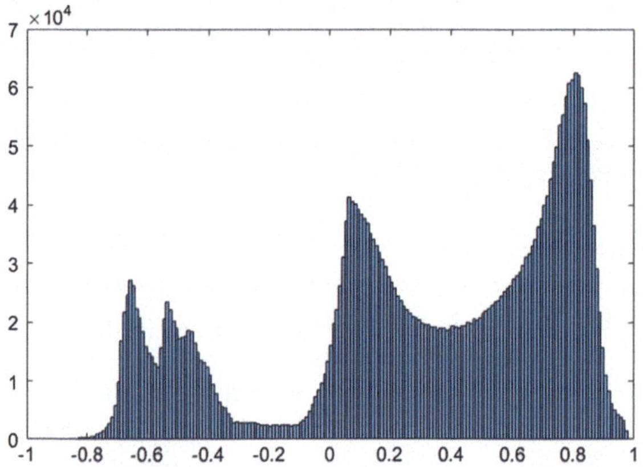

Fig. 3 Histogram of the 30 m OLI NDVI image

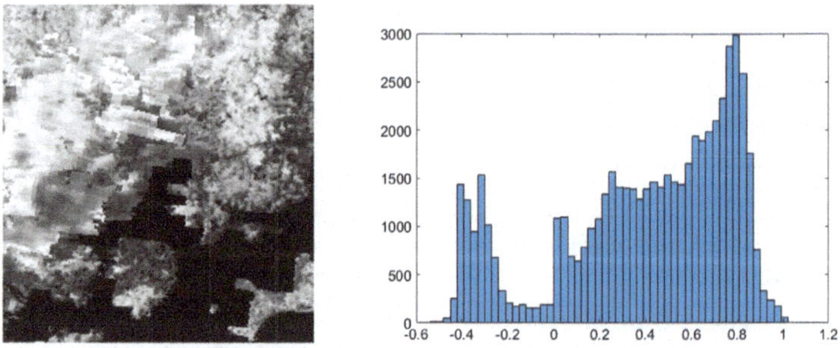

(a) 240m MOD13 Q1 image (b) Histogram of 240m MOD13 Q1 image

Fig. 4 MOD13 Q1 image of the study area and the corresponding histogram

3 Methodology

How to build NDVI downscaling model based on fractal IFS function? This question includes the following points: First, how to determine the sensitive factors affecting the spatial distribution and scale effect of NDVI for NDVI; secondly, how to use this sensitive factor to establish the vertical scale conversion factor r function in IFS, and then determine the IFS function to achieve NDVI downscaling; finally, how to evaluate downscaling conversion results. For these points, the solution is described below.

3.1 Identify Sensitive Factors

According to the above description, the water body is an important parameter affecting the spatial distribution and scale effect of NDVI; so, it can be determined that the pixel water parameter is one of the important dynamic factors of NDVI scale conversion. In addition, Wen [29] gave a method for albedo conversion from small-scale images to large-scale images, and used the pixel topographical influence factors to correct the converted results. The results demonstrate that the method was effective for albedo scale conversion under rugged terrain. Considering the close relationship between the surface reflectivity and the surface albedo, and the surface reflectance is the basic parameter for calculating NDVI, the topographic factor parameter can be determined as one of the important kinetic factors for NDVI scale conversion. Therefore, it is determined that the important dynamic factors in NDVI spatial distribution and scale conversion are pixel water parameters and topographic factors.

3.2 Determine the Vertical Conversion Factor R Function and Establish the IFS Function

Referring to Kim [26], IFS formula (1), horizontal transformation formula (2), and vertical transformation formula (3) for large-scale surface parameter pixel downscaling are obtained as follows. The IFS formula is calculated by pixel-by-pixel sliding. Get the full image downscaling results:

$$\mathrm{IFS}^{i,j}|_{n,m}\left(x^{i}, y^{j}, s^{ij}\right) = \left(p_{n}\left(x^{i}\right), q_{m}\left(y^{j}\right), I_{n,m}\left(x^{i}, y^{j}, s^{ij}\right)\right) \quad (1)$$

$$\begin{cases} p_{n}\left(x^{i}\right) = x_{n-1}^{i} + \alpha\left(x^{i} - x_{0}^{i}\right) \\ q_{m}\left(y^{i}\right) = y_{m-1}^{j} + \alpha\left(y^{j} - y_{0}^{j}\right) \end{cases} \quad (2)$$

$$I_{n,m}\left(x^{i}, y^{j}, s^{ij}\right) = \left(e_{n,m}x^{i} + f_{n,m}y^{j} + g_{n,m}x^{i}y^{j} + r_{1}\left(x^{i}, y^{j}\right)s^{ij} + k_{n,m}\right) \times r_{2}\left(x^{i}, y^{j}\right) \quad (3)$$

where $\mathrm{IFS}^{i,j}|_{n,m}(x^i, y^j, s^{ij})$ represents the surface parameter of the pixel at the (i, j) location when the large-scale pixel of the surface parameter is downscaled to the small-scale image of the $n \times m$ dimension, x^i, y^j, and s^{ij}, respectively, correspond to the x-direction coordinate $p_n(x^i)$, y-direction coordinate $q_m(y^j)$, and surface parameter values $I_{n,m}(x^i, y^j, s^{ij})$ of the three-dimensional data of the pixel, x^i_{n-1} and x^i_0, respectively, represent the x-direction starting coordinate of the (i, j) pixel in the $n \times m$-dimensional small-scale image and the x-direction starting coordinate of the large-scale pixel, α represents the downscaling ratio (small-scale/large-scale, which is less than or equal to 1), $e_{n,m}$, $f_{n,m}$, and $g_{n,m}$ $k_{n,m}$ are functions of the x- and y-coordinates of the lower left corner and upper right corner of the large-scale pixel, the downscaled surface parameter data and the vertical scale conversion surface function, $r_1(x^i, y^j)$ and $r_2(x^i, y^j)$, respectively, represent two different vertical conversion factors in the vertical scale conversion surface function. The parameters or factors not represented in the formula can be referred to the paper [26], which will not be explained here. Generally, the $p_n(x^i)$ and $q_m(y^j)$ coordinates of the (i, j) pixel are obtained by largescale divided equally to $\frac{1}{\alpha}$ parts, and the $I_{n,m}(x^i, y^j, s^{ij})$ calculation is the key. In formula 3, $r_2(x^i, y^j)$ is the same as the $r_1(x^i, y^j)$ function, but their argument coefficients are different.

For NDVI, $g_{n,m}$, $e_{n,m}$, $f_{n,m}$, and $k_{n,m}$ represent the functions of (n, m) pixel of downscaled NDVI image. And based on the special downscaled NDVI 3-D values of four-corner pixels, $g_{n,m}$, $e_{n,m}$, $f_{n,m}$, and $k_{n,m}$ could be calculated as formulas 4–11.

$$g_{n,m} = \frac{s_{n-1,m-1} - s_{n-1,m} - s_{n,m-1} + s_{n,m} - R_1^g}{x_0 y_0 - x_N y_0 - x_0 y_M + x_N y_M} \tag{4}$$

$$e_{n,m} = \frac{s_{n-1,m-1} - s_{n,m-1} - g_{n,m}(x_0 y_0 - x_N y_0) - R_1^e}{x_0 - x_N} \tag{5}$$

$$f_{n,m} = \frac{s_{n-1,m-1} - s_{n-1,m} - g_{n,m}(x_0 y_0 - x_0 y_M) - R_1^f}{y_0 - y_M} \tag{6}$$

$$k_{n,m} = s_{n,m} - e_{n,m} x_N - f_{n,m} y_M - g_{n,m} x_N y_M - R_1^k \tag{7}$$

Furthermore,

$$R_1^g = r_1\left(x^i_{n-1}, y^j_{m-1}\right) s^{ij}_{0,0} - r_1\left(x^i_n, y^j_{m-1}\right) s^{ij}_{N,0} - r_1\left(x^i_{n-1}, y^j_m\right) s^{ij}_{0,M} + r_1\left(x^i_n, y^j_m\right) s^{ij}_{N,M} \tag{8}$$

$$R_1^e = r_1\left(x^i_{n-1}, y^j_{m-1}\right) s^{ij}_{0,0} - r_1\left(x^i_n, y^j_{m-1}\right) s^{ij}_{N,0} \tag{9}$$

$$R_1^f = r_1\left(x^i_{n-1}, y^j_{m-1}\right) s^{ij}_{0,0} - r_1\left(x^i_{n-1}, y^j_m\right) s^{ij}_{0,M} \tag{10}$$

$$R_1^k = r_1\left(x^i_n, y^j_m\right) s^{ij}_{N,M} \tag{11}$$

Therefore, the calculation of $r_1(x^i, y^j)$ function is significant, and $r_1(n, m)$ $(0 \le r_1 \le 1)$ is used to adjust the NDVI surface roughness. The following focuses on the establishment of the vertical transformation formula for NDVI, that is, the determination of the r function (containing $r_1(x^i, y^j)$ and $r_2(x^i, y^j)$).

Based on the above sensitivity factors, a vertical conversion factor r function can be constructed:

$$r = \gamma \times S_{\text{water}} + \beta \times s + \delta \tag{12}$$

where: S_{water} represents the pixel water parameter, s represents the topographic information, taking into account the magnitude of the r function, the Normalized Difference Water Index (NDWI), slope (calculated by DEM image), respectively, represent the water body effect and topographic influence in the pixel γ and β are the coefficients of two parameters, respectively, and δ represents the adjustment constant. Two different orders of magnitude r are as follows:

$$r_1 = \gamma_1 \times S_{\text{water}} + \beta_1 \times s + \delta_1 \tag{13}$$

$$r_2 = \gamma_2 \times S_{\text{water}} + \beta_2 \times s + \delta_2 \tag{14}$$

For NDVI, the γ, β, and δ coefficients can be calculated by the linear regression between high-resolution NDVI image and its NDWI, slope images.

After the r function is constructed, formula 1–3 can be solved in combination with other known conditions, and NDVI downscaling can be achieved.

3.3 Evaluation of Downscaling Results

Referring to the study by Kim [26], the downscaling results accuracy will be evaluated using statistical indicators such as maximum value, minimum value, variance value, and standard deviation value (compared to high-resolution NDVI images). Besides, the histograms of downscaling NDVI and true NDVI imageries are drawn and compared, and their correlation coefficient is calculated. With those indexes, the accuracy of downscaling imagery and methodology could be validated.

4 Experiment and Result Analysis

4.1 Experiment

Since vegetation growth is closely related to topographic factors and NDWI, they are used as ancillary data to estimate the r function. The estimation of the r function

Table 2 Estimation results of parameters $(\gamma_1, \beta_1, \delta_1)$

Parameters	γ_1	β_1	δ_1
Values	-0.9445	0.1762	0.9834

Fig. 5 30 m downscaled image of MOD13 Q1

is based on the linear regression relationship between the auxiliary data and the OLI NDVI image. Based on the normalization of the three image data, the estimation results of the parameters $(\gamma_1, \beta_1, \delta_1)$ can be obtained by linear regression calculation, as shown in Table 2.

The linear regression results were evaluated. It was found that the significance level index P_Value = 0 (requirement: P_Value <0.05), indicating that the linear relationship is significant; and the correlation coefficient between the fitted line and the actual data is 0.9925, indicating that the fitting effect is very good. Therefore, the estimated result of the parameters $(\gamma_1, \beta_1, \delta_1)$ is acceptable, and the r_1 function is expressed as $r_1 = -0.9445 \times S_{water} + 0.1762 \times s + 0.9834$.

Furthermore, based on the downscaling formulas in Chapter 3.2, the MOD13 Q1 image of Xiamen was directly downscaled by $\frac{1}{8}$ multiples, and the 30 m downscaled NDVI was obtained as Fig. 5. Then, the histograms of processed NDVI images were drawn in Figs. 6 and 7, and the statistics and correlation coefficient of the NDVI images were calculated in Table 3. Based on these, the downscaled results could be evaluated and the MOD13 Q1 image could be validated.

4.2 Result Analysis

By analyzing Figs. 2, 3, 4, 5, 6 and 7 and Table 3, it is found that:

Compared with the real 30 m OLI NDVI image, the 30 m downscaling MOD13 Q1 image has smaller differences in maximum value, minimum value, mean value,

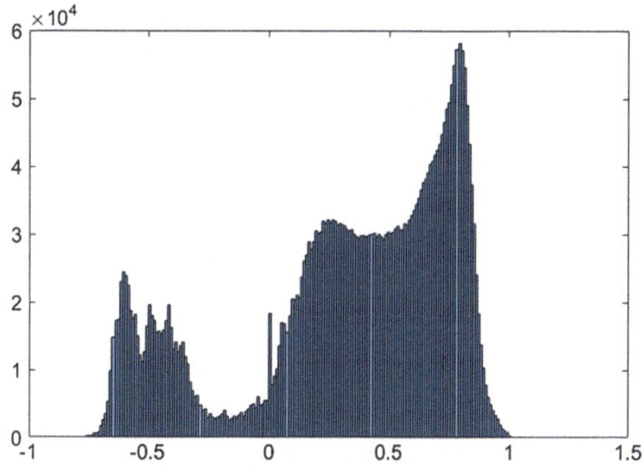

Fig. 6 Histogram of 30 m downscaled MOD13 Q1

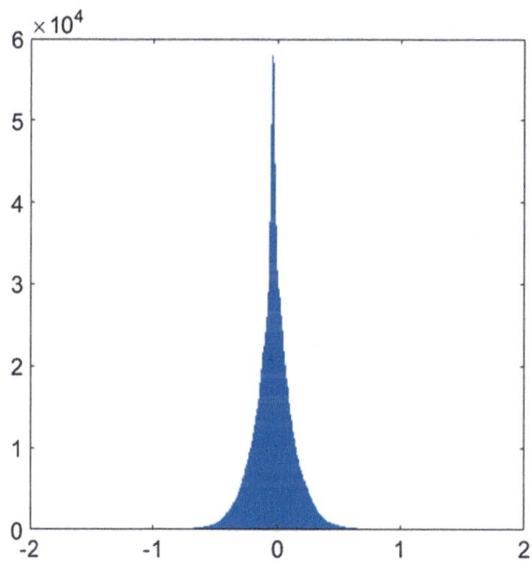

Fig. 7 Histogram of the difference image between Figs. 5 and 2 (Fig. 5–Fig. 2)

Table 3 Statistics of original and processed NDVI images

NDVI images	Maximum	Minimum	Mean	Variance	Correlation coefficient
MOD13 Q1	0.999100	−0.561500	0.397400	0.001100	–
30 m downscaled MOD13 Q1	0.999950	−0.999793	0.326412	0.206495	0.937449
OLI NDVI	0.984436	−0.961210	0.299418	0.227313	
Difference image (Fig. 5–Fig. 2)	1.672050	−1.088930	−0.029594	0.027916	–

(Note: the correlation coefficient is the one between the downscaled MOD13 Q1 and the OLI NDVI.)

and variance. The correlation coefficient between the two images is 0.93, which is highly correlated. It can be considered that the overall quality of the NDVI image obtained by downscaling of MOD13 Q1 is good, indicating that the overall quality of MOD13 Q1 is good. Comparing Figs. 4b and 6, there is a certain similarity between the distribution patterns of the two images, which indicates that the downscaling image retains the spatial distribution structure of the original image well, which proves to some extent that the original image MOD13 Q1 is of good quality. In addition, comparing Figs. 6 and 3, it is found that the downscaling NDVI image has higher proportion in the vicinity of the zero value (mainly artificial features) than the real image. In the range of 0.2–0.6, the difference is larger. The downscaling image has generally higher proportion in this value range, and the histogram is smoother, indicating that the image recognition of the NDVI difference is not high. Meanwhile, referring to the correlation between Fig. 6 and Fig. 4(b), it is known that MOD13 Q1 also has these problems within the above-mentioned range of values. Analysis of the original MOD13 Q1 image shows that the image is a 16-day NDVI composite product, and each pixel takes the maximum value of NDVI within 16 days as the result of the product release. Therefore, the histogram has a certain degree of reasonableness in the larger value area. At the same time, the histogram distribution of the difference image indicates that the pixel values are distributed in the range of $[-1,1]$, and the distribution pattern is low on both sides and high in the middle (about 0 value), which also indicates the downscale image and the real image is highly consistent.

According to the above analysis, it is considered that the overall quality of the MOD13 Q1 downscaling image is good, indicating that the overall quality of MOD13Q1 is good. While the NDVI ranges from 0.2 to 0.6, the MOD13Q1 has an overestimation and the difference recognition of the NDVI is low, which should be paid attention in practical applications.

4.3 Discussion

There are still some issues to be solved in the study:

1. The processing of abnormal seawater pixels of the MOD13 Q1 product was relatively simple, which affected the accuracy of downscaling conversion results, and it is necessary to find a better processing method later.
2. In the study, MOD13 Q1 image of Xiamen was directly downscaled by $\frac{1}{8}$ multiples to obtain 30 m NDVI, which may affect the quality of the results. In order to obtain more accurate downscaling results, if the resolution of the low-resolution image is too different from the resolution of the target resolution image (such as downscaling from 240 m MOD13 Q1 to 30 m NDVI), a hierarchical downscaling method should be adopted. First, the low-resolution surface parameter image can

be downscaled to an intermediate-resolution image, and then the intermediate-resolution image will be further downscaled to the target resolution image, which can largely guarantee the accuracy of the result.

3. When estimating the *r* function, more types of ancillary data can be considered for inclusion to improve the accuracy of the downscaling conversion function and to obtain higher quality downscaling products.
4. The time phase of the OLI image is not fully consistent with that of the MOD13 Q1 product. MOD13 Q1 product is a 16-day integrated image, whereas the OLI NDVI image, as well as the upscaled NDVI image, is taken on a specific date in the 16-day range. Thus, the OLI image is limited in time phase resolution (16 days). To achieve more accurate validation, consideration will be given in follow-up studies to cover the 16-day range of the MOD13 product by using images from different sensors with different time phases.
5. There is geometric inconsistency between the OLI image and the MOD13 Q1 image. Differences in imaging geometry should be taken into account when cross-validating the land surface parameters of different sensors if those parameters do not have a physical meaning, such as the NDVI. Such consideration is not required for parameters with physical meaning (e.g., leaf area index) because the parameter values of sensors with different spatial resolutions can be compared directly in that case. Normalization can be achieved by calculating the bidirectional reflectance distribution function (BRDF) of typical land types in the study area and then correcting the land surface reflectance of the OLI NDVI image (the implementation procedures was mentioned in our previous study [30]). Follow-up research will focus on integrating this aspect.

5 Conclusions

Based on fractal IFS, the downscaling models of RS land surface parameters can be established. The models are merged with more ancillary data, which relate to the scale effects of land surface parameters. Therefore, they are benefit for obtaining accurate downscaling results.

In summary, although the breadth and depth of fractal IFS application in the establishment of RS land surface parameters downscaling model are still insufficient, the inherent physical meaning and dynamic process expression advantages of this method make it have great application potential, which needs further digging. This method is expected to become a new universal method for quantitative downscaling of RS land surface parameters and provide new research methods.

Acknowledgements This work was supported by the National Natural Science Foundation of China "Coupling of NDVI's up-scaling and downscaling fusing with ground objects classification" (No. 41601350), the Natural Science Foundation of Fujian Province, China "Research on NDVI's scaling fusing with ground objects classification" (No. 2017J05069), and the "Scientific Research Climbing Plan" Project from Xiamen University of Technology "Spatial Distribution Estimation and Dynamic Monitoring of Soil Organic Matter Based on Multi-source and Heterogeneous Data" (No. XPDKT19010).

References

1. Huesca M, Litago J, Palaciosorueta A et al (2009) Assessment of forest fire seasonality using MODIS fire potential: a time series approach[J]. Agric For Meteorol 149(11):1946–1955
2. Leon JR, Van Leeuwen WJ, Casady GM et al (2012) Using MODIS-NDVI for the modeling of post-wildfire vegetation response as a function of environmental conditions and pre-fire restoration treatments[J]. Remote Sens 4(3):598–621
3. Zhang B, Zhang L, Xie D et al (2015) Application of synthetic NDVI time series blended from landsat and MODIS data for grassland biomass estimation[J]. Remote Sens 8(1):10
4. Nestola E, Calfapietra C, Emmerton CA et al (2016) Monitoring grassland seasonal carbon dynamics, by integrating MODIS NDVI, proximal optical sampling, and Eddy covariance measurements[J]. Remote Sens 8(3):260
5. Du L, Tian Q, Yu T et al (2013) A comprehensive drought monitoring method integrating MODIS and TRMM data[J]. Int J Appl Earth Obs Geoinf 23(1):245–253
6. Kim Y (2013) Drought and elevation effects on MODIS vegetation indices in northern Arizona ecosystems[J]. Int J Remote Sens 34(14):4889–4899
7. Veron SR, Paruelo JM (2010) Desertification alters the response of vegetation to changes in precipitation[J]. J Appl Ecol 47(6):1233–1241
8. Gao X, Huete AR, Didan K et al (2003) Multisensor comparisons and validation of MODIS vegetation indices at the semiarid Jornada experimental range[J]. IEEE Trans Geosci Remote Sens 41(10):2368–2381
9. Fensholt R, Sandholt I, Stisen S et al (2006) Evaluating MODIS, MERIS, and VEGETATION vegetation indices using in situ measurements in a semiarid environment[J]. IEEE Trans Geosci Remote Sens 44(7):1774–1786
10. Geng L, Ma M, Yu W et al (2014) Validation of the MODIS NDVI products in different land-use types using in situ measurements in the Heihe River Basin[J]. IEEE Geosci Remote Sens Lett 11(9):1649–1653
11. Liang S (2009) Quantitative Remote sensing[M]. (trans: Fan WJ). Science Press, Beijing 180–183. (in Chinese)
12. Gao F, Masek J, Schwaller M et al (2006) On the blending of the landsat and MODIS surface reflectance: predicting daily landsat surface reflectance[J]. IEEE Trans Geosci Remote Sens 44(8):2207–2218
13. Zhu XL, Chen JM, Gao F et al (2010) An enhanced spatial and temporal adaptive reflectance fusion model for complex heterogeneous regions[J]. Remote Sens Environ 114(11):2610–2623
14. Huang B, Zhang HK, Song HH et al (2013) Unified fusion of remote sensing imagery: generating simultaneously high-resolution synthetic spatial-temporal spectral earth observations[J]. Remote Sens Lett 4:561–569
15. Huang B, Zhang HK (2014) Spatio-temporal reflectance fusion via unmixing: accounting for both phenological and land-cover changes[J]. Int J Remote Sens 35(16):6213–6233
16. Wang Q, Shi W, Wang L (2014) Allocating classes for soft-then-hard subpixel mapping algorithms in units of class[J]. IEEE Trans Geosci Remote Sens 52(5):2940–2959
17. Wang Q, Shi W, Atkinson PM et al (2015) Downscaling MODIS images with area-to-point regression kriging[J]. Remote Sens Environ 166:191–204
18. Wang Q, Atkinson PM, Shi W (2015) Indicator cokriging-based subpixel mapping without prior spatial structure information[J]. IEEE Trans Geosci Remote Sens 53(1):309–323
19. Wang Q, Atkinson PM, Shi W (2015) Fast sub-pixel mapping algorithms for sub-pixel resolution change detection[J]. IEEE Trans Geosci Remote Sens 53(4):1692–1706

20. Shi W, Wang Q (2015) Soft-then-hard sub-pixel mapping with multiple shifted images[J]. Int J Remote Sens 36(5):1329–1348
21. Zhang RH, Tian J, Li ZL et al (2008) Spatial scaling and information fractal dimension of surface parameters used in quantitative remote sensing[J]. Int J Remote Sens 29:5145–5159
22. Zhang RH, Tian J, Li ZL et al (2010) Principles and methods for the validation of quantitative remote sensingproducts[J]. Sci China Earth Sci 53:741–751
23. Luan HJ, Tian QJ, Gu XF et al (2013) Establishing continuous scaling of NDVI based on fractal theory and GEOEYE-1 image[J]. J Infrared Millimeter Waves 32(6): 538–544, 549. (in Chinese)
24. Luan HJ, Tian QJ, Yu T et al (2015) Establishing continuous spatial scaling model of NDVI on fractal theory and five-index estimation system[J]. J Remote Sens 19(1):116–125 (in Chinese)
25. Wu L, Qin Q, Liu X et al (2016) Spatial up-scaling correction for leaf area index based on the fractal theory[J]. Remote Sens 8(3):197
26. Kim G, Barros AP (2002) Downscaling of remotely sensed soil moisture with a modified fractal interpolation method using contraction mapping and ancillary data[J]. Remote Sens Environ 83:400–413
27. Chen Y, Chen L (2005) Fractal geometry, 2nd edn [M]. Earthquake Press, Beijing, pp 95–98, 49-51. (in Chinese)
28. Xie HP, Sun HQ (1997) The study on bivariate fractal interpolation functions and creation of fractal interpolated surfaces[J]. Fractals 5(4):625–634
29. Wen JG, Liu Q, Liu QH et al (2009) Scale effect and scale correction of land-surface albedo in rugged terrain[J]. Int J Remote Sens 30(20):5397–5420
30. Luan H, Chen R, Zhang X et al (2017) Bidirectional reflectance distribution function estimation of typical ground objects in Xiamen and its availability in validation of normalized-difference vegetation index[C]. Int Congr Image Signal Process IEEE 625–631. (Datong)

An Autonomous Navigation Method with InSAR-Aided INS

Shuai Jiang, Yalong Pang, Luyuan Wang, Jiyang Yu, Bowen Cheng, Zongling Li, Liang Hao, and Cuilian Wang

Abstract The existing navigation systems have lots of shortcomings, the lack of cross-line resolution lead to the low precision of terrain-aided navigation(TAN) systems. Satellite signals access is necessary for Global Navigation Satellite System (GNSS). Scene matching-based navigation systems require optical image matching, without all-day or all-weather features, and these navigation systems are also affected by seasonal changes. Synthetic aperture radar (SAR) systems cannot inverse platform attitude without three-dimensional localization. Combined with the TAN and scene matching-based navigation systems, a method of InSAR/INS-integrated navigation system based on the interferograms matching is proposed to solve the above problems. The proposed method will employ fundamental interferograms characteristics, which are sensitive to terrain and platform attitude, to provide highprecision navigation. InSAR/INS-integrated navigation system is an autonomous navigation system with high accuracy, and the efficiency, robustness, and reliability of this system are improved compared to the traditional navigation systems.

Keywords InSAR-aided INS · Integrated navigation · Interferograms matching

1 Introduction

The technique of guiding the aircraft from the starting point to the destination is called navigation technology, and navigation system should have the ability to provide information such as the speed, position, attitude, and heading of the aircraft. Navigation plays an important role in the fields of modern warfare and long-range navigation of unmanned aircraft vehicles. In the current environment of strong electromagnetic interference, high-precision, high reliability, high stability without interference, all-day and all-weather working autonomous navigation system is an important research direction in the field of navigation.

The existing navigation systems include celestial navigation system, Loran-C navigation system, Doppler navigation system, satellite navigation system, and inertial

S. Jiang · Y. Pang · L. Wang (✉) · J. Yu · B. Cheng · Z. Li · L. Hao · C. Wang
Beijing Institute of Spacecraft System Engineering, 100094 Beijing, China
e-mail: 770796377@qq.com

L. Wang et al. (eds.), *Proceedings of the 6th China High Resolution Earth Observation Conference (CHREOC 2019)*, Lecture Notes in Electrical Engineering 657,
https://doi.org/10.1007/978-981-15-3947-3_46

navigation system (INS) [1]. INS has some advantages, such as extremely wide frequency band, short-term stability, and it does not need to exchange information with surroundings. Besides, INS can track and reflect any motorized motion of the vehicle, and has a very smooth output. Therefore, INS becomes an indispensable part of the existing navigation system. However, the INS output is obtained by the accelerometer and the gyroscope, so INS errors will accumulate over time, and the long persistence navigation precision is poor. At present, most of the navigation systems are based on INS and supplemented by other navigation systems. The information obtained through multisensors is merged by kalman filtering to correct the long-term accumulation error of INS and achieves the purpose of accurate navigation. At present, the integrated navigation mainly includes terrain-assisted navigation (TAN) system, GPS/INS-integrated navigation system, and scene matching-aided INS navigation system.

The TAN-integrated navigation system [2] is usually applied in the middle guidance stage, which has the advantages of autonomous, concealed, reliable, and all-weather. However, this system, without cross-direction resolution, has the disadvantages of low navigation accuracy, and it is not applicable when the terrain is flat. In addition, aircraft equipped with such navigation systems requires complex path planning to avoid fluctuate terrain areas.

The Global Navigation Satellite System-aided INS (GNSS/INS)-integrated navigation system [3] has a long development time, and it has the advantages of mature technology, high-precision, and long persistence. GNSS such as Beidou and global positioning system (GPS) are external source guidance, both of which rely on satellite signals. Nevertheless, the satellite signals are easily susceptible to electromagnetic interference, and a low power electromagnetic interference device can disturb satellite signals. Therefore, it is necessary to study the autonomous navigation systems that are not susceptible to external interference.

The scene matching-aided navigation system [2] is applied in the middle and terminal guidance stage. This system needs to match the heterogeneous images, which is greatly affected by the timeliness of the reference image. Besides, the navigation area requires extra information such as typical landmarks. In addition, optical images matching-assisted INS navigation system does not have the all-day and all-weather characteristics, so it is not applicable under wicked weather conditions such as night and rainy day.

Synthetic aperture radar-aided INS (SAR/INS)-integrated navigation system [4] is one of the scene matching-aided navigation system, and it is an autonomous navigation system. Its technology is relatively mature, but the navigation accuracy is poor; the heterogeneous image matching is not available without typical landmark; and the matching result is greatly affected by the timeliness of the reference image. Similar to TAN navigation system, complicated path planning is required, which is not conducive to completing the fast launching task.

Combined with the TAN and scene-based navigation systems, this article proposes a method of InSAR(Interferometric SAR) [5]/INS-integrated navigation system based on the interferograms matching to solve the above problems. The proposed method will employ fundamental interferogram characteristics, which are sensitive to terrain and platform attitude, to provide a high-precision navigation system. The computational efficiency of interferograms matching will also provide significantly improved performance compared to digital elevation model (DEM) matching. The interferograms are sensitive to platform roll angle, and the variant of roll angle is reflected in the interferograms, so the roll angle is inversed to modify the attitude error and then the accuracy of navigation is improved. At the same time, the interferograms are sensitive to rugged terrains, and the variant of rugged terrain is reflected in the interferograms, so it has advantages both in terms of robustness and reliability based on interferograms matching. InSAR/INS-integrated navigation system is an autonomous navigation system with high accuracy, and the efficiency, robustness, and reliability of InSAR/INS navigation system are improved compared to the traditional navigation systems.

2 Theory and Advantages of InSAR/INS-Integrated Navigation System

2.1 The Introduction of InSAR-Aided INS-Integrated Navigation System

Considering the shortcomings of the existing integrated navigation, we propose an InSAR-aided INS-integrated navigation method based on interferograms matching. Figure 1 shows the flowchart of InSAR/INS-integrated navigation system [6, 7]. The steps of the InSAR/INS-integrated navigation are detailed in the following parts. The actual and referenced interferograms are generated by the InSAR system firstly. Secondly, the actual and referenced interferograms are matched by scale-invariant feature transform (SIFT)algorithm, and the location shifts in azimuth and range

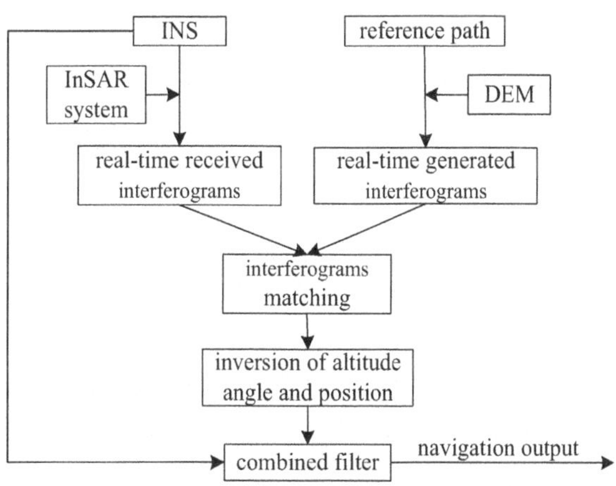

Fig. 1 Flowchart of InSAR-aided INS-integrated navigation

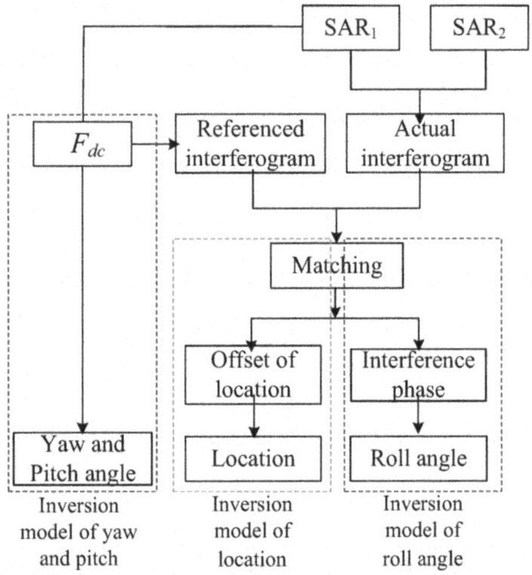

Fig. 2 Flowchart of inversion of observations

direction are obtained. Then, the position and attitude offsets are inversed according to the inversion model. Finally, the navigation outputs are obtained according to the combined filter of position and attitude information.

Figure 2 shows the inversion model of observations, including position and attitude. Aiming at the couple of position and attitude of the platform, the attitude and position inversion models will be established, and the inversion of position and attitude are realized [8].Three inversion models are established: Yaw and pitch angles will be inversed using the Doppler centroid estimated from the SAR image. Through analyzing the influence of the platform position error on the location shift [9], the inversion model of position is established; An inversion model for roll angle will be established by analyzing the effect of roll angle on the interferometric phase.

2.2 TheAdvantages of InSAR/INS-Integrated Navigation System

The InSAR system is sensitive to the terrain height and the attitude of the platform, so InSAR/INS navigation system can avoid the unfavorable terrain fluctuation, which can improve the precision of the navigation greatly. The InSAR/INS-integrated navigation system is used in the middle guidance stage and it has the particular advantages.

1. It has the three-dimensional positioning ability

Without other constraints and auxiliary information, a single SAR image cannot use its own image points to achieve accurate three-dimensional positioning, but the InSAR system is a dual-antenna system, which can observe the same regions from different positions and different angles. The phase information of the measurement area is obtained by complex conjugate multiplication of the two images, and then the elevation information in the measurement area can be obtained.

2. It depends on the interferograms matching

The interference phase is sensitive to terrain and attitude of the platform. A small attitude angle changes and terrain fluctuations can be reflected on the fringe information. Interferograms matching is effective with higher precision, when the DEM is not need.

3. It can obtain the attitude information of the platform

The existing integrated navigation systems only obtain the position information of the platform. The navigation system proposed in this article can simultaneously obtain the position and attitude information. Interference phase is very sensitive to the roll angle errors, so the roll angle can be accurately inverted with the interference phase. The yaw and pitch angles can affect the Doppler center frequency, so the yaw and pitch angles of the platform can be inverted by estimating the Doppler center frequency of the SAR data. This navigation method increases the three-dimensional attitude measurement and it can greatly improve the navigation accuracy.

4. It is not easily constrained by terrain and seasonal changes

The TAN systems do not have cross-direction resolution, so it is invalid with huge terrain undulation. The scene-assisted navigation system is also unavailable in the terrain with large relief, and it is severely affected by seasonal changes. In the navigation system mentioned in this article, the interference phase is sensitive to the terrain, and the terrain relief information can be well reflected in the interferograms, so it is still available when the terrain is undulating. In addition, the interferograms are seldom affected by the season.

5. It is an autonomous navigation system

Compared with GPS/INS-integrated navigation systems, InSAR/INS-integrated navigation is an autonomous navigation system, which is not easily interfered by the external environment. Therefore, it can reduce the probability of being discovered by the enemy, and increase the penetration capability.

6. It has the all-day and all-weather features

Compared with optical scene matching-aided navigation system, InSAR system has all-day and all-weather features, so InSAR/INS-integrated navigation can adapt to different scenarios and weather. It can work under weather conditions such as rainy days with better adaptability.

Table 1 Simulation parameter list

λ/cm	*H*/m	*B*/m	*α*/°	R_0/m	*h*/m	θ_{sq}/°
3.125	4500	1	0	3944.78	1149.4	0

3 Experimentand Results

3.1 Simulation Parameter

The algorithms such as referenced interferograms generation, interferograms matching, observation measurement inversion, and combined filtering are as shown in Sect. 2 and Fig. 1. To verify the effectiveness of our integrated navigation method, a brief introduction to the results of the experiment is necessary.

The simulation parameters of the InSAR system selected in this section are shown in Table 1, where λ is the wavelength, *H* is the platform height, *B* is the baseline length, *α* is the baseline inclination, R_0 is the slant range, *h* is the average height of DEM, and θ_{sq} is the squint angle.

3.2 Experiment Result

Figure 3 shows the inertial measurement unit (IMU) error drift trajectory obtained by inertial system. There is a large drift in the position and attitude of the platform; the position errors in the three directions are more than 100 m and the three attitude angle errors are more than 1.2 degrees.

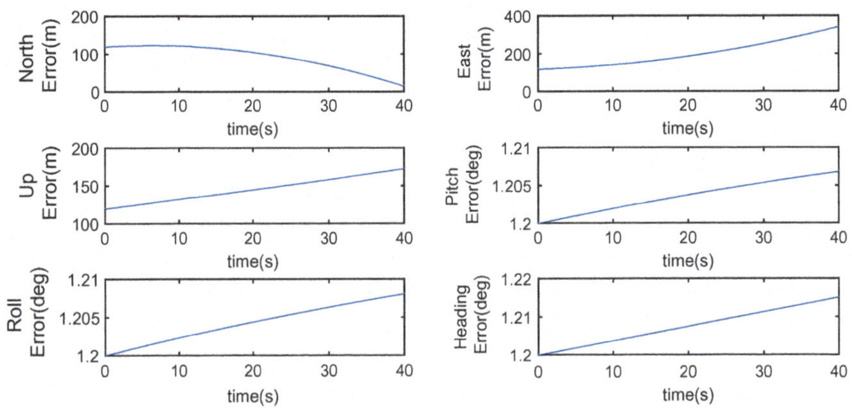

Fig. 3 Schematic diagram of IMU drift error

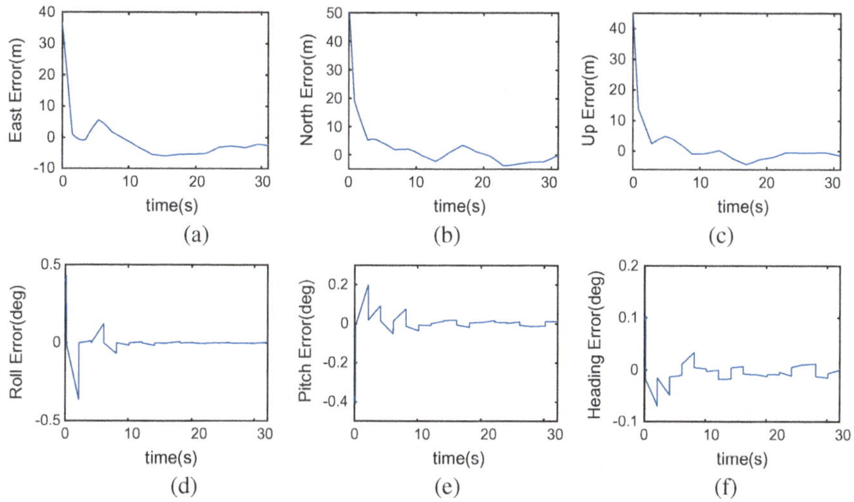

Fig. 4 Result of combined filter **a** the east error, **b** the north error, **c** the up error, **d** the roll angle error, **e** the pitch angle error, **f** the heading angle error

According to the navigation flowchart described in Fig. 1, the referenced and the actual interferograms are generated and matched. The position and attitude angle errors of the platform are inversed according to the observations inversion shown in Fig. 2. And then, the position and attitude errors are obtained by the inversion model. Finally, the result of the combined filter is shown in Fig. 4. Figure 4a shows the east error, Fig. 4b shows the north error, Fig. 4c shows the up error, Fig. 4d shows the roll angle error, Fig. 4e shows the pitch angle error, and Fig. 4f shows the heading angle error

Figure 4 shows that the position errors in the three directions are controlled within 6 m, and the three attitude angle errors of roll, pitch, and heading are controlled within 0.05 degrees after combined filtering. Compared with the errors without combined filtering shown in Fig. 3, and the IMU drift error is converged after combined filtering. The simulation results verify the feasibility of the proposed method.

4 Summary

In this article, we introduce the state of development, advantages, and disadvantages of the existing integrated navigation system firstly, and then, we propose the InSAR/INS-integrated navigation method based on interferograms matching. The principle and advantages of this integrated navigation system are introduced in detail. Finally, the feasibility of the proposed method is verified by the simulation results.

References

1. Savage P (1998) Strapdown inertial navigation integration algorithm design part 2: velocity and position algorithms[J]. J Guid Control Dyn 21(2):208–221
2. Nitti D, Bovenga F, Chiaradia M et al (2015) Feasibility of using synthetic aperture radar to aid UAV navigation[J]. Sensors 15(8):18334–18359
3. Zhou Y, Wan J, Li Z et al (2017) GPS/INS integrated navigation with BP neural network and Kalman filter[Z]. Macau, China
4. Greco M, Pinelli G, Kulpa K et al (2011) 'The study on SAR images exploitation for air platform navigation purposes'. Radar symposium (IRS), 2011 Proceedings international. Leipzig, Germany, pp 347–352
5. Rosen P, Hensley S, Joughin I et al (2002) Synthetic aperture radar interferometry. Proc IEEE 88(3):333–382
6. Shuai J, Maosheng X, Bingnan W et al (2016) The method of InSAR/INS integrated navigation[Z]. China, Guangzhou, pp 1–4
7. Shuai Jiang, Bingnan Wang, Maosheng Xiang et al (2017) The Method of the InSAR/INS integrated navigation based on fringe matching[J]. Acta Electronica Sinica 45(12):2832–2841 (in Chinese)
8. Fu X, Xiang M, Wang B et al (2017) A robust yaw and pitch estimation method for Mini-InSAR system[J]. IEEE Geosci Remote Sens Lett 14(11):2157–2161
9. Jiang S, Wang B-N, Xiang M-S et al (2018) An inversion method of the attitude for InSAR/INS integrated navigation[J]. Acta Electronica Sinica 46(3):513–519 (in Chinese)

The manufacturer's authorised representative in the EU is Springer Nature Customer Service Centre GmbH, Europaplatz 3, 69115 Heidelberg, Germany. If you have any concerns regarding our products, please contact ProductSafety@springernature.com

Printed and bound by CPI Group (UK) Ltd, Croydon, CR0 4YY
30/09/2026
02225557-0016